Sensor Signal and Information Processing II

Sensor Signal and Information Processing II

Special Issue Editors

Wai Lok Woo
Bin Gao

MDPI • Basel • Beijing • Wuhan • Barcelona • Belgrade • Manchester • Tokyo • Cluj • Tianjin

Special Issue Editors

Wai Lok Woo
Northumbria University
UK

Bin Gao
University of Electronic Science and
Technology of China
China

Editorial Office
MDPI
St. Alban-Anlage 66
4052 Basel, Switzerland

This is a reprint of articles from the Special Issue published online in the open access journal *Sensors* (ISSN 1424-8220) (available at: https://www.mdpi.com/journal/sensors/special_issues/SSIP_II).

For citation purposes, cite each article independently as indicated on the article page online and as indicated below:

LastName, A.A.; LastName, B.B.; LastName, C.C. Article Title. *Journal Name* **Year**, *Article Number*, Page Range.

ISBN 978-3-03928-270-8 (Pbk)
ISBN 978-3-03928-271-5 (PDF)

Contents

About the Special Issue Editors

Wai Lok Woo is currently a Professor of Machine Learning with Northumbria University, UK. Previously, he was the Director of Research for the Newcastle Research and Innovation Institute, and the Director of Operations for Newcastle University, UK, where he received his B. Eng. degree in electrical and electronics engineering and his M.Sc. and Ph.D. degrees in 1993, 1995, and 1998, respectively. His research interests include mathematical development of sensor signal processing and machine learning for anomaly detection, digital health, and digital sustainability. He is Associate Editor of several IEEE journals. He is interested in answering the global question of how the integration of smart sensors and machine learning advances humanity and sustains the ecosystem in the current digital transformation era. His research is funded by UK Research and Innovation.

Bin Gao is currently a Professor with the School of Automation Engineering, University of Electronic Science and Technology of China (UESTC), Chengdu, China. He received his B.Sc. degree in communications and signal processing from Southwest Jiao Tong University (2001–2005), China, MSc degree in communications and signal processing with Distinction and PhD degree from Newcastle University, UK (2006–2011). He worked as a Research Associate (2011–2013) with Newcastle university on wearable acoustic sensor technology. His research interests include electromagnetic and thermography sensing, supervised and unsupervised machine learning, wearable sensing, nondestructive testing and evaluation, and he actively publishes in these areas. He has coordinated several research projects from the National Natural Science Foundation of China.

Preface to "Sensor Signal and Information Processing II"

Smart sensors are revolutionizing the world of system design in everything from sports cars to assembly lines. These new sensors have abilities that leave their predecessors in the dust! They not only measure parameters efficiently and precisely, but they also have the ability to enhance and interrupt those measurements, thereby transforming raw data into truly useful information. The concept of a smart sensor was first introduced by NASA in the process of developing a spaceship and formed a product in 1979. Smart sensors have the ability to automatically calibrate, compensate, and collect data. Its capability determines that smart sensors have high accuracy and resolution, high stability and reliability, and good adaptability. Compared with traditional sensors, they have a high performance price ratio. Early smart sensors are processed and converted from the output signal of the sensor to the microprocessor for operation. In the 1980s, the smart sensor mainly focused on microprocessors and integrated the sensor signal conditioning circuit, microelectronic computer memory, and interface circuit to a chip, so that the sensor has a certain AI. In the 1990s, intelligent measurement technology was further improved, so that the sensor could achieve miniaturization and have the function of self-diagnosis.

Fast forwarding to 2020, sensor signal and information processing (SSIP) (https://www.mdpi.com/journal/sensors/special_issues/SSIP_II) has become an overarching field of research focusing on the mathematical foundations and practical applications of signal processing algorithms that learn, reason, and act. It bridges the boundary between theory and application, developing novel theoretically inspired methodologies targeting both longstanding and emergent signal processing applications. The core of SSIP lies in its use of nonlinear and non-Gaussian signal processing methodologies combined with convex and nonconvex optimization. SSIP encompasses new theoretical frameworks for statistical signal processing (e.g., deep learning, latent component analysis, tensor factorization, Bayesian methods) coupled with information theoretical learning, and novel developments in these areas specialized in the processing of a variety of signal modalities including audio, bio-signals, multiphysics signals, images, multispectral, and video, among others. In recent years, many signal processing algorithms have incorporated some forms of computational intelligence as part of their core framework in problem solving. These algorithms have the capacity to generalize and discover knowledge for themselves and learn new information whenever unseen data are captured. The focus of the book will be on a broad range of sensors, signal, and information processing involving the introduction and development of new advanced theoretical and practical algorithms.

Wai Lok Woo, Bin Gao
Special Issue Editors

Article

A Deep-Learning-Driven Light-Weight Phishing Detection Sensor

Bo Wei [1,*][iD], **Rebeen Ali Hamad** [1], **Longzhi Yang** [1][iD], **Xuan He** [2,3], **Hao Wang** [4], **Bin Gao** [5][iD] **and Wai Lok Woo** [1][iD]

[1]   Department of Computer and Information Sciences, Northumbria University, Newcastle upon Tyne NE1 8ST, UK; rebeen.hamad@northumbria.ac.uk (R.A.H.); longzhi.yang@northumbria.ac.uk (L.Y.); wai.l.woo@northumbria.ac.uk (W.L.W.)

[2]   School of Sino-Dutch Biomedical & Information Engineering, Northeastern University, Shenyang 110169, China; hexuan@bmie.neu.edu.cn

[3]   Neusoft Research of Intelligent Healthcare Technology, Co. Ltd., Shenyang 110169, China

[4]   Automation College, Chongqing University of Posts and Telecommunications, Chongqing 400065, China, wanghao@cqupt.edu.cn

[5]   School of Automation Engineering, University of Electronic Science and Technology of China, Chengdu 610054, China; bin_gao@uestc.edu.cn

*   Correspondence: bo.wei@northumbria.ac.uk

Received: 10 September 2019; Accepted: 28 September 2019; Published: 30 September 2019

Abstract: This paper designs an accurate and low-cost phishing detection sensor by exploring deep learning techniques. Phishing is a very common social engineering technique. The attackers try to deceive online users by mimicking a uniform resource locator (URL) and a webpage. Traditionally, phishing detection is largely based on manual reports from users. Machine learning techniques have recently been introduced for phishing detection. With the recent rapid development of deep learning techniques, many deep-learning-based recognition methods have also been explored to improve classification performance. This paper proposes a light-weight deep learning algorithm to detect the malicious URLs and enable a real-time and energy-saving phishing detection sensor. Experimental tests and comparisons have been conducted to verify the efficacy of the proposed method. According to the experiments, the true detection rate has been improved. This paper has also verified that the proposed method can run in an energy-saving embedded single board computer in real-time.

Keywords: phishing detection; cyber security; deep learning

---

1. Introduction

A phishing website is a common social engineering method that mimics trustful uniform resource locators (URLs) and webpages to gather users' sensitive and confidential information, such as user names, passwords, credit card information, etc. Figure 1 shows one example of a phishing website imitating the popular website facebook.com. It replaces "oo" with the unnoticeable "00". The webpage looks exactly the same as the official Facebook, but the phishing one will keep the username and passwords of victims and forward them to attackers. The phishing website issue is becoming increasingly severe. According to the latest phishing activity trends report from the Anti-Phishing Working Group (APWG) [1], 138,328 phishing websites were reported in the fourth quarter of 2018. The report also indicates the increasing trend of detection difficulty because attackers are trying to use multiple redirection techniques in order to make the malicious URLs obscure. There was a $48-million financial loss due to phishing in the US in 2018, only based on the cases reported to the Federal Bureau of Investigation (FBI) [2].

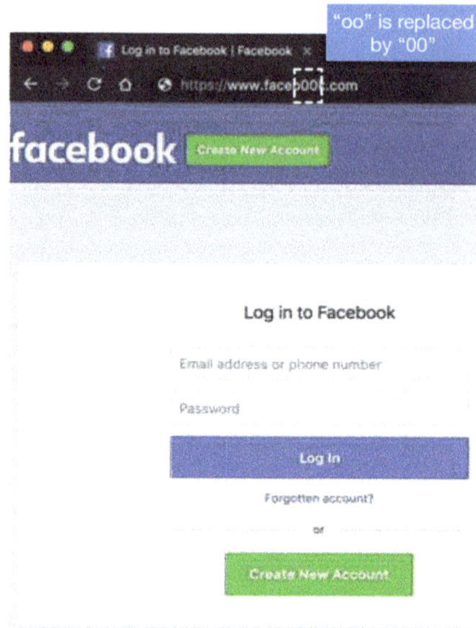

Figure 1. One example of a phishing website imitating the popular website facebook.com.

As shown in Figure 2, malicious URL recognition is relatively easy for cyber security experts since they have sufficient experience in the relevant areas. However, it is extremely difficult for normal users who usually do not pay much attention when accessing one URL. Therefore, the research community takes advantage of the expert knowledge of cyber security and designs machine-based automatic phishing URL detection. The most popular method to detect a phishing website is the use of a phishing URL tank. The URLs in that tank will be recognised as phishing URLs. Phishing URL tanks are maintained by antiphishing organisations to provide live antiphishing databases. There are several famous antiphishing organisations, such as phishtank [3], Joewein [4], hphosts [5], Malware Domains List [6], etc. Due to the rapidly increasing number of phishing websites, antiphishing organisations require comprehensive contributions from the whole community. To maintain up-to-date phishing URL tanks, they need users, including individuals and organisations, to report phishing websites manually. The URLs are fairly accurate because of this manual involvement, but there are still drawbacks in that the human effort introduces delay and extra maintenance labour costs. These handcraft list-based phishing website detection could effectively prevent further harm, but this may fail to promote warnings before its URL is reported by one user and placed in the phishing tank.

Conventional machine learning techniques have been introduced to the phishing website detection domain [7,8]. As shown in Figure 3, with the help of experts in cyber security, URLs and websites are first analysed to conduct feature selection from the malicious websites. Next, machine learning experts use the features, along with their labels, to construct a training set and take advantage of classical supervised machine learning algorithms to develop a phishing detection model. Many conventional methods, e.g., support vector machine (SVM), k-nearest neighbours algorithm (kNN), etc., have been explored to fully utilise these features. Deep learning is also incorporated in the phishing detection domain, motivated by its recent rapid development and many successful applications [9,10]. Different from classical machine learning methods involving an explicit handcrafted feature selection process, the machine learning experts can use the data directly without the knowledge from the cyber security experts (shown in Figure 3).

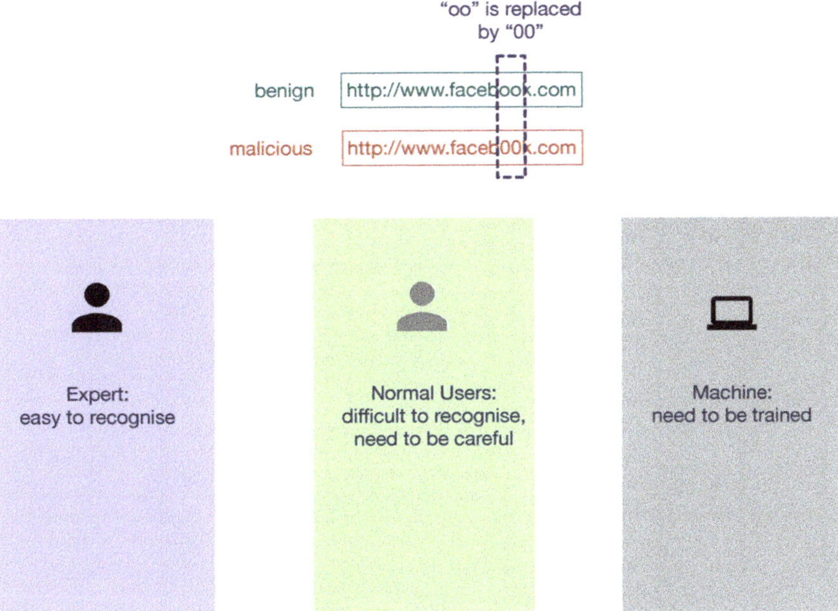

Figure 2. Difficulties to recognise malicious URLs for experts, normal users, and machines.

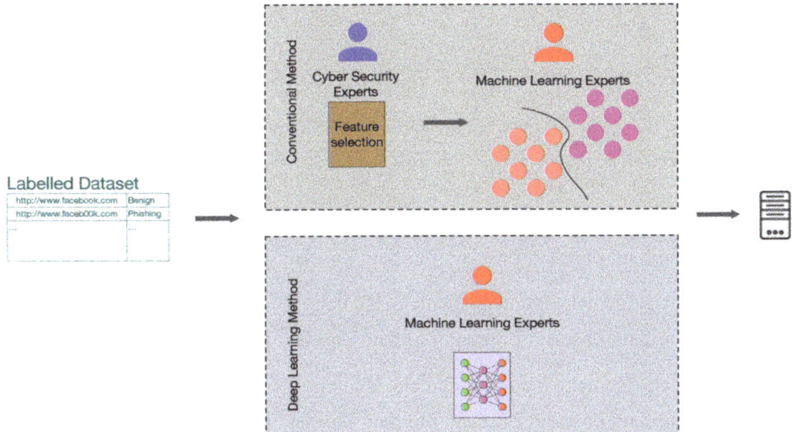

Figure 3. The phishing detection methods using classical machine learning methods and deep learning techniques.

This paper designs and implements a light-weight phishing detection sensor. The paper proposes an innovative deep learning model to enable accurate and efficient phishing detection using URLs of websites. Different from previous research works, this research also investigates the feasibility of our proposed deep learning model in resource-constrained computing devices. Furthermore, this research work implements a prototype of a deep-learning-driven light-weight phishing detection sensor in one embedded single board computer, which shows the feasibility of the integration of our method into one wireless router. This work also uses a large volume of benign and malicious URLs to construct the training set and evaluate the efficacy of the proposed model.

In summary, the main contributions of this paper are:

- This research paper proposes a novel character-level multi-spatial deep learning model to detect malicious URLs. The popular convolutional neural networks (CNN) have been explored to improve detection performance.
- This research paper also integrates the proposed model in one single board computer to enable an energy-saving and efficient phishing website sensor. As far as is known, this paper is the first to discuss the feasibility of the usage of resource-constrained computing devices to enable a phishing website detection sensor.
- This paper has conducted extensive evaluations to show the performance of the proposed method and the efficiency of our prototype.

This paper first introduces related work in Section 2. Section 3 shows the background and motivations of the proposed method. The proposed method is introduced in Section 4, and this paper evaluates the performance of the proposed model in Section 5. Section 6 shows the details of the implementation. Finally, Section 7 concludes our work.

2. Related Works

The common method to detect phishing websites is the use of blacklists to include all the reported URLs of phishing websites. This method requires largely manual efforts from the whole community. As introduced in Section 1, the blacklists are mainly maintained by antiphishing organisations. Some popular antiphishing organisations are phishtank [3], Joewein [4], hphosts [5], Malware Domains List [6], etc. Whitelists can also be created to exclude the websites that users trust. Some methods are also proposed, aiming to facilitate the labelling process for list-based phishing website detection. Cao et al. proposed a method to automatically update the whitelist from the users' familiar websites [11]. Jain et al. designed a hyperlink-based phishing detection mechanism to update the whitelist [12]. Sharifi et al. took advantage of search engines to evaluate the legitimacy of websites and create an up-to-date blacklist accordingly [13].

Machine learning has been extensively used in the phishing detection domain. Phishing detection can be classified as a supervised machine learning problem. A large number of phishing websites on blacklists can be analysed and researched by the cyber security and machine learning community. Features from two main components of a website are commonly used for phishing detection. At first, the attackers usually imitate legitimate URLs to lure users into entering phishing websites, so researchers have focused on the analysis of URL for phishing detection. Additional to URLs, the documents implemented to display one website, such as HTML, CSS, and Javascript documents, are also explored for phishing detection. Amrutkar et al. use multiple features from HTML, CSS, and javascript documents from websites to detect the phishing contents [7]. That work also investigates the website features from smart phones and aims to realise real-time malicious website detection on mobile devices. Rule-based features from URLs were explored to detect phishing internet banking webpages [14]. Natural language processing techniques are also explored in [15] to determine a malicious URL, and the authors use seven traditional classifiers along with selected features from URLs to enable an antiphishing system. Zhang et al. [16] and Xiang et al. [8] proposed Cantita and its augmented version Cantita+, which also extracted features from the contents of websites and used multiple machine learning algorithms.

Recently, deep-learning-based methods have been introduced in the phishing website detection domain. Jiang et al. used convolutional neural network (CNN) techniques, a popular model in deep learning, to detect malicious URLs [17]. One deep learning model using word embedding and CNN has also been proposed to detect malicious URLs, file paths, and registry keys [9]. Le et al. proposed URLNet to use CNN for analysing both word-level features and character-level features for malicious URL detection [10]. Yang et al. applied multiple features for detecting phishing URLs [18]. The deep learning technique has also been introduced into phishing email detection [19].

Different from the previous works, this paper proposes a new deep learning model and further investigates the feasibility of enabling an energy-saving phishing website sensor with the integration of the deep learning model in a resource-constrained computing device.

3. Background and Motivations

List-based phishing website detection is the most common method currently. The lists created by this method can offer labelled training sets, which is an essential prerequisite for the future use of machine-learning-based detection methods. Two URL lists are normally produced by list-based phishing website detection methods, i.e., a blacklist and a whitelist. Figure 4 shows the general mechanism of list-based phishing URL detection methods. The antiphishing companies use the reports from the community to create one blacklist and one whitelist. The computing devices use these two lists to detect malicious websites. The whitelist contains the user-trusted URLs. In contrast, when one URL is on the blacklist, it is recognised as a malicious URL. However, with the list-based method there remains an ongoing challenge of the detection of unknown URLs. It is difficult to classify an unknown URL that is not on any list. The common policy is to recognise that as a benign URL. If a new malicious website uses this unknown URL, the false negative could potentially harm users. Attackers take advantage of this loophole and keep changing URLs for their phishing websites to ensure the new URLs are not on the blacklist.

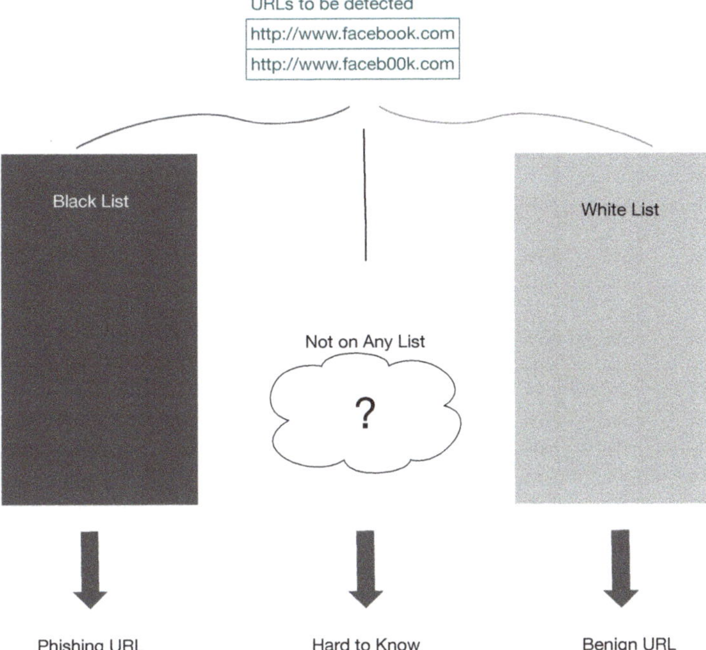

Figure 4. The mechanism of list-based phishing URL detection methods.

Phishing website detection is modelled as a supervised machine learning problem. Components from websites, such as URL, HTML, etc., are used as the training data for building a model to conduct malicious website detection. Classifiers play a vital role in supervised machine learning methods. There are several classical and popular supervised machine learning algorithms, such as kNN, SVM, etc. that have already been used for malicious website detection applications. Figure 5 shows the general process of a classical supervised learning-based malicious website detection method. The feature selection process is an initial and essential step for these classical classifiers. Informative features can help improve the detection performance, but excellent feature selection needs the expert knowledge from a cyber security perspective. Furthermore, it is always difficult to decide the best features for a particular application. Feature selection may cause a drastic loss of valuable information.

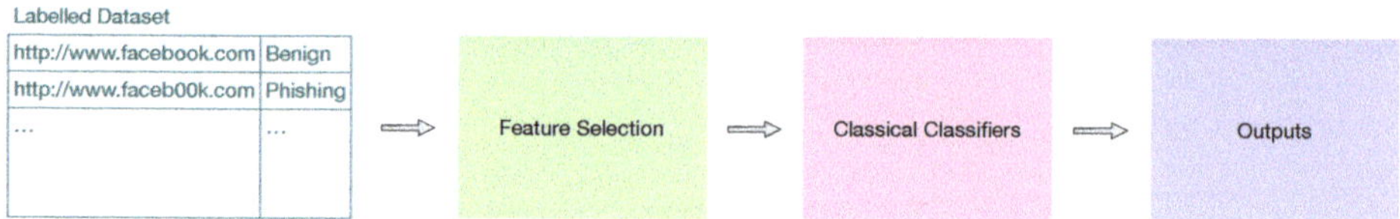

Figure 5. The general process of a classical supervised learning-based malicious website detection method.

Recently, the use of deep learning has improved the performance of many applications in image processing [20], computer vision [21], acoustic classification [22], natural language processing [23], etc. Many research works also utilise deep learning techniques in malicious website detection. Additional to the significant performance improvement, deep learning has the advantage of being featureless. As shown in Figure 6, the deep-learning-based methods do not require feature selection. The unprocessed data could be used to train a model without any extra effort, and deep learning algorithms will help select the best patterns for the final decision. Motivated by these facts, this paper also designs a deep learning model and uses unprocessed URLs to derive a deep-learning-based light-weight phishing detection sensor for inference.

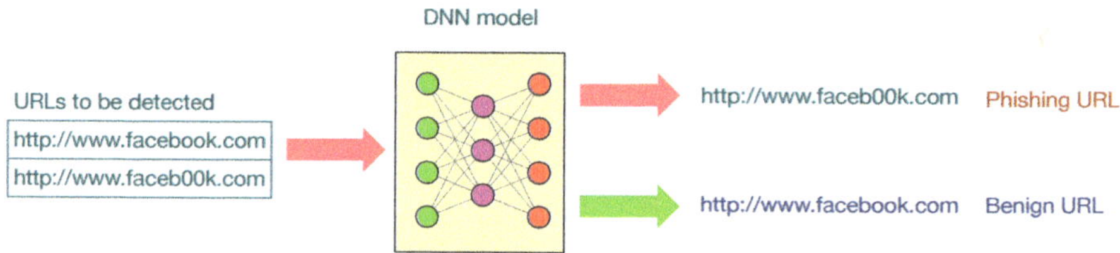

Figure 6. Deep-learning-based malicious website detection.

To enable the light-weight phishing detection sensor, another question this paper would like to address in this paper is "Can the proposed deep learning model be integrated into a resource-constrained computing device?" The paper aims to design a phishing detection sensor to achieve accurate and efficient phishing detection. By applying the designed system, it is not necessary to install antiphishing software on every single computing device and Internet of Things (IoT) device. Only the designed sensor is required for one household or office between the devices and the router. The proposed model can also be implemented into the router directly due to its computational efficiency. To summarise, this paper implements a phishing detection prototype sensor with the integration of the proposed deep learning methods.

This section will give the details of the proposed deep-learning-based phishing URL detection method. Figure 7 shows an overview of the proposed method.

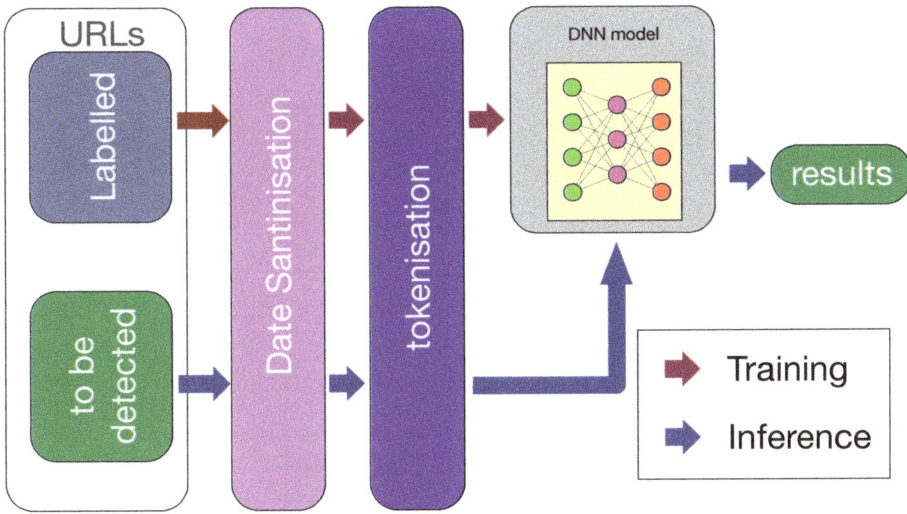

Figure 7. System structure.

4. Method

The first step of the proposed method is data sanitisation. In this step, the common URL prefixes, such as http://, https:// and www, are deleted to prevent the impact of URL presentations of the different datasets on phishing URL recognition performance. Without pruning prefixes, the inconsistency of URL formats can easily affect the quality of the model. For example, all the URLs in some phishing URL datasets contain the http prefix, which means that the trained model will falsely classify all of the URLs with the http prefix as phishing. The shorter representation will also accelerate the inference, which is also a main considerable factor for resource-constrained devices.

The tokeniser is used to vectorise each character in URLs. Character-level tokenisation is used instead of word-level analysis because URLs usually use words without any meaning. More information is contained at the character level. The attackers also mimic the URLs of authentic websites by changing several characters that are not noticeable. For example, they may change facebook.com to faceb00k.com, replacing "oo" with "00". The character-level tokenisation helps find this mimic information, improving the performance of malicious URL detection.

This paper proposes an innovative deep neural network for malicious URL detection. As shown in Figure 8, the proposed Deep Neural Network (DNN) model has the following layers: (1) embedding layers; (2) convolutional layers; (3) concatenation layer (4) dropout layers; (5) dense layers; (6) sigmoid layers. Table 1 shows the configuration of the layers of the proposed deep-layered model. The output dimension of the word embedding layer, the number of filters, and the kernel size of the convolutional layers, the rate of the dropout layer and the number of units of the dense layers are shown. Here are the details for each type of layer in the configuration.

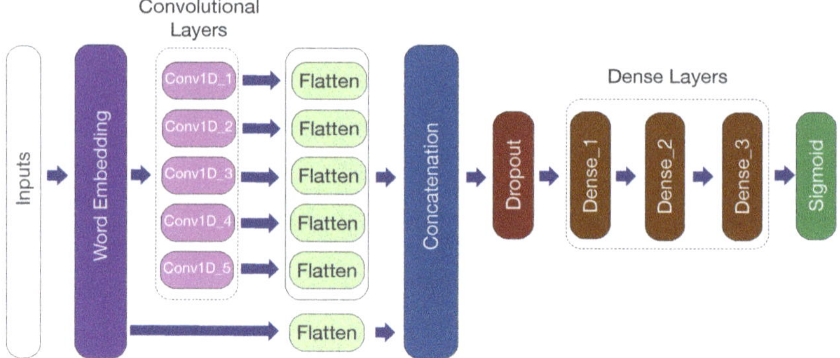

Figure 8. Structure of the proposed DNN model.

Table 1. Architecture configuration of the proposed DNN model.

	Output Dimension	
Word Embedding	32	
	Number of Filters	**Kernel Size**
Conv1D_1	256	2
Conv1D_2	256	3
Conv1D_3	256	4
Conv1D_4	256	5
Conv1D_5	256	10
	Dropout Rate	
Dropout	0.5	
	Number of Units	
Dense_1	128	
Dense_2	128	
Dense_3	128	

Embedding layer: The embedding layer is usually used in the first layer of the DNN structure for a Natural Language Processing(NLP) problem. Additional to the tokenisation, the embedding layer will return a vector. Figures 9 and 10 show examples of the simple one hot encoding and the used word embedding. Different from one hot word using binary representations for each word, the coefficients in the vector returned from the embedding layer are able to indicate the relations among characters, which can help improve the performance of NLP-related research questions. The proposed network uses embedding word configuration.

Convolutional layers: Following the embedding layer, five convolutional layers are used. For each convolutional layer, the kernel, a.k.a. a convolutional filter, is applied to extract the most useful features and remove unnecessary information. The element-wise multiplication and the summary operations occur between the filter and the relevant part of data, and the filter slides through the data to generate the features. Instead of a common sequential structure of convolutional neural networks, parallel convolutional layers are used. Each layer considers one window size of consecutive characters and extracts features from them. The rectified linear unit (ReLU) activation function is also used following each convolutional layer. The output from each convolutional layer is then flattened and subsequently concatenated.

Concatenation layer: This layer is used to concatenate the features from previous layers for further processing. Different from simply concatenating the outputs from convolutional layers, the output from the embedding layers are also combined. In addition, the output from the embedding layer (without the convolutional filtering) preserves the original information of content that can be used to detect malicious URLs as well.

Dropout layer: Dropout layer is a regularisation technique that is used to prevent overfitting during the training phase [24]. Neurons are randomly selected and ignored by the dropout layer during the training phase. Those ignored neurons are temporally removed on the forward pass, and their weights are not updated on the backward pass.

Dense layers: A dense layer is a fully connected feedback layer that equips the proposed model with the more capabilities for extracting the informative features. Following the dropout layer, three dense layers are used to analyse the patterns from the concatenation layer. One ReLU activation function also follows each dense layer.

Sigmoid layer: The sigmoid function is used in this layer to determine the malicious URLs. The range of the output from a sigmoid function is between 0 and 1, which is used in the final layer of the proposed model to show the prediction probability.

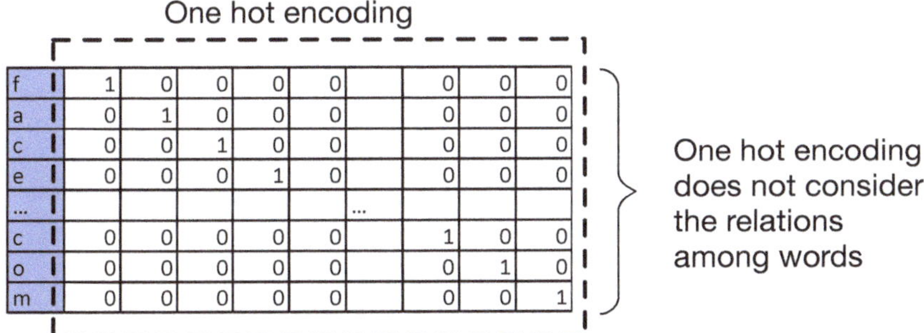

Figure 9. Example of one hot encoding.

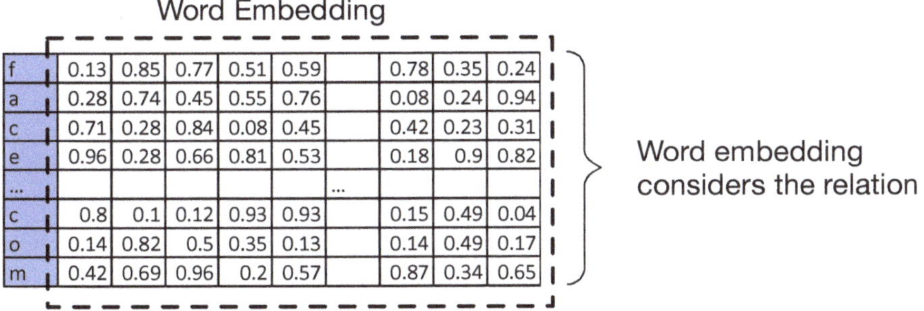

Figure 10. Example of word embedding.

5. Evaluation

This section discusses the performance of the proposed model. As discussed, the configuration of the proposed model is shown in Table 1. A PC with a Graphics Processing Unit (GPU) is used to train and evaluate the model. The computer used has an Intel Core i7 8 core CPU 3.60 GHz processor, 16 GB memory, and Nvidia GeForce GTX 1060 6 GB GPU. A total of 1,523,966 URLs were used, where 999,996 were legitimate URLs and 523,970 were phishing URLs. The legitimate URLs are from the list of Alexa top 1 million sites [25], hphosts [5], Joewein [4], malwaredomains [26], and phishtank [3]. Before using them, repeated URLs were removed to construct a dataset. The dataset was randomly split into a training set and a test set. The percentage of testing instances was 10%. The true detection rate was used as the accuracy metric, i.e., the ratio between the number of correct detected instances and the total number of instances.

Using the proposed model can achieve an 86.630% true detection rate. Many similar deep-learning-based URL detection models use similar structures but configurations with different numbers of dense layers and convolutional layers. Therefore, in the following, the effect of the dense layers, convolutional layers, and concatenation of the output from the embedding layer will be discussed.

First, the effect of the dense layers is shown in Table 2. It is expected that increasing the number of dense layers can improve performance, so we first investigate the effect of the number of dense layers. The proposed model has 4 dense layers (3 dense layers plus the sigmoid layer). Table 2 shows that the true detection rate gradually increases with the increasing number of dense layers. The proposed method can achieve an 86.630% true detection rate. With 1, 2, and 3 dense layers, the true detection rates are 86.537%, 86.538%, and 86.542%, respectively.

Table 2. The effect of the dense layers.

	Accuracy
Proposed	**86.630%**
1 Dense Layer	86.537%
2 Dense Layers	86.538%
3 Dense Layers	86.542%

Table 3 shows the effect of convolutional layers. A similar observation was also found here. Th increasing number of convolutional layers can help improve the performance the URL-based phishing detection. When using 1, 2, 3, and 4 convolutional layers, the deep learning model can achieve true detection rates of 85.401%, 85.832%, 86.169%, and 86.439%, respectively. The true detection rate of the proposed method is highest at 86.630%.

Table 3. The effect of the convolutional layers.

	Accuracy
Proposed	**86.630%**
1 Convolutional Layer	85.401%
2 Convolutional Layers	85.832%
3 Convolutional Layers	86.169%
4 Convolutional Layers	86.439%

In this paper, we also propose to concatenate the output from the word embedding layer to enable the dense layers to have the unprocessed information as well. Performance improvement can also be found using this strategy, as shown in Table 4. Without the concatenation of the output from the embedding layer, the true detection rate drops from 86.630% to 83.472%.

Table 4. The effect of the concatenation of the output from the embedding layer.

	Accuracy
Proposed	**86.630%**
No Concatenation	83.472%

6. Prototype Implementation

The proposed method is implemented by integrating the proposed deep-learning-based method into resource-constrained devices. In this work, Raspberry Pi 3 B+ was chosen to implement our prototype. Raspberry Pi 3 B+ has a Quad core 1.4 GHz 64 bit CPU with 1 GB RAM. It is powered by

5 V power input or battery and has various Input/Output (IO) ports, such as 4 USB 2.0 ports, 40-pin general-purpose input/output (GPIO) header, and Camera Serial Interface (CSI) port. Raspberry Pi 3 B+ also supports the common network ports, such as 2.4 GHz and 5 GHz IEEE 802.11 wireless cards and the Ethernet. The abundance of network cards makes Raspberry Pi 3 B+ a good candidate for simulating a router.

Figure 11 shows the implementation process for the whole system. We first use the labelled URLs to train the DNN model by using powerful computing devices, such as GPU servers, rack servers, or cloud servers. The trained DNN model is then transferred to the intelligent WiFi router that acts as the phishing malicious URL sensor. When the intelligent WiFi router receives URL requests, it conducts phishing detection by using the integrated DNN model before requiring a domain name system (DNS) server. When the URL is recognised as malicious, the smart WiFi router will raise an alarm to the user and block the user's access to that URL.

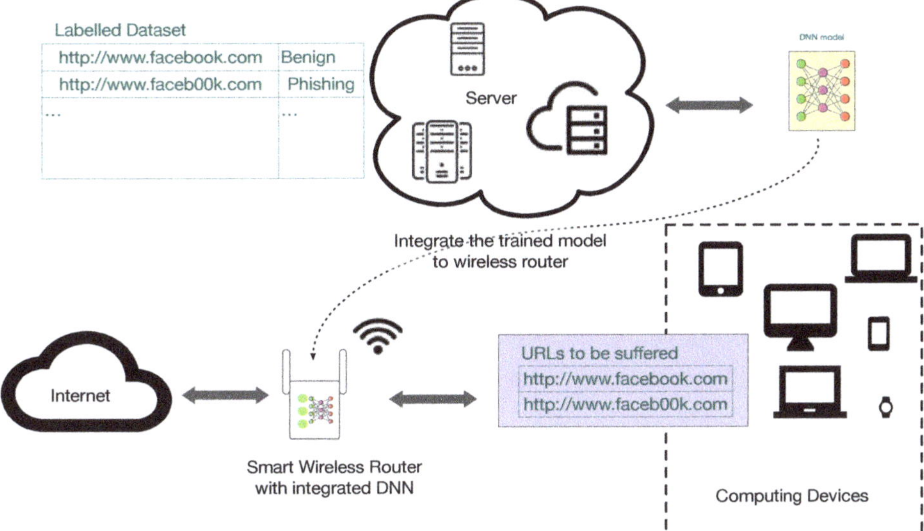

Figure 11. The implementation process.

To evaluate the efficiency of the proposed model, we measure the computational time of each step of the proposed method using Raspberry Pi. Table 5 demonstrates the execution time of data sanitisation, tokenisation, and inference using DNN. A total of 10 trials were executed, and we calculated the mean execution time for each step. The inference costs 105 ms for each URL request, which occupies most of the running time. In the meantime, the data sanitation and tokenisation take no more than 1 ms. Totally, the proposed phishing detection method uses approximately 110 ms to evaluate each URL request, which can enable real-time malicious detection.

Table 5. Execution time in the prototype.

	Execution Time (ms)
Data Sanitisation	0.0106
Tokenisation	0.1997
DNN Inference	105

The word-level word embedding method along with character-level word embedding is used in [10]. To compare that work and show the efficiency of the proposed model, a deep learning model is implemented using both word-level and character-level word embedding methods. The same convolutional layers (as shown in Figure 8) are applied on the outputs of those two word embedding layers. Two outputs from convolutional layers are concatenated for further processing by dropout, dense, and sigmoid layers to make a decision. This deep learning model was evaluated in Raspberry Pi, but an out-of-memory (OOM) error occurred when running it. In other words, there was not sufficient memory in the resource-constrained Raspberry Pi to execute the implemented deep learning model with both word-level and character-level word embedding methods. This further confirms the efficiency of the proposed light-weight model. To compare the performances, the proposed model and the model with both word-level and character-level word embedding methods are executed in the PC. The execution time of DNN inference with the proposed model is 67 ms, while the model with both word-level and character-level word embedding methods needs 96 ms. The execution time significantly reduces by 30% using the proposed model.

7. Conclusions

This paper proposed a multispatial convolutional neural network to enable an accurate and efficient phishing detection sensor. Extensive evaluations were conducted to show the performance of the proposed method. The true detection rate of the proposed method can achieve 86.63%. A prototype by using Raspberry Pi was also implemented to enable real-time phishing URL detection. With the proposed method, the execution time reduces by 30%, and real-time detection is realised in a resource-constrained device.

In the future, a webpage-content-based phishing detection model using deep learning can be proposed and implemented in a resource-constrained sensor as well.

Author Contributions: Conceptualization, B.W. and W.L.W.; methodology, B.W., H.W., and R.A.H.; validation, B.W., X.H. and B.G.; writing–original draft preparation, B.W. and R.A.H.; writing–review, X.H., L.Y. and W.L.W.; editing, W.L.W., B.G., L.Y. and H.W.; visualisation, B.W., X.H., L.Y. and H.W.

Funding: This research was funded by the National Natural Science Foundation of China (NSFC) (Grant No. 61971093 and No. 61771121), the Open Program of Neusoft Research of Intelligent Healthcare Technology, Co. Ltd. (Grant No. NRIHTOP1802), the Research and Application for Key Technologies of IoT Oriented to Smart Cities (cstc2018jszx-cyztzx0081), and the Innovation and Application for Smart Test of Supply and Demand Integration (cstc2018jszx-cyzd0404).

Conflicts of Interest: The authors declare no conflict of interest.

Abbreviations

The following abbreviations are used in this manuscript:

URL	Uniform resource locator
APWG	Anti-phishing working group
FBI	Federal Bureau of Investigation
SVM	Support vector machine
kNN	k-nearest neighbours algorithm
CNN	Convolutional neural network
ReLU	Rectified linear unit
DNS	Domain name system

References

1. Anti-Phishing Working Group (APWG). Available online: https://docs.apwg.org//reports/apwg_trends_report_q4_2018.pdf (accessed on 15 July 2019).
2. IC3 Annual Report 2018. Available online: https://pdf.ic3.gov/2018_IC3Report.pdf (accessed on 15 July 2019).
3. Phishtank. Available online: https://www.phishtank.com/ (accessed on 15 July 2019).
4. Joewein. Available online: https://joewein.net/ (accessed on 15 July 2019).
5. Hphosts. Available online: https://www.hosts-file.net/ (accessed on 15 July 2019).

6.   Malware Domains List. Available online: http://mirror1.malwaredomains.com (accessed on 15 July 2019).

7.   Amrutkar, C.; Kim, Y.S.; Traynor, P. Detecting mobile malicious webpages in real time. *IEEE Trans. Mob. Comput.* **2016**, *16*, 2184–2197. [CrossRef]

8.   Xiang, G.; Hong, J.; Rose, C.P.; Cranor, L. Cantina+: A feature-rich machine learning framework for detecting phishing web sites. *ACM Trans. Inf. Syst. Secur. (TISSEC)* **2011**, *14*, 21. [CrossRef]

9.   Saxe, J.; Berlin, K. eXpose: A character-level convolutional neural network with embeddings for detecting malicious URLs, file paths and registry keys. *arXiv* **2017**, arXiv:1702.08568.

10.   Le, H.; Pham, Q.; Sahoo, D.; Hoi, S.C. URLNet: Learning a URL representation with deep learning for malicious URL detection. *arXiv* **2018**, arXiv:1802.03162.

11.   Cao, Y.; Han, W.; Le, Y. Anti-phishing based on automated individual white-list. In Proceedings of the 4th ACM Workshop on Digital Identity Management, Alexandria, VA, USA, 31 October 2008; pp. 51–60.

12.   Jain, A.K.; Gupta, B.B. A novel approach to protect against phishing attacks at client side using auto-updated white-list. *EURASIP J. Inf. Secur.* **2016**, *2016*, 9. [CrossRef]

13.   Sharifi, M.; Siadati, S.H. A phishing sites blacklist generator. In Proceedings of the 2008 IEEE/ACS International Conference on Computer Systems and Applications, Doha, Qatar, 31 March–4 April 2008; pp. 840–843.

14.   Moghimi, M.; Varjani, A.Y. New rule-based phishing detection method. *Expert Syst. Appl.* **2016**, *53*, 231–242. [CrossRef]

15.   Sahingoz, O.K.; Buber, E.; Demir, O.; Diri, B. Machine learning based phishing detection from URLs. *Expert Syst. Appl.* **2019**, *117*, 345–357. [CrossRef]

16.   Zhang, Y.; Hong, J.I.; Cranor, L.F. Cantina: A content-based approach to detecting phishing web sites. In Proceedings of the 16th International Conference on World Wide Web, Banff, AB, Canada, 8–12 May 2007; pp. 639–648.

17.   Jiang, J.; Chen, J.; Choo, K.K.R.; Liu, C.; Liu, K.; Yu, M.; Wang, Y. A deep learning based online malicious URL and DNS detection scheme. In *International Conference on Security and Privacy in Communication Systems*; Springer: Berlin, Germany, 2017; pp. 438–448.

18.   Yang, P.; Zhao, G.; Zeng, P. Phishing Website Detection Based on Multidimensional Features Driven by Deep Learning. *IEEE Access* **2019**, *7*, 15196–15209. [CrossRef]

19.   Fang, Y.; Zhang, C.; Huang, C.; Liu, L.; Yang, Y. Phishing Email Detection Using Improved RCNN Model With Multilevel Vectors and Attention Mechanism. *IEEE Access* **2019**, *7*, 56329–56340. [CrossRef]

20.   Donahue, J.; Jia, Y.; Vinyals, O.; Hoffman, J.; Zhang, N.; Tzeng, E.; Darrell, T. Decaf: A deep convolutional activation feature for generic visual recognition. In Proceedings of the International Conference on Machine Learning, Beijing, China, June 21– 26 2014; pp. 647–655.

21.   Szegedy, C.; Vanhoucke, V.; Ioffe, S.; Shlens, J.; Wojna, Z. Rethinking the inception architecture for computer vision. In Proceedings of the IEEE Conference on Computer Vision and Pattern Recognition, Las Vegas, NV, USA, 27–30 June 2016; pp. 2818–2826.

22.   Hinton, G.; Deng, L.; Yu, D.; Dahl, G.; Mohamed, A.R.; Jaitly, N.; Senior, A.; Vanhoucke, V.; Nguyen, P.; Kingsbury, B. Deep neural networks for acoustic modeling in speech recognition. *IEEE Signal Process. Mag.* **2012**, *29*, 82–97. [CrossRef]

23.   Collobert, R.; Weston, J. A unified architecture for natural language processing: Deep neural networks with multitask learning. In Proceedings of the 25th International Conference on Machine Learning, Helsinki, Finland, 5–9 July 2008; pp. 160–167.

24.   Srivastava, N.; Hinton, G.; Krizhevsky, A.; Sutskever, I.; Salakhutdinov, R. Dropout: A simple way to prevent neural networks from overfitting. *J. Mach. Learn. Res.* **2014**, *15*, 1929–1958.

25.   Alexa Top Sites. Available online: https://www.alexa.com/topsites (accessed on 15 February 2019).

26.   Malwaredomains. Available online: https://www.malwaredomains.com/ (accessed on 15 February 2019).

Article

A Non-Linear Filtering Algorithm Based on Alpha-Divergence Minimization

Yarong Luo, Chi Guo *, Jiansheng Zheng and Shengyong You

Global Navigation Satellite System Research Center, Wuhan University, Wuhan 430079, China;
yarongluo@whu.edu.cn (Y.L.); zjs@whu.edu.cn (J.Z.); shengyongyou@whu.edu.cn (S.Y.)
* Correspondence: guochi@whu.edu.cn

Received: 25 August 2018; Accepted: 16 September 2018; Published: 24 September 2018

Abstract: A non-linear filtering algorithm based on the alpha-divergence is proposed, which uses the exponential family distribution to approximate the actual state distribution and the alpha-divergence to measure the approximation degree between the two distributions; thus, it provides more choices for similarity measurement by adjusting the value of α during the updating process of the equation of state and the measurement equation in the non-linear dynamic systems. Firstly, an α-mixed probability density function that satisfies the normalization condition is defined, and the properties of the mean and variance are analyzed when the probability density functions $p(x)$ and $q(x)$ are one-dimensional normal distributions. Secondly, the sufficient condition of the alpha-divergence taking the minimum value is proven, that is when $\alpha \geq 1$, the natural statistical vector's expectations of the exponential family distribution are equal to the natural statistical vector's expectations of the α-mixed probability state density function. Finally, the conclusion is applied to non-linear filtering, and the non-linear filtering algorithm based on alpha-divergence minimization is proposed, providing more non-linear processing strategies for non-linear filtering. Furthermore, the algorithm's validity is verified by the experimental results, and a better filtering effect is achieved for non-linear filtering by adjusting the value of α.

Keywords: alpha-divergence; Kullback–Leibler divergence; non-linear filtering; exponential family distribution

1. Introduction

The analysis and design of non-linear filtering algorithms are of enormous significance because non-linear dynamic stochastic systems have been widely used in practical systems, such as navigation system [1], simultaneous localization and mapping [2], and so on. Because the state model and the measurement model are non-linear and the state variables and the observation variables of the systems no longer satisfy the Gaussian distribution, the representation of the probability density distribution of the non-linear function will become difficult. In order to solve this problem, deterministic sampling (such as the unscented Kalman filter and cubature Kalman filter) and random sampling (such as the particle filter) are adopted to approximate the probability density distribution of the non-linear function, that is to say, to replace the actual state distribution density function by a hypothetical one [3].

In order to measure the similarity between the hypothetical state distribution density function and the actual one, we need to select a measurement method to ensure the effectiveness of the above methods. The alpha-divergence, proposed by S.Amari, is used to measure the deviation between data distributions $p(x)$ and $q(x)$ [4]. It can be used to measure the similarity between the hypothetical state distribution density function and the actual one for the non-linear filtering. Compared with the Kullback–Leibler divergence (the KL divergence), the alpha-divergence provides more choices for measuring the similarity between the hypothetical state distribution density function and the actual one. Therefore, we use alpha-divergence as a measurement criterion to measure the similarity between the two distribution functions. Indeed, adjusting the value of parameter α in the function can ensure the interesting properties of similarity measurement. Another choice of α characterizes different learning principles, in the sense that the model distribution is more inclusive ($\alpha \to \infty$) or more exclusive ($\alpha \to -\infty$) [5]. Such flexibility enables α-based methods to outperform KL-based methods with the value of α being properly selected. The higher the similarity of the two probability distributions $p(x)$ and $q(x)$, the smaller the value of alpha-divergence will be. Then, it can be proven that in a specific range of value, $q(x)$ can fully represent the properties of $p(x)$ when the value of alpha-divergence is minimum.

Because the posterior distribution of non-linear filtering is difficult to solve, given that the posterior probability distribution is $p(x)$, we can use the probability distribution $q(x)$ to approximate the posterior probability distribution $p(x)$ of non-linear filtering. The approximate distribution $q(x)$ is expected to be a distribution with a finite moment vector. This in turn means that a good choice for the approximate distribution is from the exponential family distribution, which is a practically convenient and widely-used unified family of distributions on finite dimensional Euclidean spaces.

The main contributions of this article include:

1. We define an α-mixed probability density function and prove that it satisfies the normalization condition when we specify the probability distributions $p(x)$ and $q(x)$ to be univariate normal distributions. Then, we analyze the monotonicity of the mean and the variance of the α-mixed probability density function with respect to the parameter when $p(x)$ and $q(x)$ are specified to be univariate normal distributions. The results will be used in the algorithm implementation to guarantee the convergence.
2. We specify the probability density function $q(x)$ as an exponential family state density function and choose it to approximate the known state probability density function $p(x)$. After the α-mixed probability density function is defined by $q(x)$ and $p(x)$, we prove that the sufficient condition for alpha-divergence minimization is when $\alpha \geq 1$ and the expected value of the natural statistical vector of $q(x)$ is equivalent to the expected value of the natural statistical vector of the α-mixed probability density function.
3. We apply the sufficient condition to the non-linear measurement update step of the non-linear filtering. The experiments show that the proposed method can achieve better performance by using a proper α value.

2. Related Work

It has become a common method to apply various measurement methods of divergence to optimization and filtering, among which the KL divergence, as the only invariant flat divergence, has been most commonly studied [6]. The KL divergence is used to measure the error in the Gaussian approximation process, and it is applied in the process of distributing updated Kalman filtering [7]. The proposal distribution of the particle filter algorithm is regenerated using the KL divergence after containing the latest measurement values, so the new proposal distribution approaches the actual posterior distribution [8]. Martin et al. proposed the Kullback–Leibler divergence-based differential evolution Markov chain filter for global localization for mobile robots in a challenging environment [9], where the KL-divergence is the basis of the cost function for minimization. The work in [3] provides a better measurement method for estimating the posterior distribution to apply KL minimization to the prediction and updating of the filtering algorithm, but it only provides the proof of the KL divergence minimization. The similarity of the posterior probability distribution between adjacent sensors in the distributed cubature Kalman filter is measured by minimizing the KL divergence, and great simulation results are achieved in the collaborative space target tracking task [10].

As a special situation of alpha-divergence, the KL divergence is easy to calculate, but it provides only one measurement method. Therefore, the studies on the theory and related applications of the KL divergence are taken seriously. A discrete probability distribution of minimum Chi-square divergence is established [11]. Chi-square divergence is taken as a new criterion for image thresholding segmentation, obtaining better image segmentation results than that from the KL divergence [12,13]. It has been proven that the alpha-divergence minimization is equivalent to the α-integration of stochastic models, and it is applied to the multiple-expert decision-making system [6]. Amari et al. [14] also proved that the alpha-divergence is the only divergence category, which belongs to both f-divergence and Bregman divergence, so it has information monotonicity, a geometric structure with Fisher's measurement and a dual flat geometric structure. Gultekin et al. [15] proposed to use Monte Carlo integration to optimize the minimization equation of alpha-divergence, but this does not prove the alpha-divergence minimization. In [16], the application of the alpha-divergence minimization in approximate reasoning has been systematically analyzed, and different values of α can change the algorithm between the variational Bayesian algorithm and expectation propagation algorithm. As a special situation of the alpha-divergence ($\alpha = 2q - 1$), q-entropy [17,18] has been widely used in the field of physics. Li et al. [19] proposed a new class of variational inference methods using a variant of the alpha-divergence, which is called Rényi divergence, and applied it to the variational auto-encoders and Bayesian neural networks. There are more introductions about theories and applications of the alpha-divergence in [20,21]. Although the theories and applications of alpha-divergence have been very popular, we focus on providing a theory to perfect the alpha-divergence minimization and apply it to non-linear filtering.

3. Background Work

In Section 3.1, we provide the framework of the non-linear filtering. Then, we introduce the alpha-divergence in Section 3.2, which contains many types of divergence as special cases.

3.1. Non-Linear Filtering

The actual system studied in the filtering is usually non-linear and non-Gaussian. Non-linear filtering refers to a filtering that can estimate the optimal estimation problem of the state variables in the dynamic system online and in real time from the system observations.

The state space model of non-linear systems with additive Gaussian white noise is:

$$x_k = f(x_{k-1}) + w_{k-1} \tag{1}$$

where $x_k \in R^n$ is the system state vector that needs to be estimated; w_k is the zero mean value Gaussian white noise, and its variance is $E[w_k w_k^T] = Q_k$. Equation (1) describes the state transition $p(x_k|x_{k-1})$ of the system.

The random observation model of the state vector is:

$$z_k = h(x_k) + v_k \tag{2}$$

where $z_k \in R^m$ is system measurement; v_k is the zero mean value Gaussian white noise, and its variance is $E[v_k v_k^T] = R_k$. Suppose w_k and v_k are independent of each other and the observed value z_k is independent of the state variables x_k.

The entire probability state space is represented by the generation model as shown in Figure 1. x_k is the system state; z_k is the observational variable, and the purpose is to estimate the value of state x_k. The Bayesian filter is a general method to solve state estimation. The Bayesian filter is used to calculate the posterior distribution $p(x_k|z_k)$, and its recursive solution consists of prediction steps and update steps.

Under the Bayesian optimal filter framework, the system state equation determines that the conditional transition probability of the current state is a Gaussian distribution:

$$p(x_k|x_{k-1}, z_{1:k-1}) = N(x_k|f(x_{k-1}), Q_k) \tag{3}$$

If the prediction distribution of the system can be obtained from Chapman–Kolmogorov, the prior probability is:

$$p(x_k|z_{1:k-1}) = \int p(x_k|x_{k-1}, z_{1:k-1})p(x_{k-1}|z_{1:k-1})dx_{k-1} \tag{4}$$

When there is a measurement input, the system measurement update equation determines that the measurement likelihood transfer probability of the current state obeys a Gaussian distribution:

$$p(z_k|x_k, z_{1:k-1}) = N(z_k|h(x_k), R_k) \tag{5}$$

According to the Bayesian information criterion, the posterior probability obtained is:

$$p(x_k|z_{1:k}) = \frac{p(z_k|x_k, z_{1:k-1})p(x_k|z_{1:k-1})}{p(z_k|z_{1:k-1})} \tag{6}$$

where $p(z_k|z_{1:k-1})$ is the normalized factor, and it is defined as follows:

$$p(z_k|z_{1:k-1}) = \int p(z_k|x_k, z_{1:k-1})p(x_k|z_{1:k-1})dx_k \tag{7}$$

Unlike the Kalman filter framework, the Bayesian filter framework does not demand that the update structure be linear, so it can use non-linear update steps.

In the non-linear filtering problem, the posterior distribution $p(x_k|z_{1:k})$ often cannot be solved correctly. Our purpose is to use the distribution $q(x)$ to approximate the posterior distribution $p(x_k|z_{1:k})$ without an analytical solution. Here, we use the alpha-divergence measurement to measure the similarity between the two. We propose a method that directly minimizes alpha-divergence without adding any additional approximations.

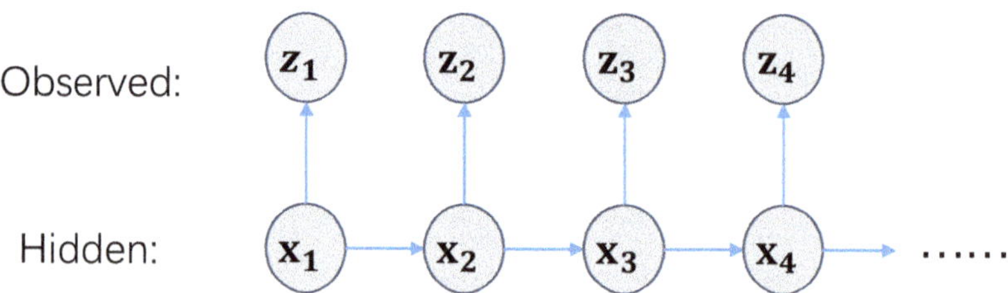

Figure 1. Hidden Markov Model (HMM).

3.2. The Alpha-Divergence

The KL divergence is commonly used in similarity measures, but we will generalize it to the alpha-divergence. The alpha-divergence is a parametric family of divergence functions, including several well-known divergence measures as special cases, and it gives us more flexibility in approximation [20].

Definition 1. *Let us consider two unnormalized distributions $p(x)$ and $q(x)$ with respect to a random variable x. The alpha-divergence is defined by:*

$$D_\alpha[p||q] = \frac{1}{\alpha(1-\alpha)} \int \alpha p(x) + (1-\alpha)q(x) - p(x)^\alpha q(x)^{1-\alpha} dx \tag{8}$$

where $\alpha \in R$, which means D_α is continuous at zero and one.

The alpha-divergence meets the following two properties:

1. $D_\alpha[p||q] \geq 0$, if and only if $p = q$, $D_\alpha[p||q] = 0$. This property can be used precisely to measure the difference between the two distributions.
2. $D_\alpha[p||q]$ is a convex function with respect to $p(x)$ and $q(x)$.

Note that the term $\int [\alpha p(x) + (1-\alpha)q(x)]dx$ disappears when $p(x)$ and $q(x)$ are normalized distributions, i.e., $\int p(x)dx = \int p(x)dx = 1$. The alpha-divergence in (8) is expressed by:

$$D_\alpha[p||q] = \frac{1}{\alpha(1-\alpha)}\left(1 - \int p(x)^\alpha q(x)^{1-\alpha} dx\right) \tag{9}$$

In general, we can get another equivalent expression of the alpha-divergence when we set $\beta = 2\alpha - 1$:

$$D_\beta[p||q] = \frac{4}{1-\beta^2} \int \frac{1-\beta}{2}p(x) + \frac{1+\beta}{2}q(x) - p(x)^{\frac{1+\beta}{2}} q(x)^{\frac{1-\beta}{2}} dx \tag{10}$$

Alpha-divergence includes several special cases such as the KL divergence, the Hellinger divergence and χ^2 divergence (Pearson's distance), which are summarized below.

- As α approaches one, Equation (8) is the limitation form of $\frac{0}{0}$, and it specializes to the KL divergence from $q(x)$ to $p(x)$ as L'Hôpital's rule is used:

$$\lim_{\alpha \to 1} D_\alpha[p||q] = \lim_{\alpha \to 1} \frac{1}{\alpha(1-\alpha)} \int \alpha p(x) + (1-\alpha)q(x) - p(x)^\alpha q(x)^{1-\alpha} dx$$
$$= \lim_{\alpha \to 1} \frac{1}{1-2\alpha} \int p(x) - q(x) - p(x)^\alpha log(p(x))q(x)^{1-\alpha} + p(x)^\alpha q(x)^{1-\alpha} log(q(x)) dx \tag{11}$$
$$= \int p(x)log\frac{p(x)}{q(x)} - p(x) + q(x)dx = KL[p||q]$$

When $p(x)$ and $q(x)$ are normalized distributions, the KL divergence is expressed as:

$$KL[p||q] = \int p(x)log\frac{p(x)}{q(x)}dx \tag{12}$$

- As α approaches zero, Equation (8) is still the limitation form of $\frac{0}{0}$, and it specializes to the dual form of the KL divergence from $q(x)$ to $p(x)$ as L'Hôpital's rule is used:

$$\lim_{\alpha \to 0} D_\alpha[p||q] = \lim_{\alpha \to 0} \frac{1}{\alpha(1-\alpha)} \int \alpha p(x) + (1-\alpha)q(x) - p(x)^\alpha q(x)^{1-\alpha} dx$$
$$= \lim_{\alpha \to 0} \frac{1}{1-2\alpha} \int p(x) - q(x) - p(x)^\alpha log(p(x))q(x)^{1-\alpha} + p(x)^\alpha q(x)^{1-\alpha} log(q(x)) dx \tag{13}$$
$$= \int q(x)log\frac{q(x)}{p(x)} + p(x) - q(x)dx = KL[q||p]$$

When $p(x)$ and $q(x)$ are normalized distributions, the dual form of the KL divergence is expressed as:

$$KL[q||p] = \int q(x)log\frac{q(x)}{p(x)}dx \tag{14}$$

- When $\alpha = \frac{1}{2}$, the alpha-divergence specializes to the Hellinger divergence, which is the only dual divergence in the alpha-divergence:

$$D_{\frac{1}{2}}[p||q] = 2\int (p(x) + q(x) - 2p(x)^{\frac{1}{2}}q(x)^{\frac{1}{2}})dx = 2\int (\sqrt{p(x)} - \sqrt{q(x)})^2 dx = 4Hel^2[p||q] \quad (15)$$

where $Hel[p||q] = \frac{1}{2}\int (\sqrt{p(x)} - \sqrt{q(x)})^2 dx$ is the Hellinger distance, which is the half of the Euclidean distance between two random distributions after taking the difference of the square root, and it corresponds to the fundamental property of distance measurement and is a valid distance metric.

- When $\alpha = 2$, the alpha-divergence degrades to χ^2-divergence:

$$\begin{aligned}
D_2[p||q] &= \frac{-1}{2}\left(\int 2p(x) - q(x) - \frac{p(x)^2}{q(x)}dx\right) \\
&= \frac{1}{2}\left(\int \frac{p(x)^2 + q(x)^2 - 2p(x)q(x)}{q(x)}dx\right) = \frac{1}{2}\int \frac{(p(x) - q(x))^2}{q(x)}dx
\end{aligned} \quad (16)$$

In the later experiment, we will adapt the value of α to optimize the distribution similarity measurement.

4. Non-Linear Filtering Based on the Alpha-Divergence

We first define an α-mixed probability density function, which will be used in the non-linear filtering based on the alpha-divergence minimization. Then, we show that the sufficient condition for the alpha-divergence minimization is when $\alpha \geq 1$ and the expected value of the natural statistical vector of $q(x)$ is equivalent to the expected value of the natural statistical vector of the α-mixed probability density function. At last, we apply the sufficient condition to the non-linear measurement update steps for solving the non-linear filtering problem.

4.1. The α-Mixed Probability Density Function

We first give a definition of a normalized probability density function called the α-mixed probability density function, which is expressed as $p_\alpha(x)$.

Definition 2. *We define an α-mixed probability density function:*

$$p_\alpha(x) = \frac{p(x)^\alpha q(x)^{(1-\alpha)}}{\int p(x)^\alpha q(x)^{(1-\alpha)}dx} \quad (17)$$

We can prove that when both $p(x)$ and $q(x)$ are univariate normal distributions, then $p_\alpha(x)$ is still the Gaussian probability density function.

Suppose that $p(x) \sim N(\mu_p, \sigma_p^2)$ and $q(x) \sim N(\mu_q, \sigma_q^2)$, so the probability density functions can be expressed as follows:

$$p(x) = \frac{1}{\sqrt{2\pi}\sigma_p}exp\left\{-\frac{(x - \mu_p)^2}{2\sigma_p^2}\right\} \quad \text{and} \quad q(x) = \frac{1}{\sqrt{2\pi}\sigma_q}exp\left\{-\frac{(x - \mu_q)^2}{2\sigma_q^2}\right\} \quad (18)$$

Then we can combine these two functions with parameter α:

$$\begin{aligned}
p(x)^\alpha q(x)^{(1-\alpha)} &= (2\pi\sigma_p^2)^{-\frac{\alpha}{2}}(2\pi\sigma_q^2)^{-\frac{1-\alpha}{2}}exp\left\{-\frac{\alpha(x - \mu_p)^2\sigma_q^2 + (1 - \alpha)(x - \mu_q)^2\sigma_p^2}{2\sigma_p^2\sigma_q^2}\right\} \\
&= \frac{S_\alpha}{\sqrt{2\pi}\sigma_\alpha}exp\left\{-\frac{(x - \mu_\alpha)^2}{2\sigma_\alpha^2}\right\}
\end{aligned} \quad (19)$$

where $\mu_\alpha = \frac{\alpha\mu_p\sigma_q^2 + (1-\alpha)\mu_q\sigma_p^2}{\alpha\sigma_q^2 + (1-\alpha)\sigma_p^2}$ is the mean of the α-mixed probability density function; $\sigma_\alpha^2 = \frac{\sigma_q^2\sigma_p^2}{\alpha\sigma_q^2 + (1-\alpha)\sigma_p^2}$ (which can be reduced to $\frac{1}{\sigma_\alpha} = \alpha\frac{1}{\sigma_p} + (1-\alpha)\frac{1}{\sigma_q}$) is the variance of the α-mixed probability density function; S_α is a scalar factor, and the expression is as follows:

$$
\begin{aligned}
S_\alpha &= (2\pi\sigma_\alpha^2)^{\frac{1}{2}}(2\pi\sigma_p^2)^{-\frac{\alpha}{2}}(2\pi\sigma_q^2)^{-\frac{1-\alpha}{2}}\exp\left\{-\frac{\alpha(1-\alpha)(\mu_p-\mu_q)^2}{2[\alpha\sigma_q^2 + (1-\alpha)\sigma_p^2]}\right\} \\
&= (2\pi\sigma_\alpha^2)^{\frac{\alpha+1-\alpha}{2}}(2\pi\sigma_p^2)^{-\frac{\alpha}{2}}(2\pi\sigma_q^2)^{-\frac{1-\alpha}{2}}\exp\left\{-\frac{\alpha(1-\alpha)(\mu_p-\mu_q)^2}{2[\alpha\sigma_q^2 + (1-\alpha)\sigma_p^2]}\right\} \\
&= \left(\frac{\sigma_q^2}{\alpha\sigma_q^2 + (1-\alpha)\sigma_p^2}\right)^{\frac{\alpha}{2}}\left(\frac{\sigma_p^2}{\alpha\sigma_q^2 + (1-\alpha)\sigma_p^2}\right)^{\frac{1-\alpha}{2}}\exp\left\{-\frac{\alpha(1-\alpha)(\mu_p-\mu_q)^2}{2[\alpha\sigma_q^2 + (1-\alpha)\sigma_p^2]}\right\}
\end{aligned}
\tag{20}
$$

Therefore, $p_\alpha(x)$ is a normalized probability density function, satisfying the normalization conditions $\int p_\alpha(x)dx = 1$. It is clear that the product of two Gaussian distributions is still a Gaussian distribution, which will bring great convenience to the representation of probability distribution of the latter filtering problem.

At the same time, we can get that the variance of $p_\alpha(x)$ is σ_α^2, which should satisfy the condition that its value is greater than zero. We can know by its denominator when $\sigma_q^2 \geq \sigma_p^2$, the value of α can take any value on the real number axis; when $\sigma_q^2 < \sigma_p^2$, the scope of α is $\alpha < \frac{\sigma_p^2}{\sigma_p^2 - \sigma_q^2}$. Then, it is easy to know that the closer σ_p^2 is to σ_q^2, the greater the range of values of α.

In addition, the influence of the mean and the variance of the two distributions on the mean and variance of the α-mixed probability density function can be analyzed to facilitate the solution of the algorithm latter. As for the variance, when $\sigma_q^2 > \sigma_p^2$, σ_α^2 decreases with the increase of α; when $\sigma_q^2 = \sigma_p^2$, it can be concluded that $\sigma_\alpha^2 = \sigma_q^2 = \sigma_p^2$; when $\sigma_q^2 < \sigma_p^2$, σ_α^2 increases with the increase of α. As for the mean value, when $\sigma_q^2 = \sigma_p^2$, $\mu_\alpha = (\mu_p - \mu_q)\alpha + \mu_q$; if $\sigma_q^2 \neq \sigma_p^2$, $\mu_\alpha = \frac{\mu_p\sigma_q^2 - \mu_q\sigma_p^2}{\sigma_q^2 - \sigma_p^2} + \frac{(\mu_q - \mu_p)\sigma_q^2\sigma_p^2}{(\sigma_q^2 - \sigma_p^2)^2\alpha + (\sigma_q^2 - \sigma_p^2)\sigma_p^2}$. It is clear that if $\mu_p > \mu_q$, then μ_α increases with the increase of α; if $\mu_p < \mu_q$, then μ_α decreases with the increase of α. The summary of the properties is shown in Table 1.

Table 1. The monotonicity of the mean μ_α and the variance σ_α^2 of the α-mixed probability density function.

	$\sigma_q^2 < \sigma_p^2$	$\sigma_q^2 = \sigma_p^2$	$\sigma_q^2 > \sigma_p^2$
	σ_α^2 **Increases with the Increase of** α	$\sigma_\alpha^2 = \sigma_q^2 = \sigma_p^2$	σ_α^2 **Decreases with the Increase of** α
$\mu_p > \mu_q$		μ_α increases with the increase of α	
$\mu_p = \mu_q$		$\mu_\alpha = \mu_p = \mu_q$	
$\mu_p < \mu_q$		μ_α decreases with the increase of α	

The monotonicity of the mean μ_α and the variance σ_α^2 with respect to α is shown in Figure 2.

It is clear that when $\mu_p < \mu_q$ and $\sigma_q^2 > \sigma_p^2$, μ_α decreases with the increase of α and σ_α^2 decreases with the increase of α; when $\mu_p < \mu_q$ and $\sigma_q^2 < \sigma_p^2$, μ_α decreases with the increase of α and σ_α^2 increases with the increase of α; when $\mu_p > \mu_q$ and $\sigma_q^2 > \sigma_p^2$, μ_α increases with the increase of α and σ_α^2 decreases with the increase of α; when $\mu_p > \mu_q$ and $\sigma_q^2 < \sigma_p^2$, μ_α increases with the increase of α and σ_α^2 increases with the increase of α.

When $\alpha \in (0, 1)$, the α-mixed probability density function is the interpolation function of $p(x)$ and $q(x)$, so its mean value and the variance are all between $p(x)$ and $q(x)$, as shown in Figure 2, and its image curve is also between them.

The above analysis will be used in the algorithm implementation of the sufficient condition in the non-linear filtering algorithm.

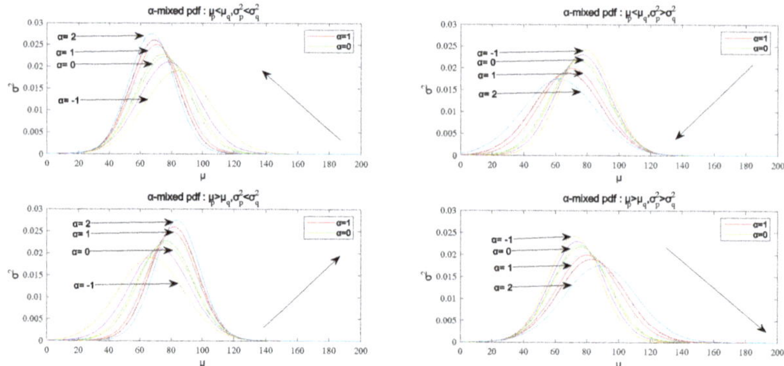

Figure 2. The monotonicity of the mean μ_α and the variance σ_α^2 with respect to α.

4.2. The Alpha-Divergence Minimization

In the solving process of the alpha-divergence minimization, either the posterior distribution itself or the calculation of the maximized posterior distribution is complex, so the approximate distribution $q(x)$ with good characterization ability is often used to approximate the true posterior distribution $p(x)$. As a result, a higher degree achieves better approximation. Here, we restrict the approximate distribution $q(x)$ to be an exponential family distribution; denote $p_e(x)$, with good properties, defined as follows:

$$p_e(x) = h(x)exp\left\{\phi^T(\theta)u(x) + g(\phi(\theta))\right\} \tag{21}$$

Here, θ is a parameter set of probability density function; $c(x)$ and $g(\phi(\theta))$ are known functions; $\phi(\theta)$ is a vector composed of natural parameters; $u(x)$ is a natural statistical vector. $u(x)$ contains enough information to express the state variable x in the exponential family distribution completely; $\phi(\theta)$ is a coefficient parameter that combines $u(x)$ based on parameter set θ.

In the non-linear filtering, assume the exponential family distribution is $p_e(x)$; arbitrary function is $p(x)$, and we use $p_e(x)$ to approximate $p(x)$, measuring the degree of approximation by the alpha-divergence. Therefore, the alpha-divergence of $p(x)$ relative to $p_e(x)$ is obtained, defined as:

$$\begin{aligned}
J = D_\alpha[p\|p_e] &= \frac{1}{\alpha(1-\alpha)}[1 - \int p(x)^\alpha p_e(x)^{1-\alpha}] \\
&= \frac{1}{\alpha(1-\alpha)}\left\{1 - \int p(x)^\alpha[h(x)exp(\phi^T(\theta)u(x) + g(\phi(\theta)))]^{1-\alpha}\right\}
\end{aligned} \tag{22}$$

We state and prove in Theorem 1 that the alpha-divergence between the exponential family distribution and the probability density function of arbitrary state variable is minimum, if and only if the expected value of the natural statistical vector in the exponential family distribution is equal to the expected value of the natural statistical vector in the α-mixed probability state density function. In Corollary 1, given $\alpha = 1$, the equivalence condition can be obtained in the case of $KL[p\|q]$. In Corollary 2, we conclude that the specialization of the exponential family distribution is obtained after being processed by the Gaussian probability density function.

Theorem 1. *The alpha-divergence between the exponential family distribution and the known state probability density function takes the minimum value; if and only if $\alpha \geq 1$, the expected value of the natural statistical vector in the exponential family distribution is equal to the expected value of the natural statistical vector in the α-mixed probability state density function, that is:*

$$E_{p_e}\{u(x)\} = E_{p_\alpha}\{u(x)\} \tag{23}$$

Proof of Theorem 1. Sufficient conditions for J minimization are that the first derivative and the second derivative satisfy the following conditions:

$$\frac{\partial J}{\partial \phi(\theta)} = 0 \quad and \quad \frac{\partial^2 J}{\partial \phi(\theta)^2} > 0 \tag{24}$$

First, we derive Equation (22) with respect to $\phi(\theta)$, and according to the conditions in the first derivative, the outcome is:

$$
\begin{aligned}
\frac{\partial J}{\partial \phi(\theta)} &= \frac{-1}{\alpha(1-\alpha)} \int p(x)^{\alpha}(1-\alpha)p_e(x)^{-\alpha}p_e(x)\left\{u(x) + \left(\frac{\partial g(\phi(\theta))}{\partial \phi(\theta)}\right)\right\}dx \\
&= \frac{-1}{\alpha} \int p(x)^{\alpha}p_e(x)^{1-\alpha}\left\{u(x) + \left(\frac{\partial g(\phi(\theta))}{\partial \phi(\theta)}\right)\right\}dx \\
&= -\frac{1}{\alpha} \int p(x)^{\alpha}p_e(x)^{1-\alpha}u(x)dx - \frac{1}{\alpha} \int p(x)^{\alpha}p_e(x)^{1-\alpha}\left(\frac{\partial g(\phi(\theta))}{\partial \phi(\theta)}\right)dx
\end{aligned}
\tag{25}
$$

Let the above equation be equal to zero, then:

$$\frac{\partial g(\phi(\theta))}{\partial \phi(\theta)} = -\int \frac{p(x)^{\alpha}p_e(x)^{1-\alpha}}{\int p(x)^{\alpha}p_e(x)^{1-\alpha}dx}u(x)dx = -\int p_{\alpha}(x)u(x)dx \tag{26}$$

In addition, since $p_e(x)$ is a probability density function, it satisfies the normalization condition:

$$\int p_e(x)dx = \int h(x)exp\left\{\phi^T(\theta)u(x) + g(\phi(\theta))\right\}dx = 1 \tag{27}$$

Derive $\phi(\theta)$ in the above equation, and the outcome is:

$$\frac{\partial}{\partial \phi(\theta)}p_e(x) = \int p_e(x)u(x)dx + \frac{\partial g(\phi(\theta))}{\partial \phi(\theta)} = 0 \tag{28}$$

The first item of Equation (23) can be obtained from Equations (26) and (28), which is the existence conditions of the stationary point for J.

To ensure that Equation (24) can minimize Equation (22), which means the stationary point is also its minimum point, we also need to prove that the second derivative satisfies the condition. Derive $\phi(\theta)$ in Equation (25); the outcome is:

$$
\begin{aligned}
\frac{\partial^2 J}{\partial \phi(\theta)^2} &= \frac{\partial}{\partial \phi(\theta)}\left\{-\frac{1}{\alpha}\int p(x)^{\alpha}p_e(x)^{1-\alpha}u(x)dx - \frac{1}{\alpha}\int p(x)^{\alpha}p_e(x)^{1-\alpha}\left(\frac{\partial g(\phi(\theta))}{\partial \phi(\theta)}\right)dx\right\} \\
&= -\frac{1-\alpha}{\alpha}\int p(x)^{\alpha}p_e(x)^{1-\alpha}\left\{u(x) + \left(\frac{\partial g(\phi(\theta))}{\partial \phi(\theta)}\right)\right\}u(x)dx \\
&\quad - \frac{1}{\alpha}\int p(x)^{\alpha}p_e(x)^{1-\alpha}\frac{\partial^2 g(\phi(\theta))}{\partial \phi(\theta)^2}dx \\
&\quad - \frac{1-\alpha}{\alpha}\int p(x)^{\alpha}p_e(x)^{1-\alpha}\left\{u(x) + \left(\frac{\partial g(\phi(\theta))}{\partial \phi(\theta)}\right)\right\}\frac{\partial g(\phi(\theta))}{\partial \phi(\theta)}dx \\
&= -\frac{1}{\alpha}\int p(x)^{\alpha}p_e(x)^{1-\alpha}\frac{\partial^2 g(\phi(\theta))}{\partial \phi(\theta)^2}dx \\
&\quad - \frac{1-\alpha}{\alpha}\int p(x)^{\alpha}p_e(x)^{1-\alpha}\left\{u(x)^2 + 2u(x)\left(\frac{\partial g(\phi(\theta))}{\partial \phi(\theta)}\right) + \left(\frac{\partial g(\phi(\theta))}{\partial \phi(\theta)}\right)^2\right\}dx \\
&= -\frac{1}{\alpha}\frac{\partial^2 g(\phi(\theta))}{\partial \phi(\theta)^2}\int p(x)^{\alpha}p_e(x)^{1-\alpha}dx + \frac{\alpha-1}{\alpha}\int p(x)^{\alpha}p_e(x)^{1-\alpha}\left\{u(x) + \frac{\partial g(\phi(\theta))}{\partial \phi(\theta)}\right\}^2dx \\
&= -\frac{1}{\alpha}\frac{\partial^2 g(\phi(\theta))}{\partial \phi(\theta)^2}\int p_{\alpha}(x)dx + \frac{\alpha-1}{\alpha}\int p(x)^{\alpha}p_e(x)^{1-\alpha}\left\{u(x) + \frac{\partial g(\phi(\theta))}{\partial \phi(\theta)}\right\}^2dx
\end{aligned}
\tag{29}
$$

For the first item, it is easy to prove $\frac{\partial^2 g(\phi(\theta))}{\partial \phi(\theta)^2} < 0$, and the proof is as follows.

It can be known from Equation (21):

$$g(\phi(\theta)) = -\log \int h(x)exp\left\{\phi^T(\theta)u(x)\right\}dx \tag{30}$$

The gradient of Equation (30) with respect to the natural parameter vector is as follows:

$$\frac{\partial g(\phi(\theta))}{\partial \phi(\theta)} = -\int \frac{h(x)exp\left\{\phi^T(\theta)u(x)\right\}}{\int h(x)exp\left\{\phi^T(\theta)u(x)\right\}dx}u(x)dx =$$
$$= -\int \frac{h(x)exp\left\{\phi^T(\theta)u(x)\right\}}{exp\left\{-g(\phi(\theta))\right\}dx}u(x)dx = -\int p_e(x)u(x)dx \tag{31}$$

Then, consider the matrix formed by its second derivative with respect to the natural parameter vector:

$$\frac{\partial^2 g(\phi(\theta))}{\partial \phi^i(\theta)\partial \phi^j(\theta)} = -\frac{\partial}{\partial \phi^j(\theta)}\int \frac{h(x)exp\left\{\phi^T(\theta)u(x)\right\}}{\int h(x)exp\left\{\phi^T(\theta)u(x)\right\}dx}u^i(x)dx$$
$$= -\frac{\partial}{\partial \phi^j(\theta)}\frac{\int h(x)exp\left\{\phi^T(\theta)u(x)\right\}u^i(x)dx}{\int h(x)exp\left\{\phi^T(\theta)u(x)\right\}dx}$$
$$= -\frac{\int h(x)exp\left\{\phi^T(\theta)u(x)\right\}u^i(x)dx\int h(x)exp\left\{\phi^T(\theta)\right\}u(x)dx}{(\int h(x)exp\left\{\phi^T(\theta)u(x)\right\}dx)^2} \tag{32}$$
$$+ \frac{\int h(x)exp\left\{\phi^T(\theta)u(x)\right\}u^i(x)dx\int h(x)exp\left\{\phi^T(\theta)u(x)\right\}u^j(x)dx}{(\int h(x)exp\left\{\phi^T(\theta)u(x)\right\}dx)^2}$$
$$= -\left\{\int p_e(x)u^i(x)u^j(x)dx - \int p_e(x)u^i(x)dx\int p_e(x)u^j(x)dx\right\}$$

According to the definition of the covariance matrix, the content in the bracket is the covariance matrix of the natural parameter vector with respect to the exponential family probability density function $p_e(x)$, and for arbitrary probability density distribution $p_e(x)$, the variance matrix is a positive definite matrix, so $\frac{\partial^2 g(\phi(\theta))}{\partial \phi(\theta)^2} < 0$; and when $\alpha > 0$, the first item is greater than zero.

The integral of the second item is the secondary moment, so $\alpha \geq 1$ or $\alpha < 0$, and the second item is greater than zero.

To sum up, when $\alpha \geq 1$, $\frac{\partial^2 J}{\partial \phi(\theta)^2} > 0$. $\square$

Corollary 1. *(See Theorem 1 of [3] for more details) When $\alpha = 1$, $p_\alpha(x) = p(x)$, $D_\alpha[p||q]$ turns into $KL[p||q]$. We can obtain the above theorem under the condition of $KL[p||q]$ and obtain the approximate distribution by minimizing the KL divergence, which also proves that the stationary point obtained when the first derivative of its KL divergence is equal to zero also satisfies the condition that its second derivative is greater than zero. The corresponding expectation propagation algorithm is shown as follows:*

$$E_{q(x)}\left\{u(x)\right\} = E_{p(x)}\left\{u(x)\right\} \tag{33}$$

Corollary 2. *(See Corollary 1.1 of [3] for more details) When the exponential family distribution is simplified as the Gaussian probability density function, its sufficient statistic for $u(x) = (x, x^2)$, we use the mean and variance of Gaussian probability density function, and the expectation of the corresponding propagation algorithm can use the moment matching method to calculate, so the first moment and the second moment are defined as follows:*

$$m = E_{p(x)}\left\{x\right\} \quad and \quad M = E_{p(x)}\left\{xx^T\right\} \tag{34}$$

The corresponding second central moment is defined as follows:

$$P = M - mm^T = E_{p(x)} \left\{ (x - m)(x - m)^T \right\} \tag{35}$$

The complexity of Theorem 1 lies in that both sides of Equation (23) depend on the probability distribution of $q(x)$ at the same time. The $q(x)$ that satisfies the condition can be obtained by repeated iterative update on $q(x)$. The specific process is shown in Algorithm 1:

Algorithm 1 Approximation of the true probability distribution $p(x)$.

Input: Target distribution parameter of $p(x)$; damping factor $\epsilon \in (0,1)$; divergence parameter $\alpha \in [1, +\infty)$; initialization value of $q(x)$

Output: The exponential family probability function $q(x)$

1: Calculate the α-mixed probability density function $p_\alpha(x)$
2: According to Equation (23), we get a new $q(x)$ using the expectation propagation algorithm described in Corollary 1, and the new $q(x)$ is denoted as $q'(x)$
3: Revalue the $q(x)$ as

$$q(x) = \frac{q(x)^\epsilon q'(x)^{1-\epsilon}}{\int q(x)^\epsilon q'(x)^{1-\epsilon} dx} \tag{36}$$

4: **while** $KL[p||q] > 0.01$ **do**
5:      Calculate the KL divergence of the old $q'(x)$ and the new $q(x)$
6: **end while**

In the above algorithms, we need to pay attention to the following two problems: giving an initial value of $q(x)$ and selecting damping factors. As for the first problem, we can know that when $\sigma_q^2 < \sigma_p^2$, the value range of α is $\alpha < \frac{\sigma_p^2}{\sigma_p^2 - \sigma_q^2}$, according to the analysis of the α-mixed probability density function in Section 4.1. Although the value of α is greater than one, the value range of α is limited under the condition that σ_p^2 is unknown in the initial state; when $\sigma_q^2 \geq \sigma_p^2$, the value of α can take any value on the whole real number axis, so the initial value we can choose is relatively larger, making $\sigma_q^2 \geq \sigma_p^2$ and $\mu_q > \mu_p$. When the value of α is greater than one, the mean value of the α-mixed probability density function will decrease, and the variance will also decrease, as shown in the upper left of Figure 2.

As for the second question, when $\alpha \in (0,1)$, the α-mixed probability density function is the interpolation function of $p(x)$ and $q(x)$ according to the analysis in Section 4.1. The value range in $(0,1)$ of damping factor ϵ is quite reasonable because the two probability density functions are interpolated when the value range of ϵ is in $(0,1)$, and the new probability density function is between the two. According to Equation (36), the smaller of ϵ, the closer the new $q(x)$ to the old $q(x)$; the larger of ϵ, the closer the new $q(x)$ to $q'(x)$. The mean value and the variance of $q'(x)$ is smaller than the real $p(x)$ according to the analysis of the first question. Then, we will continue to combine new $q(x)$ with $p(x)$ to form a α-mixed probability density function. Similarly, we clarify that the mean value and the variance of the new $q(x)$ are larger than $p(x)$, so the value of ϵ we choose should be as close as possible to one.

The convergence of the algorithm can be guaranteed after considering the above two problems, and we can get $q(x)$ that meets the conditions. It can be known from Theorem 1 that the approximation $q(x)$ of $p(x)$ can be obtained to ensure it converges on this minimum point after repeated iterative updates.

4.3. Non-Linear Filtering Algorithm Based on the Alpha-Divergence

In the process of non-linear filtering, assuming that a priori and a posteriori probability density functions satisfy the Assumed Density Filter (ADF), then define the prior parameter as $\theta_k^- = \{m_k^-, P_k^-\}$;

the corresponding distribution is prior distribution $q(x_k; \theta_k^-)$; define the posterior parameter as $\theta_k^+ = \{m_k^+, P_k^+\}$, then the corresponding distribution is posterior distribution $q(x_k; \theta_k^+)$.

The prediction of the state variance can be expressed as follows:

$$p(x_k|z_{1:k-1}) = \int p(x_k|x_{k-1}, z_{1:k-1})dx_{k-1} \tag{37a}$$

$$\theta_k^- = arg \min_{\theta} D_\alpha[p(x_k|z_{1:k-1})||q(x_k; \theta)] \tag{37b}$$

The corresponding first moment about the origin $f(x_{k-1}) = \int x_k p(x_k|z_{1:k-1})dx_k$ of $p(x_k|z_{1:k-1})$ can be obtained from Equation (37a).

By Corollary 2, when the alpha-divergence is simplified to the KL divergence, the corresponding mean value and variance are:

$$m_k^- = \int f(x_{k-1})q(x_{k-1}; \theta_{k-1}^+)dx \tag{38a}$$

$$P_k^- = \int f(x_{k-1})f(x_{k-1})^T N(x_{k-1}|m_{k-1}^+, P_{k-1}^+)dx - m_k^- m_k^{-T} + Q_k \tag{38b}$$

Here, the prior distribution $q(x_k; \theta_k^-)$ can be obtained.
Similarly, the update steps of the filter can be expressed as follows:

$$p(x_k|z_{1:k}) = \frac{p(z_k|x_k, z_{1:k-1})q(x_k; \theta_k^-)}{\int p(x_k|x_k, z_{1:k-1})q(x_k; \theta_k^-)dx_k} \tag{39a}$$

$$\theta_k^+ = arg \min_{\theta} D_\alpha[p(x_k|z_{1:k})||q(x_k; \theta)] \tag{39b}$$

It is clear according to Theorem 1:

$$E_{q(x_k; \theta_k^+)} \{u(x)\} = E_{p_\alpha(x)} \{u(x)\} = \int p_\alpha(x)u(x)dx = \int u(x)\frac{p(x_k|z_{1:k})^\alpha q(x_k; \theta_k^+)^{1-\alpha}}{\pi(x)}\pi(x)dx$$

$$\approx \sum_{i=1}^{N} u(x^i)\frac{[p(z_k|x_k^i, z_{1:k-1})q(x_k^i; \theta_k^-)]^\alpha q(x_k^i; \theta_k^+)^{(1-\alpha)}/\pi(x^i)}{\sum_j [p(z_k|x_k^j, z_{1:k-1})q(x_k^j; \theta_k^-)]^\alpha q(x_k^j; \theta_k^+)^{(1-\alpha)}/\pi(x^j)} \tag{40}$$

Here, $x^i \sim iid\pi_t(x_t), i = 1, \cdots, N$, π_t is the proposal distribution. We choose the proposal distribution as a priori distribution $q(x_k; \theta_k^-)$. We define $w^i = [p(z_k|x_k^i, z_{1:k-1})q(x_k^i; \theta_k^-)]^\alpha q(x_k^i; \theta_k^+)^{1-\alpha}/\pi(x^i)$, $W = \sum_j w^j$, so:

$$E_{q(x_k; \theta_k^+)} \{u(x)\} \approx \frac{1}{W} \sum_{i=1}^{N} w^i u(x^i) \tag{41}$$

An approximate calculation of the mean value and the variance for $q(x_k; \theta_k^+)$ is conducted:

$$m_k^+ = \frac{1}{W} \sum_{i=1}^{N} w^i x^i \tag{42a}$$

$$P_k^+ = \frac{1}{W} \sum_{i=1}^{N} w^i(x^i - m_k^i)(x^i - m_k^i)^T \tag{42b}$$

Since Equation (40) contains $q(x_k; \theta_k^+)$ on both sides of the equation, we must use Algorithm 1 to conduct the iterative calculation to get the satisfied posterior distribution $q(x_k; \theta_k^+)$.

If $\alpha = 1$, the above steps can be reduced to a simpler filtering algorithm, as shown in [3].

In this process, we do not use the integral operation of the denominator in Equation (39a), but use the Monte Carlo integral strategy proposed in [15], as shown in Equation (40). We cannot conduct resampling, which greatly reduces the calculation.

5. Simulations and Analysis

According to Theorem 1, when $\alpha \geq 1$, the non-linear filtering method we proposed is feasible theoretically. In the simulation experiment, the algorithm is validated by taking different values when $\alpha \geq 1$. We name our proposed method as AKF and compare it with the traditional non-linear filtering methods such as EKF and UKF.

We choose the Univariate Nonstationary Growth Model (UNGM) [22] to analyze the performance of the proposed method. The system state equation is:

$$x(k) = 0.5x(k-1) + \frac{2.5x(k-1)}{1+x^2(k-1)} + 8cos(1.2(k-1)) + w(k) \tag{43}$$

The observation equation is:

$$y(k) = \frac{x^2(k)}{20} + v(k) \tag{44}$$

The equation of state is a non-linear equation including the fractional relation, square relation and trigonometric function relation. $w(k)$ is the process noise with the mean value of zero and the variance of Q. The relationship between the observed signal $y(k)$ and state $v(k)$ in the measurement equation is also non-linear. $v(k)$ is the observation noise with the mean value of zero and the variance of R. Therefore, this system is a typical system with non-linear states and observations, and this model has become the basic model for verifying the non-linear filtering algorithm [22,23].

In the experiment, we set Q = 10, R = 1 and set the initial state as $p(x(1)) = N(x(1); 0, 1)$.

First, we simulate the system. When $\alpha \geq 1$, the values of α are right for the experiments; here, the value of α is two, and the entire experimental simulation time is T = 50. The result of the state estimation is shown in Figure 3, and it can be seen that the non-linear filtering method we proposed is feasible; the state value can be estimated well during the whole process, and its performance is superior to EKF and UKF in some cases.

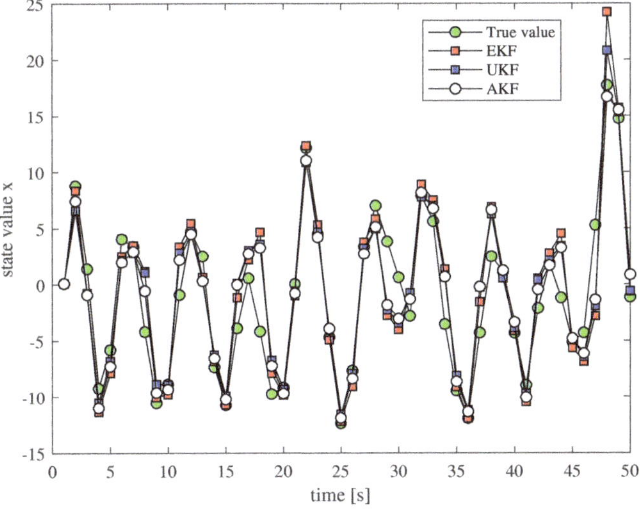

Figure 3. State estimation comparison of different non-linear filtering methods.

Second, in order to measure the accuracy of state estimation, the difference between the real state value at each moment and the estimated state value can be calculated to obtain the absolute value; thus, the absolute deviation of the state estimation at each moment is obtained, namely:

$$RMS(k) = |x_{real}(k) - x_{estimated}(k)| \tag{45}$$

As shown in Figure 4, we can see that the algorithm error we proposed is always relatively small where the absolute value deviation is relatively large. It can be seen that our proposed method performs better than other non-linear methods.

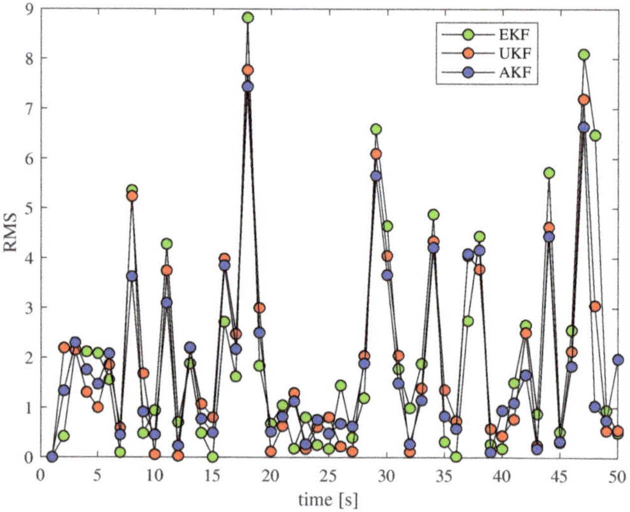

Figure 4. RMS comparison at different times.

In order to measure the overall level of error, we have done many simulation experiments. The average error of each experiment is defined as:

$$RMSE(k) = \frac{1}{T} \sum_{k-1}^{T} RMS(k) \tag{46}$$

The experimental results are shown in Table 2. We can see that when the estimation of T time series is averaged, the error mean of each AKF is minimum, which indicates the effectiveness of the algorithm, and the filtering accuracy of the algorithm is better than the other two methods under the same conditions. Because the UNGM has strong nonlinearity and we set the variance to the state noise as 10, which is quite large, so the performance differences between EKF, UKF and AKF are rather small.

Table 2. Average errors of experiments.

	1	2	3	4	5	6	7
EKF	1.6414	1.8434	1.8245	1.7749	1.6666	1.3255	⋯
UKF	1.5400	1.7703	1.6688	1.6387	1.6241	1.2243	⋯
AKF	1.4819	1.5921	1.4710	1.4694	1.4389	1.1222	⋯

Then, we analyze the influence of the initial value on the filtering results by modifying the value of process noise. As can be seen from Table 3, AKF's performance becomes more and more similar to EKF/UKF as the Q becomes smaller.

Table 3. Influence of the variance Q of state equation noise on experimental error.

Q	0.05	0.1	1	10
EKF	0.2256	0.2950	0.7288	1.7827
UKF	0.2222	0.3002	0.7396	1.6222
AKF	0.2167	0.2767	0.7144	1.5244

In the end, we analyze the performance of the whole non-linear filtering algorithm by adjusting the value of α through 20 experiments. In order to reduce the influence of the initial value on the experimental results, we take Q = 0.1 and then average the 20 experimental errors. The result is shown in Figure 5. We can see that the error grows as α grows in this example, as the noise is relatively small.

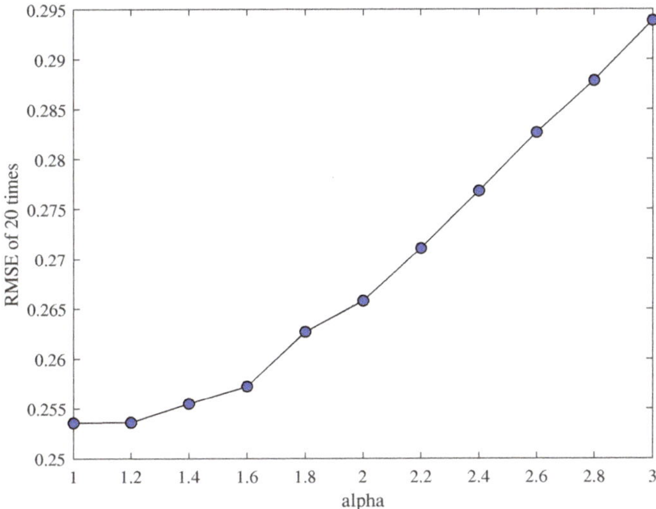

Figure 5. The error changes as α changes.

6. Conclusions

We have first defined the α-mixed probability density function and analyzed the monotonicity of the mean and the variance under different α values. Secondly, the sufficient conditions for α to find the minimum value have been proven, which provides more methods for measuring the distribution similarity of non-linear filtering. Finally, a non-linear filtering algorithm based on the alpha-divergence minimization has been proposed by applying the above two points to the non-linear filtering. Moreover, we have verified that the validity of the algorithm in one-dimensional UNGM.

Although the filtering algorithm is effective, the alpha-divergence is a direct extension of the KL divergence. We can try to verify that the minimum physical meaning of the alpha divergence is equivalent to the minimum physical meaning of the KL divergence in a further study. The algorithm should be applied to more practical applications to prove its effectiveness. Meanwhile, we can use more sophisticated particle filtering techniques, such as [24,25], to make the algorithm more efficient. Furthermore, the alpha-divergence method described above is applied to uni-modal approximations, but more attention should be paid to multi-modal distributions, which are more difficult and common in practical systems. Furthermore, it is worth designing a strategy to automatically learn the appropriate α values.

Author Contributions: Y.L. and C.G. conceived of and designed the method and performed the experiment. Y.L. and S.Y. participated in the experiment and analyzed the data. Y.L. and C.G. wrote the paper. S.Y. revised the paper. J.Z. guided and supervised the overall process.

Funding: This research was supported by a grant from the National Key Research and Development Program of China (2016YFB0501801).

Acknowledgments: The authors thank Dingyou Ma of Tsinghua University for his help.

Conflicts of Interest: The authors declare no conflict of interest.

References

1. Grewal, M.S.; Andrews, A.P. Applications of Kalman filtering in aerospace 1960 to the present [historical perspectives]. *IEEE Control Syst.* **2010**, *30*, 69–78.
2. Durrant-Whyte, H.; Bailey, T. Simultaneous localization and mapping: Part I. *IEEE Robot. Autom. Mag.* **2006**, *13*, 99–110. [CrossRef]
3. Darling, J.E.; Demars, K.J. Minimization of the Kullback Leibler Divergence for Nonlinear Estimation. *J. Guid. Control Dyn.* **2017**, *40*, 1739–1748. [CrossRef]
4. Amari, S. *Differential Geometrical Method in Statistics*; Lecture Note in Statistics; Springer: Berlin, Germany, 1985; Volume 28.
5. Minka, T. *Divergence Measures and Message Passing*; Microsoft Research Ltd.: Cambridge, UK, 2005.
6. Amari, S. Integration of Stochastic Models by Minimizing α-Divergence. *Neural Comput.* **2007**, *19*, 2780–2796. [CrossRef] [PubMed]
7. Raitoharju, M.; García-Fernández, Á.F.; Piché, R. Kullback–Leibler divergence approach to partitioned update Kalman filter. *Signal Process.* **2017**, *130*, 289–298. [CrossRef]
8. Mansouri, M.; Nounou, H.; Nounou, M. Kullback–Leibler divergence-based improved particle filter. In Proceedings of the 2014 IEEE 11th International Multi-Conference on Systems, Signals & Devices (SSD), Barcelona, Spain, 11–14 February 2014; pp. 1–6.
9. Martin, F.; Moreno, L.; Garrido, S.; Blanco, D. Kullback–Leibler Divergence-Based Differential Evolution Markov Chain Filter for Global Localization of Mobile Robots. *Sensors* **2015**, *15*, 23431–23458. [CrossRef] [PubMed]
10. Hu, C.; Lin, H.; Li, Z.; He, B.; Liu, G. Kullback–Leibler Divergence Based Distributed Cubature Kalman Filter and Its Application in Cooperative Space Object Tracking. *Entropy* **2018**, *20*, 116. [CrossRef]
11. Kumar, P.; Taneja, I.J. Chi square divergence and minimization problem. *J. Comb. Inf. Syst. Sci.* **2004**, *28*, 181–207.
12. Qiao, W.; Wu, C. Study on Image Segmentation of Image Thresholding Method Based on Chi-Square Divergence and Its Realization. *Comput. Appl. Softw.* **2008**, *10*, 30.
13. Wang, C.; Fan, Y.; Xiong, L. Improved image segmentation based on 2-D minimum chi-square-divergence. *Comput. Eng. Appl.* **2014**, *18*, 8–13.
14. Amari, S. Alpha-Divergence Is Unique, Belonging to Both f-Divergence and Bregman Divergence Classes. *IEEE Trans. Inf. Theory* **2009**, *55*, 4925–4931. [CrossRef]
15. Gultekin, S.; Paisley, J. Nonlinear Kalman Filtering with Divergence Minimization. *IEEE Trans. Signal Process.* **2017**, *65*, 6319–6331. [CrossRef]
16. Hernandezlobato, J.M.; Li, Y.; Rowland, M.; Bui, T.D.; Hernandezlobato, D.; Turner, R.E. Black Box Alpha Divergence Minimization. In Proceedings of the International Conference on Machine Learning, New York, NY, USA, 19–24 June 2016; pp. 1511–1520.
17. Tsallis, C. Possible Generalization of Boltzmann-Gibbs Statistics. *J. Stat. Phys.* **1988**, *52*, 479–487. [CrossRef]
18. Tsallis, C. Introduction to Nonextensive Statistical Mechanics. *Condens. Matter Stat. Mech.* **2004**. [CrossRef]
19. Li, Y.; Turner, R.E. Rényi divergence variational inference. In Proceedings of the Advances in Neural Information Processing Systems, Barcelona, Spain, 5–10 December 2016; pp. 1073–1081.
20. Amari, S.I. *Information Geometry and Its Applications*; Springer: Berlin, Germany, 2016.
21. Nielsen, F.; Critchley, F.; Dodson, C.T.J. *Computational Information Geometry*; Springer: Berlin, Germany, 2017.
22. Garcia-Fernandez, Á.F.; Morelande, M.R.; Grajal, J. Truncated unscented Kalman filtering. *IEEE Trans. Signal Process.* **2012**, *60*, 3372–3386. [CrossRef]

23. Li, Y.; Cheng, Y.; Li, X.; Hua, X.; Qin, Y. Information Geometric Approach to Recursive Update in Nonlinear Filtering. *Entropy* **2017**, *19*, 54. [CrossRef]
24. Martino, L.; Elvira, V.; Camps-Valls, G. Group Importance Sampling for particle filtering and MCMC. *Dig. Signal Process.* **2018**, *82*, 133–151. [CrossRef]
25. Salomone, R.; South, L.F.; Drovandi, C.C.; Kroese, D.P. Unbiased and Consistent Nested Sampling via Sequential Monte Carlo. *arXiv* **2018**, arXiv:1805.03924.

Article

A Weak Selection Stochastic Gradient Matching Pursuit Algorithm

Liquan Zhao [1,*] , **Yunfeng Hu** [1] and **Yanfei Jia** [2]

[1] Key Laboratory of Modern Power System Simulation and Control & Renewable Energy Technology, Ministry of Education (Northeast Electric Power University), Jilin 132012, China; huyunfeng22@163.com
[2] College of Electrical and Information Engineering, Beihua University, Jilin 132013, China; jia_yanfei@163.com
* Correspondence: zhao_liquan@163.com; Tel.: +86-150-4320-1901

Received: 20 April 2019; Accepted: 15 May 2019; Published: 21 May 2019

Abstract: In the existing stochastic gradient matching pursuit algorithm, the preliminary atomic set includes atoms that do not fully match the original signal. This weakens the reconstruction capability and increases the computational complexity. To solve these two problems, a new method is proposed. Firstly, a weak selection threshold method is proposed to select the atoms that best match the original signal. If the absolute gradient coefficients were greater than the product of the maximum absolute gradient coefficient and the threshold that was set according to the experiments, then we selected the atoms that corresponded to the absolute gradient coefficients as the preliminary atoms. Secondly, if the scale of the current candidate atomic set was equal to the previous support atomic set, then the loop was exited; otherwise, the loop was continued. Finally, before the transition estimation of the original signal was calculated, we determined whether the number of columns of the candidate atomic set was smaller than the number of rows of the measurement matrix. If this condition was satisfied, then the current candidate atomic set could be regarded as the support atomic set and the loop was continued; otherwise, the loop was exited. The simulation results showed that the proposed method has better reconstruction performance than the stochastic gradient algorithms when the original signals were a one-dimensional sparse signal, a two-dimensional image signal, and a low-rank matrix signal.

Keywords: compressed sensing; low rank matrix; stochastic gradient; weak selection method; reliability verification strategy; reconstruction performance

---

1. Introduction

Compressed sensing (CS) [1–4] has been receiving considerable attention. The main premise of CS theory is that the reconstruction of a high-dimensional sparse (or compressive) original signal from a low-dimensional linear measurement vector under the measurement matrix should satisfy the restricted isometry property (RIP) [5]. At present, CS is divided into the following three core aspects: Sparse representation of the signal, nonrelated linear measurements, and signal reconstruction. The sparse representation of the signal is used as the design basis for the over-complete dictionary [6,7] with the capability of sparse representation, such as discrete cosine transform (DCT), wavelet transform (WT), and Fourier transform (FT). These functions are used as the sparse representation of the signal, where they obtain a fine effect. Unrelated linear measurement is used to design the measurement matrix [8] that satisfies the RIP condition. The commonly used measurement matrices include the Gaussian random matrix, the Bernoulli random matrix, and the partial Hadamard matrix. In this study, we focused mainly on the signal reconstruction.

Signal reconstruction methods can be divided into two categories: Those based on the minimized l_1-norm problem, and the greedy pursuit algorithm based on the minimized l_0-norm problem.

Those in the first category include methods such as the basis pursuit (BP) [9] algorithm and its optimization algorithm, the gradient projection for sparse reconstruction algorithm (GPSR) [10], the iterative threshold (IT) [11], the interior point method [12], and the Bergman iteration (BT) [13] method. These algorithms are generally used to solve the convex optimization problems. The convex optimization algorithms have a better reconstruction performance and theoretical performance guarantees; however, they are sensitive to noise and usually suffer from heavy computational complexity when processing large signal reconstruction problems. The second category includes methods such as the matching pursuit (MP) [14], orthogonal matching pursuit (OMP) [15], regularized OMP (ROMP) [16], and stage-wise OMP (StOMP) [17]. These algorithms offer much faster running times than the convex optimization methods, but they lack comparable strong reconstruction guarantees. Greedy pursuit algorithms, such as subspace pursuit (SP) [18], compressive sampling matching pursuit (CoSaMP) [19,20], and iterative hard threshold (IHT) [21] algorithms, have faster running times and essentially the same reconstruction guarantees, but these algorithms are only suitable for one-dimensional (1D) signals in compressed sensing.

Several algorithms that are deemed suitable for a 1D signal and multidimensionality signals have been proposed. Ding et al. [22] and Rantzer et al. [23] proposed the forward selection method for sparse signal and low rank matrix reconstruction problems. The algorithm iteratively selects each nonzero element or each rank-one matrix. Wassell et al. [24] proposed a more general sparse basis based on previous studies [22,23]. Liu et al. [25] proposed the forward-backward method, where the atoms can be completely added or removed from the set. Bresler et al. [26] extended this algorithm beyond the quadratic loss studied in Liu et al. [25]. Soltani et al. [27] proposed an improved CoSaMP algorithm for a more general form objective function. Bahmani et al. [28] used the gradient matching pursuit (GradMP) algorithm to solve the reconstruction problem of large-scale class signals with sparsity constraints based on the CoSaMP algorithm. However, for large-scale class signal reconstruction problems, the GradMP algorithm needs to compute the full gradient of the objective function, which greatly increases the computation cost of the algorithm. Therefore, Needell et al. [29] proposed a stochastic version of the GradMP algorithm that was called the StoGradMP algorithm. Compared with the GradMP algorithm, the StoGradMP algorithm randomly selects an index and computes its associated gradient at each iteration. This operation is extremely effective for large-scale signal recovery problem.

Although the StoGradMP algorithm effectively reduces the computational cost of the algorithm, its reconstruction capability still needs improvement. In the StoGradMP algorithm, the atomic selection method of the fixed number (namely, selecting $2K$ atoms to complete the expansion of the preliminary atomic set at each round of iterations) leads to a preliminary atomic set of the existing atoms that cannot be fully matched with the original signal. When these atoms are added to the candidate atomic set, the accuracy of the least square solution and the inaccuracy of the support atomic set are affected, which then weakens the reconstruction capability of the signal and increases the computational complexity of the StoGradMP algorithm. Therefore, in this study, we created a weak selection threshold method to select the atoms that best match the original signal, thereby completing the expansion of the preliminary atomic set with a more flexible atom selection. This method improves the reconstruction performance of the algorithm. The combination of the two reliability guarantee methods ensures the correctness and effectiveness of the proposed algorithm, identifies the support atomic set, and calculates the transition estimation of the original signal. Finally, we established different original signal environments to verify the reconstruction performance of the proposed method.

The layout of this paper is as follows. Section 2 introduces the CS theory for signal reconstruction and low-rank matrix reconstruction. The StoGradMP algorithm is described in Section 3. The proposed method, with the weak selection threshold method and the reliability verification strategy of the stochastic gradient algorithm, are outlined in Section 4. The simulation results and the discussion are provided in Section 5, and the conclusion is drawn in Section 6.

2. Compressed Sensing Theory

CS theory supposes that signal x is an n-length signal. It is said to be a K-sparse signal (or compressive) if x can be well approximated using K coefficients under some nonrelated linear measurements. According to the CS theory, such a signal can be acquired by the following linear random projection:

$$u = \Phi x + \varepsilon \tag{1}$$

where $\Phi \in R^{m \times n}$, $u \in R^{m \times 1}(m \ll n)$, and $\varepsilon \in R^{m \times 1}$ are the measurement matrix [30], the observation vector, and the noise signal, respectively; u contains nearly all the information of the sparse signal x. According to Equation (1), the dimensionality of u is much lower than the dimensionality of x. This problem is an underdetermined problem, which shows that Equation (1) has an infinite number of solutions. It is difficult to reconstruct the sparse signal vector x from u. However, according to the literature [5,31], a sufficient condition for exact signal reconstruction is that the sensing matrix Φ should satisfy the RIP condition. The RIP condition is described in Definition 1.

Definition 1. *For each integer $K = 1, 2, \ldots,$ define the restricted isometry constant δ_K of the sensing matrix Φ as the smallest number, such that holds for all K -sparse signal vectors $x \in R^{n \times 1}$ with $\|x\|_0 = K$.*

$$(1 - \delta_K)\|x\|_2^2 \leq \|\Phi x\|_2^2 \leq (1 + \delta_K)\|x\|_2^2 \tag{2}$$

Assuming that the original signal x is sparse in compressed sensing, then x can be reconstructed by solving the following optimization problem:

$$\min_{x \in R^{n \times 1}} \frac{1}{2m}\|u - \Phi x\|_2^2 \ \text{Subject to} \ \|x\|_0 \leq K \tag{3}$$

where m is the number of the measurement value and $\|.\|_2^2$ denotes the square of the 2-norm of the noise signal estimate vector. Here, K controls the sparsity of the solutions to Equation (3).

The low-rank matrix reconstruction problem can be similarly formulated. We obtain the observation vector u_i, which can be described as

$$u_i = \Phi_i X + \varepsilon_i \tag{4}$$

where $i = 1, 2, \ldots, m$, the size of measurement matrix Φ_i is an $m \times n_1$; the unknown signal matrix $X \in R^{n_1 \times n_2}$ is assumed to be a low rank matrix; and ε_i is the measurement noise. According to Equation (4), the matrix X can be reconstructed by the solving the following optimization:

$$\min_{X \in R^{n_1 \times n_2}} \frac{1}{2m}\|u - \Phi X\|_2^2 \ \text{Subject to} \ \text{rank}(X) \leq R \tag{5}$$

where m is the number of the measurements; u, Φ, and X are the observation signal, measurement matrix, and low-rank matrix signal, respectively; and R controls the rank level of the solution to Equation (5).

To analyze Equations (3) and (5), we first define a more general notion of sparsity. Given the sparse basis $\Psi = \{\psi_1, \psi_2, \ldots, \psi_n\}$, which consists of the vectors ψ_i

$$x = \sum_{i=1}^{n} \alpha_i \psi_i = \Psi \alpha \tag{6}$$

where $\alpha_i = \langle x, \psi_i \rangle = \psi_i^T x$ is the projection coefficient of the original sparse signal x and $K \ll n$. x is sparse with respect to the sparse basis Ψ if the number of nonzero entries are much lower than the length of signal x; that is, $K \ll n$. The sparse basis Ψ can be explained respectively:

(1) For sparse signal reconstruction, the sparse basis Ψ could be a finite set, such as $\Psi = \{\psi_i\}_{i=1}^n$, where ψ_i is the basic vector in the Euclidean space.

(2) For low-rank matrix reconstruction, the sparse basis Ψ could be an infinite set, such as $\Psi = \{\varphi_i v_i\}_{i=1}^\infty$, where $\varphi_i v_i$ are the unit-norm rank-one matrices.

This notion is sufficiently general to address several important sparse models, such as the group sparsity and low ranks [28,32]. Therefore, we can describe Equations (3) and (5) using Equations (7) and (8), respectively:

$$\underbrace{\min_x \frac{1}{M} \sum_{i=1}^M f_i(x)}_{F(x)} \text{ Subject to } \|x\|_{0,\Psi} \leq K \tag{7}$$

$$\underbrace{\min_X \frac{1}{M} \sum_{i=1}^M f_i(X)}_{F(X)} \text{Subject to } \text{rank}(X)_\Psi \leq R \tag{8}$$

where $f_i(x)$ is the smooth function, which can be a non-convex function; $\|x\|_{0,\Psi}$ controls the sparsity level of signal; $f_i(X)$ is also the smooth function with respect to the low rank matrix, which is the non-convex function; and $\text{rank}(X)_\Psi$ determines the rank level of the low rank matrix X. In particular, $\|x\|_{0,\Psi}$ is the smallest number of atoms in Ψ, such that the original signal x can be described by

$$\|x\|_{0,\Psi} = \min_x \left\{ K : x = \sum_{i \in |E|} \alpha_i \psi_i, |E| = K \right\} \tag{9}$$

where $|E|$ denotes the number of nonzero entries in the original signal x.

According to Equations (7) and (8), the reconstruction problem of the sparse signal and the low rank matrix need to be separately explained.

For the sparse signal reconstruction, the sparse basis Ψ consists of n basic vectors, each of size n in Euclidean space. This problem can be regarded as a special case of Equation (7), where $f_i(x) = (u_i - \langle \varphi_i, x \rangle)^2$ and $M = m$. In this case, we need to decompose the observation signal u into a non-overlapping vector u_{b_i} of size b. The matrix Φ_{b_i} is the sub-matrix of size $b_i \times n$, which consists of partial row vectors in the measurement matrix Φ.

According to Equations (3) and (7), the smooth function is $F(x) = \frac{1}{2m}\|u - \Phi x\|_2^2$. Therefore, the smooth function $F(x)$ can be written as

$$F(x) = \frac{1}{M} \sum_{i=1}^M \frac{1}{2b} \|u_{b_i} - \Phi_{b_i} x\|_2^2 = \frac{1}{M} \sum_{i=1}^M f_i(x) \tag{10}$$

where $M = m/b$, representing the number of the sub-matrix M, is an integer. Consequently, each sub-function $f_i(x)$ can be treated as $f_i(x) = \frac{1}{2b}\|u_{b_i} - \Phi_{b_i}\|_2^2$. In this case, each sub-function $f_i(x)$ accounts for a collection of observations of size b, rather than only one observation. Thus, when we randomly spilt the smooth function $F(x)$ into multiple sub-functions $f_i(x)$ and block the measurement matrix Φ into multiple sub matrices Φ_{b_i}, the computation of the stochastic gradient in the stochastic gradient methods is benefitted.

For the low-rank matrix reconstruction problem, according to the explanation provided in (2) of this section, we know that the sparse basis Ψ consists of infinitely several unit-norm rank-one matrices. According to Equations (5) and (8), the smooth function can be represented as $f_i(X) = (u_i - \langle \Theta_i, X \rangle)^2$. Therefore, the smooth function $F(X)$ can be written as

$$F(X) = \frac{1}{M}\sum_{i=1}^{M} f_i(X) = \frac{1}{M}\sum_{i=1}^{M}\left(\frac{1}{2b}\sum_{j=(i-1)\times b+1}^{ib}\left(u_j - \langle \Phi_j, X\rangle\right)^2\right)$$
$$\triangleq \frac{1}{M}\sum_{i=1}^{M}\frac{1}{2b}\|u_{b_i} - \Phi_{b_i} * X\|_2 \tag{11}$$

where $M = m/b$ is the number of block matrix in the sensing matrix Φ and M is an integer. Similarly, each function $f_i(X)$ accounts for a collection of observations u_{b_i} of size b, rather than only one observation.

3. StoGradMP Algorithm

The CoSaMP [19] algorithm has become popular for reconstructing sparse or compressive signals from their linear non-adaptive measurement. According to the relevant literature, we know that the CoSaMP algorithm is fast for small-scale signals with low dimensionality, but for a large-scale signal with high dimensionality, the reconstruction accuracy and the robustness of the algorithm are considered poor and not ideal. Regarding the shortcomings of the CoSaMP algorithm, Bahmani et al. [28] summarized the idea of the CoSaMP algorithm and proposed a gradient matching pursuit (GradMP) algorithm to solve the reconstruction problem of large-scale class signals with sparsity constraints. However, for large-scale class signals, the GradMP algorithm needs to compute the full gradient of the objective function $F(x)$, which greatly increases the computational cost of the algorithm. Therefore, after the GradMP algorithm, Needell et al. proposed a stochastic version of the GradMP algorithm called StoGradMP [29], which does not need to compute the full gradient of $F(x)$. Instead, at each round of iterations, an index $i \in [M]$ is randomly selected and its associated gradient $f_i(x)$ is computed. This operation is effective for handling the large-scale signal recovery problem, as gradient computation is often prohibitively expensive. To better analyze the StoGradMP algorithm, its block diagram is shown in Figure 1.

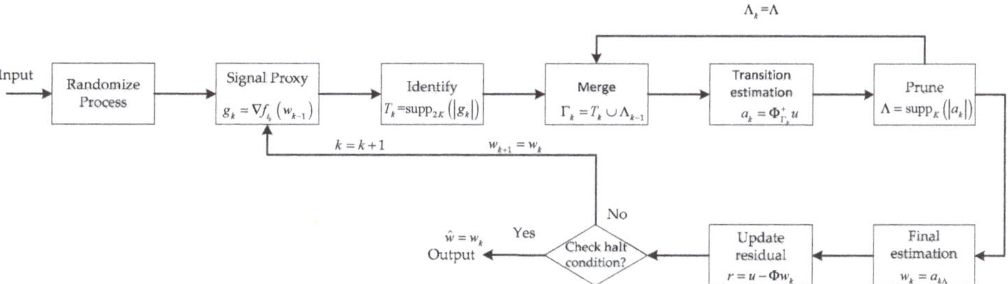

Figure 1. Block diagram of the StoGradMP algorithm.

The StoGradMP algorithm is described in Algorithm 1, where the steps at each iteration are shown below. Because the reconstruction process of the sparse original signal and the low-rank matrix are almost identical, for the sake of simplicity, we express the above two original signals using w in the subsequent explanations.

Randomize process: Randomly determine an index i_k with probability $p(i_k)$, where k is the loop index and $i_k \in M$. Then, compute its associated block matrix Φ_{b_i} and smooth functions.

Signal proxy: Compute the gradient g_k of the smooth function, where g_k is an $n \times 1$ vector. For the low-rank matrix, g_k is an $n \times n$ matrix.

Identify: In compressed sensing, when sorting the absolute values of the gradient in descending order, the first $2K$-largest absolute values of the gradient vector are selected. Then, search the atomic index of the block sensing matrix corresponding to these coefficients. Thereafter, a preliminary atomic set T_k is formed at the k-th iteration. In the low rank matrix reconstruction, the best rank $2R$ approximation to g_k is obtained by keeping the top $2R$ singular values in the singular value decomposition (SVD).

Merge: Establish the candidate atomic set Γ_k at the k-th iteration, which consists of the preliminary atomic set T_k at the current iteration and the support atomic set Λ_k at the previous iteration.

Estimate: Calculate the transition signal a_k at the current iteration, which is obtained using a sub-optimization method. This is a least squares problem for both the compressed sensing and low-rank matrix reconstruction problems. In compressed sensing, a_k is an $n \times 1$ vector, whereas in matrix recovery, a_k is an $n \times n$ matrix.

Prune: Sorting the absolute values of the transition signal vector a_k in descending order, the first K largest components are selected in vector a_k, and the atomic index of the candidate atomic set corresponding to these components is then obtained. The support atomic set Λ_k is constructed at the current iteration. The support atomic set belongs to the candidate atomic set, $\Lambda_k \in \Gamma_k$. Similarly, in the matrix reconstruction, the best rank R approximation to a_k is obtained by retaining the top R singular values in the SVD.

Update: Update the current approximate estimation of the original signal, $w_k = a_{k\Lambda}$. Here, $\Lambda = \Lambda_k$. The position of the nonzero entries of the final estimation signal w_k is determined by the index of the support atomic set. w_k is the final estimation signal at the kth iteration and w represents the original signal, which includes the sparse signal and the low rank matrix signal.

Check: When the l_2-norm of the current residual of the estimation signal w_k is smaller than the tolerance error *tol_alg* or the loop index k is greater than the maximum number of iterations (maxIter), then the reconstruction algorithm halts the iterations and the final approximation estimation of signal $\hat{w}$ is output such that $\hat{w} = w_k$. If the halt condition is not satisfied, then the algorithm continues to execute the iterations until the halt condition is met.

The entire procedure is as shown in Algorithm 1.

Algorithm 1. StoGradMP Algorithm

Input: $K, u, \Phi, p(i), b, tol\_alg$, maxIter
Output: an approximation estimation signal $\hat{w} = w_k$
Initialize: $\hat{w} = 0, k = 0, \Lambda_k = 0, T_k = 0, \Gamma_k = 0, M$
repeat
$k = k + 1$ loop index
select the i_k with probability $p(i_k)$ randomize
$g_k = \nabla f_{i_k}(w_k)$ form signal proxy
$T_k = \text{supp}_{2K}(|g_k|)$ identify $2K$ components
$\Gamma_k = T_k \cup \Lambda_{k-1}$ merge to form candidate set
$a_k = \Phi_{\Gamma_k}^+ u$ transition estimation using least squares method
$\Lambda = \text{supp}_K(|a_k|)$ prune to obtain the support atomic set
$w_k = a_{k\Lambda}$ final signal estimation
$r = u - \Phi w_k$ update the current residual
Until halting iteration condition is true, exit loop

---

4. Proposed Algorithm

The StoGradMP algorithm takes the sparsity of the original signal as the known information and uses it to complete the expansion of the preliminary atomic set in the preliminary stage of the algorithm. The StoGradMP algorithm determines the 2K-most relevant atoms in the preliminary stage of each round of iterations, and these atoms form a preliminary atomic set. Here, K represents the numerical value of the sparsity level and rank level, which is a fixed number greater than zero. This atomic selection results in the addition of smaller relevant atoms and incorrect atoms to the preliminary atomic set, which reduces the accuracy and speed of the reconstruction algorithm, thereby affecting the reconstruction performance of the algorithm. To solve this problem, we used the weak selection threshold strategy to achieve the expansion of the preliminary atomic set at the preliminary stage of the algorithm.

The entire process explanation of the proposed algorithm is described here. First, according to Equations (10) and (11) in Section 2, we selected the index i_k with probability $p(i_k)$. This step is mainly used to randomize the measurement matrix Φ to obtain a stochastic block matrix $\Phi_{b_{i_k}}$, which is expressed by

$$I = ceil(rand \times nb) \tag{12}$$

$$b_{i_k} = b \times (I-1) + 1 : b \times I \tag{13}$$

where nb is the number of block matrices according to Equation (10), which is equal to $nb = floor(m/b)$. Here, $M = nb$. b is the number of rows of the block matrix, which is equal to $b = \min(m, K)$. When the original signal is a sparse signal, K represents the numerical value of the sparsity level. When the original signal is a low-rank matrix signal, K is the numerical value of the rank level. b_{i_k} represents the index of rows of the measurement matrix, which is randomly determined. The block matrix $\Phi_{b_{i_k}}$ is also randomly selected. Then, the stochastic gradient function $f_i(w)$ is computed. Here, w consists of the symbols used in Section 3, which represents the sparse original signal and the low rank matrix. According to Equations (10) and (11), the sub-function $f_i(w)$ is expressed as

$$f_{i_k}(w_k) = \frac{1}{2b}\|u_{b_{i_k}} - \Phi_{b_{i_k}} w_{k-1}\|_2^2 \tag{14}$$

where k is the loop index, and $u_{b_{i_k}}$ and $\Phi_{b_{i_k}}$ are the i-th block observation signal and the i-th block matrix at k iteration, respectively. From Equations (12)–(14), we know that the sub-function $f_i(w)$ is also stochastically determined, and that $f_i(w)$ belongs to $F(w)$.

When the block matrix $\Phi_{b_{i_k}}$ and the stochastic gradient function $f_i(w)$ are obtained, the gradient of sub-function $f_i(w)$ is calculated, which is expressed as

$$g_k = \nabla f_i(w_k) \tag{15}$$

where g_k is the gradient of the sub-function $f_i(w)$ at the k-th iteration, w_k is the final estimation of the original signal at the k-th iteration, and $\nabla(.)$ denotes the derivative of the sub-function $f_i(w)$. Combining Equation (13) with Equation (14), the gradient g_k can be expressed as

$$g_k = -2 \times \Phi_{b_{i_k}}^T \left(u_{b_{i_k}} - \Phi_{b_{i_k}} w_{k-1} \right) \tag{16}$$

where $(.)^T$ represents the transpose operation of the matrix.

According to Equation (16), the smaller the absolute value of the gradient, the worse the match between the selected atoms and the original signal. In the StoGradMP algorithm, $2K$ is fixed and selected as the largest gradient coefficient from the gradient vector g_k to determine the atomic index of the block matrix and form the preliminary atomic set. The selected gradient coefficients may contain some smaller gradient coefficients in the StoGradMP algorithm during some iterations. This reduces the reconstruction performance and increases the computational complexity. Therefore, to improve the reconstruction performance of the StoGradMP algorithm, we used the weak selection threshold method to complete the expansion of the preliminary atomic set T_k. This process can be described as

$$\gamma = \max\left(|g_k|\right) \tag{17}$$

$$T_k = \text{supp}_{\kappa \times \gamma}\left(|g_k|\right) \tag{18}$$

where γ is the maximum value of the absolute gradient vector $|g_k|$ at the $k-th$ iteration, $\kappa \in [0.1\ 1.0]$ is the threshold, and $\text{supp}_{\kappa \times \gamma}(.)$ represents the preliminary atomic set that satisfies the weak selection threshold condition. The gradients corresponding to the preliminary atoms satisfy the condition that their absolute values are greater than $\kappa \times \gamma$. If the threshold is greater than 1, then $\kappa \times \gamma$ is greater than all gradients of the absolute value, and the atomic set is null. This causes the weak selection threshold method to fail. A too-small threshold of κ increases the number of error atoms in the preliminary atomic set. In the low rank matrix reconstruction, the best rank approximation to g_k is obtained by maintaining the singular value at a level greater than the weak selection threshold in the SVD, where γ is the maximum singular value. The preliminary atomic set consists of the singular vectors that satisfy the weak selection threshold method.

After selecting the preliminary atomic set, we used it and the previous support atomic set Λ_{k-1} to form the current candidate atomic set, which can be expressed as:

$$\Gamma_k = T_k \cup \Lambda_{k-1} \tag{19}$$

where Γ_k, T_k, and Λ_{k-1} denote the candidate atomic index set, the preliminary atomic index set, and the support atomic index set, respectively.

After the current candidate atomic set was constructed, to ensure the correctness and effectiveness of the proposed method, we added the reliability guarantee method 1 to the proposed algorithm, that is, if

$$\text{supp}|\Lambda_{k-1}| == \text{supp}|\Gamma_k| \tag{20}$$

is true, where $\text{supp}|\Gamma_k|$ and $\text{supp}|\Lambda_{k-1}|$ are represents the size (or scale) of the current candidate atomic index set Γ_k and the previous support atomic index set Λ_{k-1}, respectively. The method is unable to select the new atoms from the block matrix to add to the candidate atomic set. At this time, the loop is exited and the estimated value of the original signal is the output. We added the sub-condition judgment in the above judgment condition to prevent the proposed method from exiting the loop in the first round of iterations. Since both the candidate atomic set and the support atomic set are empty sets in the first round of iterations, this sub-condition judgment can be expressed as follows: If $k == 1$, then the estimated signal $\hat{w}$ is equal to 0.

Although the weak selection threshold method improves the correlation of the preliminary atomic set and increases the flexibility of atom selection, it is possible that when the threshold is too small, the number of columns of the candidate atomic set is greater than the number of the rows of the candidate atomic set. This leads to an inability to obtain the transition estimation of the original signal using the least squares method because the premise of the least squares method is that the number of rows of the atomic set is greater than the number of columns of the atomic set. Therefore, before solving the least squares method, we must ensure that this condition exists. Therefore, we developed the reliability guarantee method 2, that is, if

$$\text{supp}|\Gamma_k| \leq m \tag{21}$$

then,

$$\Lambda = \Gamma_k \tag{22}$$

$$\Phi_\Lambda = \Phi(:, \Lambda) \tag{23}$$

where $\text{supp}|\Gamma_k|$ represents the number of columns of the candidate atomic matrix Γ_k at the k-th iteration. If this condition is satisfied, then we regard the candidate atomic index set Γ_k as the current support atomic index set Λ, and the atoms corresponding to the current support atomic index set Λ are used to construct the current support atomic set Φ_Λ. Conversely, if the condition is not satisfied (the number of rows is smaller than the number of columns), then the matrix $\left(\Phi_\Lambda^T \times \Phi_\Lambda\right)^{-1}$ is not inverse. If this occurs, we exit the loop and let $\hat{w} = 0$.

Next, we used the least squares method to solve the sub-optimization problem, which can be described as:

$$a_k = \Phi_\Lambda^+ u \tag{24}$$

where a_k is the transition estimation signal of the original signal, u is the observation signal, and $(\Phi_\Lambda)^+$ represents the pseudo inverse of the support atomic set Φ_Λ. To better analyze the role of the reliability guarantee method 2, Equation (24) can be written as:

$$a_k = \left(\Phi_\Lambda^T \times \Phi_\Lambda\right)^{-1} \times \Phi_\Lambda^T \times u \tag{25}$$

where $(\Phi_\Lambda)^T$ and $\left(\Phi_\Lambda^T \times \Phi_\Lambda\right)^{-1}$ represent the transpose operation and inverse operation of the matrix Φ_Λ and the matrix $\Phi_\Lambda^T \times \Phi_\Lambda$, respectively. In combination with Equations (21) and (24), we can ensure that the operation $\Phi_\Lambda^T \times \Phi_\Lambda$ is invertible.

Based on Equations (22)–(24), we observed that the support atomic set is obtained using reliability guarantee method 2 and the candidate atomic set. If the reliability guarantee method 2 is true, then the current candidate atomic set can be regarded as the support atomic set. This operation is used to obtain the final support for the signal estimation. Next, we updated the current residual and final estimation of the original signal, which is expressed as

$$r_c = u - \Phi_\Lambda a_k \tag{26}$$

$$w_k = a_{k\Lambda} \tag{27}$$

where w_k is the final estimation of the original signal at the k-th iteration, $a_{k\Lambda}$ is the reconstruction signal corresponding to the support atomic index set Λ, and r_c is the current residual.

Finally, for the different original signals, we created different stop iteration conditions if

$$\|r_c\|_l \leq tol\_alg \text{ or } k \geq \text{maxIter} \tag{28}$$

is true, where tol_alg is the tolerance error of the algorithm iteration, and maxIter is the maximum number iterations of the algorithm. Specifically, if the original signal is a sparse signal, $l = 2$, that is, the l_2-norm of the residual estimation vector, then we set tol_alg and maxIter to 1×10^{-7} and $500 \times M$, respectively. When the original signal is a low-rank matrix, then the current residual estimation is a matrix, which is obtained by conducting a Frobenius norm operation on the error matrix. Here, $l = F$. We set the tol_alg and maxIter to 1×10^{-7} and $300 \times M$, respectively. According to Equation (28), when the stop iteration condition is satisfied, the algorithm stops the iterations and the output is the final estimation of the original signal $\hat{w} = w_k$. If the halt iteration condition is not satisfied, the iteration is continued, and it updates the current final estimation for the gradient computation of the next iteration, $w_{k+1} = w_k$. It continues until the stop iteration condition is true. To better analyze the proposed algorithm, its block diagram is shown in Figure 2.

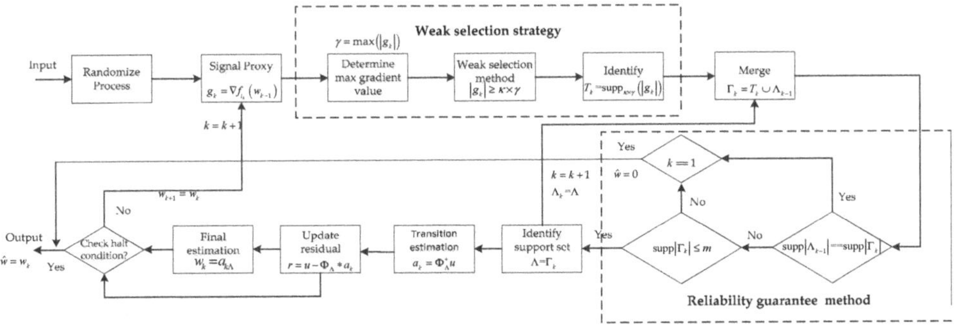

Figure 2. Block diagram of the proposed method.

The entire procedure is shown in Algorithm 2.

Algorithm 2: Proposed algorithm.

Input: $\Phi, u, p(i), b, tol\_alg, \text{maxIter}, \kappa$
Output: an approximation estimation signal $\hat{w} = w_k$
Initialize: $\hat{w} = 0, k = 0, \Lambda_k = 0, T_k = 0, \Gamma_k = 0, M$
repeat
$k = k + 1$ loop index
Select i_k from $[M]$ with probability $p(i_k)$ randomize process
$g_k = \nabla f_{i_k}(w_{k-1})$ form signal proxy
$\gamma = \max(|g_k|)$ determine the max gradient value
$T_k = \text{supp}_{\kappa \times \gamma}(|g_k|)$ weak selection method to identify the preliminary atomic set
$\Gamma_k = T_k \cup \Lambda_{k-1}$ merge to form candidate set
Reliability guarantee method 1
If $\text{supp}|\Lambda_{k-1}| == \text{supp}|\Gamma_k|$
    If $k == 1$
      $\hat{w} = 0;$
    end
    break;
end
Reliability guarantee method 2
If $\text{supp}|\Gamma_k| \le m$
    $\Lambda = \Gamma_k$ identify the support atomic set
    $\Phi_\Lambda = \Phi(:, \Lambda)$
else
    If $k == 1$
      $\hat{w} = 0;$
    end
    break;
end
$a_k = \Phi_\Lambda^+ u$ transition estimation by least squares method
$r = u - \Phi_\Lambda a_k$ update the current residual
$w_k = a_{k\Lambda}$ final signal estimation
Until halting iteration condition is true, exit loop

---

5. Discussion

We analyzed the simulation for the following experiments: 1D sparse signal reconstruction, low rank matrix reconstruction, and 2D image signal reconstruction. The reconstruction performance is an average after running the simulation 200 times using a computer with a quad-core, 64-bit processor, and 4G memory.

5.1. 1D Sparse Signal Reconstruction Experiment

In this experiment, we used a random signal with K-sparse as the original signal. The measurement matrix was randomly generated with a Gaussian distribution. We set the range of the weak selection thresholds κ to $[0.2, 0.4, 0.6, 0.8]$. The recovery error and iteration stop error of all the algorithms were set to 1×10^{-6} and 1×10^{-7}, respectively. These errors were obtained by conducting an l_2-norm operation on the error vector. The maximum number of iterations maxIter was set to $500 \times M$.

Figure 3 compares the reconstruction percentage of the proposed algorithm to the different thresholds. Figure 3 shows that when the threshold was 0.6, the reconstruction percentage of the proposed algorithm was the highest compared to the other thresholds under the same measurements and sparsity levels.

Figure 3 shows that the reconstruction percentage of the proposed algorithm was 100% for all of the sparsity and threshold levels when the number of measurements was greater than 160. Therefore, we set the range of measurements to [160 – 250] to compare the average running time of the proposed algorithm at different weak selection thresholds, as shown in Figure 4. Figure 4 shows that the average running time of the proposed algorithm was the shortest for different sparse levels when the threshold was 0.8, followed by 0.6, with very small differences between the two. Based on the analysis of Figures 3 and 4, we set the default weak selection threshold to 0.6.

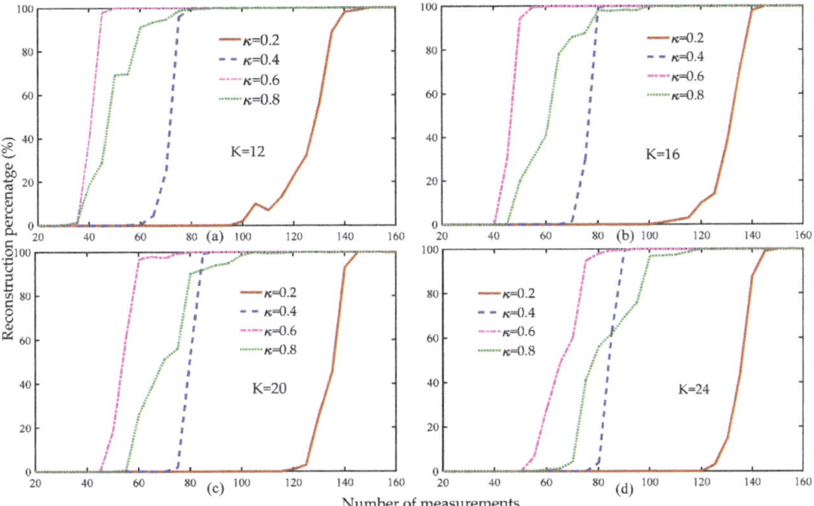

Figure 3. Reconstruction percentage of the proposed method at different weak selection thresholds ($n = 256$, $K = 12, 16, 20, 24$, $\kappa = 0.2, 0.4, 0.6, 0.8$, $m = 20 : 5 : 160$, Gaussian signal).

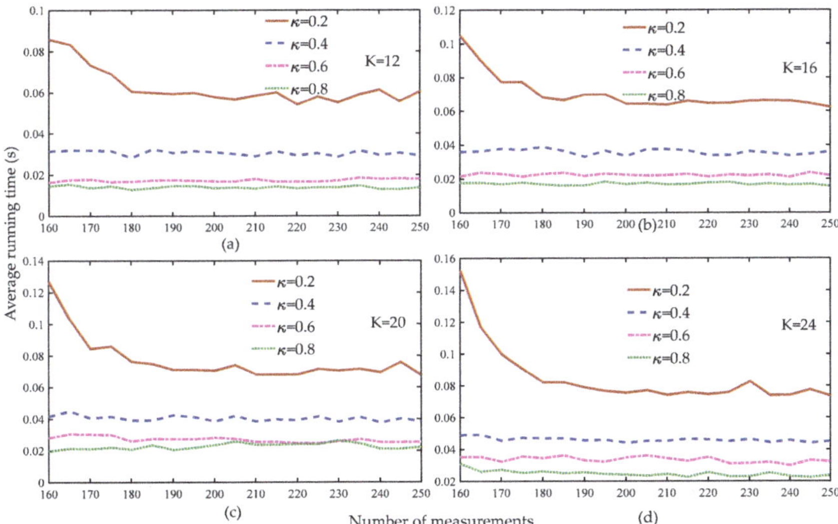

Figure 4. Average running time of the proposed method with different weak selection thresholds ($n = 256$, $K = 12, 16, 20, 24$, $\kappa = 0.2, 0.4, 0.6, 0.8$, $m = 160 : 5 : 250$, Gaussian signal).

Figure 5 compares the reconstruction percentage of the proposed algorithm using the StoGradMP algorithm. We set the sparse level to $K \in [12, 16, 20, 24]$, and the weak selection threshold to 0.6. Figure 5 shows that when the sparse level was 12, the reconstruction percentages of the proposed algorithm and the StoGradMP algorithm were nearly identical for all the measurements. When $K = 16, 20,$ or 24, the reconstruction percentage of the proposed algorithm was higher than that of the StoGradMP algorithm. When the sparse level was 24, the difference in the reconstruction percentages between the two algorithms was the largest. Therefore, we concluded that when the sparse level increases, the difference between the reconstruction percentages increases further. This means that in sparse signal reconstruction, the proposed method is more suitable for reconstruction in a larger sparsity environment compared to the StoGradMP algorithm.

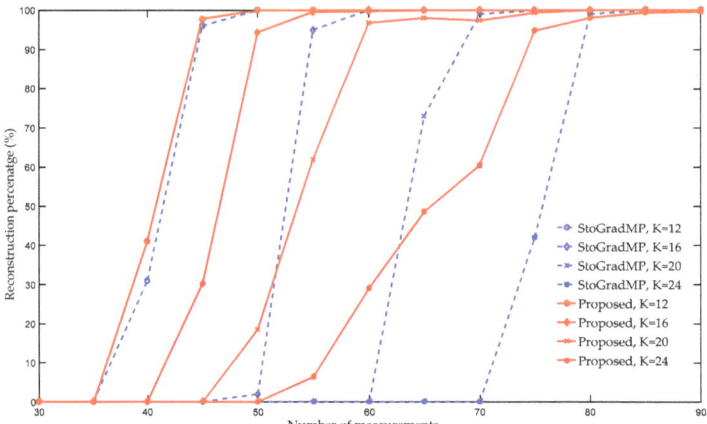

Figure 5. Reconstruction percentages of the StoGradMP and proposed algorithms ($n = 256$, $\kappa = 0.6$, $K = 12, 16, 20, 24$, $m = 20 : 5 : 90$, Gaussian signal).

Figure 6 compares the reconstruction percentage of the proposed algorithm with the StoGradMP and StoIHT algorithms. In Figure 5, the interval of measurement was set to five, and to reflect the details, we set the interval of measurement to two in Figure 4. Figure 6 shows that when $46 \leq m < 50$, the reconstruction percentages of the methods were 0%. This means that they could not complete the reconstruction. When $50 \leq m \leq 76$, the reconstruction percentage of the proposed method ranged from 0.2% to 97.6%; however, the reconstruction percentages of the StoGradMP and StoIHT algorithms were still 0%. When $72 \leq m \leq 78$, the reconstruction percentage of the StoGradMP algorithm began to increase from 0.6% to 95%. When $78 \leq m \leq 120$, the reconstruction percentages of the proposed and StoGradMP algorithms were nearly 100%. However, the StoIHT algorithm was still unable to complete reconstruction. When $120 \leq m \leq 142$, the reconstruction percentage of the StoIHT algorithm increased from 0% to 100%. When $142 \leq m$, then all the reconstruction algorithms could achieve full reconstruction. This demonstrates that the proposed method provides better reconstruction performance than the others.

Figure 7 compares the average running time. Figure 6 shows that the reconstruction percentage was 100% for all the reconstruction algorithms when the number of measurements was greater than 150. Therefore, in this simulation, we set the range of measurement to [150– 250]. Figure 7 shows that the proposed method has a shorter running time than the StoGradMP algorithm. Although the StoIHT algorithm had a shorter running time than the other algorithms, it required more measurements to achieve the same reconstruction percentage as the other algorithms.

Figure 8 compares the reconstruction percentages of the proposed algorithm and the prior improved algorithm (IStoGradMP) [33]. Both of these algorithms reconstructed the signal in an unknown sparsity environment. The main differences between the proposed algorithm and the IStoGradMP algorithm are: (1) In the preliminary atomic stage, the proposed algorithm uses the atomic matching strategy to obtain the preliminary atomic set, whereas the IStoGradMP algorithm evaluates and adjusts the estimated sparsity of the original signal to obtain the preliminary atomic set; (2) the scales of the preliminary atomic set and the support atomic set are unfixed at each iteration in the proposed algorithm, whereas the scale of the preliminary atomic set and the support atomic set are fixed at each iteration of the IStoGradMP algorithm; and (3) the support atomic set is determined by the candidate atomic set and the reliability guarantee method for the proposed method, whereas the support atomic set is determined by pruning the candidate atomic set in the IStoGradMP algorithm. We also proposed an improved StoGradMP algorithm based on the soft-threshold method [34]. This algorithm [34] requires that sparsity information of the original signal to be known, whereas the proposed method and the IStoGradMP algorithm [33] can reconstruct the signal without knowing the sparsity information. Based on the above comparative analysis, in this section, we only compared the experimental simulations in an unknown sparsity environment. We only compared the IStoGradMP and the proposed methods.

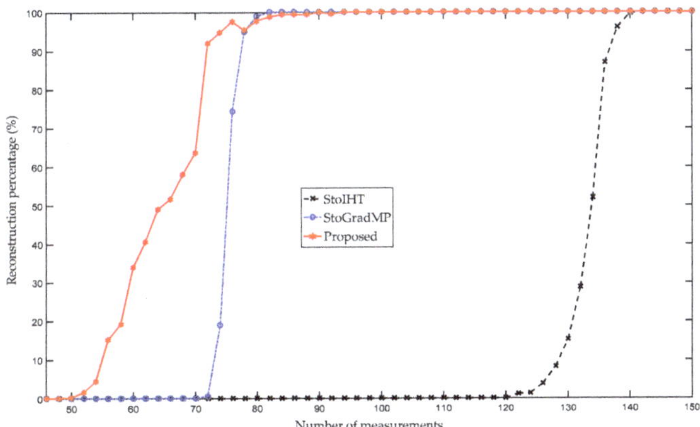

Figure 6. Reconstruction percentages of the different algorithms ($n = 256$, $K = 24$, $\kappa = 0.6$, $m = 46 : 2 : 150$, Gaussian signal).

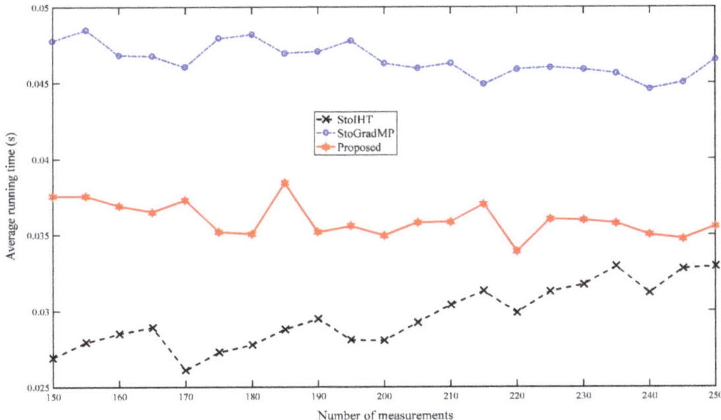

Figure 7. Average running times of the different algorithms ($n = 256$, $K = 24$, $\kappa = 0.6$, $m = 150 : 5 : 250$, Gaussian signal).

Figure 8 shows that for arbitrary measurement values, the reconstruction percentage of the proposed algorithm was higher than that of the IStoGradMP algorithm. However, we discovered that as the sparsity level increased, the gap in the reconstruction percentage between the proposed algorithm and the IStoGradMP algorithm gradually reduced. This means that the proposed method is more suitable than the IStoGradMP algorithm under smaller sparsity environments.

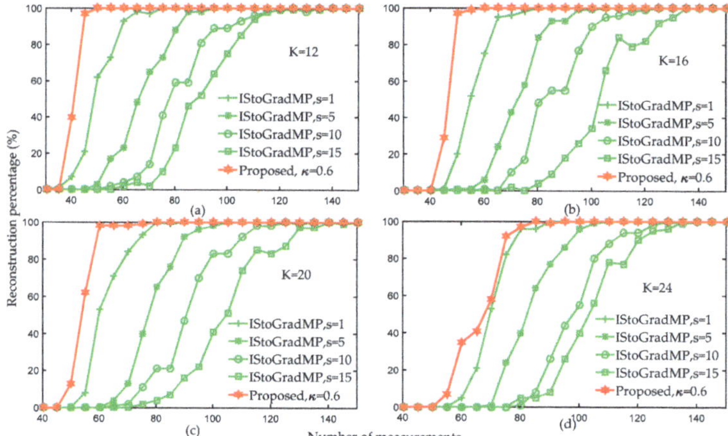

Figure 8. Reconstruction percentages of the proposed algorithm and IStoGradMP algorithm ($n = 256$, $\kappa = 0.6$, $K = 12, 16, 20, 24$, $s = 1, 5, 10, 15$, $m = 30 : 5 : 150$, Gaussian signal).

Figure 9 compares the average running times of the proposed method and the IStoGradMP method under different sparsity level conditions. Figure 8 shows that when the number of the measurements was greater than 150, the reconstruction percentage of the proposed method and the IStoGradMP was 100% for all the sparsity levels. Therefore, we set the range of the number of the measurement to $[150 - 250]$. Figure 9 shows that when the threshold of the proposed method was set to 0.6, the average running time of the proposed method was less than that for the IStoGradMP algorithm. This means that the computational complexity of the proposed algorithm was lower than that for the IStoGradMP algorithm. That is, the proposed method was faster than the IStoGradMP algorithm under the full reconstruction conditions.

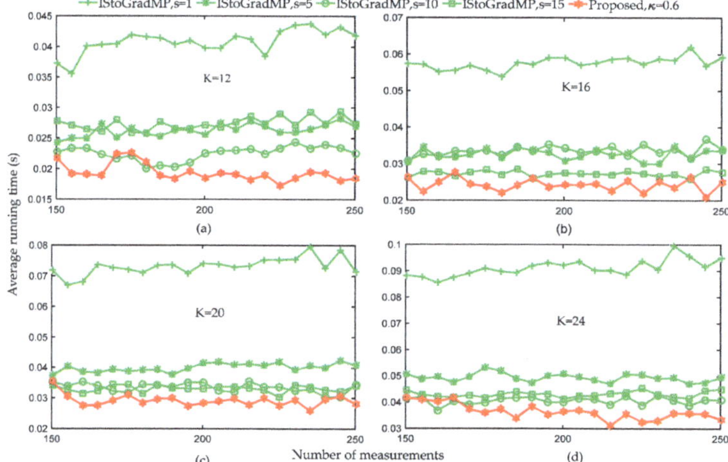

Figure 9. Average running times of the proposed algorithm and IStoGradMP algorithm ($n = 256$, $\kappa = 0.6$, $K = 12, 16, 20, 24$, $s = 1, 5, 10, 15$, $m = 150 : 5 : 250$, $\delta_K = 0.1$, Gaussian signal).

Based on the analysis of Figures 8 and 9, we conclude that for a smaller sparsity level environment, the proposed algorithm with a proper weak selection threshold has better reconstruction performance as well as a lower computational complexity than the IStoGradMP algorithm.

5.2. Low-Rank Matrix Reconstruction Experiment

In this experiment, we used the random matrix with a low-rank property as the original signal. We set the rank level of the matrix to $R = 1, 3$. The size of the low rank matrix was 10×10. The measurement matrix was randomly generated with a Gaussian distribution. The recovery error and iteration halt error of all the algorithms were set to 1×10^{-6} and 1×10^{-7}, respectively. These errors were obtained using a Frobenius norm operation on the respective error matrix. The maximum number of iterations maxIter was set to $300 \times M$.

Figure 10 compares the reconstruction percentage of the proposed method at different weak selection thresholds. Figure 10 shows that the reconstruction percentage was higher than the other algorithms when the threshold was 0.2 for the different rank levels of the matrix.

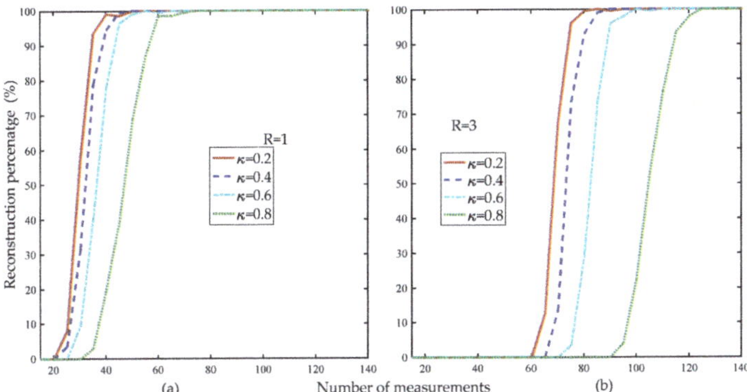

Figure 10. Reconstruction percentage of the proposed method with different weak selection thresholds ($d = 100$, $R = 1, 3$, $\kappa = 0.2, 0.4, 0.6, 0.8$, $m = 15 : 5 : 140$, random low rank matrix).

Figure 11 compares the average running times of the proposed method at different thresholds. Figure 11 shows that the smaller the weak selection threshold, the higher the reconstruction percentage, which means that a larger threshold increases the computational complexity of the proposed method. Thus, in the subsequent simulation without special instructions, the default weak selection threshold was set to 0.2.

Figure 12 compares the reconstruction percentages of the different measurements of the proposed and StoGradMP algorithms at different rank levels. We observed that the proposed method had a better reconstruction percentage than the StoGradMP algorithm at the different rank levels.

Figure 13 compares the reconstruction percentage of the proposed algorithm to the StoGradMP and StoIHT algorithms for different measurements. In Figure 12, we set the measurement interval to five, and to show details, we set the interval of measurement to two in Figure 13. Figure 13 shows that when $20 \leq m < 22$, the reconstruction percentage of all the algorithms was 0%. When $22 \leq m \leq 34$, the reconstruction percentage of the StoIHT and proposed algorithms began to increase from 0% to 88.2% and 0.6% to 87%, respectively. However, the StoGradMP algorithm struggled to complete the signal reconstruction. When $32 \leq m \leq 58$, the reconstruction percentage of the proposed algorithm ranged approximately from 87% to 100%. The reconstruction percentage of the StoIHT algorithm increased from 88.2% to 99.8%. The reconstruction percentage of the StoGradMP algorithm increased from 0.2% to 95.4%. In this measurement range, the reconstruction percentage of the proposed algorithm was higher than those of the other algorithms. When $58 \leq m$, almost all the reconstruction algorithms achieved a high probability reconstruction. Therefore, we conclude that the reconstruction percentages of the proposed method and the StoIHT method were almost the same, and higher than the StoGradMP algorithm, which existed at a lower rank level of the matrix.

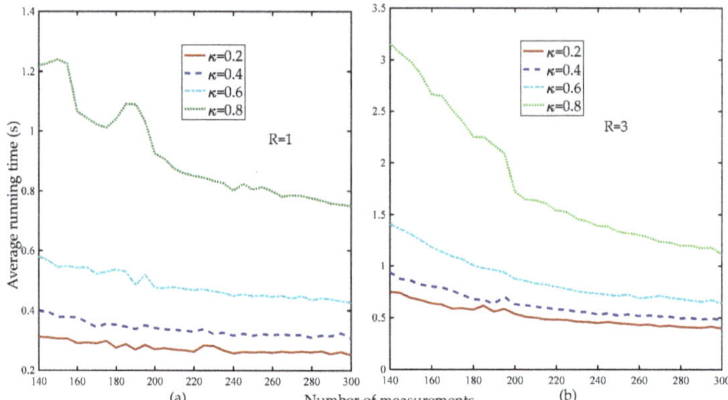

Figure 11. Average running times of the proposed method at different weak selection thresholds ($d = 100$, $R = 1, 3$, $\kappa = 0.2, 0.4, 0.6, 0.8$, $m = 200 : 5 : 300$, random low-rank matrix).

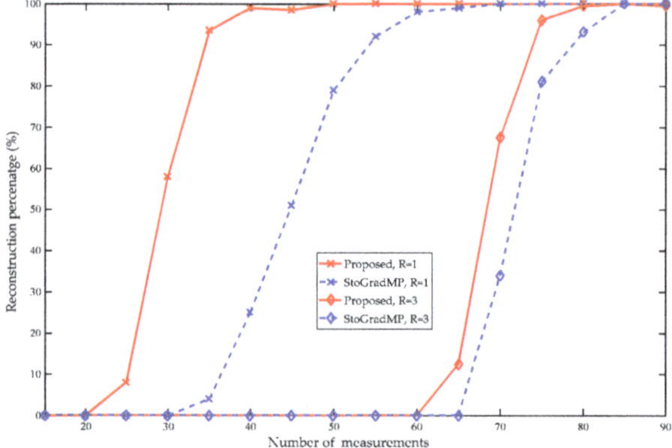

Figure 12. Reconstruction percentages of the StoGradMP and proposed algorithms ($d = 100$, $R = 1, 3$, $\kappa = 0.2$, $m = 15 : 5 : 120$, random low-rank matrix).

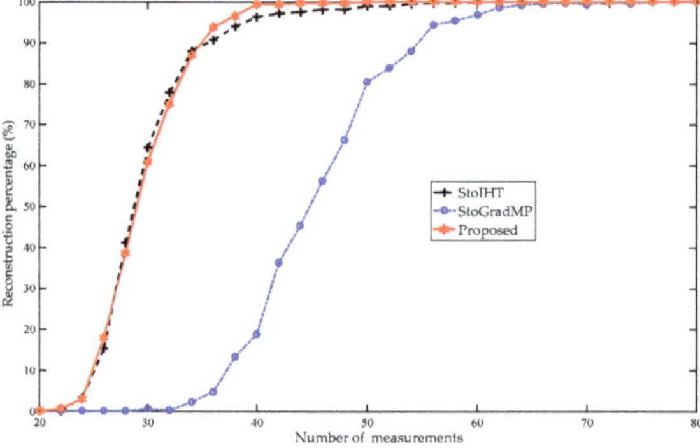

Figure 13. Reconstruction percentages of the different methods ($d = 100$, $R = 1$, $\kappa = 0.2$, $m = 20 : 2 : 70$, randomly low-rank matrix).

Figure 14 compares the average running times of the different algorithms. Figure 13 shows that when the number of measurements was more than 80, the reconstruction percentage of all the algorithms was 100%. Therefore, to better analyze the computational complexity of the different algorithms, we set the range of measurements to $[80 - 200]$ in the simulation. Figure 14 shows that the proposed algorithm had the shortest running time, followed by the StoIHT and StoGradMP algorithms. We observed that when the number of measurements increased, the running time of the StoIHT algorithm also increased, whereas the average running times of the proposed and StoGradMP algorithms tended to decrease and remain stable.

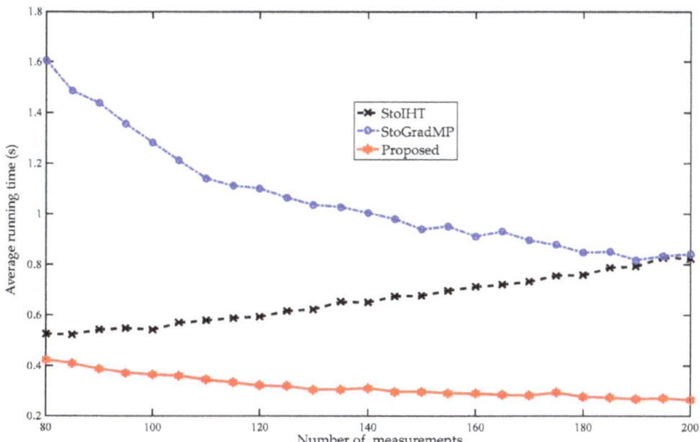

Figure 14. Average running times of the different algorithms ($d = 100, R = 1, \kappa = 0.2, m = 80 : 5 : 200$, randomly low-rank matrix).

Based on the above analysis, we conclude that the proposed algorithm with a weak selection threshold produces better reconstruction performance than the StoGradMP algorithm, as well as a lower computational complexity than the other algorithms.

5.3. 2D Image Signal Reconstruction Experiment

In this subsection, we used six 256×256 test images as the original images. We considered the image signal as a 2D signal. The test images included the following: Baboon, Boat, Cameraman, Fruits, Lena (human portrait), and Peppers. The sparse basis was a wavelet basis with sparse representation capability, and the size was 256×256. The measurement matrix was randomly generated with a Gaussian distribution, and the size was 153×256. We assumed that the sparsity was 51. The iteration halt error of the algorithm was set to 1×10^{-7}, the maximum number of iterations maxIter was set to 30, and the weak selection threshold was set to $\kappa = 0.6$.

We used the peak signal to noise ratio (PSNR) as an indicator to evaluate the reconstruction quality, which could be expressed as:

$$MSE = \frac{1}{M \times N} \sum_{i=0}^{M-1} \sum_{j=0}^{N-1} \left| \hat{x}(i,j) - x(i,j) \right|^2 \tag{29}$$

$$PSNR = 10 \times \log_{10}\left(\frac{MAX_{\hat{x}}^2}{MSE} \right) = 20 \times \log_{10}\left(\frac{MAX_{\hat{x}}}{\sqrt{MSE}} \right) \tag{30}$$

where $M = N = 256$; $\hat{x}(i,j)$ and $x(i,j)$ represent the reconstruction value and the original value of the correspondence position, respectively; MSE is the mean square error; and $MAX_{\hat{x}}$ represents the maximum value of the color of the image point. In this paper, each sample point is represented by eight bits, $MAX_{\hat{x}} = 255$. The larger the PSNR, the higher the reconstructed image quality.

Figure 15 shows the original images. Figures 16 and 17 shows the reconstructed images using the StoGradMP and proposed algorithms, respectively. Comparing the reconstructed images to the original images, we observed that the two methods successfully reconstructed the original images.

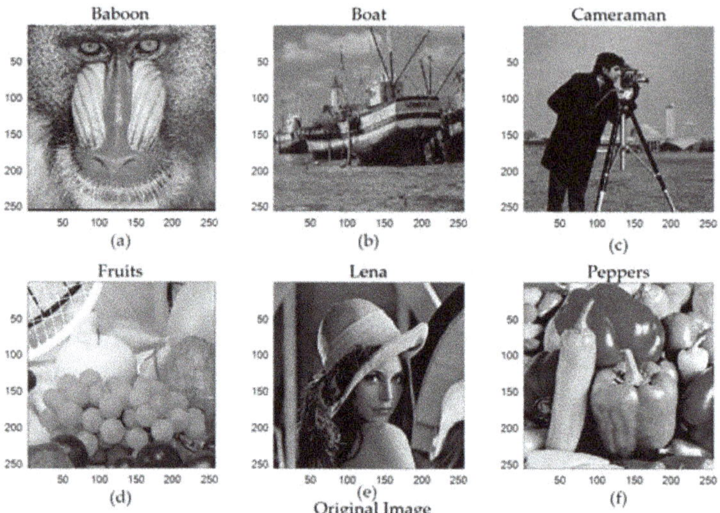

Figure 15. Original images (**a**) Baboon image, (**b**) Boat image, (**c**) Cameraman image, (**d**) Fruits image, (**e**) Lena image, (**f**) Peppers image.

Figure 16. Reconstructed images using the StoGradMP algorithm (**a**) Reconstructed Baboon image with PSNR = 16.5754 dB; (**b**) Reconstructed Boat image with PSNR = 20.9987 dB; (**c**) Reconstructed Cameraman image with PSNR = 21.8698 dB; (**d**) Reconstructed Fruits image with PSNR = 23.5299 dB; (**e**) Reconstructed Lena image with PSNR = 25.5532 dB; (**f**) Reconstructed Peppers image with PSNR = 23.7168 dB.

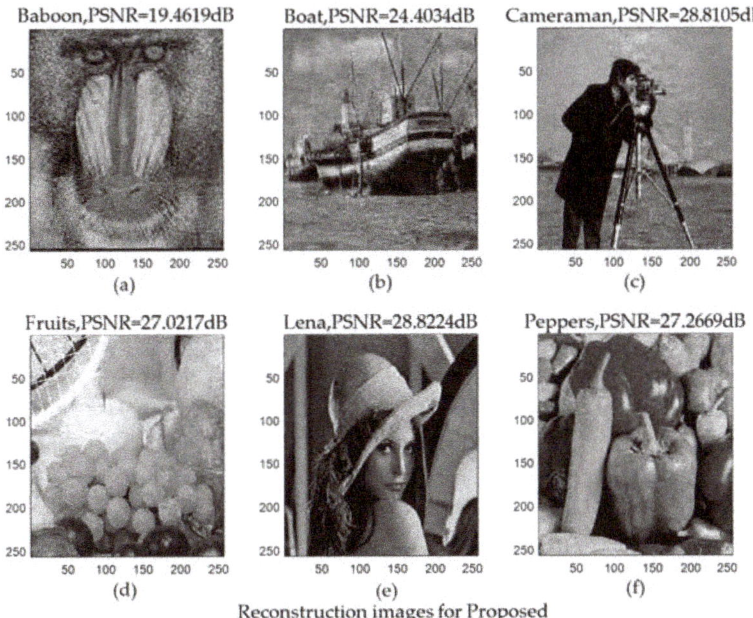

Figure 17. Reconstructed images using the proposed algorithm (**a**) Reconstructed Baboon image with PSNR = 19.4619 dB; (**b**) Reconstructed Boat image with PSNR = 24.4034 dB; (**c**) Reconstructed Cameraman image with PSNR = 28.8105 dB; (**d**) Reconstructed Fruits image with PSNR = 27.0217 dB; (**e**) Reconstructed Lena image with PSNR = 28.8224 dB; (**f**) Reconstructed Peppers image with PSNR = 27.2669 dB.

Table 1 compares the average PSNR of the StoGradMP and proposed algorithms under different test image conditions. Table 1 shows that the average PSNR of the proposed algorithm was higher than the StoGradMP algorithm for the different test images, and the average PSNR of the proposed method was higher than 3–4 dB. This shows that the reconstructed image quality of the proposed algorithm was better than the StoGradMP.

Table 1. Comparison of the average peak signal to noise ratios (PSNR) of the StoGradMP and proposed algorithms for the different test images.

	Algorithm	
Image	**StoGradMP**	**Proposed**
Baboon	16.5263 dB	19.4775 dB
Boat	20.9818 dB	24.2238 dB
Cameraman	22.3862 dB	25.7790 dB
Fruits	23.3538 dB	26.9003 dB
Lena	25.2590 dB	28.7027 dB
Peppers	23.7567 dB	27.2286 dB

Table 2 compares the average running times of the StoGradMP and proposed algorithms for the different test images. From Table 2, the average running times of the StoGradMP algorithm was longer than the proposed method for the different test images, and the average running time of the StoGradMP algorithm was more than twice that of the proposed algorithm. This means that the proposed method had lower computational complexity than the StoGradMP algorithm when images were reconstructed.

Table 2. Comparison of the average runtimes of the StoGradMP and proposed algorithms for the different test images.

Image	Algorithm	
	StoGradMP	**Proposed**
Baboon	51.04 s	19.32 s
Boat	51.45 s	20.22 s
Cameraman	51.13 s	17.94 s
Fruits	81.66 s	19.41 s
Lena	50.52 s	22.64 s
Peppers	56.88 s	19.03 s

Based on the above analysis, the proposed method has a better reconstruction performance for different test images compared to the StoGradMP algorithm, as well as a lower computational complexity than the StoGradMP algorithm.

6. Conclusions

In this paper, a novel stochastic gradient matching pursuit algorithm based on weak selection thresholds was proposed. This algorithm uses the weak selection threshold method to select the atoms that best match the original signal from the block sensing matrix and completes the expansion of the preliminary atomic set. The proposed algorithm adopts two reliability guarantee methods to identify the support atomic set and calculate the transition estimation of the original signal to ensure the correctness and effectiveness of the proposed algorithm. The proposed algorithm not only eliminates dependency on prior sparsity information of the original signal, but also increases the flexibility of the atomic selection process while improving atomic reliability. Therefore, it enhances the reconstruction accuracy and reconstruction efficiency of the proposed algorithm.

Our series of simulation results showed that the proposed method has better reconstruction performance and less computational complexity compared to the other algorithms. Future research should consider using the proposed method to process large-scale array signals, such as wireless communication signals, radar signals, and sonar signals, to enhance the useful signal, suppress noise interference, reduce the burden on sensor devices, and ensure fast real-time transmission of the array signal. The weak selection threshold was determined by setting the threshold and the maximum stochastic gradient in our proposed method. The optimal setting threshold was different for different types of signals, which affects the reconstruction performance. Our future work will consider methods to adapt the setting threshold to the signal.

Author Contributions: Conceptualization, formal analysis, investigation, and writing the original draft was done by L.Z. and Y.H. Experimental tests were done by Y.H. and Y.J. All authors have read and approved the final manuscript.

Funding: This work was supported by the National Natural Science Foundation of China (61271115) and the Foundation of Jilin Educational Committee (JJKH20180336KJ).

Conflicts of Interest: The authors declare no conflict of interest.

References

1. Laue, H.E.A. Demystifying Compressive Sensing [Lecture Notes]. *IEEE Signal Process. Mag.* **2017**, *34*, 171–176. [CrossRef]
2. Laue, H.E.A.; Du Plessis, W.P. Numerical Optimization of Compressive Array Feed Networks. *IEEE Trans. Antennas Propag.* **2018**, *66*, 3432–3440. [CrossRef]
3. Arjoune, Y.; Kaabouch, N.H.; Tamtaoui, A. Compressive sensing: Performance comparison of sparse recovery algorithms. In Proceedings of the 2017 IEEE 7th Annual Computing and Communication Workshop and Conference (CCWC), Las Vegas, NV, USA, 9–11 January 2017; pp. 1–7.

4. Giryes, R.; Eldar, Y.C.; Bronstein, A.M.; Sapiro, G. Tradeoffs between Convergence Speed and Reconstruction Accuracy in Inverse Problems. *IEEE Trans. Signal Process.* **2018**, *66*, 1676–1690. [CrossRef]

5. Chen, W.; Li, Y.L. Recovery of Signals under the Condition on RIC and ROC via Prior Support Information. *Appl. Comput. Harmon. A* **2018**, *46*, 417–430. [CrossRef]

6. Ding, X.; Chen, W.; Wassell, I.J. Joint Sensing Matrix and Sparsifying Dictionary Optimization for Tensor Compressive Sensing. *IEEE Trans. Signal Process.* **2017**, *65*, 3632–3646. [CrossRef]

7. Xu, H.; Zhang, C.; Kim, I. Coupled Online Robust Learning of Observation and Dictionary for Adaptive Analog-to-Information Conversion. *IEEE Signal Process. Lett.* **2019**, *26*, 139–143. [CrossRef]

8. Joseph, G.; Murthy, C.R. Measurement Bounds for Observability of Linear Dynamical Systems under Sparsity Constraints. *IEEE Trans. Signal Process.* **2019**, *67*, 1992–2006. [CrossRef]

9. Liu, X.; Xia, S.; Fu, F. Reconstruction Guarantee Analysis of Basis Pursuit for Binary Measurement Matrices in Compressed Sensing. *IEEE Trans. Inf. Theory* **2017**, *63*, 2922–2932. [CrossRef]

10. Wang, M.; Wu, X.; Jing, W.; He, X. Reconstruction algorithm using exact tree projection for tree-structured compressive sensing. *IET Signal Process.* **2016**, *10*, 566–573. [CrossRef]

11. Li, H.; Liu, G. Perturbation Analysis of Signal Space Fast Iterative Hard Thresholding with Redundant Dictionaries. *IET Signal Process.* **2017**, *11*, 462–468. [CrossRef]

12. Huang, X.; He, K.; Yoo, S.; Cossairt, O.; Katsaggelos, A.; Ferrier, N.; Hereld, M. An Interior Point Method for Nonnegative Sparse Signal Reconstruction. In Proceedings of the 2018 25th IEEE International Conference on Image Processing (ICIP), Athens, Greece, 7–10 October 2018; pp. 1193–1197.

13. Voronin, S.; Daubechies, I. An Iteratively Reweighted Least Squares Algorithm for Sparse Regularization. *Comput. Optim. Appl.* **2016**, *64*, 755–792.

14. Bouchhima, B.; Amara, R.; Hadj-Alouane, M.T. Perceptual orthogonal matching pursuit for speech sparse modeling. *Electron. Lett.* **2017**, *53*, 1431–1433. [CrossRef]

15. Dan, W.; Fu, Y. Exact support recovery via orthogonal matching pursuit from noisy measurements. *Electron. Lett.* **2016**, *52*, 1497–1499. [CrossRef]

16. Wang, Y.; Tang, Y.Y.; Li, L. Correntropy Matching Pursuit with Application to Robust Digit and Face Recognition. *IEEE Trans. Cybernet.* **2017**, *47*, 1354–1366. [CrossRef] [PubMed]

17. Wang, J.; Kwon, S.; Li, P.; Shim, B. Recovery of Sparse Signals via Generalized Orthogonal Matching Pursuit: A New Analysis. *IEEE Trans. Signal Process.* **2016**, *64*, 1076–1089. [CrossRef]

18. Pei, L.; Jiang, H.; Li, M. Weighted double-backtracking matching pursuit for block-sparse reconstruction. *IET Signal Process.* **2016**, *10*, 930–935. [CrossRef]

19. Satpathi, S.; Chakraborty, M. On the number of iterations for convergence of CoSaMP and Subspace Pursuit algorithms. *Appl. Comput. Harmon. Anal.* **2017**, *43*, 568–576. [CrossRef]

20. Golbabaee, M.; Davies, M.E. Inexact Gradient Projection and Fast Data Driven Compressed Sensing. *IEEE Trans. Inf. Theory* **2018**, *64*, 6707–6721. [CrossRef]

21. Huang, S.; Tran, T.D. Sparse Signal Recovery via Generalized Entropy Functions Minimization. *IEEE Trans. Signal Process.* **2019**, *67*, 1322–1337. [CrossRef]

22. Ding, Z.; Fu, Y. Deep Domain Generalization with Structured Low-Rank Constraint. *IEEE Trans. Image Process.* **2018**, *27*, 304–313. [CrossRef]

23. Grussler, C.; Rantzer, A.; Giselsson, P. Low-Rank Optimization with Convex Constraints. *IEEE Trans. Autom. Control.* **2018**, *63*, 4000–4007. [CrossRef]

24. Chen, W.; Wipf, D.; Wang, Y.; Liu, Y.; Wassell, I.J. Simultaneous Bayesian Sparse Approximation with Structured Sparse Models. *IEEE Trans. Signal Process.* **2016**, *64*, 6145–6159. [CrossRef]

25. Cong, Y.; Liu, J.; Sun, G.; You, Q.; Li, Y.; Luo, J. Adaptive Greedy Dictionary Selection for Web Media Summarization. *IEEE Trans. Image Process.* **2017**, *26*, 185–195. [CrossRef]

26. Bresler, G.; Gamarnik, D.; Shah, D. Learning Graphical Models from the Glauber Dynamics. *IEEE Trans. Inf. Theory* **2018**, *64*, 4072–4080. [CrossRef]

27. Soltani, M.; Hegde, C. Fast Algorithms for De-mixing Sparse Signals from Nonlinear Observations. *IEEE Trans. Signal Process.* **2017**, *65*, 4209–4222. [CrossRef]

28. Bahmani, S.; Boufounos, P.T.; Raj, B. Learning Model-Based Sparsity via Projected Gradient Descent. *IEEE Trans. Inf. Theory* **2016**, *62*, 2092–2099. [CrossRef]

29. Nguyen, N.; Needell, D.; Woolf, T. Linear Convergence of Stochastic Iterative Greedy Algorithms with Sparse Constraints. *IEEE Trans. Inf. Theory* **2017**, *63*, 6869–6895. [CrossRef]

30. Vehkaperä, M.; Kabashima, Y.; Chatterjee, S. Analysis of Regularized LS Reconstruction and Random Matrix Ensembles in Compressed Sensing. *IEEE Trans. Inf. Theory* **2016**, *62*, 2100–2124.
31. Wang, Q.; Qu, G. Restricted isometry constant improvement based on a singular value decomposition-weighted measurement matrix for compressed sensing. *IET Commun.* **2017**, *11*, 1706–1718. [CrossRef]
32. Shi, J.; Hu, G.; Zhang, X.; Sun, F.; Zhou, H. Sparsity-based Two-Dimensional DOA Estimation for Coprime Array: From Sum–Difference Coarray Viewpoint. *IEEE Trans. Signal Process.* **2017**, *65*, 5591–5604. [CrossRef]
33. Zhao, L.Q.; Hu, Y.F.; Liu, Y.L. Stochastic Gradient Matching Pursuit Algorithm Based on Sparse Estimation. *Electronics* **2019**, *8*, 165. [CrossRef]
34. Zhao, L.Q.; Hu, Y.F.; Jia, Y.F. Improved Stochastic Gradient Matching Pursuit Algorithm Based on the Soft-Thresholds Selection. *J. Electr. Comput. Eng.* **2018**, *2018*, 9130531.

 *sensors*

Article

Less Data Same Information for Event-Based Sensors: A Bioinspired Filtering and Data Reduction Algorithm

Juan Barrios-Avilés [†], **Alfredo Rosado-Muñoz** [*,†] , **Leandro D. Medus,**
Manuel Bataller-Mompeán and Juan F. Guerrero-Martínez

Group for Digital Design and Processing, Department of Electronic Engineering, School of Engineering,
Universitat de Valencia, Burjassot, 46100 Valencia, Spain; juan.barrios@uv.es (J.B.-A.);
leandro.medus@ext.uv.es (L.D.M.); Manuel.Bataller@uv.es (M.B.-M.); juan.guerrero@uv.es (J.F.G.-M.)
* Correspondence: alfredo.rosado@uv.es; Tel.: +34-963-543-808
† The authors contributed equally to this work.

Received: 18 September 2018; Accepted: 22 November 2018; Published: 24 November 2018

Abstract: Sensors provide data which need to be processed after acquisition to remove noise and extract relevant information. When the sensor is a network node and acquired data are to be transmitted to other nodes (e.g., through Ethernet), the amount of generated data from multiple nodes can overload the communication channel. The reduction of generated data implies the possibility of lower hardware requirements and less power consumption for the hardware devices. This work proposes a filtering algorithm (LDSI—Less Data Same Information) which reduces the generated data from event-based sensors without loss of relevant information. It is a bioinspired filter, i.e., event data are processed using a structure resembling biological neuronal information processing. The filter is fully configurable, from a "transparent mode" to a very restrictive mode. Based on an analysis of configuration parameters, three main configurations are given: weak, medium and restrictive. Using data from a DVS event camera, results for a similarity detection algorithm show that event data can be reduced up to 30% while maintaining the same similarity index when compared to unfiltered data. Data reduction can reach 85% with a penalty of 15% in similarity index compared to the original data. An object tracking algorithm was also used to compare results of the proposed filter with other existing filter. The LDSI filter provides less error (4.86 ± 1.87) when compared to the background activity filter (5.01 ± 1.93). The algorithm was tested under a PC using pre-recorded datasets, and its FPGA implementation was also carried out. A Xilinx Virtex6 FPGA received data from a 128 × 128 DVS camera, applied the LDSI algorithm, created a AER dataflow and sent the data to the PC for data analysis and visualization. The FPGA could run at 177 MHz clock speed with a low resource usage (671 LUT and 40 Block RAM for the whole system), showing real time operation capabilities and very low resource usage. The results show that, using an adequate filter parameter tuning, the relevant information from the scene is kept while fewer events are generated (i.e., fewer generated data).

Keywords: neuromorphic systems; event-based sensors; dynamic vision sensor; bioinspired event filtering; FPGA implementation; spike-based; event data reduction

---

1. Introduction

The development of event-based sensors is an important topic. Vision sensors are common [1,2] but other event-based sensors exist, especially in those areas where bioinspired sensors are developed, e.g., artificial cochleas [3] and olfactory systems [4,5]. On the one hand, the data received from the sensors consist on events (also called spikes) which greatly differ from the traditional data values received form sensors (typically, analog values). For this reason, further data processing requires special algorithms and techniques. On the other hand, new devices (network nodes) are constantly added into a laboratory or industrial communication network, increasing the volume of data transmitted. Nowadays, data transfer is increasing at a higher pace than the supported bandwidth due to the addition of advanced equipment generating and transmitting many data and causing the saturation of communication networks [6]. This is a problem, especially in those applications where real-time and low-latency are required [7]. In the case of vision sensors, which generate many data, event-based encoding techniques can be a solution so that vision sensors can be connected into an existing communication network. In frame-based cameras, it is common to use a separated ethernet network for image transmission. Event-based cameras produce data in the form of events, asynchronously [8]. Data are generated only when there is a difference in light intensity received by any of the sensors (pixels) arranged in an array. Each pixel of the camera that can sense this difference in intensity will produce an event if such difference is bigger than a threshold setting that can be adjusted. The generated event includes information about the address of the pixel in the sensor where the threshold was exceeded, together with a time stamp in order to generate a unique event, not just in space but also in time. Typically, positive or negative events are generated if the event is caused by an intensity increment or decrement, respectively. These changes in intensity are mainly caused by changes in the visual scene, which generates event data related to the scene. This behaviour is similar to a mammal brain [9], which leads to use neuromorphic systems [10] for further information processing [11], feature extraction, scene detection [12] and filtering [13,14].

Proper lighting is a key factor in traditional industrial vision systems since it is difficult to maintain a constant light due to a constantly changing environment. Traditional solutions require the use of specific lighting systems suited for specific applications [15–17]. Event-based cameras minimize light effects since only pixel intensity differences are considered and no specific light intensity is required, independently of light conditions.

Currently, applications working with event-based cameras have been mainly developed with research purposes, emulating a neuromorphic system [18–20]. However, only a few deal with the data transfer of event data [21,22]. Event-based systems have not yet achieved the desirable spread in industrial environments to benefit from their advantages. However, current event-based systems still use a high bandwidth to transmit data, higher than a typical industrial communication system could handle. According to Farabet et al. [23], an advanced event-based sensor with about 1 million neurons might generate up to 10^8 million events per second; in a relatively simple example, the authors showed an experimental test where 8 million events per second are generated. This number of events can make an event-based system require a similar bandwidth to be transmitted to frame-based vision, making their advantages overshadowed and conventional machine vision systems (frame-based) being still used in industry environments. Nowadays, event-based processing techniques are focused in producing better data for pattern recognition in neuromorphic systems [24,25] and machine learning [26] rather than event data pre-processing which could ease the task of data transmission and further machine learning or other classification, prediction or recognition algorithms due to more clear data. In [27], a filtering algorithm is proposed, aiming a similar goal to our proposed work. However, its complexity (based on two-layer processing with neural network processing) requires a high computational cost, not being feasible for on-chip implementation.

The main aim of this work was to design an algorithm able to filter data obtained from event-based sensors, generating fewer events while keeping relevant information from the scene. Thus, the volume of transmitted data from the sensor can be reduced, requiring less bandwidth, lower energy consumption and less storage, which are very important issues for data transmission in communication networks and data storage. Additionally, not only reduced data transfer is required but also real-time response needs to be provided, which means that a low complexity, yet effective, algorithm must be developed. Some works are focused on developing and improving systems for data exchange between two or more bioinspired devices [21,22], transmitting original sensor data to the processing unit, i.e., a neuromorphic system typically composed of spiking neurons.

Taking the above into consideration, an algorithm was designed and tested for processing and filtering data from event-based sensors. For this reason, we call it "Less Data Same Information (LDSI)". Since it is valid for different event-based sensors, we focused on event-based cameras. This technique is based on how biological neurons work, i.e., acquired data consist of on–off spike sequences. This algorithm is fully configurable, with the main goal of providing adjustable results of filtering and data reduction depending on the final application. The use of this filter reduces the volume of data received by a neuromorphic system for classification, prediction, or any other application. Several of the factors inherent to industrial vision systems are considered: events generated by unit of time, noise, size of the image, and strong light changes, among others.

The used materials and existing techniques are detailed in Section 2. Section 3 details the proposed LDSI bioinspired algorithm, with results provided in Section 4, including real-time performance with FPGA implementation. Finally, Sections 5 and 6 discuss the results and provide conclusions, respectively.

2. Materials and Methods

Common standard platforms and tools in the event-processing field are used. The proposed algorithm must be compatible with a wide range of existing devices, both for event-based sensors as the input, and event-based processing units. Under this guideline, Address Event Representation (AER) for event transmission was used [28]. For data visualization, jAER software was used [29]. Figure 1 shows the two main approaches for development and testing of the algorithm. Initially, the algorithm was tested offline for event-based data from a database. It was developed in C++ language and performance evaluation and visualization was directly done in jAER (Figure 1a). Once the algorithm was developed, real-time performance was verified in an online real environment using an event-based camera connected to an FPGA where the algorithm was implemented and the results were transmitted to a PC in AER format, for final jAER visualization (Figure 1b). Specifically, the implementation was done in a Virtex-6 XC6VLX240T-1FFG1156 FPGA used in the ML605 evaluation board by Xilinx. The camera was a Dynamic Vision System (DVS) from Inilabs [2] connected to the FPGA through its parallel port [30]. The parallel port is a 15-line AER bus: 7-lines for Y-axis address, next 7-lines for the X-axis address of the active pixel, and one line for the polarity. The read was controlled by two extra lines ("REQ" and "ACK") for transaction control. In addition, the FPGA was connected through a serial port to a PC; the connection, baud-rate (921,600 bps) and data protocol were made to be compatible with jAER software, where the result of the algorithm was verified.

Address-Event Representation is an efficient and universal method of transmitting event data. It was proposed by Sivilotti in 1991 [31] and, since then, it has been widely adopted in the neuromorphic hardware field. With this type of encoding, each device has its own event space defined and it transmits information only in the case of state changes in any of the sensor receptors (pixels, in the case of an event-based camera). As an example, for a silicon retina, the event space will be the whole pixel matrix, where every pixel is an independent event source. Upon a threshold event in a pixel, the information about the change is encoded into a numerical value, typically, XY coordinates of the changing pixel. Thus, as only significant changes generate new data, the amount of information that the retina generates is several orders of magnitude lower when compared to a frame-based vision camera where all pixel values are transmitted every new frame regardless of pixel intensity change, generating redundant data. Using AER, the areas of interest (areas where the image has changed) is automatically identified since only data from this area are generated. For instance, in a ball intercept task, the average event stream is 20 kEvts/s corresponding to a 40 kB/s streaming speed. Using a frame-based camera with the same time resolution would require 6.6 MB/s of raw data stream [32]. The AER communication is suitable for low latency systems. In theory, all event sources are completely independent and asynchronous, and generate an AER data packet immediately after receiving the event. In practical applications, simultaneous event collisions are common and the event source usually includes an event management and scheduling algorithm to prevent data loss.

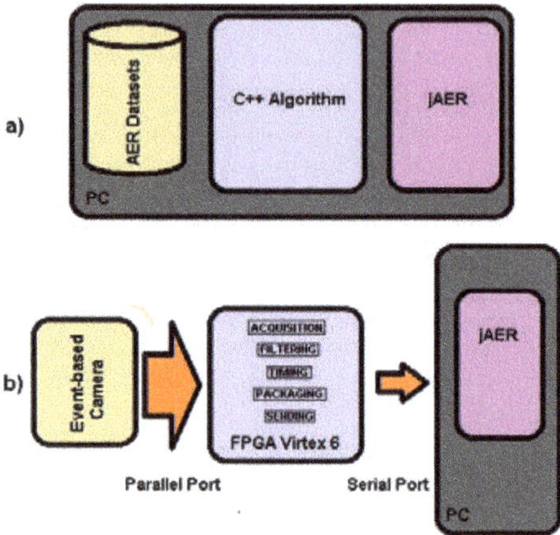

Figure 1. Procedures used to develop and test the LDSI algorithm: (**a**) offline configuration, where, from left to right, AER dataset recorded with event-based cameras was read and applied to the PC programmed LDSI algorithm whose results were provided to jAER for visualization; and (**b**) online testing, where an event-based camera was connected through its parallel port to an FPGA where the LDSI algorithm was computed and output data were properly encapsulated and sent to a PC for visualization and data-logging through a serial port.

Nowadays, there exist several protocols to encapsulate AER data. However, two main consolidated protocols are commonly used: AER1.0 and AER2.0 [26]. Currently, a new protocol version (AER3.0) is being tested for more complex and flexible data transfer between event-based devices [33]. In this application, we used AER1.0 for a dvs128 format, compatible with jAER. AER1.0 requires fewer bytes for the frame construction and, therefore, less transmission time. The protocol frame was built as follows:

1. The first bit of the first byte is used to align the data (always "1") at reception.
2. The following seven bits are the Y-axis coordinate of the pixel location.
3. The first bit of the second byte represents the polarity of the event, increment or decrement of the measured magnitude.
4. The next seven bits represent the X-axis coordinate of the pixel location.
5. Finally, the next four bytes contain the time elapsed (μs) since the last event generated (timestamp).

The operating protocol used in this case was the same as for an "edvs128" camera [34]. Proper data encapsulation was performed by the FPGA in the case of online operation.

3. Less Data Same Information (LDSI) Algorithm with Event-Based Encoding

This work proposed a novel algorithm, not only filtering noise generated in event-based cameras, but also reducing the number of redundant or irrelevant data. The proposed LDSI algorithm has a neuromorphic basis since it is based on spiking cells similar to those described by Izhikevich [9]. Specifically, it was inspired by the bipolar cells of the retina. However, the goal of this work was not to emulate a neuromorphic system but take advantage of some biological neurons concepts to reduce data transmission without loss of information. The defined model and its comparison to a biological neuron are shown in Figure 2. The layer-based model for event processing can be associated with sensory units in the sensory layer **Slayer** (pixels in case of a camera) which act as the dendrites feeding data to the nucleus (**Dlayer**) also forwarding information to synaptic terminals (**Alayer**). Each synaptic terminal in **Alayer** produces a final output represented in the output layer **Player** which can be considered as the next **Dlayer** in a successive chain of neurons.

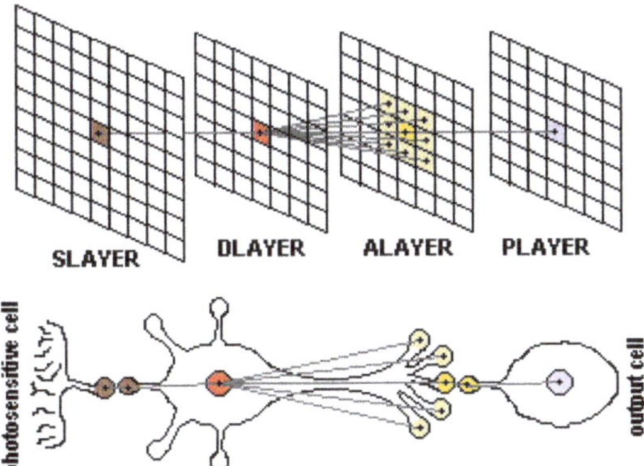

Figure 2. (**Top**) Interrelation between layers created in the LDSI algorithm. Lines between layers show the interconnections and data flow from the events generated in the input sensory layer (**Slayer**) corresponding to an event-based sensor (DVS camera in this case) to the output layer **Player** with filtered data. (**Bottom**) Equivalence of the proposed layer model in the LDSI algorithm with a biological neuron.

Thus, the model defines a single neuron composed of two units associated with the nucleus (**Dlayer**) and the axon or synaptic terminals **Alayer**, being $M \times N$ units in size. These units are arranged in two layers forming a neuronal-like structure. Each layer is defined by a bidimensional matrix of units identified by its xy coordinates in the matrix (Figure 2). Each unit in **Dlayer** and **Alayer** receive events from the input layer **Slayer** ($(M+2) \times (N+2)$ matrix size) and modifies their internal potential values, similar to biological neurons. A unit D_{xy} in **Dlayer** receives input events from the same xy position in the event generation layer (e.g., a sensory layer in an event sensor, or the output of a preceding layer). Then, the unit modifies its internal potential $\vartheta_D(x, y)$, which can be associated to the potential of the nucleus in a biological cell. Simultaneously, the units in **Alayer** modify its internal potential $\vartheta_A(x, y)$ due to input events received in **Dlayer** units located in xy, and the vicinity. Each unit in a layer modifies its internal potential and, when potential in both **Dlayer** and **Alayer** is above a threshold, the unit in **Alayer** generates an output event, reflected in **Player**, which has the same structure and size as the input layer ($(M+2) \times (N+2)$). This approach allows this LDSI filter to be included between already existing event processing modules since the **Player** output can be interpreted as the original input layer. This approach is the same as in other processing areas where different filters may be added as pre-processing.

The LDSI filter can define the number of neighbour units from **Dlayer** affecting a unit A_{xy} in **Alayer**, which is defined by the *Depthlevel*, **DL** $\in \mathbb{N}$ parameter (Figure 3). This effect resembles a receptive field affecting potential in units nearby the generation of an event.

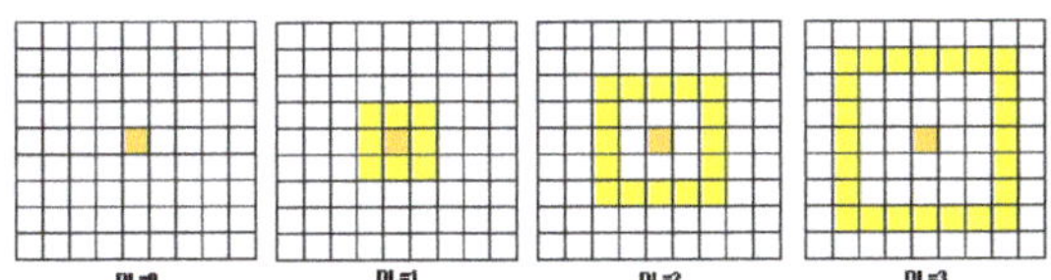

Figure 3. Interconnections existing between **Dlayer** and **Alayer** according to the DL parameter value.

In addition to **DL** already explained, the following parameters related to the units in the layers are defined:

- **Excitation level in Dlayer (ELD)**: Magnitude of the potential that a unit in the xy unit of **Dlayer** increases when an event is received from the unit located in the same xy unit in **Slayer**.
- **Excitation level in Alayer (ELA)**: The potential increment in the xy unit of **Alayer** due to an event in the same xy unit of **Dlayer**.
- **Excitation level in Alayer neighbouring units (ELAN)**: When an event is produced in an xy unit of **Dlayer**, ELAN corresponds to the potential increment of units in **Alayer** the vicinity of the xy unit. The number of affected neighbour units varies according to the **DL** value.
- **Threshold potential level in Dlayer (TPD)**: Defines the minimum value of excitation required for a certain unit in **Dlayer** to generate an output event.
- **Threshold potential level in Alayer (TPA)**: Defines the minimum value of excitation required for a certain unit in **Alayer** to generate an output event.
- **Decrement of potential in Dlayer (DPD)**: The value of potential to be decremented in **Dlayer** once MTR has elapsed.
- **Decrement of potential in Alayer (DPA)**: The value of potential to be decremented in **Alayer** once MTR has elapsed.

For event-based systems, delay between events affect how potential in a unit changes. The following parameters concerning time delay of events are defined:

- **Actualtimestamp (AT)**: The timestamp of the actual event present in a certain connection.
- **Lasttimestamp (LT)**: The timestamp of the previous event received in a certain connection.
- **Deltatime (DT)**: Time difference between the actual and the previous event coming from a certain connection. If this value is higher than MTR, the potential in the unit is decreased.
- **Maximum time to remember (MTR)**: Defines the maximum time between two events that the potential value in unit from layers **Dlayer** and **Alayer** can remain before being degraded. This parameter can be associated to a forgetting factor in the unit.

Upon an input event in the xy position of **Slayer**, the potential in the xy unit in **Dlayer** and the xy and **DL** neighbouring units in **Alayer** is increased by its corresponding excitation value. Equation (1) shows how potential $\vartheta_D(x,y)$ in an xy unit of **Dlayer** changes and Equation (2) shows the calculated potential $\vartheta_A(x,y)$ in an xy unit in **Alayer**, as a function of the above defined parameters. It is important to note that, for the same input events, potential in each layer evolves in a different form. In case of **Dlayer**, Equation (1) gives the mathematical description, and Equation (2) in the case of **Alayer**.

In case of **Dlayer**, potential is increased only when the event is received by exactly the same xy unit in **Slayer**; for **Alayer**, the potential is increased when an event exist in the xy unit in **Slayer**, or a **DL** neighbouring position. Potential is decreased if no event is received after a certain time defined by MTR, with zero limit, i.e., potential cannot be negative. When conditions are met, an output event is generated and, immediately, potential goes to zero until new events arrive. To generate an output spike in a xy unit in **Alayer**, it is important to note that this two-layer model requires that both potential in the xy unit from **Dlayer** and xy unit from **Alayer** are above TPD and TPA thresholds, respectively.

$$\vartheta_D(x,y)_{t+1} = \begin{cases} \vartheta_D(x,y)_t + ELD, & \text{if event in } D(x,y) \\ \vartheta_D(x,y)_t - DPD, & DT \geq MTR \\ 0, & (\vartheta_D(x,y)_t \geq TPD) \\ & AND\ (\vartheta_A(x,y)_t \geq TPA) \\ 0, & (\vartheta_D(x,y)_t - DPD) \leq 0 \\ \vartheta_D(x,y)_t, & no\ event\ AND\ DT < MTR \end{cases} \tag{1}$$

$$\vartheta_A(x,y)_{t+1} = \begin{cases} \vartheta_A(x,y)_t + ELA, & \text{if event in } D(x,y) \\ \vartheta_A(x,y)_t + ELAN, & \text{if event in } DL \\ & \quad vicinity\ of\ D(x,y) \\ & \quad AND\ DT < MTR \\ \vartheta_A(x,y)_t - DPA, & DT \geq MTR \\ 1, & (\vartheta_D(x,y)_t \geq TPD) \\ & \quad AND\ (\vartheta_A(x,y)_t \geq TPA) \\ 0, & \vartheta_A(x,y)_t = 1 \\ 0, & (\vartheta_A(x,y)_t - DPA) \leq 0 \\ \vartheta_A(x,y)_t, & no\ event\ AND\ DT < MTR \end{cases} \tag{2}$$

Figure 4 shows an example for the behaviour of the LDSI algorithm. In this case, **Slayer** units located in xy and $x(y+1)$ are generating events, which impact in the potential of several units in **Dlayer** and **Alayer**. According to the behaviour described in Equations (1) and (2), the figure shows the potential evolution for an xy unit in **Dlayer**, and the xy unit and all surrounding units in **Alayer**, assuming **DL** $= 1$. It is important to note that units in **Dlayer** are only affected by events in the same xy position while units in **Alayer** are affected by events in the xy unit and surrounding positions, with different potential increase depending on the event location. Additionally, for each new incoming event in **Slayer**, the time difference between the current time and the time of last event is evaluated; potential for all units in **Dlayer** and **Alayer** is decreased in the case this inter-event time is higher than MTR. When the potential of the xy units in both **Dlayer** and **Alayer** is above their respective threshold TPD and TPA, an output event is generated in the xy unit of **Player**.

The proposed model in this algorithm has two main characteristics: first, those events distant in time have a negative impact on output event generation since they decrease the potential; and, second, those spatially distant events do not contribute to potential increase in units located far from a **DL** distance. These facts allow discarding and providing low consideration to unexpected events as spurious noise events. This event processing model resembles how the mammalian brain continuously receives many data in the form of events, and, depending on the connections of the neurons and their excitation levels (strength), an output event is generated. Algorithm 1 shows how each input event from **Slayer** is processed in the subsequent layers.

Test Methodology

All parameters are integer values. Experimentally, we determined that the range of ELD, ELA, ELAN, TPD, TPA, DPD and DPA parameters should be kept between 0 and 10. Otherwise, a high computational cost is required without extra benefits. The MTR parameter is given as time units. Depending on the given values to all the parameters, the results can be adjusted to different levels of filtering. Despite the multiple possibilities and parameter value combinations, three main parameter sets were defined to provide weak, medium or restrictive level of filtering.

The LDSI algorithm was developed in two stages. First, we tested the algorithm in an "offline" environment, programming the algorithm in C++ language with the purpose of analysing its behaviour when applying different AER data from various scenes already pre-recorded as data files. The second stage was the LDSI "online" implementation where the algorithm was embedded in an FPGA so that the device obtained data from a real event-based camera, applied the LDSI algorithm and sent the resulting data in the proper format for jAER PC software reading and visualization.

For the algorithm development in the "offline" environment, we used a pre-recorded dataset publicly available on websites from other groups working with AER data processing [29,35]. The algorithm was applied with different datasets, each showing a different scene, and all of them with different noise levels, sizes and quantities of event data per unit of time. The goal of this test was to analyse the algorithm performance under different conditions and obtain enough data under different input event conditions to analyse parameter interrelations obtaining different output results.

For the "online" LDSI algorithm implementation in hardware, the algorithm was migrated from C++ to VHDL so that a more optimized computation was obtained in terms of parallelization, speed of operation and logic resource usage. It was tested in a Virtex 6 FPGA to verify the performance and compare results with the "offline" algorithm by connecting the FPGA with jAER software. The results of the "online" and "offline" LDSI show that the output events generated with the algorithm implemented in the FPGA were correct and coherent with those generated by the PC software implementation.

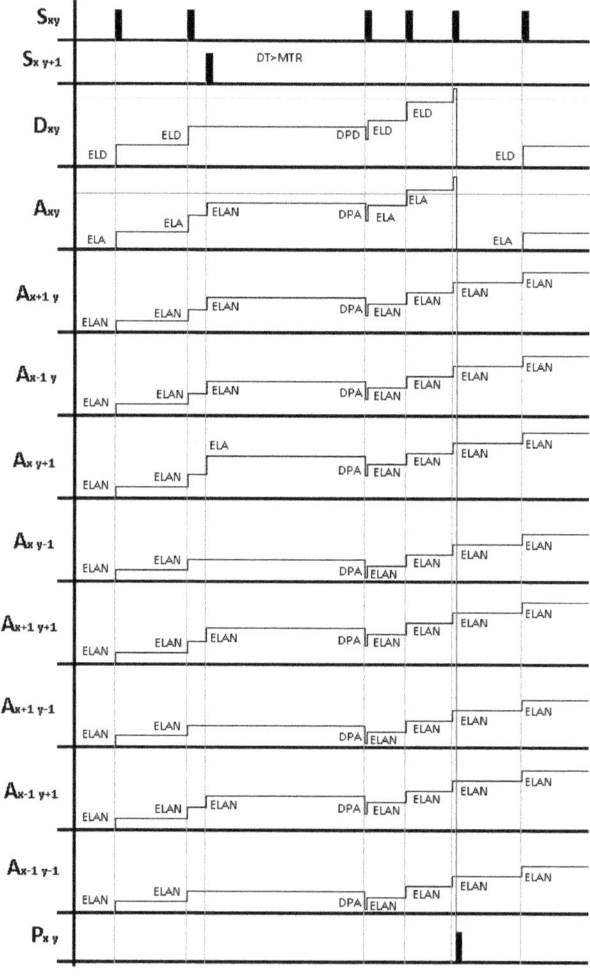

Figure 4. Event processing example in case of input events in two units of **Slayer**, xy and $x(y + 1)$. The potential value in each unit increases with input events and, when above a threshold, an output event is generated. If no events exist during a time defined by "MTR", the potential is decreased. In the case of **Dlayer** layer, only received events from the same xy unit in the previous layer increase its potential. For **Alayer**, events received from neighbour units also increase the potential. In each case, a different potential value can be defined. An output event (valued "1") is generated when the xy units in **Dlayer** and **Alayer** are above their respective threshold, simultaneously.

Algorithm 1 LDSI algorithm: Computation is performed when events exist in **Slayer**. Output events in the filter are only generated if potential is above threshold.

Require: Event-based inputs from the sensory layer **Slayer**
Ensure: Events addresses inside **Slayer**

1: **if** $Slayer_{INPUT\_EVENT}$ in (x, y) **then**
2:    {– Potential increase}
3:    $DT = AT - LT$
4:    $D_{(x,y)} = D_{(x,y)} + ELD$
5:    $A_{(x,y)} = A_{(x,y)} + ELA$
6:    **for** $i = -DL$ to DL **do**
7:      **for** $j = -DL$ to DL **do**
8:        $A_{(x+i,y+j)} = A_{(x+i,y+j)} + ELAN$
9:      **end for**
10:   **end for**
11:   **if** $DT < MTR$ **then**
12:     {– Potential decrease}
13:     **for all** $i = 0$ to $((M - 2) * DL)$ **do**
14:       **for all** $j = 0$ to $((N - 2) * DL)$ **do**
15:         **if** $D_{(i,j)} >= DPD$ **then**
16:           $D_{(i,j)} = D_{(i,j)} - DPD$
17:         **else**
18:           $D_{i,j} = 0$
19:         **end if**
20:         **if** $A_{(i,j)} >= DPA$ **then**
21:           $A_{(i,j)} = A_{(i,j)} - DPA$
22:         **else**
23:           $A_{(i,j)} = 0$
24:         **end if**
25:       **end for**
26:     **end for**
27:   **end if**
28:   **if** $D_{(x,y)} >= TPD$ AND $A_{(x,y)} >= TPA$ **then**
29:     $P_{(x,y)} = EVENT$
30:     $D_{(x,y)} = 0$
31:     $A_{(x,y)} = 0$
32:   **else**
33:     $P_{(x,y)} = NULL$
34:   **end if**
35:   $LT = AT$
36: **end if**

---

4. Results

A real scene of a slowly moving hand was captured from a DVS event camera. The same scene is compared when no filtering is applied (Figure 5a), using the existing jAER "background activity filter" (Figure 5b), and the LDSI filter using a medium level of filtering (Figure 5c). This figure shows that the LDSI result not only provided smoother and better edge definition but also reduced the produced data (generated events) compared to the original: 7974 kBytes for the LDSI versus 13,348 kBytes for jAER filter and 15,796 kBytes for the unfiltered.

To obtain a complete evaluation of the LDSI algorithm, 260 sets of parameters were tested, iterating each combination 100,000 times. Output data were analysed through a range of parameter combinations to find an interrelation between the parameters and the filter output data. The results produced by the algorithm varied from a high data reduction level and noise removal, including some removal of data from the area of interest for certain parameter values, to a "transparent mode" where most of the incoming data were transferred to the output. Thus, the parameter configuration allows the filter to be tuned according to the application or user requirements, from a very low to a highly restrictive mode. In any case, the LDSI never generated more output events than the input events and it never blurred or deformed the scene. In some cases, it was possible to obtain zero output events, i.e., null data output from the filter.

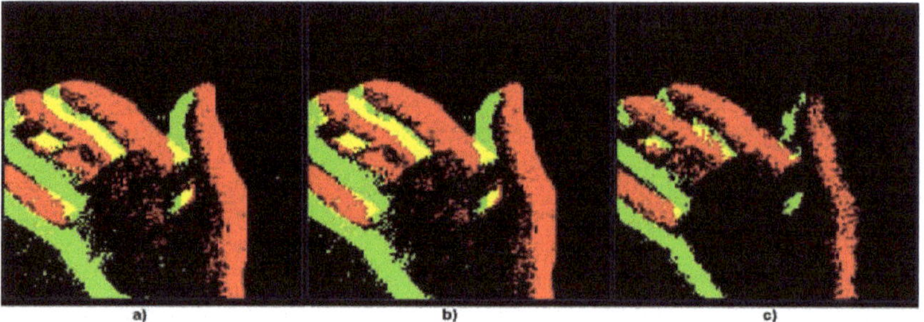

Figure 5. Comparison among the original DVS camera data, jAER "background activity filter" and the proposed LDSI algorithm: (**a**) original events from event-based camera with no filter, where noise and repetitive data were generated, mainly at the object borders; (**b**) result after applying the background noise jAER built-in filter with restrictive parameters; and (**c**) events produced after applying the LDSI algorithm with parameters selected for a compromise between data reduction and loss of main data of the scene (medium filter).

Figures 6–10 show the interrelation between some parameters and how they affect the final result of produced events. To discard non-sense resulting data as zero events at the output, some obvious combinations of parameters were not considered. Some parameters influenced noise removal while others influenced output data production.

Figure 6 shows the number of output events generated by the LDSI, depending on the TPD threshold and ELD values in **Dlayer**, while keeping constant the rest of the parameters. If ELD value is close to TPD value, an output event is more likely to be produced upon input events and then, input and output events will be very similar. For this reason, ELD values higher than TPD were discarded since that combination produces exactly the same data as the original, converting the algorithm into a repeater. On the other hand, if the ELD value is much lower than TPD, the filter will be very restrictive and fewer data will be produced, but the noise will not necessarily be discarded, as it will also discard valid events from the scene.

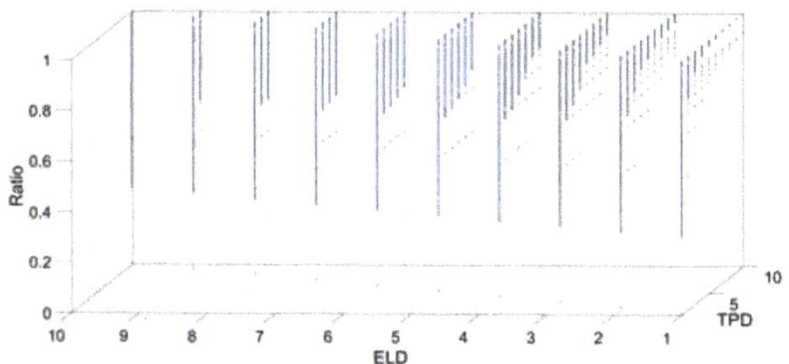

Figure 6. Output events generated by the LDSI algorithm depending on ELD and TPD values. An increase of events is produced when ELD increases. In the left corner of the graph, it is possible to see how low values of ELD at high values of TPD restricts the event production.

Figure 7 shows the relation between ELA and TPA (excitation and threshold in **Alayer**). As seen, different variations and combinations of ELA and TPA do not provide a significant modification in the output events. However, the behaviour is greatly affected by TPD and ELD since events arriving to **Alayer** come from **Dlayer** and, thus, only in the case of low values of ELA and TPA, the generated events are reduced.

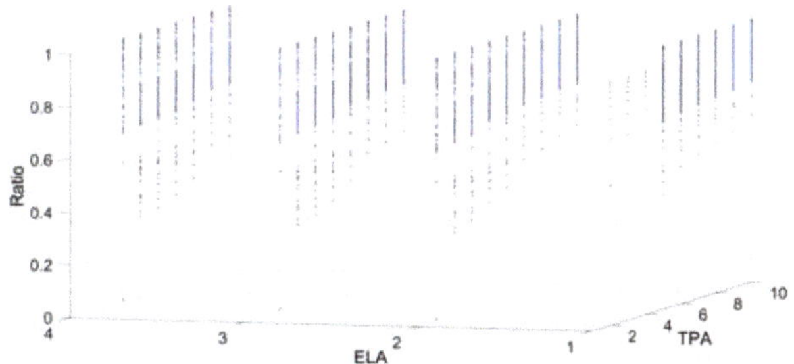

Figure 7. Outputs events as a function of the excitation level (ELA) and threshold (TPA) in **Alayer**. A value of TPA higher than ELA reduces the production of events, thus filtering noise but also some loss of data in the main scene appears.

Regarding the relationship between the threshold TPA and the excitation level of neighbours ELAN, Figure 8 shows that produced events are increased in the case of a high value in both parameters. The result indicates that, as ELAN contributes to the potential, a high value increases the possibility of producing output events since the potential in a certain unit xy increases faster.

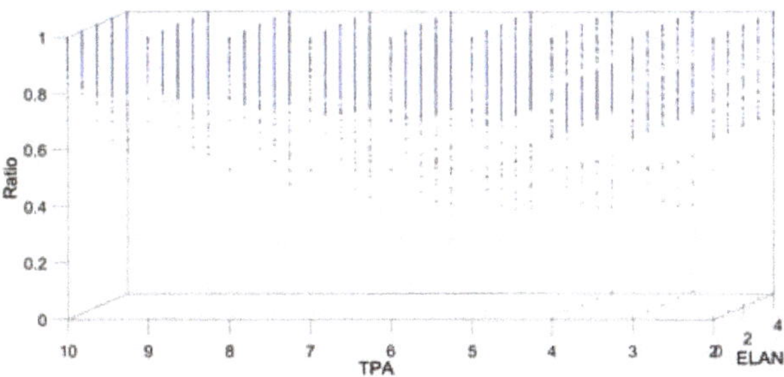

Figure 8. Relationship between the excitation level in neighbours (ELAN) and its associated threshold TPA. The production of output events increases in the case of high TPA and high ELAN.

Figure 9 shows the relationship between the threshold level TPD and the decrement level in absence of input events (DPD), in **Dlayer**. In this case, we can observe that, for the same TPD value, a reduction in produced events appears when DPD increases. This is a desired effect; it reduces potential if no input events appear. Thus, potential is highly reduced due to high values of DPD and, then, fewer output events are generated.

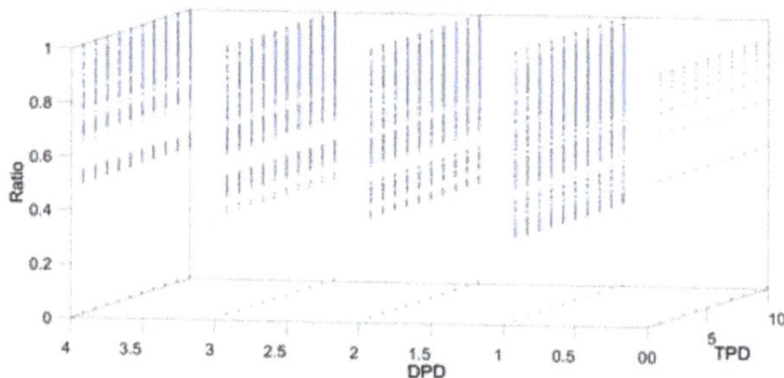

Figure 9. Relationship between the decrement potential level (DPD) and the threshold (TPD) in **Dlayer**. No direct relationship in event reduction among these values is found.

Concerning the behaviour of **Alayer** in relation to the potential decrease due to the absence of input events (DPD), Figure 10 shows the output events produced depending on TPA and DPA values. In this case, a similar effect with fewer variations than in **Dlayer** appears. A slight decrease in generated data is observed when DPD increases.

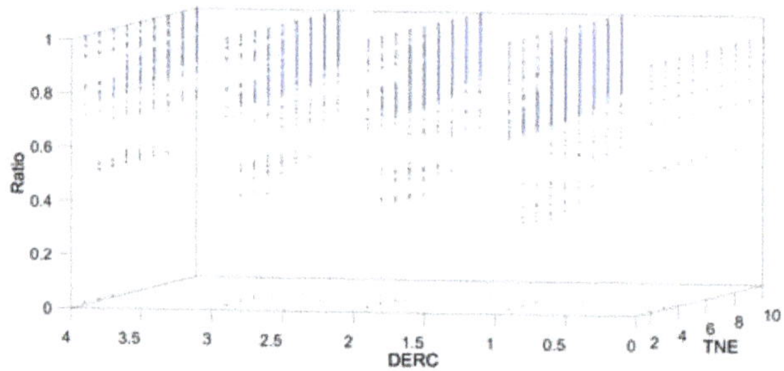

Figure 10. Relationship between the potential level (DPA) and the threshold (TPA) in **Alayer**. Only in case of low DPA or TPA values, the number of output events is reduced.

Despite the parameters of the filter are fully configurable, after analysing the results, the following conclusions can be obtained:

1. For applications where a high ratio of noise versus the main data is present, the LDSI algorithm has better performance with lower values of MTR.
2. Regardless of the quantity of noise in relation to the relevant data, it is important to define low values of ELA but preferably higher than ELAN, and TPA higher than both ELA and ELAN, which improves the noise removal.
3. When the noise is not a problem and the goal is to obtain a clear distinction of edges from the object in the scene, it is recommended to increase the value of ELD and, at same proportion, DPD and DPA.
4. Finally, it is necessary to avoid configuration parameters where results could be predictable such as ELD and ELA being equal to zero, which will produce zero events. On the contrary, TPD and TPA values lower than ELD and ELA, respectively, will produce the same output data as the input.

These statements are not mathematical facts because they depend on multiple variables such as the ratio between noise and main data of the scene, speed of the objects moving through the scene, hot pixels, size of the sensor, fast change of light (intensity), etc. However, after several tests, it was possible to realize that, under similar conditions, the result of the LDSI algorithm are consistent.

Figure 11 shows the output events under three different LDSI parameterization (weak, medium and restrictive) in comparison with the original input event data, for five different sequences (500 ms each). Five frames of a handwritten letter "L" smoothly moving side to side are displayed. Sequence (a) represents the original input events without filter; Sequence (b) represents the LDSI result with a data reduction of 33% with respect to the original image and parameters adjusted as a weak filter mode; Sequence (c) shows a 50% data reduction with medium filtering options; and Sequence (d) is the resulting sequence produced with a restrictive LDSI parameter configuration, obtaining 85% data reduction. In the case that elimination of noise close to the relevant scene information is required, a more restrictive parameter configuration is recommended, with high values of TPD and TPA with respect to ELD, ELA and ELAN, together with a low MTR value.

Table 1 shows the parameter values corresponding to weak, medium and restrictive data reduction LDSI filter parameter values, together with the number of generated data for each case. A low output data reduction implies weak filtering with low noise removal while a high data reduction means high filtering with the risk of losing relevant features in the scene. In any case, the type of scene greatly affects the results and, thus, parameters must be chosen according to the target scene or features to be extracted.

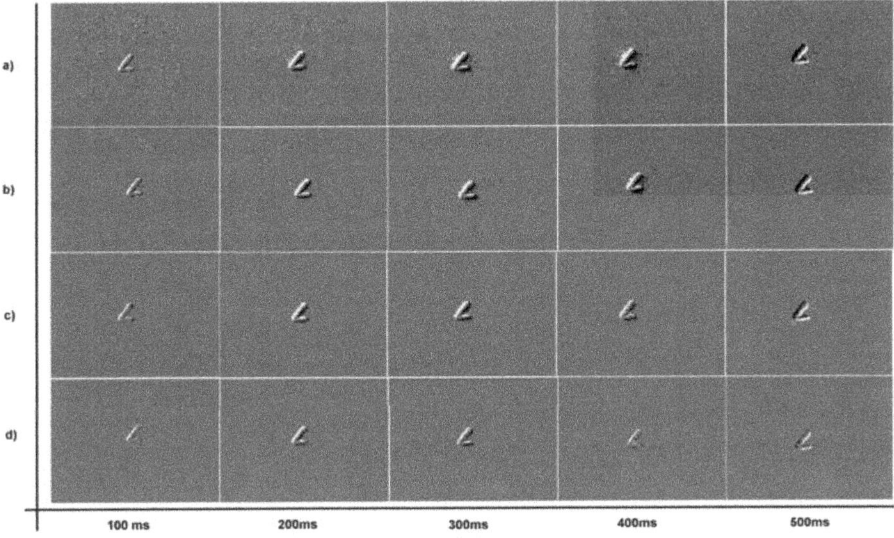

Figure 11. Image sequence of 500 ms (100 ms per image): (**a**) original data without LDSI; (**b**) low data reduction and weak noise removal; (**c**) medium data reduction and medium noise removal; and (**d**) high data reduction with restrictive parameters and high noise removal.

Table 1. Parameter values for different LDSI filter configuration and the associated number of events generated for the image shown in Figure 11. The original number of events from source with no filtering is also shown.

Parameter	Original	Weak	Medium	Restrictive
MTR	-	400	400	400
ERCO	-	2	5	1
TCE	-	3	5	2
TNE	-	2	6	8
ERCN	-	2	4	4
DERP	-	3	1	1
ERNC	-	3	4	4
DERC	-	3	1	1
Total events	25,596	17,196	8616	3769

FPGA Hardware Implementation

Concerning hardware implementation, the LDSI filter was implemented in a Xilinx Virtex6 6vlx240tlff1156-1l device. The implemented system included serial port connectivity to a PC for jAER data exchange so that pre-recorded events could be sent to the FPGA for LDSI filtering and then returned back to the PC for jAER analysis and visualization. Additionally, a DVS camera can be connected to the FPGA for real-time data event input. The FGPA applies the LDSI algorithm and filtered data are sent to the PC. The system uses a DVS camera communication protocol to receive real events from a 128 $\times$ 128 camera, and the LDSI filtering algorithm with 126 $\times$ 126 layer size according to the structure described in previous sections.

The total logic resource occupation for the camera communication protocol, the LDSI algorithm, AER packet creation and PC transmission was 671 LUTs, which is an impressively low value. In part, the reduced logic occupation is due to internal block RAM use for parameter storage: 40 internal FPGA block RAM were used. No additional FPGA resources were used for computation. Concerning the speed of operation, we used the on board 50 MHz clock but the implementation results showed that the system could run up to 177 MHz, providing enough speed for real time camera event processing and AER output generation.

In addition, the low computational complexity of the proposed algorithm would allow the use of low cost hardware device, such as simple microprocessor or microcontrollers. In fact, we also tested the system using an ATMega microcontroller and results were satisfactory up to a certain number of events per second. In the case of the FPGA, it can be guaranteed that all possible events for a 128 $\times$ 128 camera can be processed in real time.

5. Discussion

Concerning the results obtained for the hardware implementation, using the same LDSI parameters, the same scene was compared between results generated by the FPGA and the jAER software under PC, providing the same results in the FPGA and PC software, which validates the correct FPGA implementation.

Additionally, to verify the LDSI behaviour, once the LDSI algorithm has proven its capability for reduction of output data from an event-based scene, it is also necessary to verify that the relevant information in the scene is retained by the LDSI output data. Despite it can be clearly observed visually using jAER, a test pilot was conducted to formally verify this assumption. In this test, the original event data generated by a DVS camera and the event data produced by the LDSI algorithm after processing the same events were converted into a sequence of images generating a frame every 100 ms. A standard similitude algorithm typically used in industrial machine vision was applied, the algorithm provided the percentage of similitude between the LDSI output image values and a pattern initially shown to the algorithm as the master pattern. Thus, the original input data and all different LDSI configurations resulted in a similitude value compared to the initial pattern. The similitude value provides a comparison between the initial pattern and the LDSI configuration; the aim in this case was not to reach a high percentage value rather than a comparison among them. Figure 12 shows the result of the similitude algorithm applied to nine consecutive time sequences (100 ms each) for the same scene. This algorithm was based on the Mean Structural SIMilarity (MSSIM) index [36]. The similitude algorithm was evaluated for the original unfiltered data and three LDSI parameter configurations corresponding to those shown in Table 1 (weak, medium and restrictive). As seen, high data reduction was obtained while similitude results did not greatly differ from all four cases. However, higher similitude values in medium and weak LDSI parameterization with high data reduction were found when compared with the original unfiltered data (Figure 12). This fact shows that not only data reduction and denoising was obtained, but also the quality of image was increased when the LDSI filter was applied, e.g., edges were better defined.

Table 2 shows the data reduction values for the similitude analysis. As expected, the similitude ratio decreases with higher data reduction. Beyond 85% data reduction, the recognition was unsatisfactory with a high variation from one case to another (high standard deviation). However, data reduction up to 33% (weak filter) can be achieved while maintaining similar recognition results as the original data. Even in the case of 66% data reduction, the recognition ratio only drops 6.6% compared to original data. In summary, we can state that a high data reduction can be achieved in the case of weak and medium LDSI filter configurations (33% and 66% data reduction compared to the original, respectively), while maintaining a high similitude ratio.

A second test was conducted by comparing the LDSI filter result with other filtering algorithms. In this case, we used the jAER background filter algorithm. Tuned to provide the best performance, Figure 13 shows how the background filter does not remove all spurious events. The LDSI filter provides fewer spurious events and makes the relevant scene more defined.

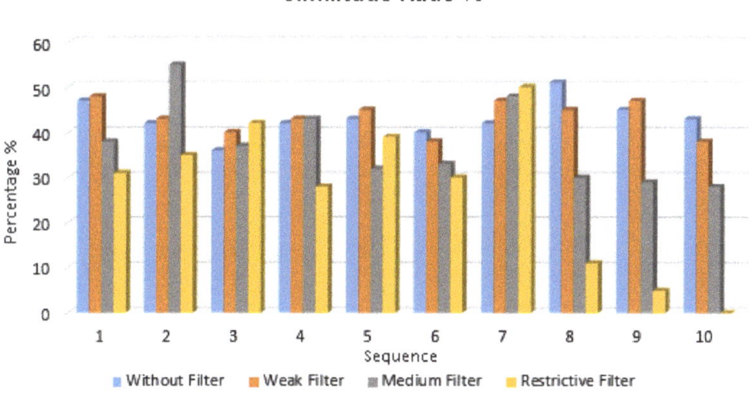

Figure 12. Similitude algorithm results for unfiltered input event data and three LDSI configurations: weak, medium and restrictive. As different scenes appear over time, the similitude ratio slightly changes. Low and medium LDSI configurations provide similar results to the original, at a higher data reduction.

Table 2. Data reduction and similitude test for original DVS camera data (unfiltered), and different LDSI parameters.

	Size (bytes)	Reduction (%)	Simil. Ratio (%)
Original data	153,576	-	43.02 ± 4.01
Weak filter	103,176	32.82	43.04 ± 3.68
Medium filter	51,696	66.34	37.64 ± 8.92
Restrictive filter	22,614	85.28	27.5 ± 16.52

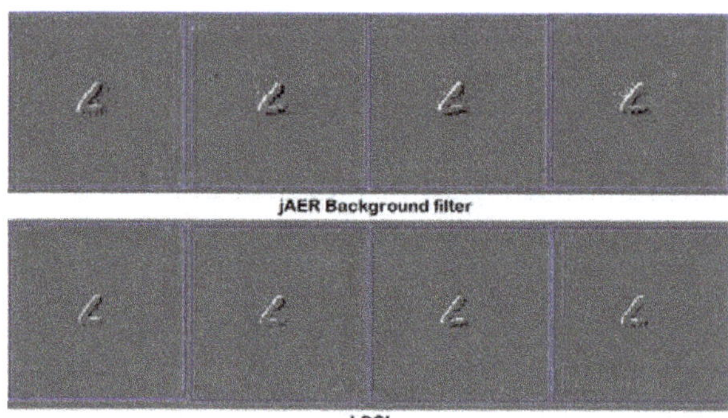

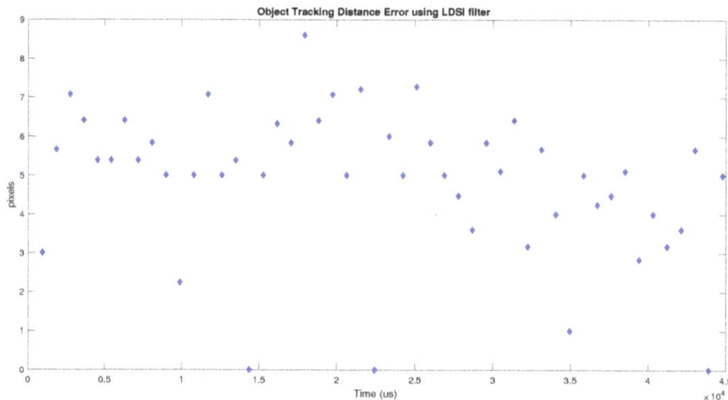

Figure 13. Filtering result for the background filter algorithm included in jAER software (**top**) and LDSI result (**bottom**).

To provide numerical results in the comparison between the LDSI and background activity filters, we applied the jAER built-in tracking algorithm, which provides the position of a moving object. For each filter, we obtained the error distance between the actual object position and the position given by the tracking algorithm. Figure 14 shows the results provided by the LDSI algorithm. Figure 15 shows the results obtained by the background activity filter. The error distance was calculated as the Euclidean distance in number of pixels. The obtained values were 4.86 ± 1.87 for LDSI, and 5.01 ± 1.93 for the background activity filters.

Figure 14. Error values in object tracking (actual position vs. tracking algorithm position) when LDSI filter is used.

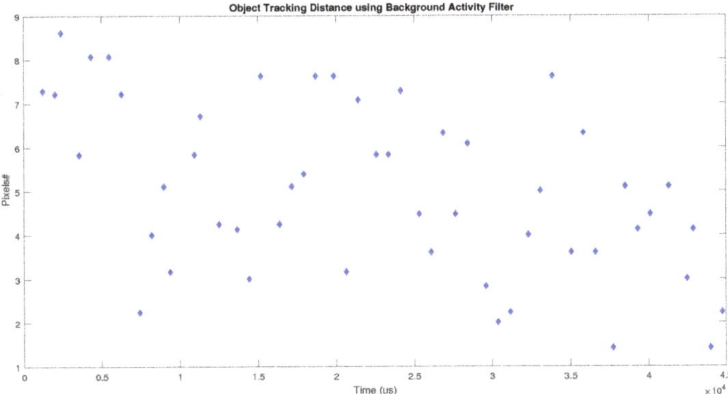

Figure 15. Error values in object tracking (actual position vs. tracking algorithm position) when the jAER built-in background activity filter is used.

Figure 14 shows a more regular behaviour along time and, in some cases, the error is zero. However, Figure 15 shows that the background activity filter provides a more irregular behaviour, being affected by those events not removed from the scene. Since the LDSI removes more noisy events, it allows the tracking algorithm to provide more accurate and regular results.

6. Conclusions

A bioinspired filter for event-based systems is proposed, based on layers of units inspired in biological neurons and their interconnections. We can conclude that the initial aim was achieved, i.e., a configurable filter providing a range of options in the result, from transparent to zero output event mode. It was demonstrated that reducing the amount of data and noise with different levels depends not only on the parameter values, but also on the interaction among them and the types of input data (fast scenes, high intensity level change, etc.). For that reason, the LDSI algorithm can be configured to be adapted to different situations due to its parameters.

Filtering algorithms for event-based cameras are seldom reported, and they are designed for specific applications. The LDSI filter is compared to a commonly used jAER software filter, providing improved results, especially in complex situations with fast moving objects in the scene. The LDSI filter can be used in any scenario where the input consists on a matrix sensor generating events, not only image sensors but also other event sensors, such as auditory [3], distance measurement or magnetic compass [34], olfactory systems [4,5], and tactile [37] sensors.

The filter was tested with a DVS camera showing that noise reduction is improved, with a reduction in produced data, too. Produced data can reach 30% event data reduction compared with the original event-based generated image with an improvement in the scene definition and noise reduction. Usually, event-based cameras produce fewer data than conventional frame-based cameras and the proposed algorithm obtains a higher reduction. This reduction ratio becomes very important when data have to be transmitted from the camera to other processing systems as neuromorphic devices or Spiking Neural Networks (SNN), especially in cases where the same communication channel has to be shared among multiple devices, e.g., an ethernet-based communication. In addition to better scene definition, the LDSI allows a reduced AER data flow, which can reduce data congestion in communication channels.

Even though multiple combinations of parameters are possible, an analysis is done to provide a guideline on the parameter configurations leading to different output filter events. The filter is fully configurable; in this work, three parameter sets were proposed: low data reduction keeping all image features (weak filter), medium data reduction with some possible loss of information (medium filter) and strong data reduction with some loss of information (restrictive filter). We recommend watching the videos included as Supplementary Materials to evaluate the performance of LDSI filter.

Finally, LDSI hardware implementation shows that low resource usage is required, being an option for event-based sensor processing as on-board filtering before transmitting data to other devices or feeding data to neuromorphic system. This is possible due to the specific LDSI design having in mind that it can be used as an intermediate processing block, fully compatible with AER input data from an event-based sensor and generating AER output data as if data were generated by the sensor, thus being "transparent" to devices receiving data. Furthermore, its simplicity would allow its implementation in low cost microcontrollers.

Supplementary Materials: Different video recordings were created to appreciate the filter performance under different filter parameters and scenes: Raw data from event-based camera without filtering https://youtu.be/ogpwNKAEWF0. LDSI algorithm output with a weak level of filtering https://youtu.be/ynNhRVe_SG4. LDSI algorithm output with a medium level of filtering https://youtu.be/XQbcV-ANoh0. LDSI algorithm output with a high level of filtering https://youtu.be/fmL8TPMGNZw. jAER tracking on event data with jAER background filter: https://youtu.be/7nptQ9EMWc8. jAER tracking with LDSI: https://youtu.be/YgIgVSa5JNY. Tracking algorithm results for low, medium and restrictive filter parameters in the case of the LDSI and background activity filters: https://www.youtube.com/playlist?list=PLgZjGCa1fp1EhRM6ctrWxv3jTyfnndXRJ.

Author Contributions: Conceptualization, A.R.-M.; Data curation, L.D.M.; Formal analysis, J.B.-A.; Investigation, J.B.-A.; Methodology, A.R.-M.; Project administration, A.R.-M.; Resources, M.B.-M.; Software, J.B.-A.; Supervision, A.R.-M.; Validation, J.B.-A.; Writing—original draft, J.B.-A. and M.B.-M.; and Writing—review and editing, A.R.-M. and J.F.G.-M.

Funding: This research received no external funding.

Conflicts of Interest: The authors declare no conflict of interest.

References

1. Pardo, F.; Boluda, J.A.; Vegara, F. Selective Change Driven Vision Sensor with Continuous-Time Logarithmic Photoreceptor and Winner-Take-All Circuit for Pixel Selection. *IEEE J. Solid-State Circ.* **2015**, *50*, 786–798. [CrossRef]

2. Brandli, C.; Berner, R.; Yang, M.; Liu, S.C.; Delbruck, T. A 240 × 180 130 dB 3 us Latency Global Shutter Spatiotemporal Vision Sensor. *IEEE J. Solid-State Circ.* **2014**, *49*, 2333–2341. [CrossRef]

3. Liu, S.C.; van Schaik, A.; Minch, B.A.; Delbruck, T. Asynchronous Binaural Spatial Audition Sensor with 2 × 64 × 4 Channel Output. *IEEE Trans. Biomed. Circ. Syst.* **2014**, *8*, 453–464. [CrossRef]

4. Vanarse, A.; Osseiran, A.; Rassau, A. An Investigation into Spike-Based Neuromorphic Approaches for Artificial Olfactory Systems. *Sensors* **2017**, *17*, 2591. [CrossRef] [PubMed]

5. Schmuker, M.; Nawrot, M.; Chicca, E. Neuromorphic Sensors, Olfaction. In *Encyclopedia of Computational Neuroscience*; Jaeger, D., Jung, R., Eds.; Springer: New York, NY, USA, 2015; pp. 1991–1997. [CrossRef]

6. Moyne, J.R.; Tilbury, D.M. The Emergence of Industrial Control Networks for Manufacturing Control, Diagnostics, and Safety Data. *Proc. IEEE* **2007**, *95*, 29–47. [CrossRef]

7. Decotignie, J.D. Ethernet-Based Real-Time and Industrial Communications. *Proc. IEEE* **2005**, *93*, 1102–1117. [CrossRef]

8. Berner, R.; Brandli, C.; Yang, M.; Liu, S.C.; Delbruck, T. A 240 × 180 10 mW 12 us latency sparse-output vision sensor for mobile applications. In Proceedings of the 2013 Symposium on VLSI Circuits, Kyoto, Japan, 12–14 June 2013.

9. Izhikevich, E.M. Simple model of spiking neurons. *IEEE Trans. Neural Netw.* **2003**, *14*, 1569–1572. [CrossRef] [PubMed]

10. Furber, S.B.; Lester, D.R.; Plana, L.A.; Garside, J.D.; Painkras, E.; Temple, S.; Brown, A.D. Overview of the SpiNNaker System Architecture. *IEEE Trans. Comput.* **2013**, *62*, 2454–2467. [CrossRef]

11. Rigi, A.; Baghaei Naeini, F.; Makris, D.; Zweiri, Y. A Novel Event-Based Incipient Slip Detection Using Dynamic Active-Pixel Vision Sensor (DAVIS). *Sensors* **2018**, *18*, 333. [CrossRef] [PubMed]

12. Rios-Navarro, A.; Cerezuela-Escudero, E.; Dominguez-Morales, M.; Jimenez-Fernandez, A.; Jimenez-Moreno, G.; Linares-Barranco, A. Real-time motor rotation frequency detection with event-based visual and spike-based auditory AER sensory integration for FPGA. In Proceedings of the 2015 International Conference on Event-based Control, Communication, and Signal Processing (EBCCSP), Krakow, Poland, 17–19 June 2015; pp. 1–6. [CrossRef]

13. Serrano-Gotarredona, R.; Serrano-Gotarredona, T.; Acosta-Jimenez, A.J.; Linares-Barranco, B. An arbitrary kernel convolution AER-transceiver chip for real-time image filtering. In Proceedings of the 2006 IEEE International Symposium on Circuits and Systems, Island of Kos, Greece, 21–24 May 2006. [CrossRef]

14. Rivas-Perez, M.; Linares-Barranco, A.; Jimenez-Fernandez, A.; Civit, A.; Jimenez, G. AER spike-processing filter simulator: Implementation of an AER simulator based on cellular automata. In Proceedings of the International Conference on Signal Processing and Multimedia Applications, Seville, Spain, 18–21 July 2011; pp. 1–6.

15. Espínola, A.; Romay, A.; Baidyk, T.; Kussul, E. Robust vision system to illumination changes in a color-dependent task. In Proceedings of the 2011 IEEE International Conference on Robotics and Biomimetics, Phuket, Thailand, 7–11 December 2011; pp. 521–526. [CrossRef]

16. Lin, W.K.; Uang, C.M.; Wang, P.C.; Ho, Z.S. LED strobe lighting for machine vision inspection. In Proceedings of the 2013 International Symposium on Next-Generation Electronics, Kaohsiung, Taiwan, 25–26 February 2013; pp. 345–346. [CrossRef]

17. Kim, H.; Cho, K.; Kim, S.; Kim, J. Color mixing and random search for optimal illumination in machine vision. In Proceedings of the 2013 IEEE/SICE International Symposium on System Integration, Kobe, Japan, 15–17 December 2013; pp. 907–912. [CrossRef]

18. Camuñas-Mesa, L.A.; Serrano-Gotarredona, T.; Linares-Barranco, B. Event-driven sensing and processing for high-speed robotic vision. In Proceedings of the 2014 IEEE Biomedical Circuits and Systems Conference (BioCAS) Proceedings, Lausanne, Switzerland, 22–24 October 2014; pp. 516–519. [CrossRef]

19. Delbruck, T.; Pfeiffer, M.; Juston, R.; Orchard, G.; Müggler, E.; Linares-Barranco, A.; Tilden, M.W. Human vs. computer slot car racing using an event and frame-based DAVIS vision sensor. In Proceedings of the 2015 IEEE International Symposium on Circuits and Systems (ISCAS), Lisbon, Portugal, 24–27 May 2015; pp. 2409–2412. [CrossRef]

20. Linares-Barranco, A.; Gomez-Rodriguez, F.; Jimenez-Fernandez, A.; Delbruck, T.; Lichtensteiner, P. Using FPGA for visuo-motor control with a silicon retina and a humanoid robot. In Proceedings of the 2007 IEEE International Symposium on Circuits and Systems, New Orleans, LA, USA, 27–30 May 2007; pp. 1192–1195. [CrossRef]

21. Partzsch, J.; Mayr, C.; Vogginger, B.; Schüffny, R.; Rast, A.; Plana, L.; Furber, S. Live demonstration: Ethernet communication linking two large-scale neuromorphic systems. In Proceedings of the 2013 European Conference on Circuit Theory and Design (ECCTD), Dresden, Germany, 8–12 September 2013. [CrossRef]

22. Fasnacht, D.B.; Whatley, A.M.; Indiveri, G. A serial communication infrastructure for multi-chip address event systems. In Proceedings of the 2008 IEEE International Symposium on Circuits and Systems, Seattle, WA, USA, 18–21 May 2008; pp. 648–651. [CrossRef]

23. Farabet, C.; Paz, R.; Perez-Carrasco, J.; Zamarreno, C.; Linares-Barranco, A.; LeCun, Y.; Culurciello, E.; Serrano-Gotarredona, T.; Linares-Barranco, B. Comparison Between Frame-Constrained Fix-Pixel-Value and Frame-Free Spiking-Dynamic-Pixel ConvNets for Visual Processing. *Front. Neurosci.* **2012**, *6*, 32. [CrossRef] [PubMed]

24. Camunas-Mesa, L.; Zamarreno-Ramos, C.; Linares-Barranco, A.; Acosta-Jimenez, A.J.; Serrano-Gotarredona, T.; Linares-Barranco, B. An Event-Driven Multi-Kernel Convolution Processor Module for Event-Driven Vision Sensors. *IEEE J. Solid-State Circ.* **2012**, *47*, 504–517. [CrossRef]

25. Zhao, B.; Ding, R.; Chen, S.; Linares-Barranco, B.; Tang, H. Feedforward Categorization on AER Motion Events Using Cortex-Like Features in a Spiking Neural Network. *IEEE Trans. Neural Netw. Learn. Syst.* **2015**, *26*, 1963–1978. [CrossRef] [PubMed]

26. Jimenez-Fernandez, A.; del Bosh, J.L.F.; Paz-Vicente, R.; Linares-Barranco, A.; Jiménez, G. Neuro-inspired system for real-time vision sensor tilt correction. In Proceedings of the 2010 IEEE International Symposium on Circuits and Systems, Paris, France, 30 May–2 June 2010; pp. 1394–1397. [CrossRef]

27. Padala, V.; Basu, A.; Orchard, G. A Noise Filtering Algorithm for Event-Based Asynchronous Change Detection Image Sensors on TrueNorth and Its Implementation on TrueNorth. *Front. Neurosci.* **2018**, *12*, 118. [CrossRef] [PubMed]

28. Boahen, K.A. Point-to-point connectivity between neuromorphic chips using address events. *IEEE Trans. Circ. Syst. II Analog Digit. Signal Process.* **2000**, *47*, 416–434. [CrossRef]

29. Open Source Software, a. User guide jAER: Java Tools for AER Neuromorphic Processing. Available online: https://inivation.com/support/software/jaer/ (accessed on 15 September 2018).

30. AER Parallel Protocol Communication. Available online: https://www.ini.uzh.ch/~amw/scx/std002.pdf (accessed on 15 September 2018).

31. Sivilotti, M.A. Wiring Considerations in Analog VLSI Systems, with Application to Field-Programmable Networks. Ph.D. Thesis, California Institute of Technology, Pasadena, CA, USA, 1991.

32. Delbruck, T.; Lang, M. Robotic goalie with 3 ms reaction time at 4event-based dynamic vision sensor. *Front. Neurosci.* **2013**, *7*, 223. [CrossRef] [PubMed]

33. AER Protocol Definition. Available online: http://inilabs.com/support/software/fileformat/ (accessed on 15 September 2018).

34. Barrios-Aviles, J.; Iakymchuk, T.; Rosado-Munoz, A.; Frances-Villora, J.V.; Bataller-Mompean, M.; Guerrero-Martinez, J.F. Event-based encoding from digital magnetic compass and ultrasonic distance sensor for navigation in mobile systems. In Proceedings of the 2016 IEEE 14th International Conference on Industrial Informatics (INDIN), Poitiers, France, 19–21 July 2016; pp. 640–645. [CrossRef]

35. MNIST Data Base. Available online: http://www2.imse-cnm.csic.es (accessed on 15 September 2018).

36. Wang, Z.; Bovik, A.C.; Sheikh, H.R.; Simoncelli, E.P. Image quality assessment: from error visibility to structural similarity. *IEEE Trans. Image Process.* **2004**, *13*, 600–612. [CrossRef] [PubMed]

37. Sorgini, F.; Massari, L.; D'Abbraccio, J.; Palermo, E.; Menciassi, A.; Petrovic, P.B.; Mazzoni, A.; Carrozza, M.C.; Newell, F.N.; Oddo, C.M. Neuromorphic Vibrotactile Stimulation of Fingertips for Encoding Object Stiffness in Telepresence Sensory Substitution and Augmentation Applications. *Sensors* **2018**, *18*, 261. [CrossRef] [PubMed]

Article

A Regularized Weighted Smoothed L_0 Norm Minimization Method for Underdetermined Blind Source Separation

Linyu Wang, Xiangjun Yin, Huihui Yue * and Jianhong Xiang

College of Information and Communication Engineering, Harbin Engineering University, Harbin 150001, China; wanglinyu@hrbeu.edu.cn (L.W.); yinxiangjun@hrbeu.edu.cn (X.Y.); xiangjianhong@hrbeu.edu.cn (J.X.)
* Correspondence: yuehuihui@hrbeu.edu.cn

Received: 22 October 2018; Accepted: 30 November 2018; Published: 4 December 2018

Abstract: Compressed sensing (CS) theory has attracted widespread attention in recent years and has been widely used in signal and image processing, such as underdetermined blind source separation (UBSS), magnetic resonance imaging (MRI), etc. As the main link of CS, the goal of sparse signal reconstruction is how to recover accurately and effectively the original signal from an underdetermined linear system of equations (ULSE). For this problem, we propose a new algorithm called the weighted regularized smoothed L_0-norm minimization algorithm (WReSL0). Under the framework of this algorithm, we have done three things: (1) proposed a new smoothed function called the compound inverse proportional function (CIPF); (2) proposed a new weighted function; and (3) a new regularization form is derived and constructed. In this algorithm, the weighted function and the new smoothed function are combined as the sparsity-promoting object, and a new regularization form is derived and constructed to enhance de-noising performance. Performance simulation experiments on both the real signal and real images show that the proposed WReSL0 algorithm outperforms other popular approaches, such as SL0, BPDN, NSL0, and L_p-RLSand achieves better performances when it is used for UBSS.

Keywords: image reconstruction; nullspace measurement matrix; regularized least squares problem; smoothed L_0-norm; sparse signal recovery; UBSS; weighted function

---

1. Introduction

The problem that UBSS [1,2] needs to address is how to separate multiple signals from a small number of sensors. The essence of this problem is to solve the optimal solution of the undetermined linear system of equations (ULSE). Fortunately, as a new undersampling technique, compressed sensing (CS) [3–5] is an effective way to solve ULSE, which makes it possible to apply CS to UBSS.

The model of CS is shown in Figure 1. According to this figure, it can be see that CS boils down to the form,

$$\mathbf{y} = \mathbf{\Phi x} + \mathbf{b}, \tag{1}$$

where $\mathbf{\Phi} = [\boldsymbol{\phi}_1, \boldsymbol{\phi}_2, ..., \boldsymbol{\phi}_n] \in \mathbb{R}^{m \times n}$ is a sensing matrix with the condition of $m \ll n$ and $\boldsymbol{\phi}_i \in \mathbb{R}^m$, $i = 1, 2, ..., n$, which can be further represented as $\mathbf{\Phi} = \boldsymbol{\psi}\boldsymbol{\varphi}$, while $\boldsymbol{\psi}$ is a random matrix, and $\boldsymbol{\varphi}$ is the sparse basis matrix. $\mathbf{y} \in \mathbb{R}^m$ is the vector of measurements. Moreover, $\mathbf{b} \in \mathbb{R}^m$ denotes the additive noise.

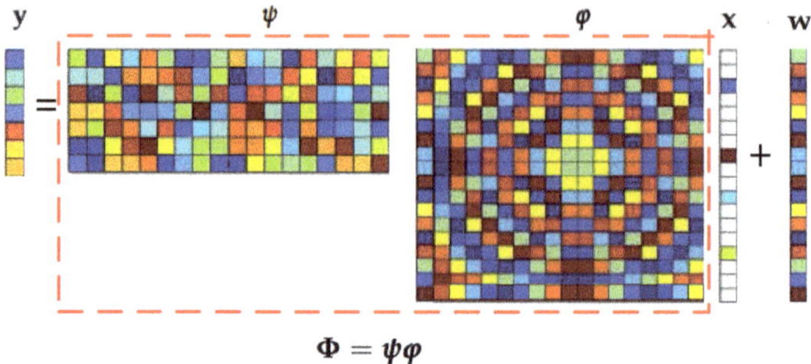

Figure 1. Frame of compressed sensing (CS).

To solve the ULSE in Equation (1), we try to recover the sparse signal $\mathbf{x}$ from the given $\{\mathbf{y}, \mathbf{\Phi}\}$ by CS. According to CS, this problem is transformed into solving the L_0-norm minimization problem.

$$(P_0) \quad \arg\min_{\mathbf{x}\in\mathbb{R}^n} ||\mathbf{x}||_0, \text{ s.t. } ||\mathbf{\Phi}\mathbf{x} - \mathbf{y}||_2^2 \leq \epsilon. \tag{2}$$

where ϵ denotes error. This rather wonderful attempt is actually supported by a brilliant theory [6]. Based on this theory, in the noiseless case, it is proven that the sparsest solution is indeed a real signal when $\mathbf{x}$ is sufficiently sparse and $\mathbf{\Phi}$ satisfies the restricted isometry property (RIP) [7]:

$$1-\delta_K \leq \frac{||\mathbf{\Phi}\mathbf{x}||_2^2}{||\mathbf{x}||_2^2} \leq 1+\delta_K, \tag{3}$$

where K is the sparsity of signal $\mathbf{x}$ and $\delta_K \in (0,1)$ is a constant. In Equation (2), the L_0-norm is nonsmooth, which leads an NP-hard problem. In practice, two alternative approaches are usually employed to solve the problem [8]:

- Greedy search by using the known sparsity as a constraint;
- The relaxation method for the P_0.

For greedy search, the main methods are based on greedy matching pursuit (GMP) algorithms, such as orthogonal matching pursuit (OMP) [9,10], stage-wise orthogonal matching pursuit (StOMP) [11], regularized orthogonal matching pursuit (ROMP) [12], compressive sampling matching pursuit (CoSaMP) [13], generalized orthogonal matching pursuit (GOMP) [14,15], and subspace pursuit (SP) [16,17] algorithms. The objective function of these algorithms is given by:

$$\arg\min_{\mathbf{x}\in\mathbb{R}^n} \frac{1}{2} ||\mathbf{\Phi}\mathbf{x} - \mathbf{y}||_2^2, \text{ s.t. } ||\mathbf{x}||_0 \leq K. \tag{4}$$

As shown in the above equation, the features of GMP algorithms can be concluded as:

- Using sparsity as prior information;
- Using the least squares error as the iterative criterion.

The advantage of GMP algorithms is that the computational complexity is low, but the reconstruction accuracy is not high in the noise case.

At present, the relaxation method for P_0 is widely used. The relaxation method is mainly divided into two categories: the constraint-type algorithm and the regularization method. The constraint-type algorithm can also be divided into L_1-norm minimization methods and smoothed L_0-norm minimization methods. The representative algorithm of the former is the BPalgorithm [18], and the latter is the smoothed L_0-norm minimization (SL0) algorithm. For the SL0 algorithm, the objective function can be expressed as:

$$(P_F) \quad \arg\min_{\mathbf{x}\in\mathbb{R}^n} F_\sigma(\mathbf{x}), \quad \text{s.t. } ||\mathbf{\Phi x} - \mathbf{y}||_2^2 \leq \epsilon.$$

$$\lim_{\sigma\to0} F_\sigma(\mathbf{x}) = \lim_{\sigma\to0} \sum_{i=1}^{n} f_\sigma(x_i) \approx ||\mathbf{x}||_0. \tag{5}$$

where $F_\sigma(\mathbf{x})$ is a smoothed function, which approximates the L_0-norm when $\sigma \to 0$. Compared with L_1 or L_p, a small σ is selected to make the function close to L_0-norm [8]; therefore, $F_\sigma(\mathbf{x})$ are closer to the optimal solution.

Based on the idea of approximation, Mohimani used a Gauss function to approximate the L_0-norm [19], which is described as:

$$f_\sigma(x_i) = 1 - \exp(-\frac{x_i^2}{2\sigma^2}). \tag{6}$$

According to the equation, we can know:

$$f_\sigma(x_i) \approx \begin{cases} 1 & \text{if } x_i \gg \sigma \\ 0 & \text{if } x_i \ll \sigma. \end{cases} \tag{7}$$

when σ is a small enough positive value, the Gauss function is almost equal to the L_0-norm. Furthermore, the Gauss function is differentiable and smoothed; hence, it can be optimized by optimization methods such as the gradient descent (GD) method. Zhao proposed another smoothed function: the hyperbolic tangent (tanh) [20],

$$f_\sigma(x_i) = \frac{\exp(\frac{x_i^2}{2\sigma^2}) - \exp(-\frac{x_i^2}{2\sigma^2})}{\exp(\frac{x_i^2}{2\sigma^2}) + \exp(-\frac{x_i^2}{2\sigma^2})}. \tag{8}$$

This smoothed function makes a closer approximation to the L_0-norm than the Gauss function, as shown in [19], with the same σ; hence, it performs better in sparse signal recovery. Indeed, a large number of simulation experiments confirmed this view.

Another relaxation method is the regularization method. For CS, sparse signal recovery in the noise case is a very practical and unavoidable problem. Fortunately, the regularization method makes the solution of this problem possible [21,22]. The regularization method can be described as a "relaxation" approach that tries to solve the following unconstrained recovery problem:

$$(P_v) \arg\min_{\mathbf{x}\in\mathbb{R}^n} \frac{1}{2} || \mathbf{\Phi x} - \mathbf{y}||_2^2 + \lambda v(\mathbf{x}), \tag{9}$$

where $\lambda > 0$ is the parameter that balances the trade-off between the deviation term $|| \mathbf{\Phi x} - \mathbf{y}||_2^2$ and the sparsity regularizer $v(\mathbf{x})$. The sparse prior information is enforced via the regularizer $v(\mathbf{x})$, and a proper $v(\mathbf{x})$ is crucial to the success of the sparse signal recovery task: it should favor sparse solutions and make sure the problem P_v can be solved efficiently in the meantime.

For regularization, various sparsity regularizers have been proposed as the relaxation of the L_0-norm. The most popular algorithms are the convex L_1-norm [22,23] and the nonconvex L_p-norm to the pth power [24,25]. In the noiseless case, the L_1-norm is equivalent to the L_0-norm, and the L_1-norm is the only norm with sparsity and convexity. Hence, it can be optimized by convex optimization methods. However, according to [8], in the noisy case, the L_1-norm is not exactly equivalent to the L_0-norm, so the effect of promoting sparsity is not obvious. Compared to the L_1-norm, the nonconvex L_p-norm to the pth power makes a closer approximation to the L_0-norm; therefore, L_p-norm minimization has a better sparse recovery performance [8].

In view of the above explanation, in this paper, a compound inverse proportional function (CIPF) function is proposed as a new smoothed function, and a new weighted function is proposed to promote sparsity. For the noise case, a new regularization form is derived and constructed to enhance de-noising performance. The experimental simulation verifies the superior performance of this algorithm in signal and image recovery, and it has achieved good results when applied to UBSS.

This paper is organized as follows: Section 2 introduces the main work of this paper. The steps of the ReRSL0algorithm and the selection of related parameters are described in Section 3. Experimental results are presented in Section 4 to evaluate the performance of our approach. Section 5 verifies the effect of the proposed weighted regularized smoothed L_0-norm minimization (WReSL0) algorithm in UBSS. Section 6 concludes this paper.

2. Main Work of This Paper

In this paper, based on the P_F in Equation (9), we propose a new objective function, which is given by:

$$\underset{\mathbf{x} \in \mathbb{R}^n}{\arg \min} \mathbf{W} H_\sigma(\mathbf{x}), \quad \text{s.t.} \ ||\mathbf{\Phi x} - \mathbf{y}||_2^2 \leq \epsilon. \tag{10}$$

According to this equation, We not only propose a smoothed function approximating the L_0-norm, but also propose a weighted function to promote sparsity. This section focuses on the relevant contents of $\mathbf{W} = [w_1, w_2, ... w_n]^T$ and $H_\sigma(\mathbf{x})$.

2.1. New Smoothed Function: CIPF

According to [26], some properties of the smoothed functions are summarized in the following:

Property: Let $f : \mathbb{R} \to [-\infty, +\infty]$ and, define $f_\sigma(r) \approx f_\sigma(r/\sigma)$ for any $\sigma > 0$. The function f has the property, if:

(a) f is real analytic on (r_0, ∞) for some r_0;
(b) $\forall r \geq 0, f''(r) \geq -\epsilon_0$, where $\epsilon_0 > 0$ is some constant;
(c) f is convex on $\mathbb{R}$;
(d) $f(r) = 0 \leftrightarrow r = 0$;
(e) $\lim\limits_{r \to +\infty} f(r) = 1$.

It follows immediately from **Property** that $\{f_\sigma(r)\}$ converges to the L_0-norm as $\sigma \to 0^+$, i.e.,

$$\lim\limits_{\sigma \to 0+} f_\sigma(r) = \begin{cases} 0 & \text{if } r = 0 \\ 1 & \text{otherwise.} \end{cases} \tag{11}$$

Based on **Property**, this paper proposes a new smoothed function model called CIPF, which satisfies **Property** and better approximates the L_0-norm. The smoothed function model is given as:

$$f_\sigma(r) = 1 - \frac{\sigma^2}{\alpha r^2 + \sigma^2}. \tag{12}$$

In Equation (12), α denotes a regularization factor, which is a large constant. By experiments, the factor α is determined to be 10, which is a good result of the simulation. σ represents a smoothed factor, and when it is smaller, it will make the proposed model closer to the L_0-norm. Obviously,

$$\lim\limits_{\sigma \to 0} f_\sigma(r) = \begin{cases} 0, & r = 0 \\ 1, & r \neq 0 \end{cases} \text{ or approximately } f_\sigma(r) \approx \begin{cases} 0, & |r| \ll \sigma \\ 1, & |r| \gg \sigma \end{cases} \text{ is satisfied. Let:}$$

$$H_\sigma(\mathbf{x}) = \sum_{i=1}^n f_\sigma(x_i) = n - \sum_{i=1}^n \frac{\sigma^2}{\alpha x_i^2 + \sigma^2} \tag{13}$$

where $H_\sigma(\mathbf{x}) \approx ||\mathbf{x}||_0$ for small values of σ, and the approximation tends to equality when $\sigma \to 0$.

Figure 2 shows the effect of the CIPF model approximating the L0-norm. Obviously, the CIPF model makes a better approximation.

In conclusion, the merits of the CIPF model can be summarized as follows:

- It closely approximates the L_0-norm;
- It is simpler in form than that in the Gauss and tanh function models.

These merits make it possible to reduce the computational complexity on the premise of ensuring the accuracy of sparse signal reconstruction, which is of practical significance for sparse signal reconstruction.

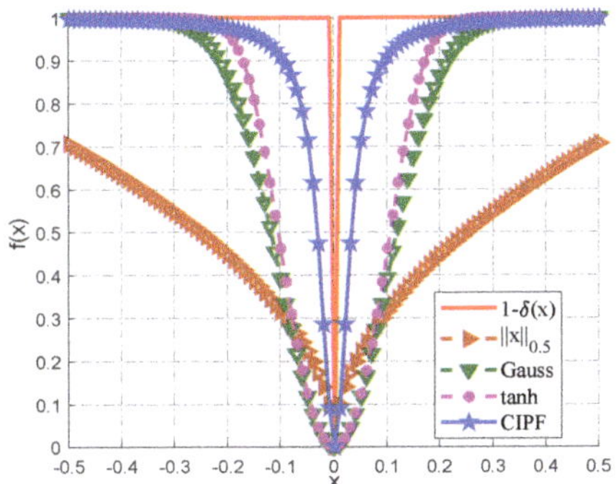

Figure 2. Different functions used in the literature to approximate the L_0-norm; some of them are plotted in this figure, and the $L_{0.5}$-norm is displayed for comparison. CIPF, compound inverse proportional function.

2.2. New Weighted Function

Candès et al. [27] proposed the weighted L_1-norm minimization method, which employs the weighted norm to enhance the sparsity of the solution. They provided an analytical result of the improvement in the sparsity recovery by incorporating the weighted function with the objective function. Pant et al. [28] applied another weighted smoothed L_0-norm minimization method, which uses a similar weighted function to promote sparsity. The weighted function can be summarized as follows:

- Candès et al.: $w_i = \begin{cases} \frac{1}{|x_i|} & x_i \neq 0 \\ \infty & x_i = 0 \end{cases}$;
- Pant et al.: $w_i = \frac{1}{|x_i|+\zeta}$, ζ is a small enough positive constant.

From the two weighted functions, we can find a phenomenon: a large signal entry x_i is weighted with a small w_i; on the contrary, a small signal entry x_i is weighted with a large value w_i. By analysis, the large w_i forces the solution $\mathbf{x}$ to concentrate on the indices where w_i is small, and by construction, these correspond precisely to the indices where $\mathbf{x}$ is nonzero.

Combined with the above idea, we propose a new weighted function, which is given by:

$$w_i = e^{-\frac{|x_i|}{\sigma}}, \ \text{s.t. } i = 1, 2, ..., n. \tag{14}$$

As for Candès et al., when the signal entry is zero or close to zero, the value of w_i will be very large, which is not suitable for computation by a computer. Although Pant et al. noticed the problem and improved the weighted function to avoid it, the constant ζ depends on experience. Actually, the proposed weighted function can avoid the two problems. Moreover our weighted function can be satisfied with the phenomenon. When the small signal entry x_i can be weighted with a large w_i and a large signal entry x_i can be weighted with a small w_i, this can make the large signal entry and small signal entry closer. In this way, the direction of optimization can be kept as consistent as possible, and the optimization process tends to be more optimal. Therefore, the proposed weighted function can have a better effect.

3. New Algorithm for CS: WReSL0

3.1. WReSL0 Algorithm and Its Steps

Here, in order to analyze the problem more clearly, we rewrite Equation (10) as follows:

$$\arg\min_{\mathbf{x}\in\mathbb{R}^n} \mathbf{W}H_\sigma(\mathbf{x}), \quad \text{s.t. } ||\mathbf{\Phi x} - \mathbf{y}||_2^2 \le \epsilon.$$

where $H_\sigma(\mathbf{x}) = \mathbf{I} - \frac{\sigma^2}{\alpha\mathbf{x}^2+\sigma^2}$ ($\mathbf{I} \in \mathbb{R}^N$ is a unit vector) is a differentiable smoothed accumulated function. The weighted function $\mathbf{W} = \mathrm{e}^{-\frac{|\mathbf{x}|}{\sigma}}$. Therefore, we can obtain the gradient of CIPF, which is written as:

$$\mathbf{G} = \frac{\partial H_\sigma(\mathbf{x})}{\partial \mathbf{x}} = \frac{2\alpha\sigma^2\mathbf{x}}{(\alpha\mathbf{x}^2 + \sigma^2)^2} \tag{15}$$

According to Equation (15), as in [28], we can obtain:

$$\mathbf{WG} = \left(\mathrm{e}^{-\frac{|\mathbf{x}|}{\sigma}}\right)^T \frac{2\alpha\sigma^2\mathbf{x}}{(\alpha\mathbf{x}^2 + \sigma^2)^2} \tag{16}$$

Solving the problem of ULSE is to solve the optimization problem in Equation (10). As for this problem, there are many methods, such as split Bregman methods [29–31], FISTA [32], alternating direction methods [33], gradient descent (GD) [34], etc. In order to reduce the computational complexity, this paper adopts the GD method to optimize the proposed objective function.

Given σ, a small target value $\sigma_{\min}$, and a sufficiently large initial value $\sigma_{\max}$, after referring to the annealing mechanism in simulated annealing [35], this paper proposes a monotonically-decreasing sequence $\{\sigma_t | t = 2, 3, ..., T\}$, which is generated as:

$$\sigma_t = \sigma_{\max}\theta^{-\gamma(t-1)}, \quad \text{s.t. } t = 1, 2, 3, ..., T. \tag{17}$$

where $\gamma = \frac{\log_\theta(\sigma_{\max}/\sigma_{\min})}{T-1}$, θ is a constant that is larger than one, and T is the maximum number of iterations. Using such a monotonically-decreasing sequence can avoid the case of too small of a σ leading to the local optimum.

Similar to SL0, WReSL0 also consists of two nested iterations: the external loop, which begins with a sufficiently large value of σ, i.e, $\sigma_{\max}$, responsible for the gradually decreasing strategy in Equation (17), and the internal loop, which for each value of σ, finds the maximizer of $H_\sigma(\mathbf{x})$ on $\{\mathbf{x}|||\mathbf{Ax} - \mathbf{y}||_2 \le \epsilon\}$.

According to the GD algorithm, the internal loop consists of the gradient descent step, which is given by:

$$\hat{\mathbf{x}} = \mathbf{x} + \mu\boldsymbol{d}, \tag{18}$$

where $\boldsymbol{d} = \boldsymbol{g}$ and μ denotes a step size factor. This part is similar to SL0, followed by solving the problem:

$$\arg\min_{\mathbf{x}^*\in\mathbb{R}^n} ||\mathbf{x}^* - \hat{\mathbf{x}}||_2^2, \quad \text{s.t.} ||\mathbf{\Phi x}^* - \mathbf{y}||_2^2 \le \epsilon \tag{19}$$

where $\mathbf{x}^*$ denotes the optimal solution. By regularization, this form can be converted to another form as follows,

$$\underset{\mathbf{x}^* \in \mathbb{R}^n}{\arg\min} \ ||\mathbf{x}^* - \hat{\mathbf{x}}||_2^2 + \lambda ||\mathbf{\Phi}\mathbf{x}^* - \mathbf{y}||_2^2. \tag{20}$$

where λ is the regularization parameter, which is adapted to balance the fit of the solution to the data y and the approximation of the solution to the maximizer of $H_\sigma(\mathbf{x})$. Weighted least squares (WLS) can be used to solve this problem, and the solution is:

$$\mathbf{x}^* = \left[\begin{bmatrix} \mathbf{I}_n \\ \mathbf{\Phi} \end{bmatrix}^H \begin{bmatrix} \mathbf{I}_n & 0 \\ 0 & \lambda \mathbf{I}_m \end{bmatrix} \begin{bmatrix} \mathbf{I}_n \\ \mathbf{\Phi} \end{bmatrix} \right]^{-1} \begin{bmatrix} \mathbf{I}_n \\ \mathbf{\Phi} \end{bmatrix}^H \begin{bmatrix} \mathbf{I}_n & 0 \\ 0 & \lambda \mathbf{I}_m \end{bmatrix} \begin{bmatrix} \hat{\mathbf{x}} \\ \mathbf{y} \end{bmatrix}. \tag{21}$$

By calculation, Equation (21) is equivalent to:

$$\mathbf{x}^* = \left(\mathbf{I}_n + \lambda \mathbf{\Phi}^H \mathbf{\Phi} \right)^{-1} \left(\hat{\mathbf{x}} + \lambda \mathbf{\Phi}^H \mathbf{y} \right) \tag{22}$$

where $\mathbf{I}_n$ and $\mathbf{I}_m$ are both identity matrices of size $n \times n$ and $m \times m$, respectively. Therefore, we can obtain:

$$\begin{aligned}
\mathbf{x}^* - \hat{\mathbf{x}} &= \left(\mathbf{I}_n + \lambda \mathbf{\Phi}^H \mathbf{\Phi} \right)^{-1} \left(\hat{\mathbf{x}} + \lambda \mathbf{\Phi}^H \mathbf{y} \right) - \hat{\mathbf{x}} \\
&= \left(\mathbf{I}_n + \lambda \mathbf{\Phi}^H \mathbf{\Phi} \right)^{-1} \left(\hat{\mathbf{x}} + \lambda \mathbf{\Phi}^H \mathbf{y} - \left(\mathbf{I}_n + \lambda \mathbf{\Phi}^H \mathbf{\Phi} \right) \hat{\mathbf{x}} \right) \\
&= \left(\mathbf{I}_n + \lambda \mathbf{\Phi}^H \mathbf{\Phi} \right)^{-1} \left(\hat{\mathbf{x}} + \lambda \mathbf{\Phi}^H \mathbf{y} - \hat{\mathbf{x}} - \lambda \mathbf{\Phi}^H \mathbf{\Phi} \hat{\mathbf{x}} \right) \\
&= - \left(\lambda^{-1} \mathbf{I}_n + \mathbf{\Phi}^H \mathbf{\Phi} \right)^{-1} \mathbf{\Phi}^H \left(\mathbf{\Phi} \hat{\mathbf{x}} - \mathbf{y} \right)
\end{aligned}$$

According to the above analysis and derivation, we can get:

$$\mathbf{x}^* = \hat{\mathbf{x}} - \left(\lambda^{-1} \mathbf{I}_n + \mathbf{\Phi}^H \mathbf{\Phi} \right)^{-1} \mathbf{\Phi}^H \left(\mathbf{\Phi} \hat{\mathbf{x}} - \mathbf{y} \right) \tag{23}$$

The initial value of the internal loop is the maximizer of $H_\sigma(\mathbf{x})$ obtained for $\sigma_{\max}$. To increase the speed, the internal loop is repeated a fixed and small number of times (L). In other words, we do not wait for the GD method to converge in the internal loop.

According to the explanation above, we can conclude the steps of the proposed WReSL0 algorithm, which are given in Table 1. As for σ, it can be shown that function $H_\sigma(\mathbf{x})$ remains convex in the region where the largest magnitude of the component of $\mathbf{x}$ is less than σ. As the algorithm starts at the original value $\mathbf{x}^{(0)} = \mathbf{\Phi}^H (\mathbf{\Phi}\mathbf{\Phi}^H)^{-1}\mathbf{y}$, the above choice of σ_1 ensures that the optimization starts in a convex region. This greatly facilitates the convergence of the WReSL0 algorithm.

Table 1. Weighted regularized smoothed L_0-norm minimization (WReSL0) algorithm using the GD method.

- Initialization:
  (1) Set $L, \mu = \sigma / (2\alpha), \hat{\mathbf{x}}^{(0)} = \mathbf{\Phi}^H (\mathbf{\Phi}\mathbf{\Phi}^H)^{-1}\mathbf{y}$.
  (2) Set $\sigma_{\max} = \sqrt{\alpha} \max |\mathbf{x}|$, $\sigma_{\min} = 0.01$, and $\sigma_t = \sigma_{\max} \theta^{-\gamma(t-1)}$, where $\gamma = \frac{\log_\theta (\sigma_{\max}/\sigma_{\min})}{T-1}$, and T is the maximum number of iterations.
- **while** $t < T$, **do**
  (1) Let $\sigma = \sigma_t$.
  (2) Let $\mathbf{x} = \hat{\mathbf{x}}^{(t-1)}$.
  for $l = 1, 2, ..., L$
  (a) $\mathbf{x} \leftarrow \mathbf{x} - \mu \left(e^{-\frac{|\mathbf{x}|}{\sigma}} \right)^T \frac{2\alpha\sigma^2 \mathbf{x}}{(\alpha\mathbf{x}^2 + \sigma^2)^2}$.
  (b) $\mathbf{x} \leftarrow \mathbf{x} - \left(\lambda^{-1} \mathbf{I}_n + \mathbf{\Phi}^H \mathbf{\Phi} \right)^{-1} \mathbf{\Phi}^H \left(\mathbf{\Phi} \hat{\mathbf{x}} - \mathbf{y} \right)$
  (3) Set $\hat{\mathbf{x}}^{(t-1)} = \mathbf{x}$.
- The estimated value is $\hat{\mathbf{x}} = \hat{\mathbf{x}}^{(t)}$.

3.2. Selection of Parameters

The selection of parameters μ and σ will affect the performance of the WReSL0 algorithm; thus, this paper discusses the selection of these two above parameters in this section.

3.2.1. Selection of Parameter μ

According to the algorithm, each iteration consists of a descent step $x_i \leftarrow x_i - \mu\left(e^{-\frac{|x_i|}{\sigma}}\right)\frac{2\alpha\sigma^2 x_i}{(\alpha x_i^2 + \sigma^2)^2}, 1 \leq i \leq n$, followed by a projection step. If for some values of i, we have $|x_i| \gg \sigma$, then the algorithm does not change the value of x_i in that descent step; however, it might be changed in the projection step. If we are looking for a suitably large μ, a suitable choice is to make the algorithm force all those values of $\mathbf{x}$ satisfying $|x_i| \lesssim \sigma$ toward zero. Therefore, we can get:

$$x_i - \mu\left(e^{-\frac{|x_i|}{\sigma}}\right)\frac{2\alpha\sigma^2 x_i}{(\alpha x_i^2 + \sigma^2)^2} \approx 0 \tag{24}$$

and:

$$\left(e^{-\frac{|x_i|}{\sigma}}\right) \xrightarrow{x_i \to 0} 1 \tag{25}$$

Combining Equations (24) and (25), we can further obtain:

$$x_i - \mu\frac{2\alpha\sigma^2 x_i}{(\alpha x_i^2 + \sigma^2)^2} \approx 0 \tag{26}$$

By calculation, we can obtain:

$$\mu \approx \frac{\left(\alpha x_i^2 + \sigma^2\right)^2}{2\alpha\sigma^2} \xrightarrow{x_i \to 0} \frac{\sigma^2}{2\alpha} \tag{27}$$

According to the above derivation, we have come to the conclusion that $\mu \approx \frac{\sigma^2}{2\alpha}$. Therefore, we can set $\mu = \frac{\sigma^2}{2\alpha}$.

3.2.2. Selection of Parameter σ

According to Equation (17), the descending sequence of σ is generated by $\sigma_t = \sigma_{\max}\left(\frac{\sigma_{\min}}{\sigma_{\max}}\right)^{\frac{t-1}{T-1}}$ (it is obtained through simplification of Equation (17)). Parameter $\sigma_{\min}$ and parameter $\sigma_{\max}$ should be appropriately selected. The selection of $\sigma_{\min}$ and $\sigma_{\max}$ is discussed below.

For the initial value of σ, i.e., $\sigma_{\max}$, here, let $\tilde{x} = \max\{|\mathbf{x}|\}$; suppose there is a constant b, in order to make the algorithm converge quickly; let parameter $\sigma_{\max}$ satisfy:

$$H_\sigma(\tilde{x}) = 1 - \frac{\sigma_{\max}^2}{\alpha\tilde{x}^2 + \sigma_{\max}^2} \leq b \Rightarrow \sigma_{\max} \geq \left(\sqrt{\frac{1-b}{b}}\alpha\right)\tilde{x}. \tag{28}$$

From the equation, we can see that constant b satisfies $\frac{1-b}{b} \geq 0$; thus $0 < b \leq 1$, and here, we define constant b as 0.5. Hence, $\sigma_{\max} = \sqrt{\alpha}\max\{|\mathbf{x}|\}$.

For the final value $\sigma_{\min}$, when $\sigma_{\min} \to 0$, $H_{\sigma_{\min}}(\mathbf{x}) \to ||\mathbf{x}||_0$. That is, the smaller $\sigma_{\min}$, the more $H_{\sigma_{\min}}(\mathbf{x})$ can reflect the sparsity of signal $\mathbf{x}$, but at the same time, it is also more sensitive to noise; therefore, the value $\sigma_{\min}$ should not be too small. Combining [19], we choose $\sigma_{\min} = 0.01$.

4. Performance Simulation and Analysis

The numerical simulation platform is MATLAB 2017b, which is installed on a computer with a Windows 10, 64-bit operating system. The CPU of the simulation computer is the Intel (R) Core (TM) i5-3230M, and the frequency is 2.6 GHz. In this section, the performance of the WReSL0 algorithm is verified by signal and image recovery in the noise case.

Here, some state-of-the-art algorithms are selected for comparison. The parameters are selected to obtain the best performance for each algorithm: for the BPDNalgorithm [36], the regularization parameter $\lambda = \sigma_N \sqrt{2\log(n)}$; for the SL0 algorithm [19], the initial value of smoothed factor $\delta_{\max} = 2 \max\{|\mathbf{x}|\}$, the final value of smoothed factor $\delta_{\min} = 0.01$, scale factor is set as step size $L = 5$, and the attenuation factor $\rho = 0.8$; for the NSL0algorithm [20], the initial value of smoothed factor $\delta_{\max} = 4 \max\{\mathbf{x}\}$, the final value of smoothed factor $\delta_{\min} = 0.01$, the step size $L = 10$, and the attenuation factor $\rho = 0.8$; for L_p-RLSalgorithm [24], the number of iterations $T = 80$, the norm initial value $p_1 = 1$, the norm final value $p_T = 0.1$, the initial value of regularization factor $\epsilon_1 = 1$, the final value of regularization factor $\epsilon_T = 0.01$, and the algorithm termination threshold $E_t = 10^{-25}$; for the WReSL0 algorithm, the initial value of smoothed factor $\sigma_{\max} = \sqrt{c}\max\{|\mathbf{x}|\}$, the final value of smoothed factor $\sigma_{\min} = 0.01$, the iterations $T = 30$, the step size $L = 5$, and the regularization parameter $\lambda = 0.1$. All experiments are based on 100 trials.

4.1. Signal Recovery Performance in the Noise Case

In this part, we discuss signal recovery performance in the noise case. We add noise **b** to the measurement vector **y**; moreover, $\mathbf{b} = \delta_N \Omega$, Ω is randomly formed and follows the Gaussian distribution of $\mathcal{N}(0, 1)$. For signal recovery under noise conditions, we evaluate the performance of algorithms by the normalized mean squared error (NMSE) and the CPU running time (CRT). NMSE is defined as $||x - \hat{x}||_2 / ||x||_2$. CRT is measured with *tic* and *toc*. In order to analyze the de-noising performance of the WReSL0 algorithm in context closer to the real situation, we constructed a certain signal as an experimental object in the experiments in this section. The signal is given by:

$$\begin{cases} \mathbf{x}_1 = \alpha_1 \sin(2\pi f_1 T_s \mathbf{t}) \\[2mm] \mathbf{x}_2 = \beta_1 \cos(2\pi f_2 T_s \mathbf{t}) \\[2mm] \mathbf{x}_3 = \alpha_2 \sin(2\pi f_3 T_s \mathbf{t}) \\[2mm] \mathbf{x}_4 = \beta_2 \cos(2\pi f_4 T_s \mathbf{t}) \\[2mm] \mathcal{X} = \mathbf{x}_1 + \mathbf{x}_2 + \mathbf{x}_3 + \mathbf{x}_4 \end{cases} \tag{29}$$

where $\alpha_1 = 0.2$, $\alpha_2 = 0.1$, $\beta_1 = 0.3$, and $\beta_2 = 0.4$. $f_1 = 50$ Hz; $f_2 = 100$ Hz; $f_3 = 200$ Hz; and $f_4 = 300$ Hz. Here, **t** is a sequence with $\mathbf{t} = [1, 2, 3, ..., n]$, and T_s is sampling interval with the value of $\frac{1}{f_s}$. f_s is the sampling frequency with the value of 800 Hz. The object that needs to be reconstructed can be expressed as:

$$\mathbf{y} = \Phi \mathbf{x} + \delta_N \Omega. \tag{30}$$

where $\mathbf{x} \in \mathbb{R}^n$ is a sparse signal in the frequency domain, and it is the Fourier transform expression of $\mathcal{X}$, $\mathbf{y} \in \mathbb{R}^m$. Here, let $n = 128$, $m = 64$. Moreover, Φ can be represented as $\Phi = \psi\varphi$; here, ψ is a randn matrix generated by a Gaussian distribution, and φ is a sparse basis matrix generated by Fourier transform. Here, φ can be given by Fourier $\mathbf{I}_{n \times n}$, and $\mathbf{I}_{n \times n}$ is a unit matrix. This target signal $\mathcal{X}$ is sparse in Fourier space; hence, the signal $\mathcal{X}$ can be recovered from given $\{\mathbf{y}, \Phi\}$ by CS recovery methods.

Figure 3 shows the signal recovery effect. Obviously, BPDN and SL0 do not perform well, while NSL0, L_p-RLS and the proposed WReSL0 perform quite well. This verifies that the regularization mechanism has a good de-noising effect. Figure 4 shows the frequency spectrum of the recovered signal by the selected algorithms. The spectrum of the signal recovered by our proposed WReSL0 algorithm is almost the same as the original signal, while other algorithms fail to achieve this effect.

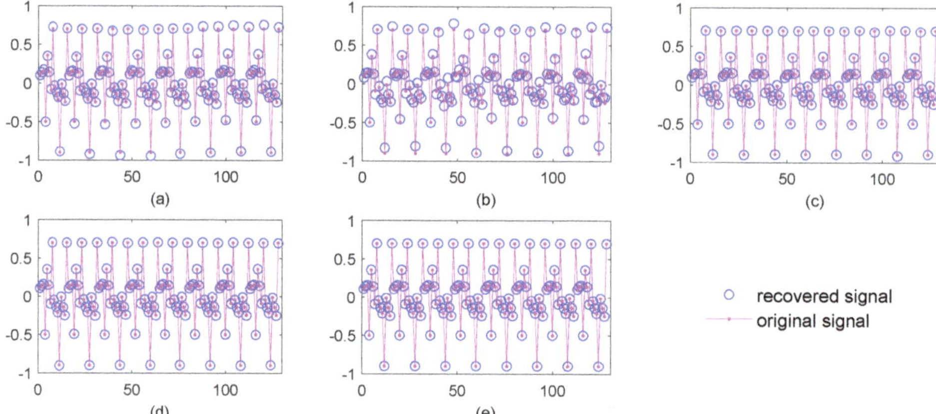

Figure 3. Signal recovery effect by BPDN, SL0, NSL0, L_p-RLS, and weighted regularized smoothed L_0-norm minimization (WReSL0) when noise intensity $\delta_N = 0.2$. (**a**) signal recovery by the BPDN algorithm; (**b**) signal recovery by the SL0 algorithm; (**c**) signal recovery by NSL0 algorithm; (**d**) signal recovery by the L_p-RLS algorithm; (**e**) signal recovery by the WReSL0 algorithm.

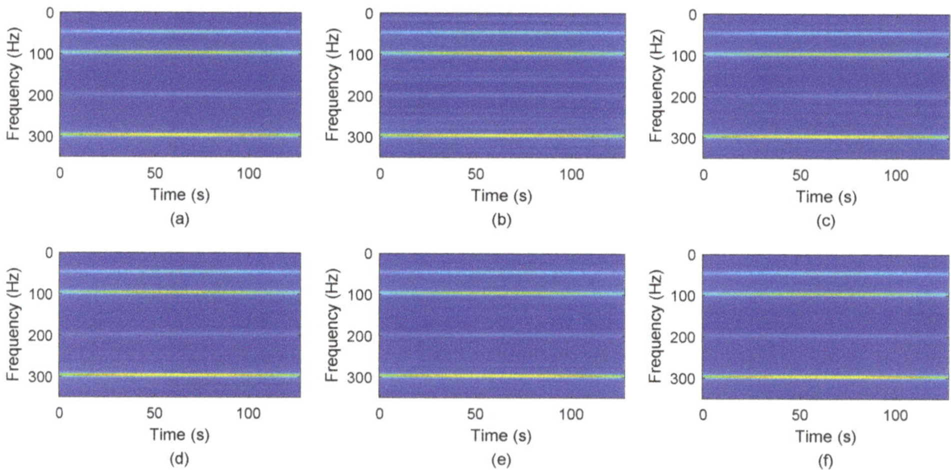

Figure 4. Frequency spectrum analysis of the original signal and the signal recovered by BPDN, SL0, NSL0, L_p-RLS, and WReSL0 when noise intensity $\delta_N = 0.2$. (**a**) original signal; (**b**) signal recovery by the BPDN algorithm; (**c**) signal recovery by the SL0 algorithm; (**d**) signal recovery by the NSL0 algorithm; (**e**) signal recovery by the L_p-RLS algorithm; (**f**) signal recovery by the WReSL0 algorithm.

Table 2 shows the CRT of all algorithms. The n changes according to a given sequence $[170, 220, 270, 320, 370, 420, 470, 520]$. From the table, for any n, SL0 has the shortest computation time, followed by WReSL0, NSL0, and L_p-RLS, and BPDN has the longest computation time. The BPDN algorithm is generally implemented by the quadratic programming method, and the computational complexity of this method is very high, thus resulting in a large increase in the overall computation time of the algorithm. Furthermore, in L_p-RLS, the iterative process adopts the conjugate gradient method with high complexity, while NSL0 and WReSL0 do not. Compared with NSL0, WReSL0 is more prominent in the decrease of computation time.

Table 2. Signal CPU running time (CRT) analysis for BPDN, SL0, NSL0, L_p-RLS, and the proposed WReSL0 with signal length changes according to the sequence [170,220,270,320,370,420,470,520] when $\delta_N = 0.2$.

Signal Length (n)	CPU Running Time (Seconds)				
	BPDN	**SL0**	**NSL0**	**Lp-RLS**	**WReSL0**
170	0.195	0.057	0.091	0.194	0.063
220	0.289	0.139	0.230	0.350	0.142
270	0.495	0.229	0.426	0.505	0.291
320	0.767	0.320	0.639	0.712	0.509
370	1.059	0.456	0.926	0.982	0.892
420	1.477	0.613	1.133	1.491	1.017
470	1.941	0.796	1.478	2.118	1.344
520	2.619	1.038	2.089	2.910	1.882

The performance of each algorithm under different noise intensities is shown in Figure 5. When $\delta_N = 0$, SL0 outperforms other algorithms, but with the increase of δ_N, the effect of SL0 becomes worse and worse. This result further illustrates that the traditional constrained sparse recovery algorithm does not have the performance of anti-noising. For BPDN, NSL0, L_p-RLS, and WReSL0, they all applied the regularization mechanism, and they are indeed superior to SL0 in the noise case. Therefore, the proposed WReSL0 in this paper has the best de-noising performance.

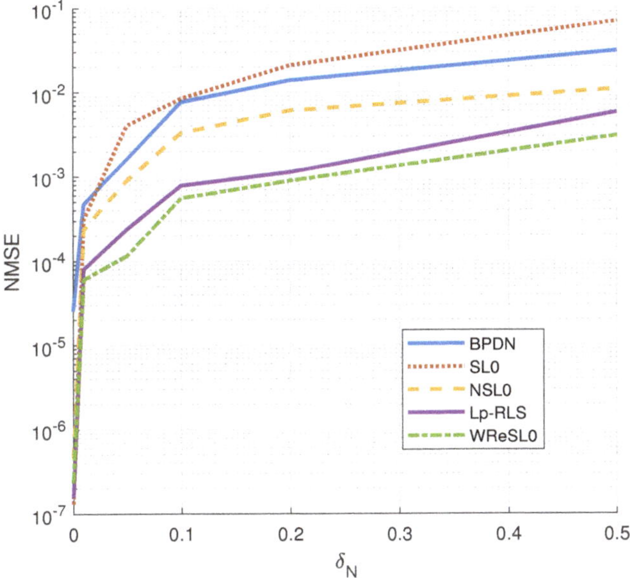

Figure 5. NMSE analysis by BPDN, SL0, NSL0, L_p-RLS, and WReSL0 when noise intensity δ_N changes according to the sequence [0, 0.1, 0.2, 0.3, 0.4, 0.5].

4.2. Image Recovery Performance in the Noise Case

Real images are considered to be approximately sparse under some proper basis, such as the DCT basis, DWT basis, etc. Here, we choose the DWT basis to recover these images. We compare the recovery performances based on the four real images in Figure 6: boat, Barbara, peppers, and Lena. The size of these images is 256×256; the compression ratio (CR; defined as m/n) is 0.5; and the noise δ_N equals 0.01. We still choose SL0, BPDN, NSL0, and L_p-RLS to make comparisons. For image recovery, the object of image processing is given by:

$$\mathbf{Y} = \mathbf{\Phi}\mathbf{X} + \mathbf{B} \tag{31}$$

Here, $\mathbf{Y}$, $\mathbf{X}$, $\mathbf{B}$ are matrices, and among these, $\mathbf{Y}, \mathbf{B} \in \mathbb{R}^{m \times n}, \mathbf{X} \in \mathbb{R}^{n \times n}$. In order to meet the basic requirements of CS, we perform the following processing:

$$\mathbf{Y}_i = \mathbf{\Phi}\mathbf{X}_i + \mathbf{B}_i \ \ s.t. \ i = 1, 2, ..., n. \tag{32}$$

where $\mathbf{Y}_i$, $\mathbf{X}_i$, $\mathbf{B}_i$ are the column vectors of $\mathbf{Y}$, $\mathbf{X}$, $\mathbf{B}$, respectively. $\mathbf{B}_i = \delta_N \mathbf{\Omega}$, $\mathbf{\Omega}$ obeys the Gaussian distribution $\mathcal{N}(0, 1)$.

To perform image recovery, we valuate it by the peak signal to noise ratio (PSNR) and the structural similarity index (SSIM). PSNR is defined as:

$$\text{PSNR} = 10\log(255^2/\text{MSE}) \tag{33}$$

where $\text{MSE} = ||x - \hat{x}||_2^2$, and SSIM is defined as:

$$\text{SSIM}(p, q) = \frac{(2\mu_p + \mu_q + c_1)(2\sigma_{pq} + c_2)}{(\mu_p^2 + \mu_q^2 + c_1)(\sigma_p^2 + \sigma_q^2 + c_2)}. \tag{34}$$

Among these, μ_p is the mean of image p, μ_q is the mean of image q, σ_p is the variance of image p, σ_q is the variance of image q, and σ_{pq} is the covariance between image p and image q. Parameters $c_1 = z_1 L$ and $c_2 = z_2 L$, for which $z_1 = 0.01, z_2 = 0.03$, and L is the dynamic range of pixel values. The range of SSIM is $[-1, 1]$, and when these two images are the same, SSIM equals one.

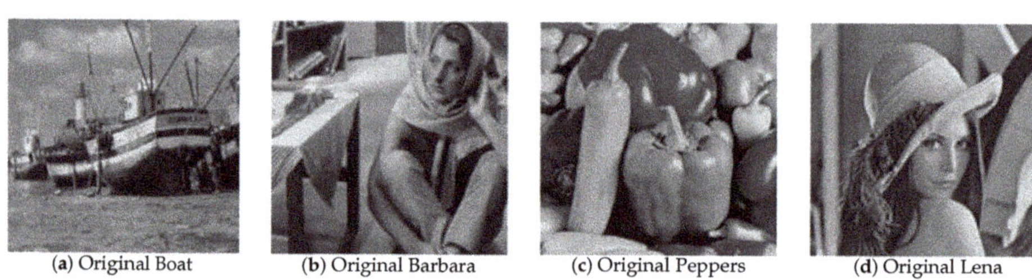

(a) Original Boat   (b) Original Barbara   (c) Original Peppers   (d) Original Lena

Figure 6. Original images: (**a**) boat; (**b**) Barbara; (**c**) peppers; (**d**) Lena.

Figure 7 shows the recovery effect of boat and Barbara with noise intensity $\delta_N = 0.01$. For boat and Barbara, the recovered images by SL0 and BPDN have obvious water ripples, while recovered images by other algorithms have no such water ripples. Similarly, for peppers and Lena, the recovered images by SL0 and BPDN are blurred compared with the recovered images by other algorithms. The NSL0, L_p-RLS, and WReSL0 algorithms are also effective at noisy image recovery. For the NSL0, L_p-RLS, and WReSL0 algorithms, their recovery effects are very similar. In order to further analyze the advantages and disadvantages of the algorithms, we analyze the PSNR and SSIM of the images recovered by these algorithms, and the results are shown in Tables 3 and 4. By observation and analysis, L_p-RLS performs better than NSL0, and at the same time, WReSL0 outperforms L_p-RLS. Hence, the WReSL0 proposed by this paper is superior to the other selected algorithms in image processing.

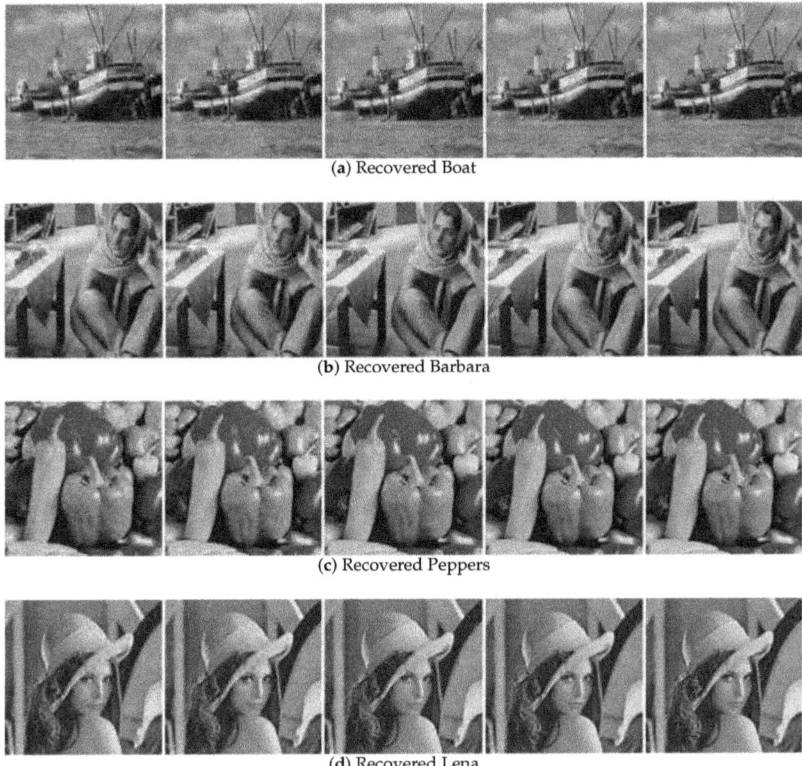

(a) Recovered Boat

(b) Recovered Barbara

(c) Recovered Peppers

(d) Recovered Lena

Figure 7. Image recovery effect by the BPDN, SL0, NSL0, L_p-RLS, and WReSL0 algorithms with noise intensity $\delta_N = 0.01$. In (**a**–**d**), from left to right, are: image recovered by the BPDN, SL0, NSL0, L_p-RLS, and WReSL0 algorithms.

Table 3. PSNR and SSIM analysis of recovered images (boat and Barbara) by SL0, BPDN, NSL0, L_p-RLS, and WReSL0.

Items	Barbara		Boat	
	PSNR (dB)	SSIM	PSNR (dB)	SSIM
SL0	27.983	0.981	26.959	0.969
BPDN	28.834	0.984	27.376	0.971
NSL0	31.296	0.991	31.247	0.988
L_p-RLS	31.786	0.992	31.797	0.989
WReSL0	32.244	0.993	32.369	0.991

Table 4. PSNR and SSIM analysis of recovered images (peppers and Lena) by SL0, BPDN, NSL0, L$_p$-RLS, and WReSL0.

Items	Peppers		Lena	
	PSNR (dB)	**SSIM**	**PSNR (dB)**	**SSIM**
SL0	28.677	0.982	30.334	0.987
BPDN	29.542	0.985	29.875	0.983
NSL0	31.373	0.991	32.639	0.993
L$_p$-RLS	33.757	0.994	34.051	0.995
WReSL0	34.231	0.996	34.653	0.997

5. Application in Underdetermined Blind Source Separation

The problem of UBSS stems from cocktail reception, which is shown in Figure 8. Suppose the source signal matrix $\mathbf{S}(t) = [s_1(t), s_2(t), ..., s_m(t)]^T$, the mixed matrix (Sensors) $\mathbf{A}$ is $m \times n$ ($m \ll n$) matrix, the Gaussian noise $\mathbf{G}(t) = [g_1(t), g_2(t), ..., g_m(t)]^T$ is generated by Gaussian distribution, and the observed mixed signal matrix $\mathbf{X}(t) = [x_1(t), x_2(t), ..., x_n(t)]^T$; therefore, the general mathematical models of UBSS can be summarized as:

$$\mathbf{X}(t) = \mathbf{A}\mathbf{S}(t) + \mathbf{G}(t) \tag{35}$$

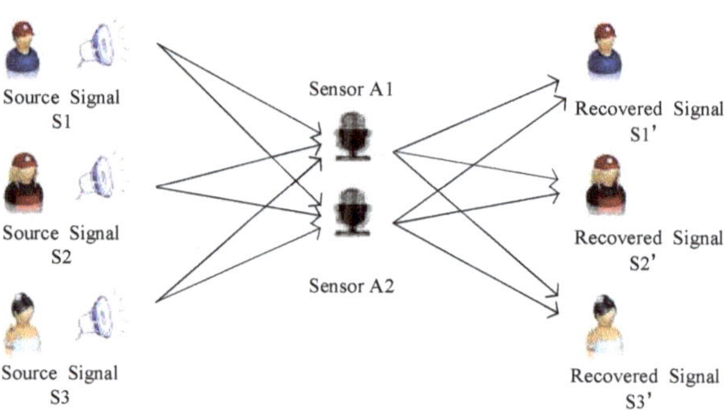

Figure 8. Schematic diagram of cocktail reception signal mixing.

In fact, each signal has L data collected; therefore, $\mathbf{X} \in \mathbb{R}^{m \times L}$, $\mathbf{A} \in \mathbb{R}^{m \times n}$ ($m \ll n$), $\mathbf{S}(t) \in \mathbb{R}^{n \times L}$, and $\mathbf{G} \in \mathbb{R}^{m \times L}$, and $\mathbf{G}$ can be represented as $\delta_N \mathbf{W}$ ($\mathbf{W}$ obeys $\mathcal{N}(0, 1)$). The purpose of UBSS is to use the mixed signal matrix $\mathbf{x}(t)$ to estimate the sof the source signal matrix $\mathbf{s}(t)$. In fact, this is the process of solving the underdetermined linear system of equations (ULSE). For this problem, we can use the two-step method to solve it, which is shown in Figure 9.

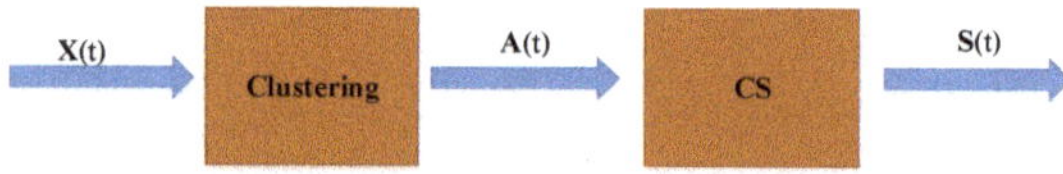

Figure 9. Schematic diagram of two-step method for UBSS.

From Figure 9, firstly, we get the mixed matrix by the clustering method and then use CS technology to separate the signal, so as to restore the original signal.

5.1. Process Analysis of CS Applied to UBSS

5.1.1. Solving the Mixed Matrix by the Potential Function Method

In this section, we choose the potential function method to solve the mixed matrix $\mathbf{A}$. To verify the performance of the proposed WReSL0 algorithm better, we choose four simulated signals and four real images to organize experiments in this section.

Suppose there are four source signals, which are:

$$\begin{cases} s_1(t) = 5\sin(2\pi f_1 t) \\[6pt] s_2(t) = 5\sin(2\pi f_2 t) \\[6pt] s_3(t) = 5\sin(2\pi f_3 t) \\[6pt] s_4(t) = 5\sin(2\pi f_4 t) \\[6pt] \mathbf{S} = [s_1(t), s_2(t), s_3(t), s_4(t)]^T \end{cases} \tag{36}$$

where $f_1 = 310$ Hz, $f_2 = 210$ Hz, $f_3 = 110$ Hz, and $f_4 = 10$ Hz. The length of each source signal s_i $(i = 1, 2, 3, 4)$ is 1024, and the sample frequency is 1024 Hz. These four signals are shown in Figure 10.

The four source images are the classic standard test images: boat, Barbara, peppers, and Lena, which are in Figure 6.

Suppose there are two sensors that receive signals and another two sensors that receive images. Mixed matrices $\mathbf{A}$ and $\mathbf{B}$ are set as:

$$\mathbf{A} = \begin{bmatrix} A_1 \\ A_2 \end{bmatrix} = \begin{bmatrix} 0.9930 & 0.9941 & 0.1092 & 0.9304 \\ 0.2116 & 0.0757 & 0.9647 & 0.3837 \end{bmatrix}$$

$$\mathbf{B} = \begin{bmatrix} B_1 \\ B_2 \end{bmatrix} = \begin{bmatrix} 0.9354 & 0.9877 & -0.6730 & 0.1097 \\ 0.3535 & 0.07846 & 0.7396 & 0.9940 \end{bmatrix} \tag{37}$$

By this mixed matrix and added Gaussian noise ($\delta_N = 0.1$), we can get the two mixed signals, which are shown in Figure 11, and the two mixed images, which are shown in Figure 12. Then, we can get the estimated mixed matrix $\hat{\mathbf{A}}$ and $\hat{\mathbf{B}}$ by clustering by the potential function method [37]. As shown in Figure 13, the potential function method can cluster well. By clustering, we get the estimated values of $\mathbf{A}$ and $\mathbf{B}$, as follows:

$$\hat{\mathbf{A}} = \begin{bmatrix} \hat{A}_1 \\ \hat{A}_2 \end{bmatrix} = \begin{bmatrix} 0.9792 & 0.9969 & 0.1097 & 0.9239 \\ 0.2028 & 0.0785 & 0.9940 & 0.3827 \end{bmatrix}$$

$$\hat{\mathbf{B}} = \begin{bmatrix} \hat{B}_1 \\ \hat{B}_2 \end{bmatrix} = \begin{bmatrix} 0.9478 & 0.9431 & -0.6483 & 0.1130 \\ 0.3476 & 0.0765 & 0.7075 & 0.9979 \end{bmatrix} \tag{38}$$

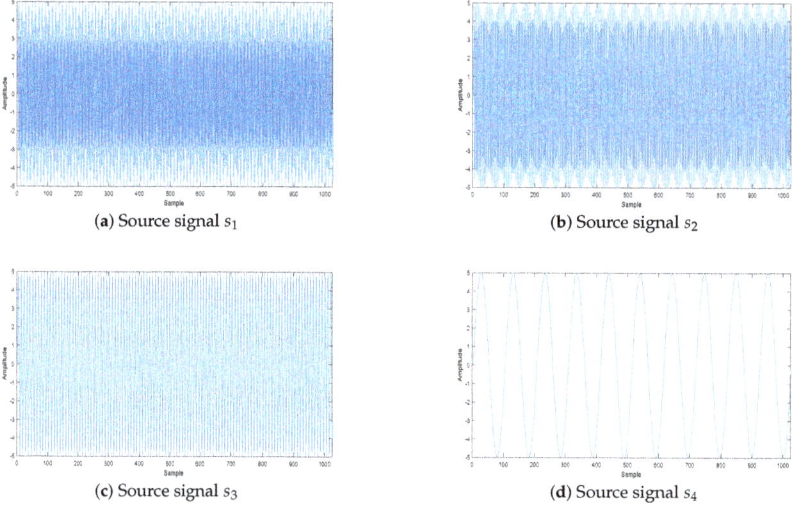

(**a**) Source signal s_1

(**b**) Source signal s_2

(**c**) Source signal s_3

(**d**) Source signal s_4

Figure 10. Source signal.

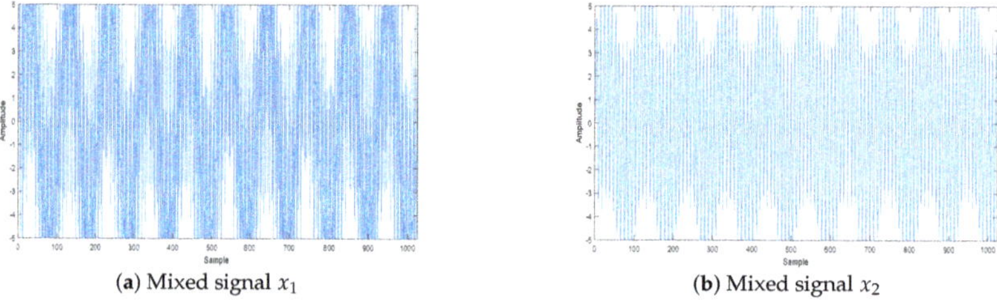

(**a**) Mixed signal x_1

(**b**) Mixed signal x_2

Figure 11. Mixed signal by sensors.

(**a**) Mixed image I_1

(**b**) Mixed image I_2

Figure 12. Mixed image by sensors.

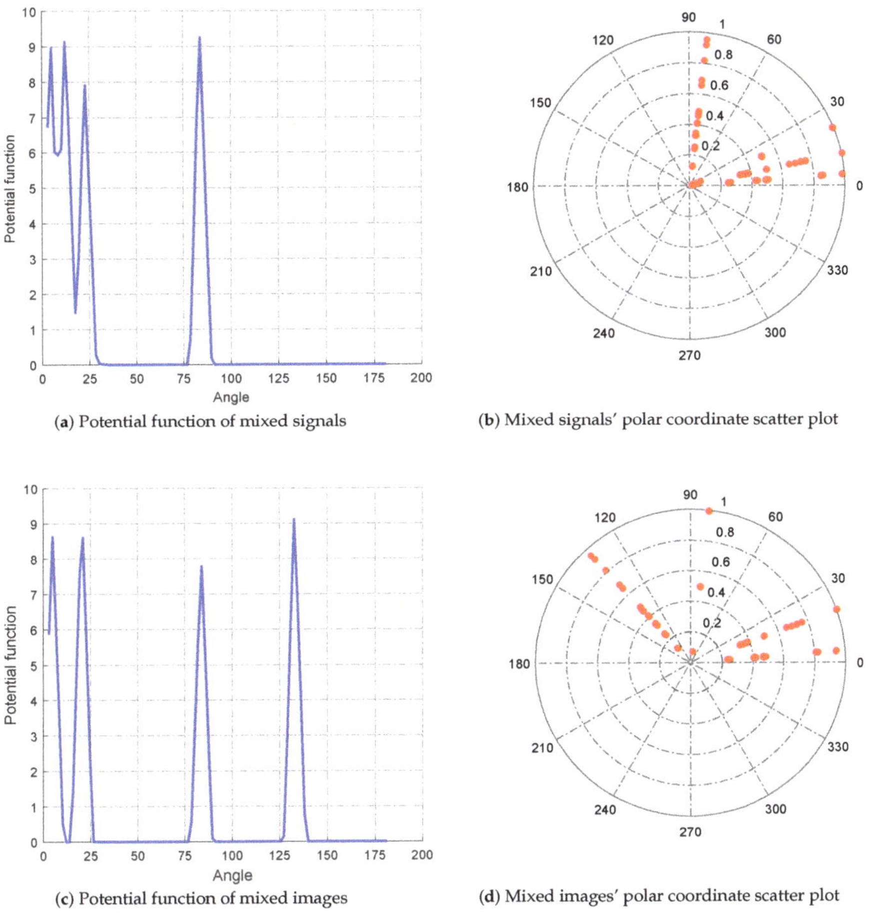

(**a**) Potential function of mixed signals

(**b**) Mixed signals' polar coordinate scatter plot

(**c**) Potential function of mixed images

(**d**) Mixed images' polar coordinate scatter plot

Figure 13. Clustering analysis.

By calculation, the error of solving the mixed matrix is $\frac{||\mathbf{A}-\hat{\mathbf{A}}||_F}{||\mathbf{A}||_F} \times 100\% = 1.763\%$ and $\frac{||\mathbf{B}-\hat{\mathbf{B}}||_F}{||\mathbf{B}||_F} \times 100\% = 3.64\%$. This error range is much smaller than the classical k-means and fuzzy c-means, thus laying a foundation for the reconstruction of compressed sensing.

5.1.2. Using CS to Separate Source Signals

The next problem is to get $\mathbf{S}(t)$ from known $\mathbf{A}(t)$ and $\mathbf{X}(t)$. Here, we solve this problem by CS. The solution process is similar to the image reconstruction process. The difference is that the sparse basis used here is the Fourier basis. Then, we apply the proposed RWeSL0 algorithm to this process. First, we transform the obtained $\mathbf{x}(t)$ into column vectors:

$$\mathbf{x}(t) = [x_1(t), x_2(t)]^T \Rightarrow \tilde{\mathbf{x}}(t) = \begin{bmatrix} x_1(t) \\ x_2(t) \end{bmatrix} \tag{39}$$

Then, we use the Fourier (for the sparse signal) or DWT (for the image) basis for sparse representation and extend the matrix and the valuated mixed matrix to obtain the sensing matrix.

$$\tilde{\mathbf{A}} = \hat{\mathbf{A}} \otimes \mathbf{I}_{L \times L}, \quad or \quad \tilde{\mathbf{B}} = \hat{\mathbf{B}} \otimes \mathbf{I}_{L \times L}$$

$$\mathbf{\Psi} = Fourier(\mathbf{I}_{L \times L})/\sqrt{L}, \quad or \quad \mathbf{\Psi} = DWT(\mathbf{I}_{L \times L})/\sqrt{L}$$

$$\tilde{\mathbf{\Psi}} = \begin{bmatrix} \mathbf{\Psi} & 0 & \cdots & 0 \\ 0 & \mathbf{\Psi} & \cdots & \cdots \\ \cdots & \cdots & \cdots & 0 \\ 0 & \cdots & 0 & \mathbf{\Psi} \end{bmatrix} \tag{40}$$

$$\mathbf{\Phi} = \tilde{\mathbf{A}}\tilde{\mathbf{\Psi}}, \quad or \quad \mathbf{\Phi} = \tilde{\mathbf{B}}\tilde{\mathbf{\Psi}}$$

For this equation, $\otimes$ denotes the Kronecker product sign, $Fourier(\cdot)$ represents the Fourier transform, and DWT represents the discrete wavelet transform. Therefore, the CS-UBSS model can be described as:

$$\hat{\mathbf{X}}(t) = \tilde{\mathbf{A}}(t)\mathbf{S}(t) + \mathbf{G}(t)$$

$$= \tilde{\mathbf{A}}(t)\tilde{\mathbf{\Psi}}\mathbf{\Theta}(t) + \mathbf{G}(t)$$

$$= \mathbf{\Phi}\mathbf{\Theta}(t) + \mathbf{G}(t)$$

$$or \tag{41}$$

$$\hat{\mathbf{X}}(t) = \tilde{\mathbf{B}}(t)\mathbf{S}(t) + \mathbf{G}(t)$$

$$= \tilde{\mathbf{B}}(t)\tilde{\mathbf{\Psi}}\mathbf{\Theta}(t) + \mathbf{G}(t)$$

$$= \mathbf{\Phi}\mathbf{\Theta}(t) + \mathbf{G}(t)$$

where $\mathbf{\Theta}$ is the Fourier transform or DWT of $\mathbf{S}(t)$, so $\mathbf{\Theta}$ is a sparse signal. As for UBSS in the images, firstly, each image matrix needs to be transformed into a row vector, then the four row vectors form a matrix $\mathbf{S}(t)$. At the same time, the sparse basis in Equation (40) needs to be replaced by DWT.

Then, we can recover the source signal by CS. In summary, the above can be described as the flowchart in Figure 14.

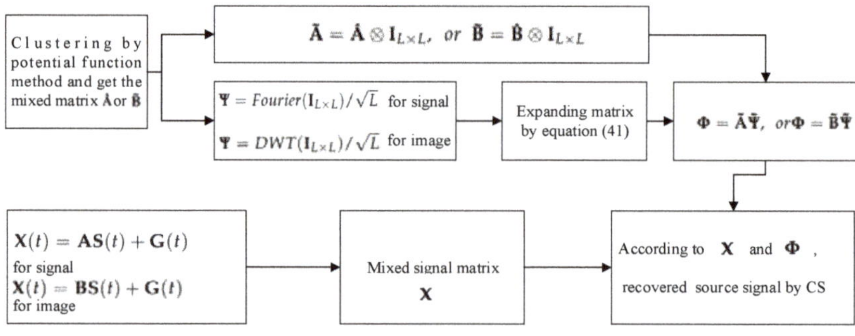

Figure 14. Flowchart of UBSS by CS.

5.2. Performance Analysis of the WReSL0 Algorithm Applied to UBSS

5.2.1. The Effect of the WReSL0 Algorithm Applied to UBSS

In this section, we evaluate the effect of the WReSL0 algorithm applied to UBSS by the separation of signals and spectrum analysis.

The effect of the separation of signals is shown in Figure 15: the source signals are well separated, and the separation signals and the original signals are very similar. Figure 16 displays the error between the original source signal and the recovered source signal. It indicates that the error between the original source signal and the recovered source signal is fairly small, and the WReSL0 algorithm can better deal with the problem of UBSS. In addition, We get the time-frequency diagram of the restored signal by short-time Fourier transform. Figure 17 is the time-frequency diagram. From this figure, we find that each signal has the same frequency as the original signal, and it also validates the rationality of the proposed algorithm for UBSS.

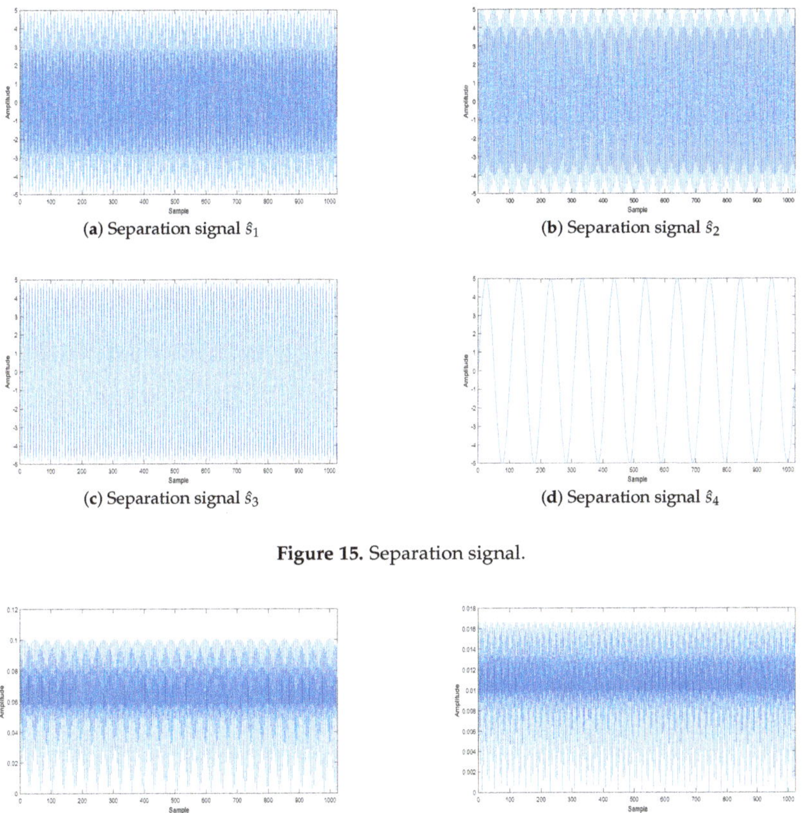

(a) Separation signal $\hat{s}_1$

(b) Separation signal $\hat{s}_2$

(c) Separation signal $\hat{s}_3$

(d) Separation signal $\hat{s}_4$

Figure 15. Separation signal.

(a) Error signal $\hat{s}_1 - s_1$

(b) Error signal $\hat{s}_2 - s_2$

Figure 16. *Cont.*

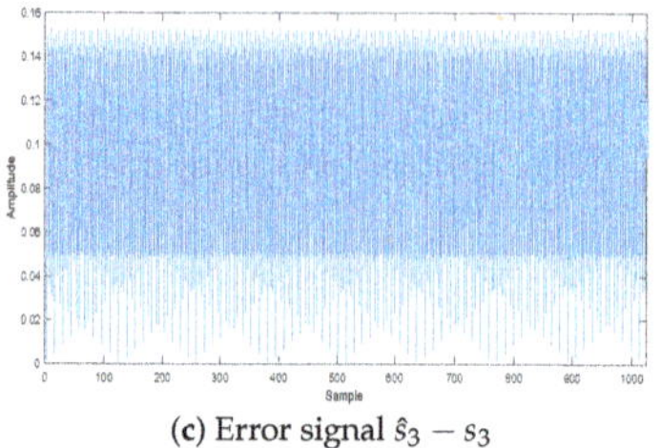

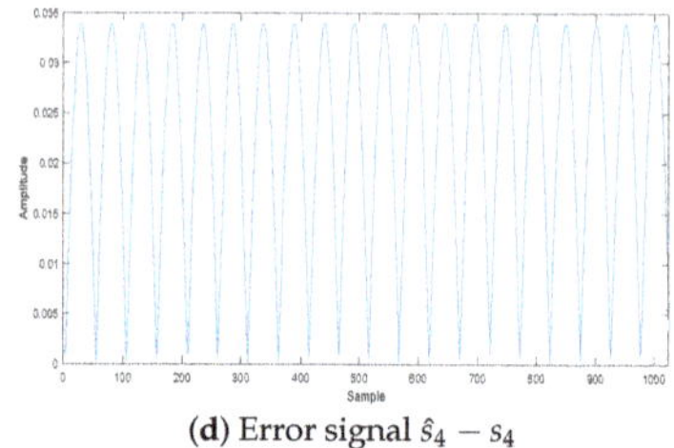

(**c**) Error signal $\hat{s}_3 - s_3$ (**d**) Error signal $\hat{s}_4 - s_4$

Figure 16. Separation signal error analysis.

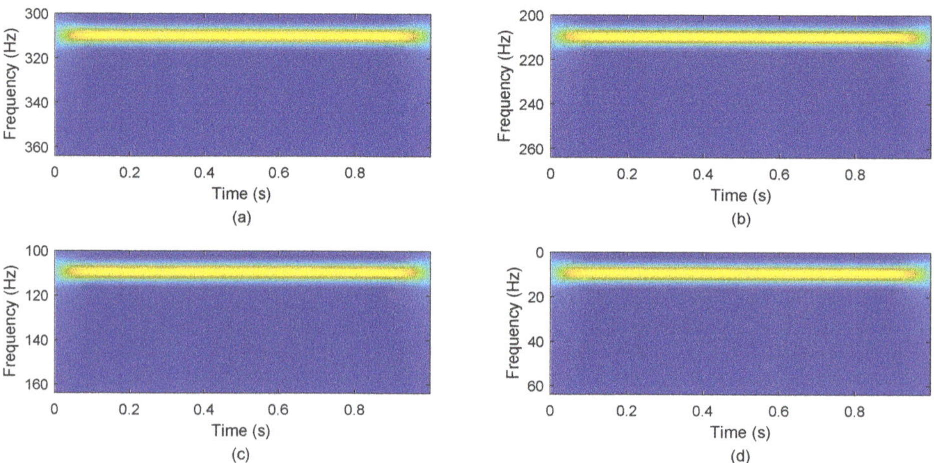

Figure 17. Separation signals' frequency spectrum. Subfigures (**a**–**d**) show the frequency spectrums of separation signals $\hat{s}_1$, $\hat{s}_2$, $\hat{s}_3$, and $\hat{s}_4$.

5.2.2. Performance Comparisons of the Selected Algorithms

Here, we use the SL0, NSL0, and L_p-RLS algorithms and the classical shortest path method (SPM) [38] to make a comparison in different noise cases. In order to analyze the situation of signal recovery clearly, we apply average SNR (ASNR) (for the signal) and average peak SNR (APSNR) (for the image) to evaluate. Let the original source signal be s_i and the recovered source signal be $\hat{s}_i$, so ANSR is defined as:

$$\text{ASNR} = \frac{1}{n} \sum_{i=1}^{n} \text{SNR}_i$$

$$\text{SNR}_i = 20 \log \frac{\|\hat{s}_i - s_i\|_2}{\|s_i\|_2,} \tag{42}$$

and PSNR is defined as:

$$\text{APSNR} = \frac{1}{n} \sum_{i=1}^{n} \text{PSNR}_i$$

$$\text{PSNR}_i = 10 \log \frac{255^2 \times M \times N}{\|\hat{s}_i - s_i\|_2} \tag{43}$$

where M and N are the width and height of the image.

The ASNR comparisons are shown in Table 5. From the table, we can see that ASNR attenuates sharply when δ_N increases from 0.15–0.2. The reason is that the error of the valuated mixed matrix $\hat{A}$ increases obviously, which leads those CS recovery algorithms to perform poorly. In fact, from this table, our proposed RWeSL0 algorithm performs well when δ_N is less than 0.15, and when δ_N is greater than 0.15, the L_p-RLS algorithm performs best, followed by our proposed RWeSL0 algorithm.

The APSNR comparisons are shown in Table 6. In this table, It is clear that APSNR is not high, and it drops greatly when δ_N increases from 0.15–0.2. From Figure 18, we can see that these separated images seem to be enveloped in mist, which leads to a low APSNR. Therefore, we will try our best to improve this problem in the future.

In summary, the CS technique can be used in UBSS and performs well especially for the signal recovery. Our proposed WReSL0 algorithm can perform well in UBSS for the signal recoverywhen the noise is small; and regarding image recovery, we will develop this in the future.

Table 5. Average SNR (ASNR) analysis for separated signals by SPM, SL0, NSL0, L_p-RLS, and the proposed WReSL0 with δ_N changing according to sequence [0,0.1,0.15,0.18,0.2] with 100 runs.

Oise Intensity (δ_N)	Error of $\hat{A}$(%)	ASNR (dB)				
		SPM	**SL0**	**NSL0**	**L_p-RLS**	**WReSL0**
0	1.763	45.443	41.576	42.324	38.412	39.993
0.1	1.763	36.788	35.278	36.034	37.091	39.295
0.15	1.763	31.407	30.754	32.930	35.332	38.975
0.18	112.6	26.355	24.063	25.437	28.305	26.650
0.2	126.3	11.201	9.974	12.358	17.549	15.581

Table 6. APSNR analysis for separated images by SPM, SL0, NSL0, L_p-RLS, and the proposed WReSL0 with δ_N changing according to the sequence [0,0.1,0.15,0.18,0.2] with 100 runs.

Noise Intensity (δ_N)	Error of $\hat{B}$(%)	APSNR (dB)				
		SPM	**SL0**	**NSL0**	**L_p-RLS**	**WReSL0**
0	3.64	16.447	19.211	20.035	16.372	18.483
0.1	3.64	15.639	16.305	17.327	15.407	17.849
0.15	3.64	13.407	14.754	14.930	14.932	17.351
0.18	133.2	9.355	11.063	11.437	10.305	11.650
0.2	142.4	5.201	5.974	6.358	3.549	5.581

(a) Separated Boat   (b) Separated Barbara   (c) Separated Peppers   (d) Separated Lena

Figure 18. Separated images: (**a**) boat; (**b**) Barbara; (**c**) peppers; (**d**) Lena.

6. Conclusions

In this paper, we propose the WReSL0 algorithm to recover the sparse signal from given $\{\mathbf{y}, \mathbf{\Phi}\}$ in the noise case. The WReSL0 algorithm is constructed under the GD method, in which the update process of $\mathbf{x}$ in the inner loop adopts the regularization mechanism to enhance the de-noising performance. As a key part of the WReSL0 algorithm, a weighted smoothed function

$\mathbf{W}^T H_\sigma(\mathbf{x})$ is proposed to promote sparsity and provide the guarantee of robust and accurate signal recovery. Furthermore, We deduced the value of μ and the initial value $\sigma_{\max}$ to ensure the optimization performance of the algorithm. Performance simulation experiments on both real signals and real images show that the proposed WReSL0 algorithm performs better than the L_1 or L_p regularization methods and the classical L_0 regularization methods. Finally, we apply the proposed WReSL0 algorithm to solve the problem of UBSS and also make comparisons with the classical SPM, SL0, NSL0, and Lp-RLS algorithms. Experiments show that this algorithm has some advanced performance. In addition, we would also like to apply the the proposed algorithm to other CS applications such as the RPCA [39], SAR imaging [40], and other de-noising methods [41].

Author Contributions: All authors have made great contributions to the work. L.W., X.Y., H.Y., and J.X. conceived of and designed the experiments; X.Y. and H.Y. performed the experiments and analyzed the data; X.Y. gave insightful suggestions for the work; X.Y. and H.Y. wrote the paper.

Funding: This research received no external funding.

Acknowledgments: This paper is supported by the National Key Laboratory of Communication Anti-jamming Technology.

Conflicts of Interest: The authors declare no conflict of interest.

References

1. Zhang, C.Z.; Wang, Y.; Jing, F.L. Underdetermined Blind Source Separation of Synchronous Orthogonal Frequency Hopping Signals Based on Single Source Points Detection. *Sensors* **2017**, *17*, 2074. [CrossRef]
2. Zhen, L.; Peng, D.; Zhang, Y.; Xiang, Y.; Chen, P. Underdetermined blind source separation using sparse coding. *IEEE Trans. Neural Netw. Learn. Syst.* **2017**, *99*, 1–7. [CrossRef] [PubMed]
3. Donoho, D.L. Compressed sensing. *IEEE Trans. Inf. Theory* **2006**, *52*, 1289–1306. [CrossRef]
4. Candès, E.J.; Wakin, M.B. An introduction to compressive sampling. *IEEE Signal Process. Mag.* **2008**, *2*, 21–30. [CrossRef]
5. Badeńska, A.; Błaszczyk, Ł. Compressed sensing for real measurements of quaternion signals. *J. Frankl. Inst.* **2017**, *354*, 5753–5769. [CrossRef]
6. Candès, E.J. The restricted isometry property and its implications forcompressed sensing. *C. R. Math.* **2008**, *910*, 589–592. [CrossRef]
7. Cahill, J.; Chen, X.; Wang, R. The gap between the null space property and the restricted isometry property. *Linear Algebra Its Appl.* **2016**, *501*, 363–375. [CrossRef]
8. Huang, S.; Tran, T.D. Sparse Signal Recovery via Generalized Entropy Functions Minimization. *arXiv* **2017**, arXiv:1703.10556.
9. Tropp, J.A.; Gilbert, A.C. Signal recovery from random measurements via orthogonal matching pursuit. *IEEE Trans. Inf. Theory* **2007**, *12*, 4655–4666. [CrossRef]
10. Determe, J.F.; Louveaux, J.; Jacques, L.; Horlin, F. On the noise robustness of simultaneous orthogonal matching pursuit. *IEEE Trans. Signal Process.* **2016**, *65*, 864–875. [CrossRef]
11. Donoho, D.L.; Tsaig, Y.; Starck, J.L. Sparse solution of underdetermined linear equations by stagewise orthogonal matching pursuit. *IEEE Trans. Inf. Theory* **2012**, *2*, 1094–1121. [CrossRef]
12. Needell, D.; Vershynin, R. Signal recovery from incompleteand inaccurate measurements via regularized orthogonal matching pursuit. *IEEE J. Sel. Top. Signal Process.* **2010**, *2*, 310–316. [CrossRef]
13. Needell, D.; Tropp, J.A. CoSaMP: Iterative signal recovery from incomplete and inaccurate samples. *Commun. ACM* **2010**, *12*, 93–100. [CrossRef]
14. Jian, W.; Seokbeop, K.; Byonghyo, S. Generalized orthogonal matching pursuit. *IEEE Trans. Signal Process.* **2012**, *12*, 6202–6216. [CrossRef]
15. Wang, J.; Kwon, S.; Li, P.; Shim, B. Recovery of sparse signals via generalized orthogonal matching pursuit: A new analysis. *IEEE Trans. Signal Process.* **2016**, *64*, 1076–1089. [CrossRef]
16. Dai, W.; Milenkovic, O. Subspace pursuit for compressive sensing signal reconstruction. *IEEE Trans. Inf. Theory* **2009**, *5*, 2230–2249. [CrossRef]

17. Goyal, P.; Singh, B. Subspace pursuit for sparse signal reconstruction in wireless sensor networks. *Procedia Comput. Sci.* **2018**, *125*, 228–233. [CrossRef]

18. Liu, X.J.; Xia, S.T.; Fu, F.W. Reconstruction guarantee analysis of basis pursuit for binary measurement matrices in compressed sensing. *IEEE Trans. Inf. Theory* **2017**, *63*, 2922–2932. [CrossRef]

19. Mohimani, H.; Babaie-Zadeh, M.; Jutten, C. A Fast Approach for Overcomplete Sparse Decomposition Based on Smoothed L0 Norm. *IEEE Trans. Signal Process.* **2009**, *57*, 289–301. [CrossRef]

20. Zhao, R.; Lin, W.; Li, H.; Hu, S. Reconstruction algorithm for compressive sensing based on smoothed L0 norm and revised newton method. *J. Comput.-Aided Des. Comput. Graph.* **2012**, *24*, 478–484.

21. Ye, X.; Zhu, W.P. Sparse channel estimation of pulse-shaping multiple-input–multiple-output orthogonal frequency division multiplexing systems with an approximate gradient L_2-SL0 reconstruction algorithm. *Iet Commun.* **2014**, *8*, 1124–1131. [CrossRef]

22. Nowak, R.D.; Wright, S.J. Gradient projection for sparse reconstruction: Application to compressed sensing andother inverse problems. *IEEE J. Sel. Top. Signal Process.* **2007**, *1*, 586–597.

23. Long, T.; Jiao, W.; He, G. RPC estimation via ℓ_1-norm-regularized least squares (L1LS). *IEEE Trans. Geosci. Remote Sens.* **2015**, *8*, 4554–4567. [CrossRef]

24. Pant, J.K.; Lu, W.S.; Antoniou, A. New improved algorithms for compressive sensing based on ℓ_p norm. *IEEE Trans. Circuits Syst. II Express Br.* **2014**, *3*, 198—202. [CrossRef]

25. Wipf, D.; Nagarajan, S. Iterative Reweighted and Methods for Finding Sparse Solutions. *IEEE J. Sel. Top. Signal Process.* **2016**, *2*, 317–329.

26. Zhang, C.; Hao, D.; Hou, C.; Yin, X. A New Approach for Sparse Signal Recovery in Compressed Sensing Based on Minimizing Composite Trigonometric Function. *IEEE Access* **2018**, *6*, 44894–44904. [CrossRef]

27. Candès, E.J.; Wakin, M.B.; Boyd, S.P. Enhancing sparsity by weighted L1 minimization. *J. Fourier Anal. Appl.* **2008**, *14*, 877–905. [CrossRef]

28. Pant, J.K.; Lu, W.S.; Antoniou, A. Reconstruction of sparse signals by minimizing a re-weighted approximate L0-norm in the null space of the measurement matrix. In Proceedings of the IEEE International Midwest Symposium on Circuits and Systems, Seattle, WA, USA, 1–4 August 2010; pp. 430–433.

29. Aggarwal, P.; Gupta, A. Accelerated fmri reconstruction using matrix completion with sparse recovery via split bregman. *Neurocomputing* **2016**, *216*, 319–330. [CrossRef]

30. Chu, Y.J.; Mak, C.M. A new qr decomposition-based rls algorithm using the split bregman method for L1-regularized problems. *Signal Process.* **2016**, *128*, 303–308. [CrossRef]

31. Hu, Y.; Liu, J.; Leng, C.; An, Y.; Zhang, S.; Wang, K. Lp regularization for bioluminescence tomography based on the split bregman method. *Mol. Imaging Biol.* **2016**, *18*, 1–8. [CrossRef]

32. Liu, Y.; Zhan, Z.; Cai, J.F.; Guo, D.; Chen, Z.; Qu, X. Projected iterative soft-thresholding algorithm for tight frames in compressed sensing magnetic resonance imaging. *IEEE Trans. Med. Imaging* **2016**, *35*, 2130–2140. [CrossRef]

33. Yang, L.; Pong, T.K.; Chen, X. Alternating direction method of multipliers for a class of nonconvex and nonsmooth problems with applications to background/foreground extraction. *Mathematics* **2016**, *10*, 74–110. [CrossRef]

34. Antoniou, A.; Lu, W.S. *Practical Optimization: Algorithms and Engineering Applications*; Springer: New York, NY, USA, 2007.

35. Samora, I.; Franca, M.J.; Schleiss, A.J.; Ramos, H.M. Simulated annealing in optimization of energy production in a water supply network. *Water Resour. Manag.* **2016**, *30*, 1533–1547. [CrossRef]

36. Goldstein, T.; Studer, C. Phasemax: Convex phase retrieval via basis pursuit. *IEEE Trans. Inf. Theory* **2018**, *64*, 2675–2689. [CrossRef]

37. Wei-Hong, F.U.; Ai-Li, L.I.; Li-Fen, M.A.; Huang, K.; Yan, X. Underdetermined blind separation based on potential function with estimated parameter's decreasing sequence. *Syst. Eng. Electron.* **2014**, *36*, 619–623.

38. Bofill, P.; Zibulevsky, M. Underdetermined blind source separation using sparse representations. *Signal Process.* **2001**, *81*, 2353–2362. [CrossRef]

39. Su, J.; Tao, H.; Tao, M.; Wang, L.; Xie, J. Narrow-band interference suppression via rpca-based signal separation in time–frequency domain. *IEEE J. Sel. Top. Appl. Earth Obs. Remote Sens.* **2017**, *99*, 1–10. [CrossRef]

40. Ni, J.C.; Zhang, Q.; Luo, Y.; Sun, L. Compressed sensing sar imaging based on centralized sparse representation. *IEEE Sens. J.* **2018**, *18*, 4920–4932. [CrossRef]
41. Li, G.; Xiao, X.; Tang, J.T.; Li, J.; Zhu, H.J.; Zhou, C.; Yan, F.B. Near—Source noise suppression of AMT by compressive sensing and mathematical morphology filtering. *Appl. Geophys.* **2017**, *4*, 581–589. [CrossRef]

Article

Detail Preserved Surface Reconstruction from Point Cloud

Yang Zhou [1,2], Shuhan Shen [1,2]* and Zhanyi Hu [1,2]

[1] National Laboratory of Pattern Recognition, Institute of Automation, Chinese Academy of Sciences, Beijing 100190, China; yang.zhou@nlpr.ia.ac.cn (Y.Z.); huzy@nlpr.ia.ac.cn (Z.H.)
[2] School of Artificial Intelligence, University of Chinese Academy of Sciences, Beijing 100049, China
* Correspondence: shshen@nlpr.ia.ac.cn

Received: 18 February 2019; Accepted: 11 March 2019; Published: 13 March 2019

Abstract: In this paper, we put forward a new method for surface reconstruction from image-based point clouds. In particular, we introduce a new visibility model for each line of sight to preserve scene details without decreasing the noise filtering ability. To make the proposed method suitable for point clouds with heavy noise, we introduce a new likelihood energy term to the total energy of the binary labeling problem of Delaunay tetrahedra, and we give its s-t graph implementation. Besides, we further improve the performance of the proposed method with the dense visibility technique, which helps to keep the object edge sharp. The experimental result shows that the proposed method rivalled the state-of-the-art methods in terms of accuracy and completeness, and performed better with reference to detail preservation.

Keywords: computer vision; 3D reconstruction; point cloud

1. Introduction

Image-based scene reconstruction is a fundamental problem in Computer Vision. It has many practical applications in fields such as entertainment industry, robotics, cultural heritage digitalization and geographic systems. Image-based scene reconstruction has been studied for decades due to its low cost data acquisition and various usages. In recent years, researchers have made tremendous progress in this field. As far as small objects under controlled conditions are concerned, the performance of current scene reconstruction methods could achieve results comparable to those generated by laser scans or structured-light based methods [1,2]. However, when it comes to large scale scenes with multi-scale objects, current reconstruction methods have some problems with the completeness and accuracy, especially when concerning scene details [3].

Scene details such as small scale objects and object edges are an essential part of scene surfaces. Figure 1 shows an example of preserving scene details in reconstructing an ancient Chinese architecture. In general, cultural heritage digitalization projects, representing scene details such as the brackets in Figure 1, are among the most important tasks. The point cloud representation is often redundant and noisy, and the mesh representation is concise but it sometimes lose some information. Therefore, preserving scene details in reconstructing multi-scale scenes has been a difficult problem in surface reconstruction. The existing surface reconstruction methods [4–7] either ignore the scene details or rely on further refinement to restore them. Firstly, this is because, compared with noise, the supportive points in such part of the scene are sparse, making it difficult to distinguish true surface points from false ones. Secondly, the visibility models and associated parameters employed in existing methods are not particularly suitable for large scale ranges, where scene details are usually compromised for overall accuracy and completeness. While the first case seems to be unsolvable due to the lack of sufficient information, in this work, we focus on the second case. In particular, we extend the work of our conference paper [8], and suggest a new method with a new visibility model for surface reconstruction.

(a)   (b)

Figure 1. An example of the application of the proposed method in the field of cultural heritage digitalization, (**a**) point cloud; (**b**) mesh.

In many previous surface reconstruction methods [4–6,9–13], visibility information that records a 3D point is seen by the views used to help to generate accurate surface meshes. To use the visibility information, assumptions of the visibility model are made so that the space between camera center and the 3D point is free-space and the space behind the point along the line of sight is full-space. However, the points are often contaminated with noise and the full-space scales are often hard to determinate. To preserve scene details without decreasing the noise filtering ability, we propose a new visibility model with error tolerance and adaptive end weights. We also introduce a new likelihood energy representing the punishment of wrongly classifying a part of space as free-space or full-space, which helps to improve the ability of the proposed method to efficiently filter noise. Moreover, we further improve the performance of the proposed method with the dense visibility technique, which helps to keep the object edge sharp. Experimental results show that the proposed method rivals the state-of-the-art methods in terms of accuracy and completeness, and performs better with reference to detail preservation.

2. Related Work

In recent years, various works have been done to advance the image-based scene reconstruction. Referring to small objects, silhouette based methods [14–18] are proposed. The silhouettes provide proper bounds for the objects, which help reduce the computing cost and yield a good model for the scene. However, good silhouettes rely on effective image segmentation, which remains a difficult task. Furthermore, silhouettes can hardly be used for large scale scenes. Volumetric methods such as space carving [19–21], level sets [22,23] and volumetric graph cut [9–11] often yield good results for small objects. However, in the case of large scale scenes, the computational and memory costs increase rapidly as scene scale grows. Consequently, this makes them unsuitable for large scale scene reconstruction. Vogiatzis et al. [9] proposed a volumetric graph cut based method to reconstruct an object by labeling voxels as inside or outside, in which a photo-consistency term is introduced to enhance the final result. Tran and Davis [10] and Lempitsky et al. [11] also exploited the same idea, the former by adding predetermined locations of possible surface as surface constraints and the latter by estimating visibility based on position and orientation of local surface patches, then optimizing on a CW-complex.

As far as outdoor scenes, uncontrollable imaging conditions and multiple scale structures make it hard to reconstruct scene surfaces. A common process of reconstructing large scale scenes is to generate the dense point cloud from calibrated images first, and then extract the scene surface. The dense point cloud can be generated through the depth fusion method [24–28] or through the feature expansion methods [29]. In depth-map fusion methods, depth-maps are usually computed independently, and then merged into one point cloud; in feature expansion methods instead, the sparse point cloud is generated first, and then expanded to points near the seed points. Usually, the depth-map fusion based methods yield a relatively denser but noisier point cloud. Once the dense point cloud is generated, scene surface can be reconstructed through poisson surface reconstruction [30] or through graph cut based methods [4,5,12,13]. In [12], firstly the dense point cloud is used to generate Delaunay tetrahedra, then a visibility model is introduced to weight the facets in Delaunay tetrahedra, and finally the inside–outside binary labeling of tetrahedra is solved and the surface is extracted; it consists of triangles between tetrahedra with different labels. The basic assumption of the visibility model in [12] is that the space between camera center and 3D point is free-space, while the space behind 3D point is full-space. Then, this typical visibility model is promoted by a refined version in [13], namely soft-visibility, to cope with noisy point clouds. Apart from using dense point clouds, Bódis-Szomorú et al. [31] also proposed a method to produce a surface mesh by fitting the meshes reconstructed by single views and a sparse point cloud. The method is parallelizable and the quality of the final meshes is comparable to that of a state-of-the-art pipeline [32].

Besides the point-cloud based methods, large scale scene reconstruction can also be achieved through volume-based methods. Häne et al. [6,33] proposed a method for scene reconstruction and object classification. The scene space is represented by voxels and each one is given a class label. The accuracy of the classification relies on the decision tree, which is made up of labeled images. The extensive works in [34,35] reduce the heavy memory cost of the method in [33] by introducing octree structure and block scheme, respectively. Savinov et al. [36] exploited an idea similar to that in [6,33], and used full multilabel ray potential and continuously inspired anisotropic surface regularization together to yield 3D semantic models. Ummenhofer and Brox [37] proposed a method that is capable of handling a billion points. This method uses an octree structure to manage the points and then reconstruct the scene using a level-set alike method.

In this paper, we extend the work of our conference paper [8], follow the idea in [5,13] and exploit the visibility model for the line of sight in the binary labeling problem of Delaunay tetrahedra. The main contributions of the proposed method are: (1) two new visibility models for the visibility information of the points on the vertices of Delaunay tetrahedra and inside them, respectively; and (2) the dense visibility technique, which exploits the visibility information of unmerged points for better performance of preserving scene details. Experimental comparison results show that the proposed method rivals the state-of-the-art methods [5,24,27,29,38] in terms of accuracy and completeness, and performs better in detail preservation.

3. Visibility Models and Energies

The pipeline of the proposed method is shown in Figure 2. The input of the proposed method is a dense point cloud generated by multi-view stereo (MVS) methods. Each point in the point cloud is attached with the visibility information recording that from which views the point is seen. Next, Delaunay tetrahedra are constructed from given input point cloud, and the scene reconstruction problem is formulated as a binary labeling problem to label tetrahedra as inside or outside. Then, an *s-t* graph is constructed with tetrahedra as the vertices and facets of tetrahedra as the edges. Finally, by minimizing the energy defined on the *s-t* graph, the vertices are separated into two parts, i.e., inside and outside, and the scene surface is extracted which consists of triangle facets lying between tetrahedra with different labels. The key issue of this process is to find a proper energy. In doing so, we introduce a new visibility model to formulate the energy of the binary labeling problem. In the following subsections, we detail the visibility models in [12,13] and our new one. Before that, we give the meaning of the symbols used in this work in Table 1 for better understanding.

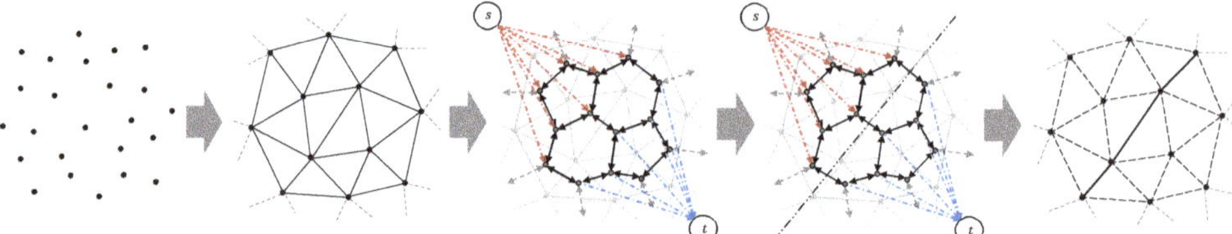

Figure 2. Pipeline of the proposed method in 2D. From left to right are the input point cloud, Delaunay tetrahedra, the *s-t* graph, the energy minimization result and the final surface mesh.

Table 1. Symbols used in this work.

Symbol	Meaning
v	line of sight
c	camera center (a 3D point)
p	3D point
T	tetrahedron
l_T	label of tetrahedron T
$D(l_T)$	unary energy of the label assignment of tetrahedron T
$W(l_{T_i}, l_{T_j})$	pair-wise energy of the label assignments of two adjacent tetrahedra
α_v	weight of a line of sight v
N_v	amount of tetrahedra intersected with a line of sight v
d	distance between point p and the intersecting point of a segment and a facet
σ	scale factor
r	the radius of the circumsphere of the end tetrahedron
$U_{out}(T)$	energy of tetrahedron T being labeled as outside
$U_{in}(T)$	energy of tetrahedron T being labeled as inside
$f(T)$	free-space support of tetrahedron T
β	constant for transferring $f(T)$
$\lambda, \lambda_{vis}, \lambda_{like}, \lambda_{qual}$	balance factor
$E, E_{vis}, E_{like}, E_{qual}, E_{vis}^{typical}$	energy
w_f	weight of a facet f
ϕ, ψ	angle

3.1. Existing Visibility Models

The typical visibility model [12] assumes that, for each line of sight v, the space between camera center c and point p is free-space; the space behind point p along the line of sight is full-space. For the Delaunay tetrahedra of the point cloud, the tetrahedra intersected by segment (c, p) should be labeled as outside; the tetrahedron right behind the point p should be labeled as inside. For example, in Figure 3a, according to the above model, tetrahedra T_1–T_5 should be labeled as outside, while tetrahedron T_6 should be labeled as inside.

For a single line of sight, the above label assignment is desirable. While taking all lines of sight into consideration, some surface part might not be handled properly, for example, that between T_2 and T_3 in Figure 3a. This scenario violates the label assignment principle described above. To punish the conflicts, the facets intersected by a line of sight are given weights (energy) of $W(l_{T_i}, l_{T_j})$. Similarly, weights for punishing bad label assignments of the first tetrahedron and last tetrahedron are $D(l_{T_1})$ and $D(l_{T_{N_v+1}})$, respectively. Therefore, the visibility energy is the sum of the penalties of all the bad label assignments in all the lines of sight, as

$$E_{vis}^{typical} = \sum_{v \in \mathcal{V}} \left[D(l_{T_1}) + \sum_{i=1}^{N_v-1} W(l_{T_i}, l_{T_{i+1}}) + D(l_{T_{N_v+1}}) \right] \tag{1}$$

where

$$D(l_{T_1}) = \begin{cases} \alpha_v & \text{if } l_{T_1} = 1 \\ 0 & \text{otherwise} \end{cases}, \quad D(l_{T_{N_v+1}}) = \begin{cases} \alpha_v & \text{if } l_{T_{N_v+1}} = 0 \\ 0 & \text{otherwise} \end{cases}, \quad W(l_{T_i}, l_{T_j}) = \begin{cases} \alpha_v & \text{if } l_{T_i} = 0 \wedge l_{T_j} = 1 \\ 0 & \text{otherwise} \end{cases}$$

where $\mathcal{V}$ denotes the visibility information set, containing all lines of sight; N_v is the number of tetrahedra intersected by a single line of sight v, indexed from the camera center c to the point p; $N_v + 1$ denotes the tetrahedron behind the point p; l_T is the label of tetrahedron T, with 1 stands for inside and 0 outside; α_v is the weight of a line of sight v, which can be the photo consistency score of point p in current view.

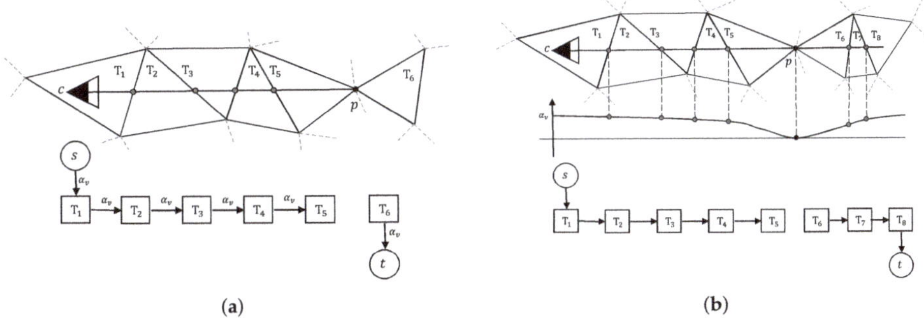

(a)   **(b)**

Figure 3. Typical visibility model and soft visibility model in 2D. (**a**) Typical visibility model; and (**b**) soft visibility model for a single line of sight v; how to assign weight (energy) to the tetrahedron containing camera center, the end tetrahedron and the facets intersected by (c, p) or its extension are shown.

The typical visibility model as well as the energy formulation in Equation (1) described above is effective in most cases, but it has several flaws in relation to dense and noisy point clouds [13]. The surfaces reconstructed by [12] tend to be overly complex, with bumps on the surface and handles inside the model. One possible solution to these problems is the soft visibility proposed in [13], as shown in Figure 3b. In the soft visibility model, the basic assumption is similar to the previous typical visibility model. However, the edge weights are multiplied by a weight factor $(1 - e^{-d^2/2\sigma^2})$, in which d represents the distance between the point p and the intersecting point. In addition, the end tetrahedron in the soft visibility model is shifted to a distance of 3σ along the line of sight.

3.2. Our Proposed Visibility Model

Although the soft visibility model is effective to filter noise points and helps to yield visually smoothed models, it sometimes performs poorly in preserving details, especially in a large scene containing some relatively small scale objects (see Experimental Results). According to our observations, this happens mainly because of the improperly chosen relaxation parameter σ and the strong constraint imposed on the end of line of sight in the tetrahedron $k\sigma$ from the point p along the line of sight. In some cases, such end tetrahedra would be free-space even though the point p is a true surface point.

To balance noise filtering and detail preserving, we propose a new visibility model, which is shown in Figure 4a. In our visibility model, we also use the relaxed visibility constraints in the soft visibility model in the space between the camera center c and the point p, i.e., the weight factor $(1 - e^{-d^2/2\sigma^2})$ is kept to ensure that the final model is not overly complex. Then, we set the end of line of sight in the tetrahedron just right behind the point p, to avoid the wrong end in the soft visibility model in the case of small scale objects. To determine the weight of the t-edge of the end tetrahedron, we compare the end tetrahedra of noisy points and true surface points on datasets with quasi-truth. Figure 4b shows a typical end tetrahedron of noisy points and that of true surface points on densely sampled surfaces in 2D space. Noise points tend to appear in somewhere a bit away from true surface, which makes the end tetrahedra (triangles in 2D) thin and long, and true surface points are often surrounded by other true surface points, which makes their end tetrahedra flat and wide. Based on the above observations, we set a weight of $\alpha_v(1 - e^{-r^2/2\sigma^2})$ to the t-edge of the end tetrahedron, where r is the radius of the circumsphere of the end tetrahedron.

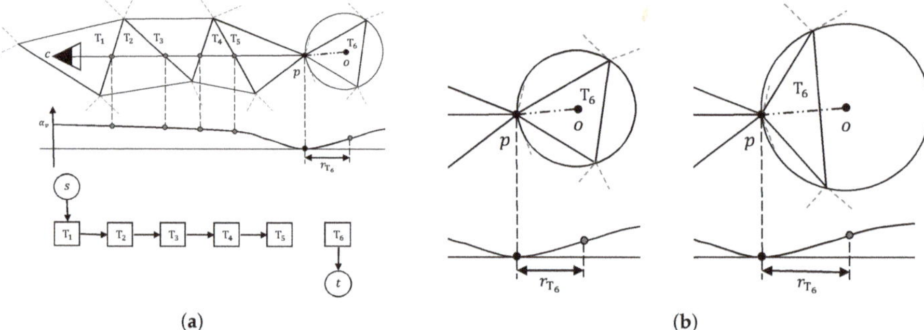

Figure 4. Our visibility model and end tetrahedra comparison in 2D. (**a**) In our visibility model, for a single line of sight v, how to assign weight (energy) to the tetrahedron containing camera center, the end tetrahedron and the facets intersected by (c, p) is shown. (**b**) From left to right: Typical end tetrahedron of noise points and that of true surface points on densely sampled surfaces.

With our new visibility model, our visibility energy is formulated as

$$E_{vis} = \sum_{v \in \mathcal{V}} [D(l_{T_1}) + \sum_{i=1}^{N_v-1} W(l_{T_i}, l_{T_{i+1}}) + D(l_{T_{N_v+1}})] \tag{2}$$

where

$$D(l_{T_1}) = \begin{cases} \alpha_v & \text{if } l_{T_1} = 1 \\ 0 & \text{otherwise} \end{cases}, \quad D(l_{T_{N_v+1}}) = \begin{cases} \alpha_v(1 - e^{-r^2/2\sigma^2}) & \text{if } l_{T_{N_v+1}} = 0 \\ 0 & \text{otherwise} \end{cases},$$

$$W(l_{T_i}, l_{T_j}) = \begin{cases} \alpha_v(1 - e^{-d^2/2\sigma^2}) & \text{if } l_{T_i} = 0 \wedge l_{T_j} = 1 \\ 0 & \text{otherwise} \end{cases}$$

Instead of setting the tolerance parameter σ to a constant manually, we set σ adaptively within 0.5–1% of the length of line of sight in our visibility model. The underlying reason is that, typically, in depth-map fusion methods, the error bound used to filter bad correspondences while generating dense point clouds is adaptively set to 0.5–1% of the depth of the point in the current view, as in [28]. This gives each point a confidence interval along the line of sight.

4. Likelihood Energy for Efficient Noise Filtering

In both the typical visibility model and our proposed one, the end of each line of sight is set in the tetrahedra right behind the point p. This practice sometimes could weaken the ability of noise filtering. When the surface is sampled very densely, unexpected handles could appear inside the model [13], or part of the surface could fail to be reconstructed, as shown in Figure 5. This is mainly due to the unbalanced links of s-edges and t-edges in the s-t graph, i.e., the s-edges are too strong to be cut as the camera centers are consistent for all lines of sight, while the t-edges are weak because their weights are scattered by the varying locations of the points. The noisier the point cloud is, the greater the gap between *s*-edges and *t*-edges becomes.

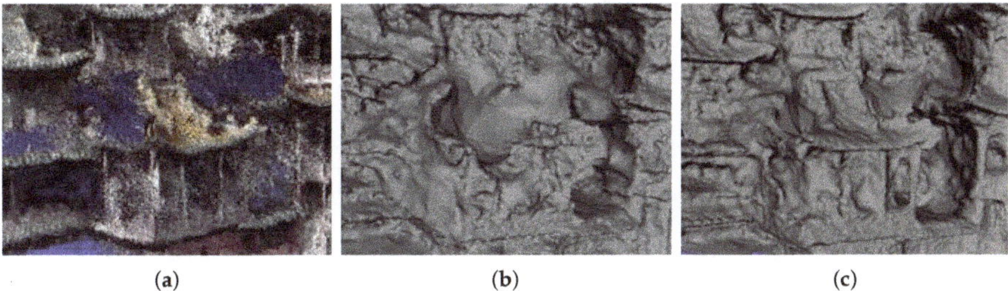

(a)   (b)   (c)

Figure 5. Surface reconstruction without and with the likelihood energy. From left to right: (**a**) point cloud with heavy noise; (**b**) reconstructed meshes without; and (**c**) with the likelihood energy.

4.1. Likelihood Energy

To solve these problems, we introduce a likelihood energy E_{like} to the total energy of the binary labeling problem. E_{like} is defined as

$$E_{like} = \sum_{i=1}^{N} D(l_{T_i}), \text{ where } D(l_{T_i}) = \begin{cases} U_{out}(T_i) & \text{if } l_{T_i} = 1 \\ U_{in}(T_i) & \text{otherwise} \end{cases} \tag{3}$$

where N is the total number of Delaunay tetrahedra. E_{like} measures the penalties of a wrong label assignment. For each tetrahedron, it is attached with two attributes that describe how likely it is to be outside or inside. If it is mistakenly labeled, a penalty is introduced, i.e., $U_{out}(T_i)$ or $U_{in}(T_i)$.

To evaluate the likelihood of the label assignment of a tetrahedron, we employ the measure *free-space support* [5], which is used to measure the emptiness of a compact space. For each line of sight v that intersects tetrahedra T, it contributes to the emptiness of T_i, thus increasing the probability of T to be outside. The free-space support $f(T)$ of tetrahedron T is computed as

$$f(T) = \sum_{v \in \mathcal{V}_T} \alpha_v, \text{ with } \mathcal{V}_T = \{v | v \cap T \neq \varnothing\} \tag{4}$$

where $\mathcal{V}_T$ is the set of lines of sight v that intersect with tetrahedron T. To evaluate $f(T)$ to adequately describe the likelihood energy, we set $U_{out}(T) = \lambda f(T)$ and $U_{in}(T) = \lambda(\beta - f(T))$, where λ is a constant used to scale the range of $f(T)$, and β is a constant greater than $f(T)$ for all tetrahedra T.

4.2. Implementation of the Likelihood Energy

For the likelihood term E_{like}, we link two edges for each vertex i in *s-t* graph. One is from source s to vertex i, with weight $U_{out}(T_i)$; the other is from vertex i to sink t, with weight $U_{in}(T_i)$. However, to reduce the complexity of the graph, we cross out the *s*-edges of all vertices, and only link the vertices to sink t whose correspondent tetrahedra have lower $f(T)$ than the 75th percentile of all $f(T)$s. Since some of the vertices in *s-t* graph have a heavy-weight edge linked to source s, we rely on the visibility constraints to label truly outside tetrahedra, instead of linking them with extra *s*-edges.

Note that the free space support threshold of 75th percentile was empirically set, as shown in Figures 6 and 7. We generated dense point clouds of four datasets, and applyiedDelaunay tetrahedralization to them. The four datasets were Fountain-P11 [3], Herz-Jesu-P8 [3], Temple-P312 [1] and scan23-P49 [2]. Then, we labeled the tetrahedra with method described in [5], and took the result as the quasi-truth. Finally, we evaluated the ratios of number of outside and inside tetrahedra in different proportions of all free-space support scores, as shown in Figure 6. In Figure 6, we can see that, generally, tetrahedra with lower $f(T)$ ($f(T)$ lower than the 75th percentile of all free-space support scores) had a higher probability to be truly inside. As shown in Figure 7, the true positive rates and false positive rates with different free-space support thresholds were evaluated. It is noteworthy that, when the free-space support threshold was set as the 75th percentile of each dataset, both true positive rate and false positive rate were reasonable. Therefore, we only linked those tetrahedra to sink t whose free-space support was lower than 75th percentile of all free-space support scores. In Figure 5, we can see that the likelihood energy is helpful for filtering noise.

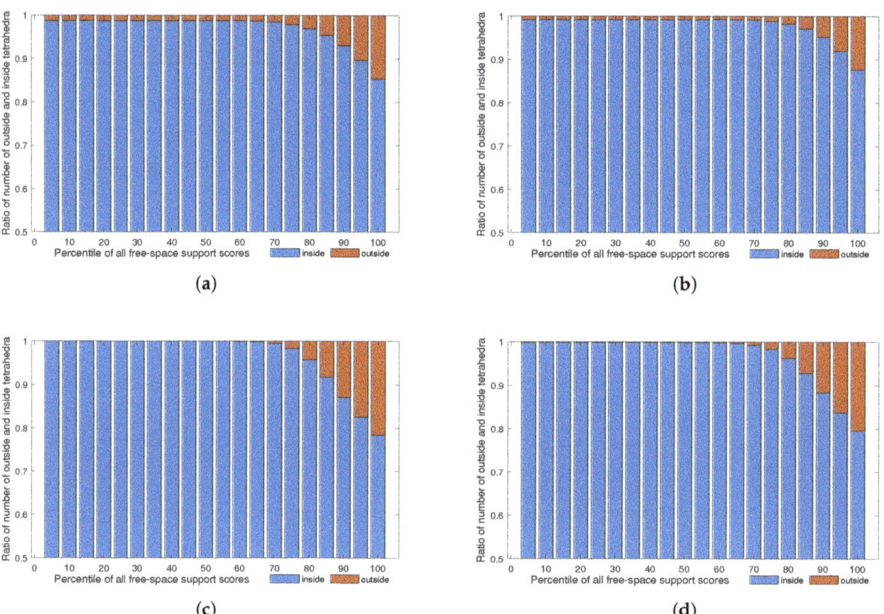

Figure 6. Free-space support analysis. The four graphs show the ratios of number of outside and inside tetrahedra in different percentiles of all free-space support scores. The four datasets are: (**a**) Fountain-P11 [3]; (**b**) Herz-Jesu-P8 [3]; (**c**) Temple-P312 [1]; and (**d**) scan23-P49 [2].

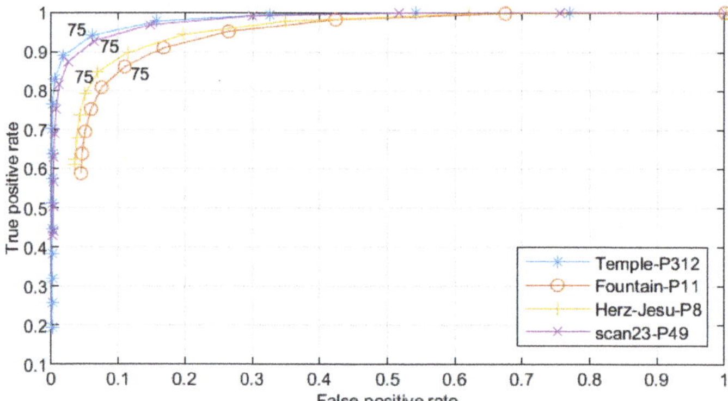

Figure 7. Free-space support threshold evaluation. The true positive rates and false positive rates in the same four datasets as in Figure 6 are evaluated by setting different free-space support thresholds.

5. Surface Reconstruction with Energy Minimization

With the likelihood energy and the proposed visibility model, the total energy of the binary labeling problem of the Delaunay tetrahedra is formulated as

$$E_{total} = E_{vis} + \lambda_{like} E_{like} + \lambda_{qual} E_{qual} \tag{5}$$

where λ_{like} and λ_{qual} are two constant balancing factors; E_{vis} and E_{like} are defined in Equations (2) and (3); and E_{qual} is the surface quality energy introduced in [13] as

$$E_{qual} = \sum_f w_f \tag{6}$$

where

$$w_f = 1 - \min\{\cos(\phi), \cos(\psi)\}, \text{ if } l_{T_1^f} \neq l_{T_2^f}$$

The total energy E_{total} defined in Equation (5) could be represented as an *s-t* graph and minimized by the Maxflow/Mincut algorithm [39]. In E_{total}, the graph construction for the visibility energy E_{vis} and the surface quality energy E_{qual} is straightforward, and the likelihood E_{like} is implemented as described in Section 4.2. Then, the energy minimization problem is solved using the Maxflow/Mincut algorithm on the *s-t* graph, and the optimal label assignment of Delaunay tetrahedra is yielded. Ultimately, a triangle mesh, which consists of triangles lying between tetrahedra with different labels, is extracted. A further optional refinement can be applied as described in [7].

6. Dense Visibility for Edge Preservation

Although the proposed visibility model in Figure 4 and the energy formulation in Equation (5) are carefully designed for preserving the scene details, the object edges sometimes still appear to be inaccurate concerning bumps and dents. When referring to the original depth maps, the depths of object edges are quite smooth. This scenario is shown in Figure 8. Figure 8a,b shows the object edges in 3D meshes reconstructed by the method in [5] and the proposed method described in the previous sections, as well as the original image and the corresponding depth map with a similar view point. We can easily see that the object edges in the reconstructed meshes failed to keep the smoothness as in the depth map. This could be due to the error of either point locations or visibility information which is introduced in the depth map fusion process. To circumvent such problem, instead of fusing the depths in matched cameras, we generate the dense point cloud simply by joining all of the 3D points recovered with depth maps and camera parameters. This could result in a much denser and more redundant point cloud, but it also contains more useful information for better surface reconstruction. To alleviate the memory and the computational cost, points are sampled and then Delaunay tetrahedra are constructed from the sampled point cloud. Instead of discarding those points unused for tetrahedralization and their visibility information, we apply a modified version of our visibility model to use their visibility information, as shown in Figure 9. To keep the ability to filter noise and select true surface points, we keep most of the visibility model in Figure 4 along the line of sight and only modify the part near the 3D point. The difference between the visibility models in Figure 9 and the one in Figure 4 is that, for a point *p* that lies in a tetrahedron, the end of the line of sight is set in the tetrahedron right behind the one that contains *p*, and the facet between them is punished with a weight multiplied by the same weight factor as in the soft visibility model. In this way, we keep the end of the line of sight close to the 3D point and the tolerance to noise. In Figure 8c, we show the surface meshes containing object edges reconstructed by the method in [5] and the proposed method with the dense visibility technique. We can infer from the results in Figure 8 that the dense visibility technique is helpful for preserving object edges.

(a)   (b)   (c)

Figure 8. Object edges in the reconstructed meshes, the original image and the depth map. From left to right: (**a**) the object edges in surface meshes reconstructed by the method in [5] (top) and our method (bottom) without the dense visibility technique; (**b**) the original image and the corresponding depth map with a similar view point; and (**c**) the object edges in surface meshes reconstructed by the method in [5] (top) and our method (bottom) with the dense visibility technique.

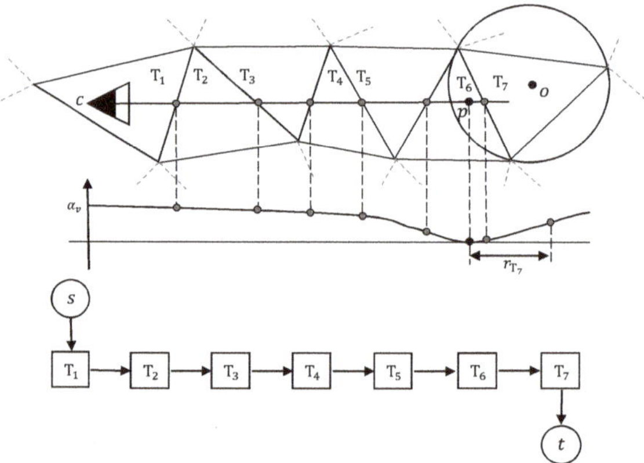

Figure 9. The modified version of our visibility model in 2D. In this visibility model, for a single line of sight v, how to assign weight (energy) to the tetrahedron containing camera center, the end tetrahedron and the facets intersected by (c, p) or its extension is shown.

7. Experimental Results

In our experiments, the input dense point cloud was generated from images with the open source library OpenMVG (http://imagine.enpc.fr/~moulonp/openMVG/) and OpenMVS (http://cdcseacave.github.io/openMVS/). Sparse point cloud was generated by OpenMVG, then densified with OpenMVS. Delaunay tetrahedralization was computed using CGAL (http://www.cgal.org/) library. Maxflow/Mincut algorithm [39] aws used. The proposed method was tested on public benchmark MVS dataset [2] and Tanks and Temples dataset [40].

We first tested the proposed method on MVS dataset [2]. MVS dataset [2] contains over one hundred scenes consisting of images depicting compact objects under controlled lighting conditions. Figure 10 shows the result on the MVS dataset [2]. The reference model is in the first column. From the second column to the last column, there are the models of Tola et al. [27], Furukawa and Ponce [29], Campbell et al. [24], Jancosek and Pajdla [5] and the proposed method, respectively. The final meshes given by Tola et al. [27], Furukawa and Ponce [29] and Campbell et al. [24] were generated by Poisson surface reconstruction method [30] and trimmed, which were provided in the benchmark MVS dataset [2]; the meshes of Jancosek and Pajdla [5] were generated by OpenMVS, which contains an reimplementation of the method in [5]. Figure 10 shows that the proposed method could reconstruct complex scenes as well as regular scenes, and it had great potential for preserving scene details. Figure 11 shows the accuracy and completeness of reconstructed meshes over twenty scenes, which are scans 1–6, 9–10, 15, 21, 23–24, 29, 36, 44, 61, 110, 114, 118 and 122. The detailed information of the 3D models evaluated on MVS dataset [2] is presented in Table 2. The evaluation method is described in [2], in which accuracy is measured as the distance from the MVS reconstruction to the reference model, and the completeness is measured from the reference model to the MVS reconstruction. In addition to the evaluation of the surface, we also evaluated the point clouds generated by the methods in [24,27,29] as well as OpenMVS, with the point clouds in [24,27,29] provided by MVS dataset [2]. We can see in Figure 11 that generally the point clouds achieved better scores than the surface meshes, which complies with the results in [2]. The underlying reason could be that the mesh representation discarded points that are redundant but close to the true surface, and simultaneously fixed unexpected gaps. Comparing the result within the surface meshes, the proposed method without the dense visibility technique was not outstanding, in terms of both accuracy and completeness. By applying the dense visibility technique, the proposed method achieved the best median accuracy and median completeness, and the second best mean accuracy and mean completeness. In odd rows of Figure 12 are the local point cloud and the surface meshes of Jancosek and Pajdla [5], the proposed method without the dense visibility technique and the proposed method; in the even rows of Figure 12 are the evaluation result (lower is better) of the corresponding local models through the method in [41]. We show the ability of the three methods to preserve thin objects and object edges. In some cases, the method in [5] failed completely in reconstructing them; even the complete ones were less accurate, both visually and quantitatively, than those provided by the proposed method with or without the dense visibility technique. In addition, compared with the method in [5], the proposed method showed a tremendous capability of preserving sharp object edges.

To better visualize the differences between the models generated by Jancosek and Pajdla [5] and the proposed method, we enlarged some parts of the meshes generated by the two methods. The enlarged views are shown in Figure 12. We also tested the proposed method on benchmark Tanks and Temples dataset [40]. Scenes in Tanks and Temples dataset [40] are realistic and contain plenty of objects with different scales in both outdoor conditions and indoor environments. We evaluated the proposed method as well as two other methods on four outdoor scenes (Barn, Ignatius, Truck and Courthouse) of the training set and the eight scenes of the intermediate set of Tanks and Temples dataset [40]. Figure 13 shows the result of the proposed method on four scenes of the training set of Tanks and Temples dataset [40]. Looking at Figure 13 from left to right, we can see the input images, the precision and recall of the model generated by the proposed method and the F-scores of the models generated by Colmap [38], Jancosek and Pajdla [5], and the proposed method with and without the dense visibility technique. The evaluation method is described in [40], in which precision is measured as the distance from the MVS reconstruction to the reference model, the completeness is measured from the reference model to the MVS reconstruction, and the F-score is the harmonic mean of precision and recall with a given threshold. Table 3 presents the evaluation result of the 3D models through the method in [42]. Since the evaluation method in [40,42] takes point clouds as the input, the meshes generated by Jancosek and Pajdla [5] and the proposed method were sampled to acquire point clouds. The point clouds yielded by Colmap [38] are provided in Tanks and Temples dataset [40]. From the evaluation result of the four scenes in Figure 13 and Table 3, we concluded similarly as for the MVS dataset [2] that the proposed method outperformed the method in [5] with the dense visibility technique, while it performd slightly worse than the method in [5] without it. The proposed method performed better than the other three methods in Barn, Truck and Courthouse and rivalled Colmap [38] in Ignatius when evaluated through the method in [40], while it achieved the best result in the four scenes when evaluated through the method in [42]. Figure 14 shows the detailed views of the proposed method and the method of Jancosek and Pajdla [5] on four scenes of the training set. As in Figure 12, we also present the local models and the evaluation result using the method in [41] in odd rows and even rows, respectively. In Figure 14, Jancosek and Pajdla's method [5] inaccurately reconstructed the edge of the roof, wrongly fixed the gap between the arm and the chest of the statue, and failed to reconstruct the rearview mirror of the truck and the light stand of the lamp, while the proposed method performed well in these parts. It is noteworthy that the proposed method also had a great ability of deducing the close form of the real surface when there were no supportive points due to the occlusion, a common phenomenon in 3D reconstruction. A good example is the arm and the chest of the statue in Figure 14. We can see that the proposed method fixed the vacant part of the chest reasonably, even though it was occluded by the arm of the statue, while the method in [5] gave a less satisfactory solution. The underlying reason is that, in the proposed visibility models, the end of a line of sight lies in the tetrahedra right behind the 3D point, which is good for separating small scale objects from other objects. Table 4 and Figure 15 present the result of the proposed method on the intermediate set of Tanks and Temples dataset [40]. The detailed information of the 3D models evaluated on the intermediate set of Tanks and Temples dataset [40] is presented in Table 5. In Table 4, we can see that the result of the proposed method and that in [5] are not outstanding. However, compared to the evaluation result of the point clouds of OpenMVG + OpenMVS, the result of both the proposed method and the method in [5] achieves a substantial boost in all scenes except M60. The underlying reason could be that the two methods simultaneously filtered most noise points and fixed unexpected gaps during the surface reconstruction process. These two behaviors had opposite effects on the evaluation result, since the former one increased the F-score while the latter one decreased it. In Table 4, we can also find that generally with the dense visibility technique the proposed method outperformed the method in [5], while without the dense visibility technique it performed worse than the method in [5]. Figure 15 shows the detailed views of the proposed method and the method of Jancosek and Pajdla [5] on eight scenes of the intermediate set. Table 4 shows the F-scores of several public methods and the proposed method on eight scenes of the intermediate set. In dealing with the thin objects such as the human legs in Francis, the horse ear in Horse, the wire in M60, the barrel in Panther, the bar in Playground and the handle in Train, Jancosek and Pajdla's method [5] either failed to reconstruct it or wrongly fixed the gaps, while the proposed method performed well in these parts. Compared to the meshes of the method in [5], the proposed method also kept distinguishing object edges such as the cloth folds in Family, the horse mouth in Horse and the doorframe in Lighthouse. Therefore, we can infer that the proposed method has a strong ability to preserve the scene details.

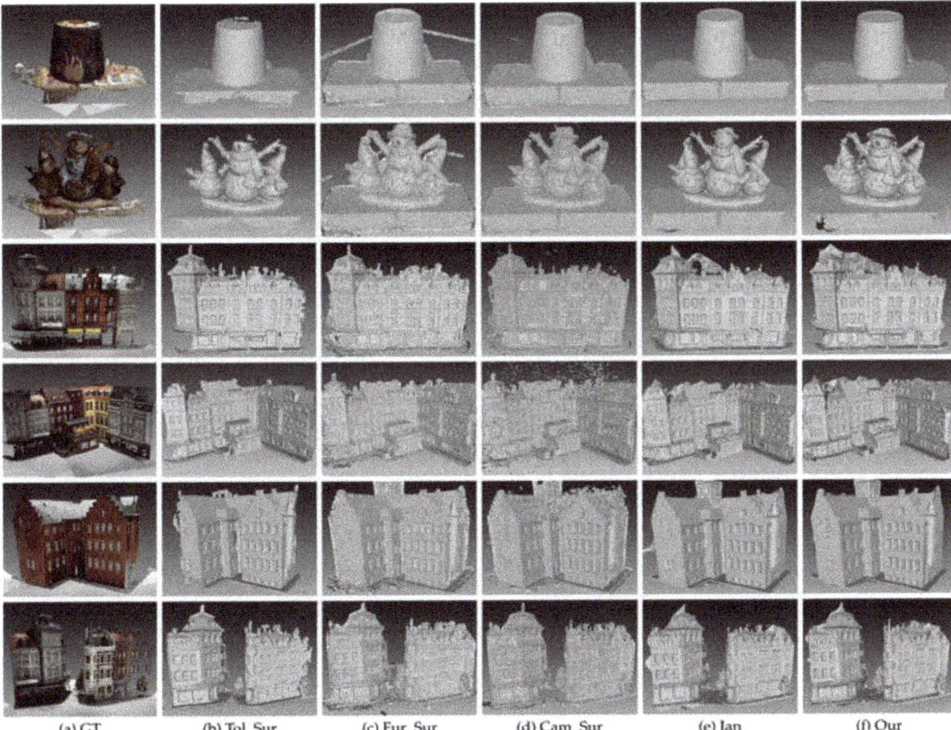

Figure 10. Result of the five methods on MVS dataset [2]. GT is the reference model; Tol is Tola et al. [27]; Fur is Furukawa and Ponce [29]; Cam is Campbell et al. [24]; Jan is Jancosek and Pajdla [5]; Our is the proposed method; and *_Sur is the surface generated by poisson surface reconstruction method [30].

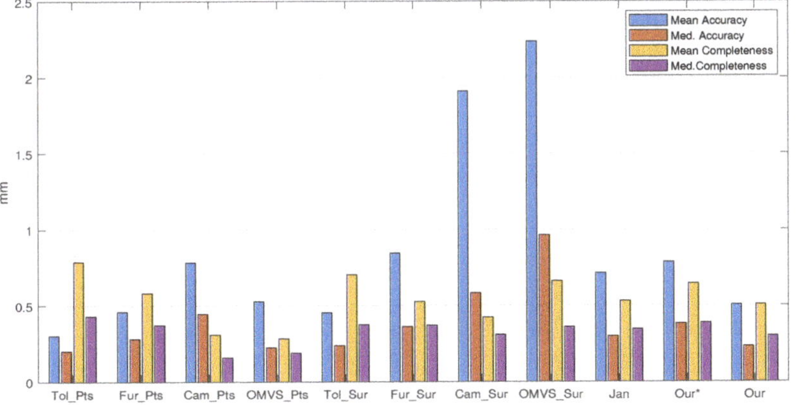

Figure 11. Quantitative evaluation (lower is better) of the methods on MVS dataset [2]. The abbreviations of the methods are given in Figure 10. In addition, OMVS is OpenMVS; Our* is the proposed method without the dense visibility technique; and *_Pts is the point cloud generated by method *.

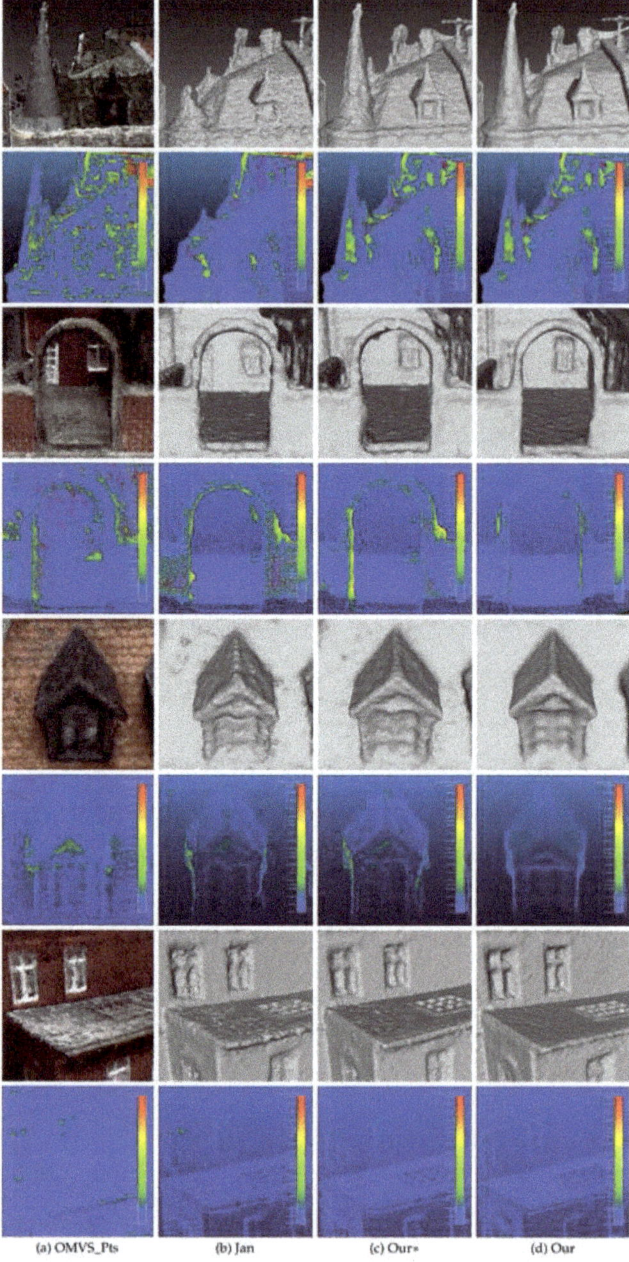

Figure 12. Detailed views of three methods on MVS dataset [2]. From left to right are: (**a**) the point cloud generated by OpenMVS; (**b**) the mesh of Jancosek and Pajdla [5]; (**c**) the mesh of the proposed method without the dense visibility technique; and (**d**) the mesh of the proposed method. In the even rows are the evaluation result (lower is better) of the corresponding local models in the odd rows through the method in [41]. The unit is mm for all numbers.

(a) input   (b) precision   (c) recall   (d) result

Figure 13. Result of the proposed method on four scenes of the training set of Tanks and Temples dataset [40]. From left to right are: (**a**) the input images; (**b**) the precision of the model generated by the proposed method; (**c**) the recall of the model generated by the proposed method; and (**d**) the evaluation result (higher is better) of the models generated by Colmap [38] and three other methods depicted in Figure 10.

Table 2. Information of the 3D models evaluated on MVS dataset [2]. *_Pts is the number of points in a 3D point cloud. *_Vtx and *_Fcs are the number of vertices and facets of a 3D mesh, respectively. The unit of all numbers is million.

SceneID	1	2	3	4	5	6	9	10	15	21	23	24	29	36	44	61	110	114	118	122
Tol_Pts	1.0	1.1	0.9	0.7	0.9	1.0	1.0	0.7	1.0	1.0	1.1	0.8	0.7	1.1	0.9	0.7	0.7	1.2	1.0	0.9
Tol_Vtx	2.1	2.2	2.1	1.8	1.9	2.3	2.6	1.7	2.3	2.1	2.3	2.4	1.1	2.0	1.5	1.6	1.6	2.1	2.1	1.8
Tol_Fcs	4.2	4.4	4.2	3.5	3.7	4.6	5.2	3.3	4.7	4.3	4.5	4.8	2.1	4.1	3.1	3.2	3.2	4.2	4.2	3.5
Fur_Pts	2.3	2.6	2.5	2.2	2.2	2.4	2.4	1.9	2.5	3.0	3.1	2.5	2.3	2.7	2.7	1.6	2.2	2.6	2.6	2.4
Fur_Vtx	1.1	1.1	1.2	0.8	0.8	0.7	1.0	0.7	2.5	2.9	2.8	1.0	2.1	1.9	1.7	0.9	1.8	1.5	1.7	1.4
Fur_Fcs	2.2	2.2	2.4	1.6	1.6	1.5	2.0	1.5	4.9	5.8	5.5	1.9	4.2	3.8	3.4	1.7	3.6	2.9	3.3	2.7
Cam_Pts	23.6	29.6	22.2	20.8	20.2	23.6	19.8	13.0	22.0	24.0	29.5	20.2	16.5	29.5	20.2	7.6	19.9	26.1	30.2	21.7
Cam_Vtx	4.2	4.6	8.1	4.8	6.8	6.7	16.0	2.6	12.0	8.9	4.1	2.6	3.3	3.2	5.1	3.7	6.3	5.1	31.2	6.1
Cam_Fcs	8.5	9.2	16.3	9.5	13.5	13.4	32.0	5.1	24.0	17.8	8.2	5.2	6.6	6.3	10.2	7.3	12.5	10.2	62.4	12.1
OMVS_Pts	11.8	11.0	12.2	10.1	11.8	10.9	9.1	8.3	9.2	10.1	12.2	9.0	7.8	11.3	9.8	8.9	8.0	13.1	8.8	8.4
Jan_Vtx	0.6	0.6	0.7	0.5	0.6	0.6	0.6	0.5	0.6	0.8	0.7	0.6	0.6	0.7	0.7	0.5	0.5	0.7	0.6	0.6
Jan_Fcs	1.3	1.2	1.4	1.1	1.2	1.2	1.2	1.0	1.3	1.6	1.4	1.2	1.2	1.5	1.5	0.9	1.0	1.3	1.2	1.1
Our*_Vtx	1.2	1.1	1.1	1.0	0.9	1.0	1.0	0.9	1.1	1.3	1.3	1.2	1.0	1.3	1.1	0.7	1.0	1.1	0.9	1.0
Our*_Fcs	2.5	2.3	2.2	2.0	1.8	2.0	2.0	1.9	2.2	2.6	2.6	2.4	2.0	2.6	2.3	1.5	1.9	2.2	1.8	1.9
Our_Vtx	1.6	1.6	1.4	1.0	1.2	1.4	1.3	1.2	1.5	1.5	1.7	1.3	1.3	1.3	1.2	0.7	1.0	1.5	1.2	1.1
Our_Fcs	3.3	3.2	2.9	2.0	2.5	2.9	2.6	2.4	3.0	3.1	3.4	2.6	2.7	2.6	2.5	1.4	2.0	3.1	2.4	2.3

Table 3. Evaluation result of the 3D models evaluated on the training set of Tanks and Temples dataset [40] through the method in [42]. M stands for million. Precision is expressed as a proportion 1:k, where k is the size of the scene divided by the standard error. The unit of all other numbers is mm.

Type	Barn					Ignatius				
	Colmap	OMVS_Pts	Jan	Our*	Our	Colmap	OMVS_Pts	Jan	Our*	Our
Pts	6.2M	35.4M	6.3M	5.8M	6.7M	1.3M	13.1M	3.5M	2.9M	3.3M
mean	19.24	17.70	10.44	11.14	10.23	2.66	3.55	2.51	2.90	2.14
95.5%<	59.93	34.75	28.17	29.42	26.04	7.96	12.15	6.33	7.69	5.15
99.7%<	221.83	181.46	118.49	120.25	113.45	39.18	35.91	38.55	38.49	34.38
Precision	1:800	1:1000	1:1400	1:1400	1:1500	1:500	1:600	1:800	1:800	1:900

Table 3. *Cont.*

Type	Courthouse					Truck				
	Colmap	OMVS_Pts	Jan	Our*	Our	Colmap	OMVS_Pts	Jan	Our*	Our
Pts	17.3M	63.4M	14.4M	13.7M	15.0M	3.8M	22.5M	3.7M	3.0M	3.5M
mean	97.56	247.94	93.15	96.84	91.88	8.07	7.29	6.65	7.13	6.47
95.5%<	315.51	597.98	302.36	311.29	294.31	24.67	23.62	21.48	23.87	21.46
99.7%<	2261.96	5108.95	2086.73	2117.38	2029.45	154.56	174.76	147.83	151.72	143.58
Precision	1:300	1:200	1:400	1:400	1:400	1:400	1:400	1:600	1:600	1:700

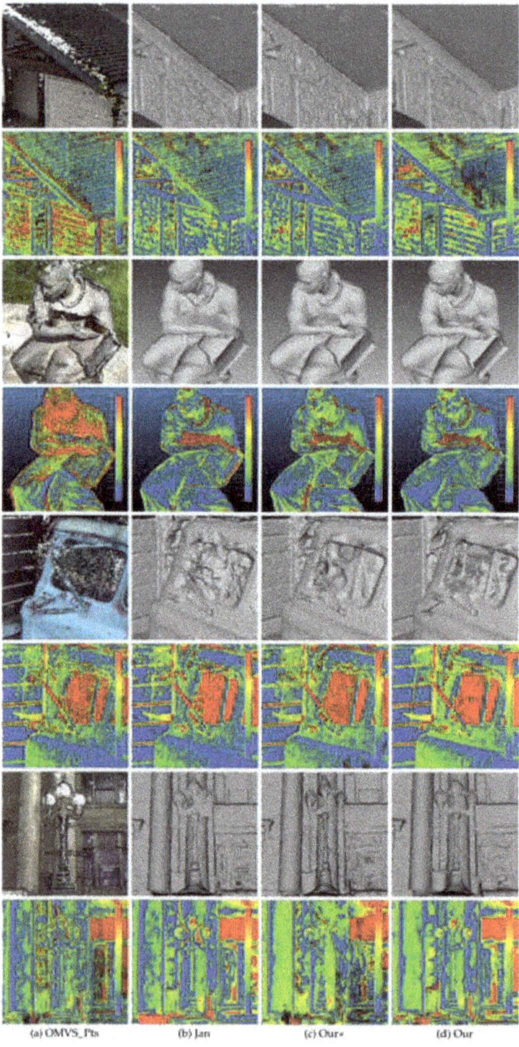

Figure 14. Detailed views of three methods on four scenes of the training set of Tanks and Temples dataset [40]. From left to right are: (**a**) the point cloud generated by OpenMVS; (**b**) the mesh of Jancosek and Pajdla [5]; (**c**) the mesh of the proposed method without the dense visibility technique; and (**d**) the mesh of the proposed method. In the even rows are the evaluation result (lower is better) of the corresponding local models in the odd rows through the method in [41]. The unit is m for all numbers.

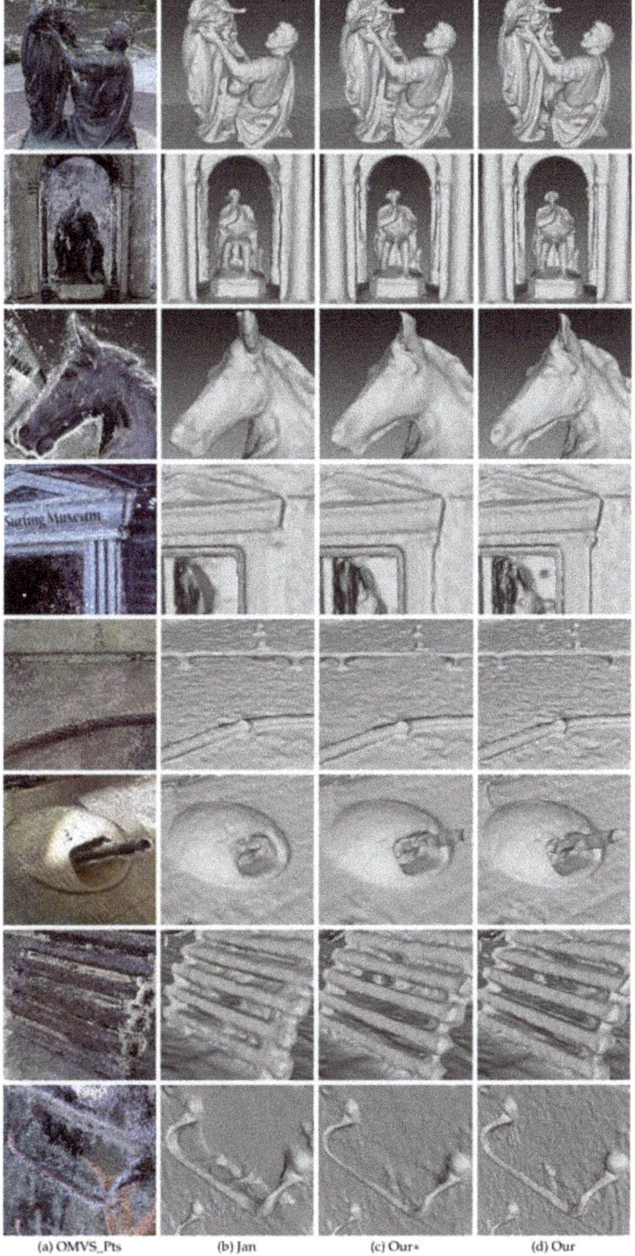

Figure 15. Detailed views of three methods on the intermediate set of Tanks and Temples dataset [40]. From top to bottom are Family, Francis, Horse, Lighthouse, M60, Panther, Playground and Train. From left to right are: (**a**) the point cloud generated by OpenMVS; (**b**) the mesh of Jancosek and Pajdla [5]; (**c**) the mesh of the proposed method without the dense visibility technique; and (**d**) the mesh of the proposed method.

Table 4. Leaderboard [1] of the methods and the result of the proposed method with respect to F-score on the intermediate set of Tanks and Temples dataset [40].

Method	Family	Francis	Horse	Lighthouse	M60	Panther	Playground	Train	Mean
PMVSNet	70.04	44.64	40.22	**65.20**	**55.08**	**55.17**	60.37	**54.29**	55.62
Altizure-HKUST	**74.60**	**61.30**	38.48	61.48	54.93	53.32	56.21	49.47	**56.22**
ACMH	69.99	49.45	**45.12**	59.04	52.64	52.37	58.34	51.61	54.82
Dense R-MVSNet	73.01	54.46	43.42	43.88	46.80	46.69	50.87	45.25	50.55
R-MVSNet	69.96	46.65	32.59	42.95	51.88	48.80	52.00	42.38	48.40
i23dMVS4	56.64	33.75	28.40	48.42	39.23	44.87	48.34	37.88	42.19
MVSNet	55.99	28.55	25.07	50.79	53.96	50.86	47.90	34.69	43.48
COLMAP	50.41	22.25	25.63	56.43	44.83	46.97	48.53	42.04	42.14
Pix4D	64.45	31.91	26.43	54.41	50.58	35.37	47.78	34.96	43.24
i23dMVS_3	56.21	33.14	28.92	47.74	40.29	44.20	46.93	37.66	41.89
OpenMVG + OpenMVS	58.86	32.59	26.25	43.12	44.73	46.85	45.97	35.27	41.71
OpenMVG + MVE	49.91	28.19	20.75	43.35	44.51	44.76	36.58	35.95	38.00
OpenMVG + SMVS	31.93	19.92	15.02	39.38	36.51	41.61	35.89	25.12	30.67
Theia-I + OpenMVS	48.11	19.38	20.66	30.02	30.37	30.79	23.65	20.46	27.93
OpenMVG + PMVS	41.03	17.70	12.83	36.68	35.93	33.20	31.78	28.10	29.66
Jan	62.69	47.44	34.52	57.94	38.67	47.06	55.26	39.90	47.94
Our*	62.46	46.68	32.61	57.66	33.66	44.25	52.40	38.25	46.00
Our	65.21	49.41	35.41	59.04	37.57	47.85	56.77	41.28	49.07

[1] https://www.tanksandtemples.org/leaderboard/.

Table 5. Information of the 3D models evaluated on the intermediate set of Tanks and Temples dataset [40]. *_Pts is the number of points in a 3D point cloud. *_Vtx and *_Fcs are the number of vertices and facets of a 3D mesh, respectively. The unit of all numbers is million.

Scene	Family	Francis	Horse	Lighthouse	M60	Panther	Playground	Train
OMVS_Pts	12.0	17.8	9.0	28.4	24.8	26.0	28.6	31.9
Jan_Vtx	2.0	1.9	1.5	2.8	5.1	4.1	5.7	4.8
Jan_Fcs	4.1	3.7	3.0	5.6	10.1	8.3	11.3	9.6
Our*_Vtx	1.6	1.2	1.2	1.7	3.6	2.9	4.1	3.2
Our*_Fcs	3.2	2.4	2.3	3.5	7.2	5.8	8.2	6.5
Our_Vtx	2.5	2.7	2.1	3.3	5.3	5.4	6.9	6.3
Our_Fcs	5.1	5.4	4.2	6.6	10.6	10.8	13.9	12.5

8. Conclusions

In this paper, we present a new surface reconstruction method. The proposed method is designed to preserve scene details while keeping the ability to filter noise. To make the proposed method efficient to filter out noise and to select true surface points, we introduce a new visibility model with error tolerance and adaptive end weights. Along with the proposed visibility model, a new likelihood energy term is added to the total energy of the binary labeling problem to promote the robustness of the proposed method to noise. Moreover, we further improve the performance of the proposed method with the dense visibility technique, which avoids the error introduced in the point cloud generation process and provide denser visibility information. We tested the proposed method on two publicly available benchmark datasets. Experimental results on different datasets show that the proposed method rivalled the state-of-the-art methods in terms of accuracy and completeness, while it could preserve scene details such as thin objects and sharp edges. Our future work will consist in segmenting the model and adding semantic knowledge to each part of the model.

Author Contributions: Conceptualization, Y.Z.; Data curation, Y.Z. and S.S.; Formal analysis, Y.Z.; Funding acquisition, S.S. and Z.H.; Investigation, Y.Z.; Methodology, Y.Z. and S.S.; Project administration, S.S. and Z.H.; Resources, Y.Z. and S.S.; Software, Y.Z. and S.S.; Supervision, S.S. and Z.H.; Validation, Y.Z.; Visualization, Y.Z.; Writing—original draft, Y.Z.; and Writing—review and editing, Y.Z., S.S. and Z.H.

Funding: This research was funded by National Natural Science Foundation of China grant numbers 61632003, 61873265 and 61573351.

Conflicts of Interest: The authors declare no conflict of interest.

References

1. Seitz, S.M.; Curless, B.; Diebel, J.; Scharstein, D.; Szeliski, R. A comparison and evaluation of multi-view stereo reconstruction algorithms. In Proceedings of the IEEE Computer Society Conference on Computer Vision and Pattern Recognition IEEE, New York, NY, USA, 17–22 June 2006; pp. 519–528.
2. Jensen, R.; Dahl, A.; Vogiatzis, G.; Tola, E.; Aanæs, H. Large scale multi-view stereopsis evaluation. In Proceedings of the IEEE Conference on Computer Vision and Pattern Recognition IEEE, Columbus, OH, USA, 23–28 June 2014; pp. 406–413.
3. Strecha, C.; von Hansen, W.; Van Gool, L.; Fua, P.; Thoennessen, U. On benchmarking camera calibration and multi-view stereo for high resolution imagery. In Proceedings of the IEEE Conference on Computer Vision and Pattern Recognition IEEE, Anchorage, AL, USA, 23–28 June 2008; pp. 1–8.
4. Sinha, S.N.; Mordohai, P.; Pollefeys, M. Multi-view stereo via graph cuts on the dual of an adaptive tetrahedral mesh. In Proceedings of the 11th IEEE International Conference on Computer Vision, Rio De Janeiro, Brazil, 14–21 October 2007; pp. 1–8.
5. Jancosek, M.; Pajdla, T. Exploiting visibility information in surface reconstruction to preserve weakly supported surfaces. *Int. Schol. Res. Not.* **2014**, *2014*, 798595. [CrossRef] [PubMed]
6. Häne, C.; Zach, C.; Cohen, A.; Pollefeys, M. Dense semantic 3d reconstruction. *IEEE Trans. Pattern Anal. Mach. Intell.* **2017**, *39*, 1730–1743. [CrossRef] [PubMed]
7. Vu, H.H.; Labatut, P.; Pons, J.P.; Keriven, R. High accuracy and visibility-consistent dense multiview stereo. *IEEE Trans. Pattern Anal. Mach. Intell.* **2012**, *34*, 889–901. [CrossRef] [PubMed]
8. Zhou, Y.; Shen, S.; Hu, Z. A New Visibility Model for Surface Reconstruction. In Proceedings of the CCF Chinese Conference on Computer Vision, Tianjin, China, 11–14 October 2017; pp. 145–156.
9. Vogiatzis, G.; Torr, P.H.; Cipolla, R. Multi-view stereo via volumetric graph-cuts. In Proceedings of the IEEE Computer Society Conference on Computer Vision and Pattern Recognition, IEEE, San Diego, CA, USA, 20–25 June 2005; pp. 391–398.
10. Tran, S.; Davis, L. 3D surface reconstruction using graph cuts with surface constraints. In Proceedings of the European Conference on Computer Vision, Graz, Austria, 7–13 May 2006; pp. 219–231.
11. Lempitsky, V.; Boykov, Y.; Ivanov, D. Oriented visibility for multiview reconstruction. In Proceedings of the European Conference on Computer Vision, Graz, Austria, 7–13 May 2006; pp. 226–238.
12. Labatut, P.; Pons, J.P.; Keriven, R. Efficient multi-view reconstruction of large-scale scenes using interest points, delaunay triangulation and graph cuts. In Proceedings of the 11th IEEE international Conference on Computer Vision, Rio De Janeiro, Brazil, 14–21 October 2007; pp. 1–8.
13. Labatut, P.; Pons, J.P.; Keriven, R. Robust and efficient surface reconstruction from range data. *Comp. Graph Forum* **2009**, *28*, 2275–2290. [CrossRef]
14. Laurentini, A. The visual hull concept for silhouette-based image understanding. *IEEE Trans. Pattern Anal. Mach. Intell.* **1994**, *16*, 150–162. [CrossRef]
15. Esteban, C.H.; Schmitt, F. Silhouette and stereo fusion for 3D object modeling. *Comp. Vis. Image Underst.* **2004**, *96*, 367–392. [CrossRef]
16. Starck, J.; Miller, G.; Hilton, A. Volumetric stereo with silhouette and feature constraints. In Proceedings of the British Machine Vision Conference, Edinburgh, UK, 4–7 September 2006; pp. 1189–1198.
17. Franco, J.S.; Boyer, E. Fusion of multiview silhouette cues using a space occupancy grid. In Proceedings of the 10th IEEE International Conference on Computer Vision, San Diego, CA, USA, 20–25 June 2005; pp. 1747–1753.
18. Guan, L.; Franco, J.S.; Pollefeys, M. 3d occlusion inference from silhouette cues. In Proceedings of the IEEE Conference on Computer Vision and Pattern Recognition, IEEE, Minneapolis, MN, USA, 17–22 June 2007; pp. 1–8.
19. Kutulakos, K.N.; Seitz, S.M. A theory of shape by space carving. *Int. J. Comp. Vis.* **2000**, *38*, 199–218. [CrossRef]
20. Broadhurst, A.; Drummond, T.W.; Cipolla, R. A probabilistic framework for space carving. In Proceedings of the 8th IEEE International Conference on Computer Vision, Vancouver, BC, Canada, 7–14 July 2001; pp. 388–393.

21. Yang, R.; Pollefeys, M.; Welch, G. Dealing with Textureless Regions and Specular Highlights-A Progressive Space Carving Scheme Using a Novel Photo-consistency Measure. In Proceedings of the 9th IEEE International Conference on Computer Vision, IEEE, Nice, France, 13–16 October 2003; pp. 576–584.
22. Jin, H.; Soatto, S.; Yezzi, A.J. Multi-view stereo reconstruction of dense shape and complex appearance. *Int. J. Comp. Vis.* **2005**, *63*, 175–189. [CrossRef]
23. Pons, J.P.; Keriven, R.; Faugeras, O. Multi-view stereo reconstruction and scene flow estimation with a global image-based matching score. *Int. J. Comp. Vis.* **2007**, *72*, 179–193. [CrossRef]
24. Campbell, N.D.; Vogiatzis, G.; Hernández, C.; Cipolla, R. Using multiple hypotheses to improve depth-maps for multi-view stereo. In Proceedings of the European Conference on Computer Vision, Marseille, France, 12–18 October 2008; pp. 766–779.
25. Goesele, M.; Curless, B.; Seitz, S.M. Multi-view stereo revisited. In Proceedings of the IEEE Computer Society Conference on Computer Vision and Pattern Recognition, New York, NY, USA, 17–22 June 2006; pp. 2402–2409.
26. Hiep, V.H.; Keriven, R.; Labatut, P.; Pons, J.P. Towards high-resolution large-scale multi-view stereo. In Proceedings of the IEEE Conference on Computer Vision and Pattern Recognition, Miami, FL, USA, 20–25 June 2009; pp. 1430–1437.
27. Tola, E.; Strecha, C.; Fua, P. Efficient large-scale multi-view stereo for ultra high-resolution image sets. *Mach. Vis. Appl.* **2012**, *23*, 903–920. [CrossRef]
28. Shen, S. Accurate multiple view 3d reconstruction using patch-based stereo for large-scale scenes. *IEEE Trans. Image Process.* **2013**, *22*, 1901–1914. [CrossRef] [PubMed]
29. Furukawa, Y.; Ponce, J. Accurate, dense, and robust multiview stereopsis. *IEEE Trans. Pattern Anal. Mach. Intell.* **2010**, *32*, 1362–1376. [CrossRef] [PubMed]
30. Kazhdan, M.; Hoppe, H. Screened poisson surface reconstruction. *ACM Trans. Graph.* **2013**, *32*, 29. [CrossRef]
31. Bódis-Szomorú, A.; Riemenschneider, H.; Van Gool, L. Superpixel meshes for fast edge-preserving surface reconstruction. In Proceedings of the IEEE Conference on Computer Vision and Pattern Recognition, Boston, MA, USA, 7–12 June 2015; pp. 2011–2020.
32. Jancosek, M.; Pajdla, T. Multi-view reconstruction preserving weakly-supported surfaces. In Proceedings of the IEEE Conference on Computer Vision and Pattern Recognition, Colorado Springs, CO, USA, 20–25 June 2011; pp. 3121–3128.
33. Hane, C.; Zach, C.; Cohen, A.; Angst, R.; Pollefeys, M. Joint 3D scene reconstruction and class segmentation. In Proceedings of the IEEE Conference on Computer Vision and Pattern Recognition, Portland, OR, USA, 23–28 June 201; pp. 97–104.
34. Blaha, M.; Vogel, C.; Richard, A.; Wegner, J.D.; Pock, T.; Schindler, K. Large-scale semantic 3d reconstruction: an adaptive multi-resolution model for multi-class volumetric labeling. In Proceedings of the IEEE Conference on Computer Vision and Pattern Recognition, IEEE, Las Vegas, NE, USA, 26 June–1 July 2016; pp. 3176–3184.
35. Cherabier, I.; Hane, C.; Oswald, M.R.; Pollefeys, M. Multi-label semantic 3d reconstruction using voxel blocks. In Proceedings of the 4th International Conference on 3D Vision, Stanford University, CA, USA, 25–28 October 2016; pp. 601–610.
36. Savinov, N.; Hane, C.; Ladicky, L.; Pollefeys, M. Semantic 3d reconstruction with continuous regularization and ray potentials using a visibility consistency constraint. In Proceedings of the IEEE Conference on Computer Vision and Pattern Recognition, Las Vegas, NE, USA, 27–30 June 2016; pp. 5460–5469.
37. Ummenhofer, B.; Brox, T. Global, dense multiscale reconstruction for a billion points. In Proceedings of the IEEE International Conference on Computer Vision, IEEE, Washington, DC, USA, 7–13 December 2015; pp. 1341–1349.
38. Schönberger, J.L.; Frahm, J.M. Structure-from-motion revisited. In Proceedings of the IEEE Conference on Computer Vision and Pattern Recognition, Las Vegas, NE, USA, 27–30 June 2016; pp. 4104–4113.
39. Goldberg, A.V.; Hed, S.; Kaplan, H.; Tarjan, R.E.; Werneck, R.F. Maximum flows by incremental breadth-first search. In Proceedings of the European Symposium on Algorithms, Saarbrücken, Germany, 5–9 September 2011; pp. 457–468.
40. Knapitsch, A.; Park, J.; Zhou, Q.Y.; Koltun, V. Tanks and Temples: Benchmarking Large-Scale Scene Reconstruction. *ACM Trans. Graph.* **2017**, *36*, 78. [CrossRef]

Sensors **2019**, *19*, 1278

41. Jurjević, L.; Gašparović, M. 3D Data Acquisition Based on OpenCV for Close-range Photogrammetry Applications. In Proceedings of the ISPRS Hannover Workshop: HRIGI 17–CMRT 17–ISA 17–EuroCOW 17, Hannover, Germany, 6–9 June 2017.
42. Sapirstein, P. Accurate measurement with photogrammetry at large sites. *J. Archaeol. Sci.* **2016**, *66*, 137–145. [CrossRef]

Article

Efficient Fiducial Point Detection of ECG QRS Complex Based on Polygonal Approximation

Seungmin Lee, Yoosoo Jeong, Daejin Park *, Byoung-Ju Yun * and Kil Houm Park *

School of Electronics Engineering, Kyungpook National University, Daegu 41566, Korea;
lsm1106@knu.ac.kr (S.L.); ysjung@ee.knu.ac.kr (Y.J.)
* Correspondence: boltanut@knu.ac.kr (D.P.); bjisyun@ee.knu.ac.kr (B.-J.Y.); khpark@ee.knu.ac.kr (K.H.P.);
Tel.: +82-53-950-5548 (D.P.)

Received: 20 November 2018; Accepted: 13 December 2018; Published: 19 December 2018

Abstract: Electrocardiogram signal analysis is based on detecting a fiducial point consisting of the onset, offset, and peak of each waveform. The accurate diagnosis of arrhythmias depends on the accuracy of fiducial point detection. Detecting the onset and offset fiducial points is ambiguous because the feature values are similar to those of the surrounding sample. To improve the accuracy of this paper's fiducial point detection, the signal is represented by a small number of vertices through a curvature-based vertex selection technique using polygonal approximation. The proposed method minimizes the number of candidate samples for fiducial point detection and emphasizes these sample's feature values to enable reliable detection. It is also sensitive to the morphological changes of various QRS complexes by generating an accumulated signal of the amplitude change rate between vertices as an auxiliary signal. To verify the superiority of the proposed algorithm, error distribution is measured through comparison with the QT-DB annotation provided by Physionet. The mean and standard deviation of the onset and the offset were stable as -4.02 ± 7.99 ms and -5.45 ± 8.04 ms, respectively. The results show that proposed method using small number of vertices is acceptable in practical applications. We also confirmed that the proposed method is effective through the clustering of the QRS complex. Experiments on the arrhythmia data of MIT-BIH ADB confirmed reliable fiducial point detection results for various types of QRS complexes.

Keywords: electrocardiogram; QRS complex; fiducial point; polygonal approximation; dynamic programming; QT-database; MIT-BIH arrhythmia database

---

1. Introduction

Electrocardiogram (ECG) signals are electronically converted signals from the depolarization and repolarization of the atria and ventricle [1]. Generally, signals are composed of a P-wave, QRS complex, and T-wave, which occurred in the depolarization of the atrium and ventricle, and the repolarization of the ventricle, respectively [2,3]. Signals are determined by cardiac activity, so the signal instantly shows an arrhythmia. Signal analysis to diagnose arrhythmia has been widely used to recognize the deformation of the signal and analyze the feature value when arrhythmia occurs [4], and it is used for monitoring such as mental stress [5] and fear [6]. The study of signal analysis is subdivided according to various techniques and their purposes. In general, the signal analysis system can be divided into the noise removal step [7], fiducial point detection step [8–10], and feature value acquisition and arrhythmia classification step [11–13]. Various applications related to signal analysis include a signal monitoring system [14], heart rate measurement, signal compression [15,16], personal authentication [17,18], and so on.

The fiducial point detection step identifies the most important fiducial points of the ECG signal, which are represented by the onset, the peak, and the offset of the P-wave, the QRS complex, and the T-wave [8,19–23]. With accurate detection of these fiducial points, the width and the interval information of ECG waveforms used as feature values can be accurately measured. Therefore, the detection of accurate fiducial points is an important research field that greatly affects all subsequent ECG signal analysis.

The R-peak, which is the peak of the QRS complex, is the easiest to detect because it has the largest amplitude value. It is used not only for measuring heart rate, but also for detecting other fiducial points. Typical R-peak detection methods are based on the Pan's method, which is approximately 99% accurate. However, accurate detection methods of the fiducial points of the QRS complex other than the R-peak have been not clearly determined. Their low detection rate and inaccuracy are problematic due to signal deformation caused by various arrhythmias. These difficulties stem from the ambiguity of the reference regions that serve as a starting point.

Among various signal compression techniques, a method using polygonal approximation compresses a signal through a small number of vertices. In this case, since the onset and offset of the waveform represent the boundary between the waveform region and the baseline region, they are well-preserved as vertices.

Figure 1 illustrates the difference between the existing fiducial point detection method and the polygonal approximation method.

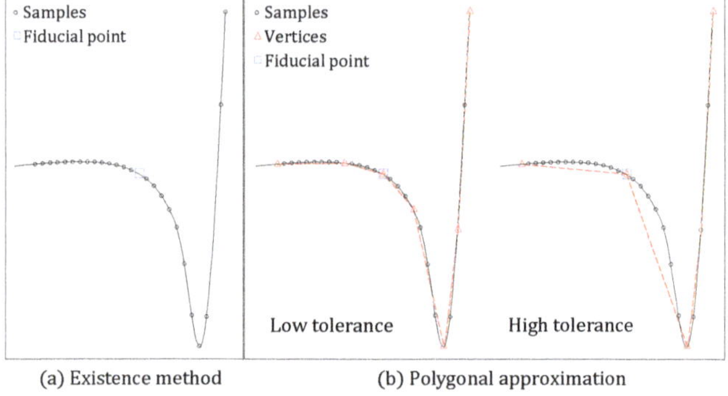

Figure 1. Motivation of the proposed work.

As shown in Figure 1a, since the existing technique uses all samples, the position of the fiducial point is ambiguous due to the samples having similar feature values nearby. Therefore, there is a high possibility of error in the threshold value, and the detection result is unreliable.

On the other hand, since the polygonal approximation uses the atomic vertices, there is a large difference in the feature values of the vertices. As shown in Figure 1b, even if the tolerance used in the approximation and the number of vertices changes, the fiducial point can be represented as a vertex of the same or similar position, so that stable detection thereof is possible.

In this paper, we propose a curvature-based vertex selection technique to solve the ambiguity of the fiducial points in the QRS complex. Our approach is roughly divided into three stages. The first stage consists of initial vertex selection using curvature-based polygonal approximation. Since the curvature value of the fiducial point is large, most of the fiducial points are represented as vertices. The second stage is an incremental vertex selection using repetitive sequential polygonal approximation, and the third stage performs an additional vertex optimization step using dynamic programming. These steps are applied for a missing case due to ambiguous curvature value.

After applying the polygonal approximation, unwanted variations of the QRS complex—such as the presence of Q-waves, S-waves, and the polarity of the waveform—remain a major problem. To mitigate the side effects of this problem, we have prepared an auxiliary signal by accumulating data from the polygonal approximation signal. This cumulative signal monotonically increases by accumulating the absolute value of the rate of change of the amplitude between the vertices of the polygonal signal. Thus, by effectively expressing and emphasizing the feature of the fiducial point, we could effectively represent changes in the QRS complex's shape and polarity. From the cumulative signal, we analyze feature values for each vertex, such as the amplitude difference between R-peak and vertex, the time difference between reference point and vertex and the angles with neighbor vertices. Then, we determine the vertex with the largest sum of these feature values as a fiducial point.

This paper is organized as follows. Section 2 briefly reviews the ECG signal composition and explains why detecting the fiducial point is relatively difficult. Section 3 introduces the curvature-based vertex selection technique proposed in this paper and shows the expected benefits when applied to ECG signals. Section 4 details our algorithm for generating the cumulative signal from the polygonal approximated signal and detecting the fiducial point therefrom. In Section 5, the performance of the proposed algorithm is verified through experiments on QT-DB [24] and MIT-BIH ADB [25], and Section 6 concludes the paper.

2. Composition of ECG Signal

The ECG signal, which consists of the P-wave, QRS complex, and T-wave, includes the corresponding onset, offset, and peak points, which are referred as fiducial points. Figure 2 shows the fiducial points and feature values of the P-wave, QRS complex, and T-wave of the ECG signal.

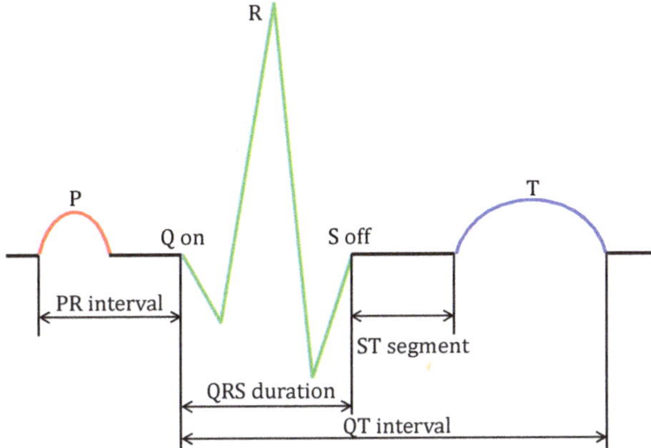

Figure 2. Composition of ECG signal.

As shown in Figure 2, the ECG signal is divided into a waveform region in which the amplitude is changed by depolarization and repolarization, and a baseline region in which no amplitude change occurs. The boundary point is a fiducial point of each waveform. However, the actual input signal contains various noise, not ideal forms, as shown in Figure 2. Typical types of noise are as follows [26–29].

1. Power line interference: various high frequency noise according to country.
2. Baseline wander: a low-frequency noise (0.15 up to 0.3 Hz). This noise results from the patient breathing and leads to a baseline shift in the signals.
3. Electrode contract noise, electrode motion artifacts, muscle contractions, electrosurgical noise, instrumentation noise, and so on.

1 and 2 are typical high- and low-frequency ECG signal noises, respectively. Since noise complicates the baseline, it is difficult to estimate baseline and thus detect the boundary with an ambiguous waveform. Most of study use the bandpass filter for suppress noises and in some case,

it uses a notch filter [30] for aiming to suppress the power line interference, such as high frequency of 50 Hz or 60 Hz.

Figure 3 shows the result of applying a high-pass and low-pass filter to suppress baseline deviation and power line interference.

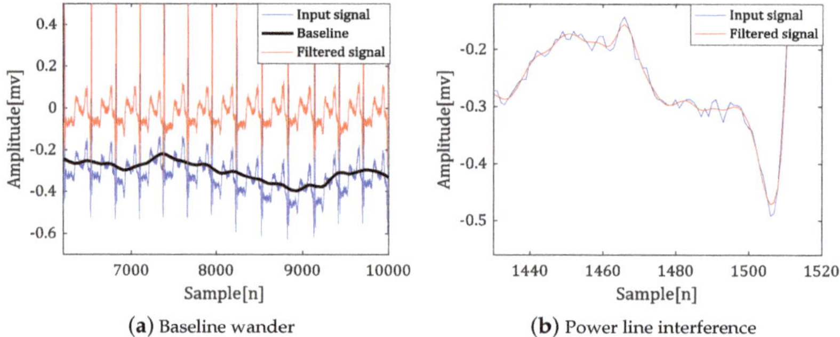

(**a**) Baseline wander

(**b**) Power line interference

Figure 3. Noises of ECG signal and filtering results.

Peaks, such as the R-peak, can be easily detected because the amplitude has a local maximum or minimum and the rate of amplitude change is large enough. In contrast, the boundary between the baseline and the waveform is still ambiguous, even when filtering is applied. This is because the amplitude change occurs slowly, and the features are similar to the surrounding samples. In this paper, we propose an effective fiducial point detection technique by emphasizing ambiguous fiducial points based on polygonal approximation.

3. Polygonal Approximation of ECG Signal

Signal approximation techniques have been widely studied by using polygonal approximation, such as sequential polygonal and cyclic polygonal approximation, and polynomial approximation, such as B-spline. However, these techniques are problematic, since they select too many vertices and errors are not minimized.

In the curvature-based vertex selection technique [31], curvature-based polygonal approximation [32], sequential polygonal approximation [33], and dynamic programming [34] are proposed as methods for minimizing vertices and resultant errors. This algorithm selects an initial vertex using a curvature-based polygonal approximation. A fiducial point with a large curvature is selected as a vertex. However, there is a problem when a fiducial point having an ambiguous curvature value is not selected as an initial vertex. To solve this problem, the sequential polygonal approximation method is adopted to select additional vertices, and the dynamic programming technique can optimize the position of the additional vertices by minimizing errors.

The algorithm flow of the curvature-based vertex selection method for the input ECG signal (S) is summarized as follows.

1. Separate the R-R section of the input signal. In this paper, we detect the R-peak by Pan's method.
2. After calculating the curvature for the separated R-R section, the curvature-based polygonal approximation technique is applied to select the initial vertices. Equation (1) represents the set of initial vertices.

$$V^I = \{v_1^I, v_2^I, \cdots, v_S^I\} \tag{1}$$

3. We apply the sequential polygonal approximation method to the interval between each initial vertex to select additional vertices. Equation (2) represents a set of $N_{V_i} - 1$ additional vertices

between the i-th initial vertex and the $i + 1$-th initial vertex, and both end vertices coincide with the two initial vertices.

$$V_i = \{v_{i,0}, \cdots, v_{i,N_{V_i}}\}$$

$$v_{i,0} = v_i^I, \qquad v_{i,N_{V_i}} = v_{i+1}^I$$

(2)

4. Dynamic programming is applied to the additional vertices to optimize their position. Equation (3) is a set of corrected vertices for the additional vertex set V_i.

$$V_i^{Opt} = \{v_{i,0}^{Opt}, \cdots, v_{i,N_{V_i}}^{Opt}\}$$

$$v_{i,0}^{Opt} = v_{i,0} = v_i^I, \qquad v_{i,N_{V_i}}^{Opt} = v_{i,N_{V_i}} = v_{i+1}^I$$

(3)

5. Repeat steps 2–4 to proceed with polygonal approximation for the entire input signal. Equation (4) represents the set of N_V vertices as the result of vertex selection.

$$V = \{v_1, \cdots, v_{N_V}\}, v_i = (v_{x_i}, v_{y_i})$$

(4)

Figure 4 shows the result of each step of the polygonal approximation.

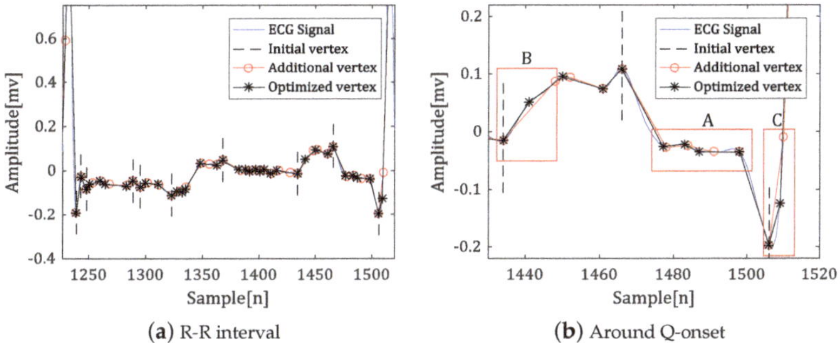

Figure 4. Additional vertex calibration results using dynamic programming.

In general, when the curvature-based polygonal approximation is applied to a pole with a large curvature value, it appears as an initial vertex, as shown in Figure 4a. However, with a smooth transition of the amplitude value near the fiducial point, such as the onset of the QRS complex and the offset of the P-wave in Figure 4b, there is a side effect wherein the fiducial point is not selected as the initial vertex. By applying additional vertex selection and correction to solve this problem, the fiducial points are efficiently represented by the vertex as shown in region A, and the similarity between the original signal and the approximated signal is preserved, as shown in B and C.

4. Fiducial Point Detection Based on Polygonal Approximation

The curvature-based vertex selection technique represents the ECG signal as a small number of vertices, and then detects the onset and offset of the QRS complex by analyzing the characteristic values of each vertex. However, it is not easy to express the characteristic value from the fiducial point because the QRS complex has various shapes based on its polarity, as well as the presence of the Q- and S-peaks.

To resolve the difficulty of extracting features from the QRS complex's ambiguous shape, robust fiducial point detection using various auxiliary signals has been proposed. In Pan's method, the R-peak detection is assisted by an auxiliary signal generated by the derivative of the signal and an average filter. In Manriquez's method, the fiducial point is detected by a threshold value of the auxiliary signal generated from the Hilbert transform. This paper is also based on the auxiliary signal, for which we propose the cumulative signal of the polygonal approximation to preserve morphological features of the vertex that represent the fiducial point.

4.1. Generate the Cumulative Signal

To acquire the cumulative signal, we first obtain the amplitude difference (V^D) for the vertex, as shown in Equation (4).

$$V^D = \{v_1^D, \cdots, v_{N_V}^D\}, \quad v_i^D = (v_{x_i}^D, v_{x_i}^D)$$
$$v_{x_i}^D = v_{x_i}, \quad v_{y_i}^D = v_{y_i} - v_{y_{i-1}}, \quad v_{y_1}^D = 0 \tag{5}$$

With the absolute value of the amplitude difference obtained using Equation (5), that value is accumulated as shown in Equation (6) to generate the cumulative signal.

$$V^{D'} = \{v_1^{D'}, \cdots, v_{N_V}^{D'}\}, \quad v_i^{D'} = (v_{x_i}^{D'}, v_{x_i}^{D'})$$
$$v_{x_i}^{D'} = v_{x_i}^D, \quad v_{y_i}^{D'} = \textstyle\sum_{k=1}^{i} |v_{y_k}^D| \tag{6}$$

This simplifies the signal as monotonically increased, even if the QRS complex appears as a downward wave or includes Q-peaks and S-peaks that appear as downward or upward waves. The vertex corresponding to the fiducial point also maintains the feature of dividing the baseline and waveform regions.

Figure 5 shows the cumulative signal results for various shapes of the polygonal approximation signals.

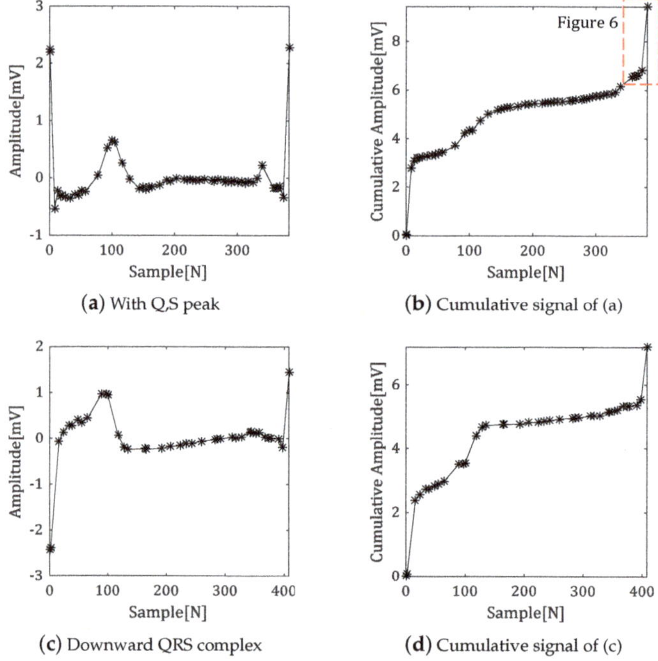

(a) With Q,S peak

(b) Cumulative signal of (a)

(c) Downward QRS complex

(d) Cumulative signal of (c)

Figure 5. Comparison of cumulative signal results according to waveform type.

As shown in Figure 5 even if the signal includes Q- or S-peaks, or the QRS complex shows a downward wave, it can be expressed as a cumulative signal of similar shape. In this case, the features are also similar, so that effective fiducial point detection is possible.

4.2. Algorithm of Fiducial Point Detection

The feature value of each vertex of the accumulated signal is analyzed to determine the fiducial point. In this paper, we propose three types of features for each vertex to determine the fiducial point: the amplitude difference between R-peak and vertex, the time difference between the reference point and vertex, and the angles with neighbor vertices.

4.2.1. Amplitude Difference between R-Peak and Vertex

Figure 6, which is magnified from red-dotted box in Figure 5b, show amplitude difference between R-peak and vertex from this cumulative signal.

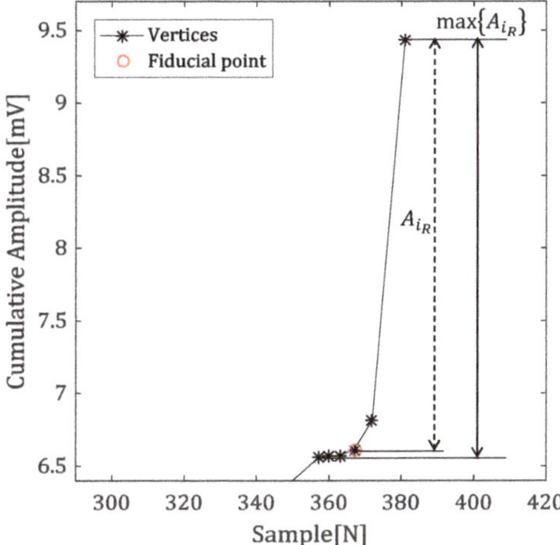

Figure 6. The amplitude difference between R-peak and the vertex in the cumulative signal.

The onset and offset of the QRS complex are the boundary points between it and the baseline region. The amplitude difference between the R-peak and the vertex in the accumulated signal is close to the maximum value near the fiducial point. Therefore, the Q-peak with a largest amplitude difference in Figure 5a become a smallest amplitude difference in the cumulative signal, which makes it easier to determine the fiducial point.

Equation (7) represents the feature value obtained by using the amplitude difference between the R-peak and the vertex.

$$A_{i_L} = v_{y_i}^{D'} - v_{y_1}^{D'}$$
$$A_{i_R} = v_{y_{N_V}}^{D'} - v_{y_i}^{D'}$$

(7)

A_{i_L} denotes an amplitude difference between the previous R-peak and vertex, which is used to detect the offset. Similarly, A_{i_R} is used to detect the onset.

4.2.2. Time Difference between Reference Point and Vertex

Figure 7 shows the time difference between the vertex which is 0.3 s away from the R-peak, and the reference point.

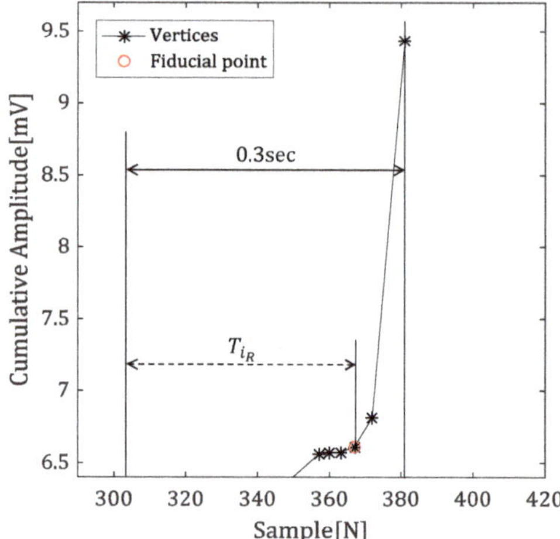

Figure 7. The time difference between the reference point and vertex in the cumulative signal.

In this paper, we suggest the time difference as a second feature value for excluding the case of detecting the onset of the P-wave. Generally, the normal width of the QRS complex is about 0.08 to 0.12 s. We use a reference point based on the point 0.3 s away from the R-peak for estimating the time difference and consider that the larger the time difference is, the more likely it is to be the fiducial point.

If the ventricular arrhythmia occurred and the width of the QRS complex is increased, fiducial point may have lower time difference feature value when there is a vertex at notch in QRS complex. However, this problem can be easily solved from the amplitude difference between previous R-peak and vertex, since the amplitude of notch and fiducial point is similar to previous R-peak and baseline, respectively.

Equation (8) represents the feature value obtained by using the time difference between the references and vertex.

$$T_{i_L} = \left(v_{x_1}^{D'} + 0.3 \times F \right) - v_{x_i}^{D'}$$

$$T_{i_R} = v_{x_i}^{D'} - \left(v_{x_{N_V}}^{D'} - 0.3 \times F \right)$$

(8)

F denotes a sampling frequency and T_{i_L} denotes a time difference between the previous R-peak and vertex, which is used to detect the offset. Similarly, T_{i_R} is used to detect the onset.

4.2.3. Angles with Neighbor Vertices

Since the fiducial point is boundary between the waveform region and the baseline region, most significant feature of vertices of fiducial point is angle between the horizontal line and straight line connecting the neighbor vertex.

Figure 8 shows the angles of the vertices corresponding to the fiducial points with the left and right vertices in the cumulative signal.

In the case of the onset, the angle with the left vertex is close to 0 degrees, and the angle with the right vertex is close to 90 degrees. In the case of the offset, it is reversed. θ_{i_L} and θ_{i_R} mean the left and right angles of the i-th vertex, respectively.

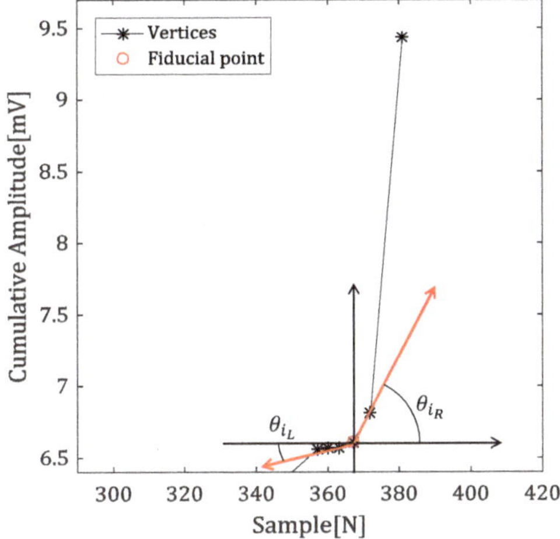

Figure 8. The angle of fiducial point in cumulative signal.

4.2.4. Detecting the Fiducial Point

Based on the feature values of the fiducial point, we can calculate the feature value of each vertex in the searching interval. Our approach provides a method to select the point with the highest probabilities of being the fiducial point by summarizing all feature values. Equations (9) and (10) represent equations used to detect the onset and offset of the QRS complex, respectively.

$$v_{Q_{on}} = argmax\{\omega_A(A_{i_R}) + \omega_T(T_{i_R}) + \omega_{\theta_C}(\theta_{i_L}) + \omega_{\theta_S}(\theta_{i_R})\}, \tag{9}$$

$$v_{S_{off}} = argmax\{\omega_A(A_{i_L}) + \omega_T(T_{i_L}) + \omega_{\theta_S}(\theta_{i_L}) + \omega_{\theta_C}(\theta_{i_R})\}, \tag{10}$$

where,

$$\omega_A(A_i) = \frac{A_i}{max(A_i)}, \quad \omega_T(T_i) = \frac{T_i}{0.3 \times F},$$

$$\omega_{\theta_C}(\theta_i) = \sqrt{cos\theta_i}, \quad \omega_{\theta_S}(\theta_i) = sin^2\theta_i$$

ω_A, ω_T, and ω_θ are weight functions for normalizing each feature value. ω_A uses the maximum value in the search interval because sensitive amplitude changes in the QRS complex may be affected according to each heartbeat. ω_T uses 0.3 s, which means the reference, and the ω_{θ_C} and ω_{θ_S} are used to have higher feature value when the baseline and waveform angles are close to 0 and 90 degree, respectively.

In the case of the baseline direction angle, the square root is added to consider that the vertex corresponding to the fiducial point may have a high value of about 30 to 40 degrees due to noise or signal distortion. On the other hand, since the waveform angle has a low possibility of distortion because of the large amplitude change, a square is added to have a low feature value for the vertices except the fiducial point.

5. Experiment and Analysis of Results

Figure 9 is the flowchart of our proposed algorithm.

In the preprocessing step, noise suppression and R-peak detection are performed. Breathing and muscle movements cause the low-frequency noise, and power noise of 30 Hz or 60 Hz causes the high-frequency noise. In this paper, these noises are suppressed using a Butterworth bandpass filter of

1–25 Hz and the most widely known Pan's method is applied to R-peak detection. The QT-DB and MIT-BIH ADB provided by Physionet are used to evaluate the performance of the proposed algorithm.

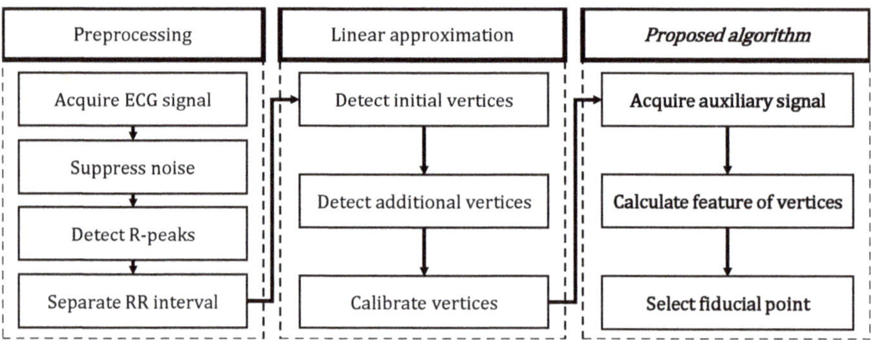

Figure 9. Algorithm flowchart.

5.1. Experiment in QT-DB

The QT-DB contains a total of 105 fifteen-minute excerpts of two-channel ECGs, which were carefully selected to avoid significant baseline wander or other artifacts. Within each record, around 30 numbers of beats were manually annotated by cardiologists, who can identify the onset, peak, and offset of the QRS complex. The proposed algorithm works on a single-channel signal, while cardiologist manually recorded one annotation considering both channel of signal simultaneously. Therefore, to compare our approach with the manual annotations on the QT-DB for each of the two single-channels, it is reasonable to choose the annotation result of the channel with less error [10]. After all the errors are obtained, the mean of total error μ and the standard deviation of total error σ are computed by averaging the intrarecording mean and standard deviation of each set of data. The standard deviation of the total error is used to measure the criterion for the algorithm's stability. In the CSE working party [35], tolerances for standard deviation of the error for the onset and offset of QRS complex are suggested as 6.5 ms and 11.6 ms, respectively.

We summarized the experimental results by our proposed method in Figure 10 according to the types of DB constituting the QT-DB.

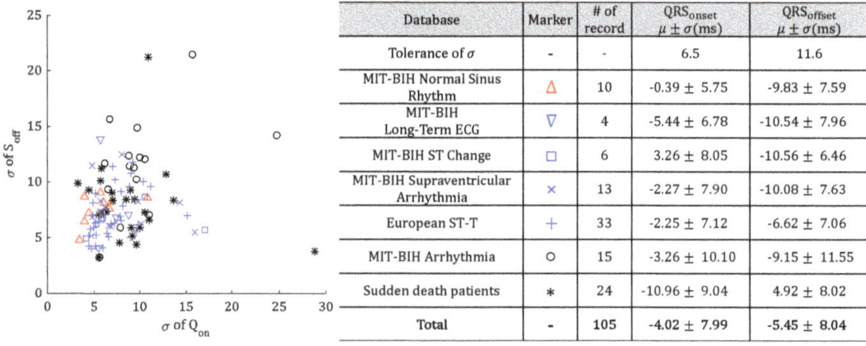

Database	Marker	# of record	QRS$_{onset}$ $\mu \pm \sigma$(ms)	QRS$_{offset}$ $\mu \pm \sigma$(ms)
Tolerance of σ	-	-	6.5	11.6
MIT-BIH Normal Sinus Rhythm	△	10	-0.39 ± 5.75	-9.83 ± 7.59
MIT-BIH Long-Term ECG	▽	4	-5.44 ± 6.78	-10.54 ± 7.96
MIT-BIH ST Change	□	6	3.26 ± 8.05	-10.56 ± 6.46
MIT-BIH Supraventricular Arrhythmia	×	13	-2.27 ± 7.90	-10.08 ± 7.63
European ST-T	+	33	-2.25 ± 7.12	-6.62 ± 7.06
MIT-BIH Arrhythmia	○	15	-3.26 ± 10.10	-9.15 ± 11.55
Sudden death patients	*	24	-10.96 ± 9.04	4.92 ± 8.02
Total	-	105	-4.02 ± 7.99	-5.45 ± 8.04

Figure 10. Distribution of standard deviation for data.

Most of the data is regarded as having satisfied the tolerance or shown on the detection result similar to the tolerance. Especially in the case of MIT-BIH normal sinus rhythm DB (represented as a red marker), it can be confirmed that the standard deviation of the total error satisfies tolerance. On the other hand, MIT-BIH arrhythmia DB and sudden-death patient DB, represented by a black marker, contains various arrhythmia heartbeats and caused large errors compared to another DB.

Table 1 shows the performance of the proposed algorithm compared to the existing detection algorithm.

Table 1. QRS segmentation performance comparison in the QT-DB.

Method	Ref	QRS Onset (ms)	QRS Offset (ms)
This work	-	-4.02 ± 7.99	-5.45 ± 8.04
Yazdani and Vesin	[36]	6.16 ± 8.3	1.5 ± 4.2
Martinez et al.	[20]	-0.2 ± 7.2	2.5 ± 8.9
Ghaffari et al.	[37]	-0.6 ± 8.0	0.3 ± 8.8
Manriquez and Zhang	[21]	-2.6 ± 7.1	0.7 ± 8.0
Manriquez and Zhang	[38]	0.58 ± 7.18	-0.95 ± 8.25
Dumont et al.	[39]	0.3 ± 6.6	-1.9 ± 8.3
Martinez et al.	[10]	4.6 ± 7.7	0.8 ± 8.7
Jane et al.	[40]	-7.82 ± 10.86	-3.64 ± 10.74
Laguna et al.	[23]	-3.6 ± 8.6	-1.1 ± 8.3
Tolerance	[35]	6.5	11.6

As with other algorithms, the standard deviation of error of the onset is out of tolerance. On the other hand, the offset satisfies the tolerance and confirmed a lower error than other algorithms.

To analyze experimental results in detail, we tried to cluster the experimental results for each data set. The experimental results of some data according to the types of QRS complex are shown in Figure 11.

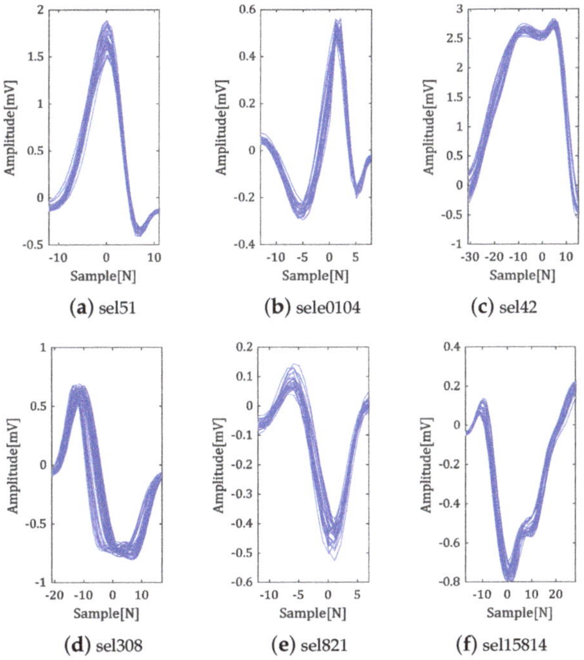

Figure 11. Clustering results.

Clustering was performed around the R-peak, and the position of the R-peak was shown as the origin. As shown in Figure 11a,b, not only the detection results of the normal type of QRS complex are stable, but also the fiducial point detection is well-performed for various types of QRS complex, such as absence of Q-peak, widen QRS complex and downward QRS complex (see Figure 11c–f).

5.2. Experiment in MIT-BIH ADB

Since MIT-BIH ADB only annotate the arrhythmia type of each heartbeat, statistical analysis for fiducial point detection is hard. We use the mean length of QR section, which start from the onset to R-peak, and RS section, which start from R-peak to the offset, as reference for onset and offset detection, respectively. According to the type of heartbeat, we separate into normal heartbeats and abnormal heartbeats, and calculate the mean and standard deviation of each type of heartbeats, respectively.

However, in the case of arrhythmia, there are various types of abnormal heartbeats. Most occur with premature atrial contraction (PAC; annotated as 'A') or premature ventricular contraction (PVC; annotated as 'V'), with the remainder very rare. For example, supraventricular premature or ectopic beat (SVP; annotated as 'S') only appears twice in record 208 over the total 107,265 heartbeats in MIT-BIH ADB. Thus, the test for abnormal heartbeat only proceeded for one of the most typical abnormal heartbeat types for records with more than 30 PACs or PVCs.

Figure 12 shows the distribution of standard deviation of fiducial points in MIT-BIH ADB.

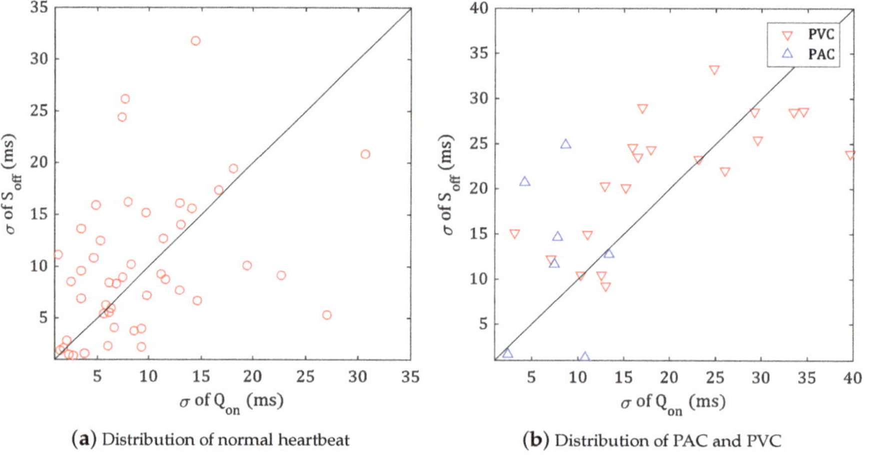

(**a**) Distribution of normal heartbeat   (**b**) Distribution of PAC and PVC

Figure 12. Distribution of standard deviation for MIT-BIH ADB.

Figure 12a is distribution of fiducial points for normal heartbeats (annotated as 'N') or other representable type of heartbeat, such as left bundle branch block (LBBB; annotated as 'L'), right bundle branch block (RBBB; annotated as 'R') or pacemaker (PM; annotated as '/'). Considering that the length of each record is 30 min, which is 60 times longer than the length of QT-DB used in Section 5.1, and most signals are unstable with various arrhythmia, it can be confirmed that the detection result of fiducial point is excellent.

Generally, since the PAC does not affect the QRS complex, the stability of the fiducial point detection is high as shown in Figure 12b. On the other hand, since PVC deforms QRS complex into various shapes, the standard deviation of width is increased.

Table 2 represents the detailed distribution of stable data in Figure 12b. In the case of PVC, the results for uniform PVC are shown.

Table 2. Detailed results for stable data in Figure 12.

Record of PAC	♮ of Beat	σ of Onset (ms)	σ of Offset (ms)	Record of PVC	♮ of Beat	σ of Onset (ms)	σ of Offset (ms)
100	33	2.45	1.59	114	43	12.64	10.48
207	106	7.89	14.60	116	109	13.04	9.29
209	383	7.50	11.61	119	444	3.21	15.14
220	94	10.85	1.28	201	198	7.09	12.23
222	212	15.28	16.78	208	992	11.02	14.96
232	1381	8.66	24.80	221	396	10.31	10.45
Average		8.77	11.78	Average		9.55	12.09

Figure 13 shows the results of the proposed algorithm for the part of the MIT-BIH ADB data in which normal heartbeat and PVC occur consecutively.

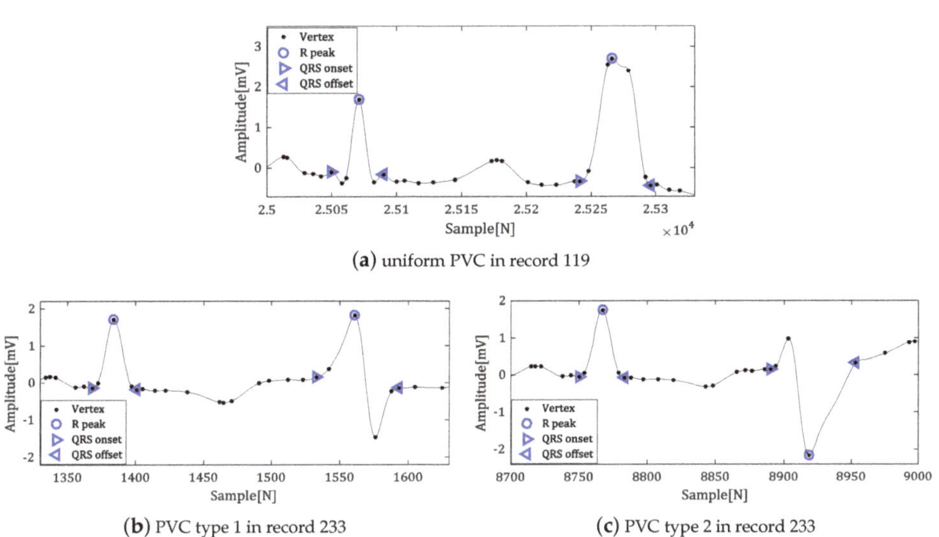

Figure 13. The result of fiducial point detection for arrhythmia data.

The proposed algorithm detected the fiducial points regardless of whether Q- or S-waves are present or not in the normal heartbeat. In addition, we could reliably detect the fiducial points for the various types of QRS complex with arrhythmia.

From this experiment, we can confirm that the proposed polygonal approximation method can effectively detect fiducial points compared with other algorithms. In addition, it obtains stable results for various types of QRS including arrhythmia. In some cases, detection error is high due to the acquisition of signals such as the V2 and V5 channel, rather than signal acquisition by the modified lead II (MLII) channel, which emphasizes the QRS complex. The distortion in the preprocessing process causes an error. Therefore, we consider that the signal acquisition clearly reflecting the shape of the QRS complex—such as the MLII channel—or the stable noise suppression technique can improve the performance of the proposed algorithm.

6. Conclusions

In this paper, the proposed algorithm focuses on mitigating the ambiguity of the fiducial point by polygonal approximation to extract the feature points of ECG signals, motivated from the characteristic that the polygonal approximation preserves the fiducial point as a vertex. Therefore, our approach resulted in better performance compared to other signal compression techniques. In addition, we propose an effective auxiliary signal for stable detection results in various types of QRS complex using these features. Experimental results show that the proposed auxiliary signal-based technique enables stable detection for various applications in real ECG databases equipped with QT-DB and MIT-BIH ADB.

In future research, the proposed method is required not only to expand the detection of P- and T-waves, but also to apply post-processing to improve the existing technique. This method can also be extended to study adaptive signal compression technique according to importance of intervals by reapplying fiducial point detection results. The trade-off between the accuracy and energy consumption could be achieved by adjusting the fiducial points compression ratio, so the extremely long-time ECG monitoring services are capable in the battery-operated wearable systems.

Author Contributions: S.L. designed the entire core architecture and performed the hardware/software implementation and experiments; Y.J. performed the simulation and measurements; D.P. has his responsibility in writing entire paper as the main corresponding author; B.-J.Y. presented the initial concept and has his role as co-corresponding author; K.H.P. is the principle investigator serving co-corresponding author.

Funding: This study was supported by the BK21 Plus project funded by the Ministry of Education, Korea (21A20131600011). This research was supported by Basic Science Research Program through the National Research Foundation of Korea (NRF) funded by the Ministry of Education (2014R1A6A3A04059410, NRF-2018R1A6A1A03025109, 2018R1A6A3A01011035). The APC was funded by NRF-2018R1A6A1A03025109.

Conflicts of Interest: The authors declare no conflict of interest. The founding sponsors had no role in the design of the study; in the collection, analyses, or interpretation of data; in the writing of the manuscript, and in the decision to publish the results.

Abbreviations

The following abbreviations are used in this manuscript:

ECG	Electrocardiogram
PVC	Premature Ventricular Contraction
PAC	Premature Atrial Contraction
SVP	Supraventricular premature
LBBB	Left Bundle Branch Block
RBBB	Right Bundle Branch Block
PM	PaceMaker
MLII	Modified Lead II

References

1. Huszar, R.J. *Basic Dysrhythmias: Interpretation & Management*; Mosby Jems/Elsevier: Maryland Heights, MO, USA, 2007.
2. Chan, H.; Chou, W.; Chen, S.; Fang, S.; Liou, C.; Hwang, Y. Continuous and online analysis of heart rate variability. *J. Med. Eng. Technol.* **2005**, *29*, 227–234. [CrossRef] [PubMed]
3. Clifford, G.D.; Azuaje, F.; McSharry, P. *Advanced Methods and Tools for ECG Data Analysis*; Artech House, Inc.: Norwood, MA, USA, 2006.
4. Oussama, B.M.; Saadi, B.M.; Zine-Eddine, H.S. Extracting Features from ECG and Respiratory Signals for Automatic Supervised Classification of Heartbeat Using Neural Networks. *Asian J. Inf. Technol.* **2015**, *14*, 53–59.
5. Salai, M.; Vassányi, I.; Kósa, I. Stress Detection Using Low Cost Heart Rate Sensors. *J. Healthc. Eng.* **2016**, *2016*, 5136705. [CrossRef] [PubMed]
6. Covello, R.; Fortino, G.; Gravina, R.; Aguilar, A.; Breslin, J.G. Novel method and real-time system for detecting the Cardiac Defense Response based on the ECG. In Proceedings of the 2013 IEEE International Symposium on Medical Measurements and Applications (MeMeA), Gatineau, QC, Canada, 4–5 May 2013; pp. 53–57.

7. Poungponsri, S.; Yu, X.H. An adaptive filtering approach for electrocardiogram (ECG) signal noise reduction using neural networks. *Neurocomputing* **2013**, *117*, 206–213. [CrossRef]

8. Pan, J.; Tompkins, W.J. A Real-Time QRS Detection Algorithm. *IEEE Trans. Biomed. Eng.* **1985**, *BME-32*, 230–236. [CrossRef]

9. Nygårds, M.E.; Sörnmo, L. Delineation of the QRS complex using the envelope of the e.c.g. *Med. Biol. Eng. Comput.* **1983**, *21*, 538–547. [CrossRef] [PubMed]

10. Martinez, J.P.; Almeida, R.; Olmos, S.; Rocha, A.P.; Laguna, P. A wavelet-based ECG delineator: Evaluation on standard databases. *IEEE Trans. Biomed. Eng.* **2004**, *51*, 570–581. [CrossRef]

11. Dokur, Z.; Ölmez, T. ECG beat classification by a novel hybrid neural network. *Comput. Methods Programs Biomed.* **2001**, *66*, 167–181. [CrossRef]

12. Hu, Y.H.; Palreddy, S.; Tompkins, W.J. A patient-adaptable ECG beat classifier using a mixture of experts approach. *IEEE Trans. Biomed. Eng.* **1997**, *44*, 891–900.

13. Tsipouras, M.G.; Fotiadis, D.I.; Sideris, D. Arrhythmia classification using the RR-interval duration signal. In Proceedings of the Computers in Cardiology, Memphis, TN, USA, 22–25 September 2002; pp. 485–488.

14. Kumar, A.; Komaragiri, R.; Kumar, M. Design of wavelet transform-based electrocardiogram monitoring system. *ISA Trans.* **2018**, *80*, 381–398. [CrossRef]

15. Kim, T.H.; Kim, S.Y.; Kim, J.H.; Yun, B.J.; Park, K.H. Curvature-based ECG signal compression for effective communication on WPAN. *J. Commun. Netw.* **2012**, *14*, 21–26. [CrossRef]

16. Mamaghanian, H.; Khaled, N.; Atienza, D.; Vandergheynst, P. Compressed Sensing for Real-Time Energy-Efficient ECG Compression on Wireless Body Sensor Nodes. *IEEE Trans. Biomed. Eng.* **2011**, *58*, 2456–2466. [CrossRef] [PubMed]

17. Israel, S.A.; Irvine, J.M.; Cheng, A.; Wiederhold, M.D.; Wiederhold, B.K. ECG to identify individuals. *Pattern Recognit.* **2005**, *38*, 133–142. [CrossRef]

18. Arteaga-Falconi, J.S.; Osman, H.A.; Saddik, A.E. ECG Authentication for Mobile Devices. *IEEE Trans. Instrum. Meas.* **2016**, *65*, 591–600. [CrossRef]

19. Lin, H.Y.; Liang, S.Y.; Ho, Y.L.; Lin, Y.H.; Ma, H.P. Discrete-wavelet-transform-based noise removal and feature extraction for ECG signals. *IRBM* **2014**, *35*, 351–361.

20. Martinez, A.; Alcaraz, R.; Rieta, J.J. Application of the phasor transform for automatic delineation of single-lead ECG fiducial points. *Physiol. Meas.* **2010**, *31*, 1467. [CrossRef]

21. Manriquez, A.I.; Zhang, Q. An algorithm for robust detection of QRS onset and offset in ECG signals. In Proceedings of the Computers in Cardiology, Bologna, Italy, 14–17 September 2008; pp. 857–860.

22. Madeiro, J.P.; Cortez, P.C.; Marques, J.A.; Seisdedos, C.R.; Sobrinho, C.R. An innovative approach of QRS segmentation based on first-derivative, Hilbert and Wavelet Transforms. *Med. Eng. Phys.* **2012**, *34*, 1236–1246. [CrossRef] [PubMed]

23. Laguna, P.; Jané, R.; Caminal, P. Automatic Detection of Wave Boundaries in Multilead ECG Signals: Validation with the CSE Database. *Comput. Biomed. Res.* **1994**, *27*, 45–60. [CrossRef]

24. Laguna, P.; Mark, R.G.; Goldberg, A.; Moody, G.B. A database for evaluation of algorithms for measurement of QT and other waveform intervals in the ECG. In Proceedings of the Computers in Cardiology, Lund, Sweden, 7–10 September 1997; pp. 673–676.

25. Moody, G.B.; Mark, R.G. The MIT-BIH Arrhythmia Database on CD-ROM and software for use with it. In Proceedings of the Computers in Cardiology, Chicago, IL, USA, 23–26 September 1990; pp. 185–188.

26. Merone, M.; Soda, P.; Sansone, M.; Sansone, C. ECG databases for biometric systems: A systematic review. *Expert Syst. Appl.* **2017**, *67*, 189–202. [CrossRef]

27. Elhaj, F.A.; Salim, N.; Harris, A.R.; Swee, T.T.; Ahmed, T. Arrhythmia recognition and classification using combined linear and nonlinear features of ECG signals. *Comput. Methods Programs Biomed.* **2016**, *127*, 52–63. [CrossRef]

28. Friesen, G.M.; Jannett, T.C.; Jadallah, M.A.; Yates, S.L.; Quint, S.R.; Nagle, H.T. A comparison of the noise sensitivity of nine QRS detection algorithms. *IEEE Trans. Biomed. Eng.* **1990**, *37*, 85–98. [CrossRef] [PubMed]

29. Berkaya, S.K.; Uysal, A.K.; Gunal, E.S.; Ergin, S.; Gunal, S.; Gulmezoglu, M.B. A survey on ECG analysis. *Biomed. Signal Process. Control* **2018**, *43*, 216–235. [CrossRef]

30. Alcaraz, R.; Hornero, F.; Rieta, J.J. Dynamic time warping applied to estimate atrial fibrillation temporal organization from the surface electrocardiogram. *Med. Eng. Phys.* **2013**, *35*, 1341–1348. [CrossRef]

31. Yun, B.J. Curvature-Based Vertex Selection for Reducing Contour Information. Ph.D. Thesis, Korea Advanced Institute of Science and Technology, Daejeon, Korea, 2002.
32. Mokhtarian, F.; Suomela, R. Robust image corner detection through curvature scale space. *IEEE Trans. Pattern Anal. Mach. Intell.* **1998**, *20*, 1376–1381. [CrossRef]
33. O'Connell, K.J. Object-adaptive vertex-based shape coding method. *IEEE Trans. Circuits Syst. Video Technol.* **1997**, *7*, 251–255. [CrossRef]
34. Bellman, R.; Dreyfus, S. *Applied Dynamic Programming*; Princeton Legacy Library, Princeton University Press: Princeton, NJ, USA, 2015.
35. The CSE Working Party. Recommendations for measurement standards in quantitative electrocardiography. *Eur. Heart J.* **1985**, *6*, 815–825.
36. Yazdani, S.; Vesin, J.M. Extraction of QRS fiducial points from the ECG using adaptive mathematical morphology. *Digit. Signal Process.* **2016**, *56*, 100–109. [CrossRef]
37. Ghaffari, A.; Homaeinezhad, M.; Akraminia, M.; Atarod, M.; Daevaeiha, M. A robust wavelet-based multi-lead electrocardiogram delineation algorithm. *Med. Eng. Phys.* **2009**, *31*, 1219–1227. [CrossRef]
38. Manriquez, A.I.; Zhang, Q. An algorithm for QRS onset and offset detection in single lead electrocardiogram records. In Proceedings of the 2007 29th Annual International Conference of the IEEE Engineering in Medicine and Biology Society, Lyon, France, 23–26 August 2007; pp. 541–544.
39. Dumont, J.; Hernandez, A.I.; Carrault, G. Parameter optimization of awavelet-based electrocardiogram delineator with an evolutionary algorithm. In Proceedings of the Computers in Cardiology, Lyon, France, 25–28 September 2005; pp. 707–710.
40. Jane, R.; Blasi, A.; Garcia, J.; Laguna, P. Evaluation of an automatic threshold-based detector of waveform limits in Holter ECG with the QT database. In Proceedings of the Computers in Cardiology, Lund, Sweden, 7–10 September 1997; pp. 295–298.

Article

Signal Amplification Gains of Compressive Sampling for Photocurrent Response Mapping of Optoelectronic Devices

George Koutsourakis [1,*] , **James C. Blakesley** [1] **and Fernando A. Castro** [1,2]

1   National Physical Laboratory (NPL), Hampton Road, Teddington, Middlesex TW11 0LW, UK
2   Advanced Technology Institute, University of Surrey, Guildford, Surrey GU2 7XH, UK
*   Correspondence: george.koutsourakis@npl.co.uk

Received: 17 May 2019; Accepted: 25 June 2019; Published: 28 June 2019

Abstract: Spatial characterisation methods for photodetectors and other optoelectronic devices are necessary for determining local performance, as well as detecting local defects and the non-uniformities of devices. Light beam induced current measurements provide local performance information about devices at their actual operating conditions. Compressed sensing current mapping offers additional specific advantages, such as high speed without the use of complicated experimental layouts or lock-in amplifiers. In this work, the signal amplification advantages of compressed sensing current mapping are presented. It is demonstrated that the sparsity of the patterns used for compressive sampling can be controlled to achieve significant signal amplification of at least two orders of magnitude, while maintaining or increasing the accuracy of measurements. Accurate measurements can be acquired even when a point-by-point scan yields high noise levels, which distort the accuracy of measurements. Pixel-by-pixel comparisons of photocurrent maps are realised using different sensing matrices and reconstruction algorithms for different samples. The results additionally demonstrate that such an optical system would be ideal for investigating compressed sensing procedures for other optical measurement applications, where experimental noise is included.

Keywords: non-destructive testing; current mapping; digital micromirror device; compressed sensing

1. Introduction

The non-uniformities of material structure and local defects can have an influence on the overall performance of optoelectronic devices, such as solar cells and photodiodes. Therefore, it is important to develop methods that provide spatially resolved information on the defects and inhomogeneities of such semiconductor devices. Light/laser beam induced current (LBIC) methods have been established for the spatial characterisation of solar cells [1], photodiodes [2,3], and other sensors and p–n junction devices [4,5]. For the realisation of current mapping, a light beam scans the device being tested, and the induced current is measured for every point. A variety of different system approaches have been proposed, making the LBIC measurement systems able to deliver spatial maps of electrical properties [6], local reflectivity [7], performance parameters [8], and material properties of optoelectronic devices. Recent implementations utilise multiple laser wavelengths that enable measurements on a larger range of samples and for different energy ranges [1,9].

Although useful and sometimes necessary, photoresponse mapping measurements are usually time-consuming, since a small spot size has to scan the entire active area of the device for a total area current map, which means the smaller the spot size, the lengthier the measurements. Focusing the laser beam on a small spot often requires elaborate optical elements and accurate alignment. A very frequent solution is to use a microscope objective lens to achieve a spot size of several micrometres [10]. The point-by-point scan is realised by an x–y stage, which means that there is always a time delay from moving from one point to the next. Continuous acquisition methods have been reported in order to accelerate this process [11]; nevertheless, this can result in small distortions of the current. The alternative option to maximise scan speed is to use piezo-electric mirror systems to guide the beam on the sample [12,13]. Spot sizes of several micrometres also mean very weak signals. For this reason, lock-in techniques for accurate current readings were introduced, even in very early systems [14], and have been used in almost every LBIC system implementation ever since. Combining all of the above features into one system is not trivial, and LBIC systems can become very complicated to realise.

The first attempt to utilise digital light processing (DLP) for current mapping was by a fast tomographic current mapping method for photovoltaic (PV), based on a digital micromirror device (DMD) for implementing the scan [15]. DLP devices utilise a DMD to create light projections [16]. Photocathode quantum efficiency mapping using a digital micromirror device has been reported, where the DMD implements the laser raster scan [17]. A DLP projector has been utilised for low-resolution spatial uniformity characterisation of solar cells [18]. A DMD-based system has also been reported for fast spectral response measurements of PV devices [19], where additional frequency modulation for each wavelength band has been reported, in order to accelerate measurements [20]. High-frequency light modulation of more than 40 GHz has also been introduced recently, with an Si light emitter embedded in a p-channel, metal oxide, semiconductor field effect transistor (PMOSFET) structure [21].

Using a DMD to apply compressed sensing (CS) current mapping of PV devices has been demonstrated in recent work [22–24]. CS current mapping has also been demonstrated by utilising an LCD (liquid crystal display) monitor to project the necessary patterns for compressive sampling [25]. The CS current mapping method is based on the CS sampling theory [26,27]. According to this theory, it is possible to reconstruct a signal from highly incomplete or inaccurate information. Compression of signals is something very common in everyday life. For instance, in JPEG image compression, most of the signal information is thrown away at the transform compression stage. Only the necessary elements for describing the image in the transform domain are kept (K elements). The image is reconstructed using these very few K elements, which provide a sparse representation of the image. The aim of CS imaging is to directly acquire the K coefficients necessary for an almost exact reconstruction of a signal. This is achieved by only acquiring $M < N$ measurements for capturing an N pixel image, where $K < M$. There are a large number of compressive sampling applications, such as CS Magnetic resonance imaging (MRI) [28], the single-pixel camera [29], CS radar imaging [30], CS confocal microscopy [31], and many more.

In previous work, we have presented the CS current mapping methodology and a design for a CS current mapping measurement system for solar cells [22,23]. In this work, the signal amplification aspects of the sampling process and technical approaches for optimised sampling are presented. Although the performance of different CS aspects (algorithms, transforms, and matrices) can be investigated by simulations, aberrations of compressive sampling due to instrumentation and optics only show in experimental investigations, such as the one presented in this work. The significant signal amplification gains of CS current mapping for optoelectronic devices are demonstrated and discussed. The optimum ways to achieve such amplification for the measured signal and the impact of sensing matrix sparsity (defined later on) on the accuracy of measurements are investigated for the first time for such an application. Three types of devices are used to illustrate that the choice of sensing matrix sparsity is dependent of sample and measurement instrumentation. The robustness of CS current mapping against long-term measurement noise is studied, and a pixel-by-pixel comparison of compressive and point-by-point sampling for current mapping is realised. Different measurement settings and samples are tested using the DMD optical system. This comparison aids in determining the most suitable occasions in which each sampling method should be realised, and presents a realistic performance evaluation of compressive sampling for this specific application. In addition, it is demonstrated that this optical setup is ideal for realistic experimental comparisons of reconstruction algorithms for optical measurement applications of compressed sensing.

2. Methodology

2.1. Experimental Layout

The optical current mapping system used in this work is based on a DMD kit and is presented in Figure 1. A single mode fibre-coupled laser source of 40 mW at a 637 nm wavelength is used. The light output of the fibre is collimated such that the beam overfills the DMD micro-mirror area. The DMD is a V-7000 module, consisting of a 1024 × 768 pixel micromirror array, each micromirror having a pixel size of 13.7 × 13.7 µm. A spatial filter is used to reject the diffracted and non-collimated components of the beam. Finally, a mirror is used for guiding the beam onto the sample, which is placed horizontally on a z-stage platform. A National Instruments PXIe-4139 source measure unit (SMU) is used for measuring the current for both cases of sampling (raster scanning and patterns). The sample is placed at the focal plane of the last lens, so that the scanning spot or the patterns are projected onto the sample. In order to apply a compressive or a point-by-point scan, a number of micromirrors are grouped together to form one pixel, and the number of grouped micromirrors depends on the selected optical resolution. The spot shape is square. The sampling methods are presented in Figure 2. As can be seen in the picture of the DMD on the right of Figure 2, not all of the active area of the DMD (1024 × 768 pixels) is used. A square 700 × 700 pixel area is used to project the patterns, in order to create a square projection. Groups of 7 × 7 micromirrors are binned together, creating projections of 100 × 100 pixels. This results in a 100 × 100 resolution of the final current maps. The sampling rate that can be achieved is 30 points or patterns per second and this sampling rate is used for all silicon samples of this work. For the organic device measured, a slower sampling rate was selected (5 samples/s), due to the slower response of the specific organic photovoltaic device [32].

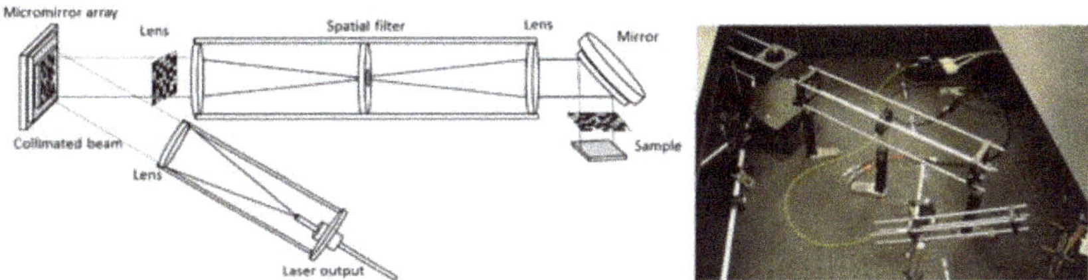

Figure 1. On the left, schematic of the optical system used for compressed sensing (CS) current mapping. On the right, a photo of the system.

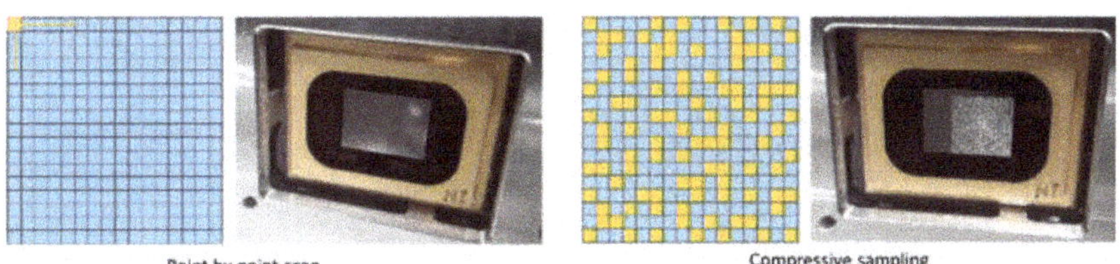

Figure 2. Schematics of the two different sampling modes used, and photos of how each is implemented on the digital micromirror device (DMD): on the left, point-by-point sampling; on the right, compressive sampling.

2.2. Compressed Sensing Current Mapping

For the application of compressive sampling, a series of binary patterns are projected onto the sample's area to be measured, and the photocurrent response of the sample is measured for each pattern. The patterns are generated by the DMD, assigning pixels (binned groups of micromirrors) as either "on" or "off", illuminating or shading different points of the sample, as can be seen in Figure 2. Similar to JPEG image compression, the sequence of patterns measures and compresses the necessary information, in order to successfully reconstruct the photocurrent response map. This is a standard procedure for optical CS imaging systems [33] that is analytically described for CS current mapping of photovoltaic (PV) devices in [34]. Compared to a point-by-point scan, fewer measurements are required in order to produce a current map of a sample.

In summary, for the application of CS current mapping, a series of binary patterns $\Phi = \{\varphi_m\}_{m=1}^{M}$ are projected onto the sample, in order to acquire a compressed representation of the signal $\mathbf{x}$, which has N elements, using $M < N$ linear measurements. Each row of the sensing matrix Φ is a binary pattern expressed as a vector, which makes Φ an $N \times M$ matrix. The current response of the PV device is measured for each pattern, populating the measurement vector $\mathbf{y}$. An underdetermined problem $\mathbf{x}$: $\mathbf{y} = \Phi\mathbf{x}$ is created, since $\mathbf{y}$ has fewer elements than $\mathbf{x}$. Random binary matrices are used in this work to produce the sensing matrix, as it has been shown that they possess the necessary properties needed for compressive sampling [35]. The discrete cosine transform (DCT) is applied as a basis to provide the sparse representation of the signal. Two different algorithms are used in this work for solving this underdetermined problem and reconstructing the current map. The first is the $\ell 1$ norm minimisation basis pursuit algorithm, included in the $\ell 1$ magic toolbox in MatLab developed by Candès and Romberg [36]. A second algorithm used is the orthogonal matching pursuit (OMP) algorithm [37]. Using one of the above algorithms, the underdetermined problem is solved, and the current map is reconstructed.

Although it is not within the scope of this work to investigate different reconstruction algorithms, the right choice of algorithm can be crucial for the successful reconstruction of the final current map. Nevertheless, given a specific algorithm, the choice of sensing matrix sparsity does not significantly affect the reconstructed image, as will be demonstrated in this work. This is shown by acquiring similar current mapping results when using different sensing matrices, for each of the two algorithms. In addition, it is demonstrated that this simple optical experimental setup is ideal for comparing CS reconstruction algorithms under real measurement conditions, and not just in simulations. Although there are a large number of reconstruction algorithms reported in the literature for use with CS, the algorithms of this work are selected due to their simplicity and known theoretical performance.

Three samples are used in this work, and an area of 1 cm by 1 cm of each sample is always measured, as well as a monocrystalline silicon (*c*-Si) reference cell, with an active area of 2 cm by 2 cm; an organic photovoltaic (OPV) cell, with an area of 1 cm by 1 cm, non-uniform performance, and a weak current; and a large multicrystalline silicon (*mc*-Si) solar cell with an area of 8 cm by 8 cm, which yields noisy measurements due to its large area. Photocurrent response measurements are acquired at short circuit conditions for all the samples. The samples are presented in Figure 3, with a random pattern projected on them using the DMD optical system. A series of such random patterns are used for compressive sampling.

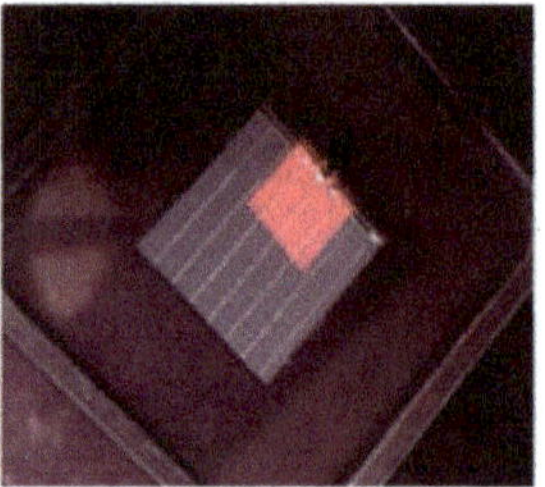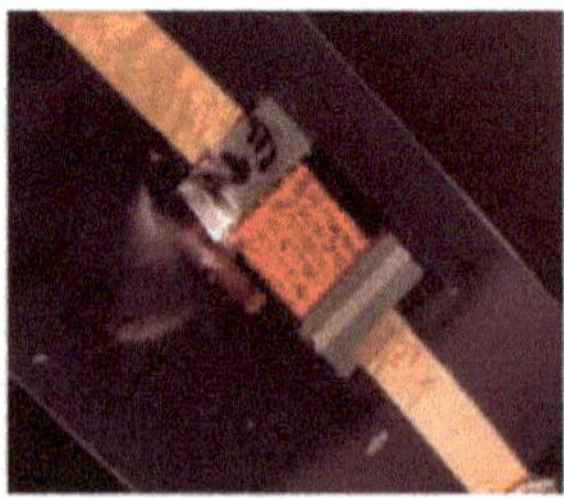

Figure 3. The three samples used in this work, with a random pattern for compressive sampling projected on them. On the left is the c-Si reference cell, in the middle is the organic photovoltaic (OPV) device, and on the right is the large mc-Si cell.

2.3. Sensing Matrix Sparsity

The impact of sensing matrix sparsity on the measurement process and on measurement accuracy can be significant. In this work, 100×100 pixel random binary sensing matrices are used, with different levels of sparsity, which means that they have a proportion of pixels in the "on" state between 1% and 99%. In this scenario, 50% means that half of the elements of the sensing matrix are in the "on" state, and the rest are in the "off" state. As a result, the projected patterns on the sample have half of their pixels bright ("on") and the other half dark ("off"). A proportion of 1% simply means that only 1% of the pixels are in the "on" state, resulting in 100 illuminated pixels for a 10,000 pixel projection. As a result, the amplitude of the current response measured, when a series of patterns (rows in a sensing matrix) is projected onto the sample ,will depend on the sparsity of the sensing matrix. This influences signal levels, and so has an impact on the measurement signal-to-noise ratio (SNR). It should be noted that "measurement SNR" is the SNR at the sampling level—the final image SNR of the reconstructed current maps will be lower, and will also depend on the artefacts inserted by the reconstruction procedure. In reality, initial sampling SNR is only one of the factors that influences the final image SNR of the reconstructed image, but it is still a very significant factor for compressive sampling, as will be shown below. Increased sparsity will mean fewer pixels in the "on" state, while reduced sparsity will mean more pixels in the "on" state. Although one could argue that regarding sparsity, 1% and 99% can be the same thing, for the sake of clarity the above convention is adopted throughout this work. This is explained in Figure 4.

It has been demonstrated in CS microscopy that sensing matrix sparsity can have an influence on CS imaging applications [38]. When using very sparse sensing matrices, the probability of having two adjacent pixels in the "on" state at the same time is small. If, in a projected pattern, there are two adjacent pixels in the "on" state simultaneously, the result may be an overlapping excited area in the sample. In CS application cases, as in the optical system of this work, due to light scattering and the diffusion of charge carriers, it may be uncertain to which of the two adjacent pixels the additional measured signal, which contributes to the global current reading of the specific pattern, is generated. Consequently, there may eventually be increased measurement noise in the final reconstructed current map, because of this uncertainty. On the other hand, with very sparse matrices the measured signal is significantly reduced. When using less sparse sensing matrices, many more pixels are in the "on" state, which results in a significant signal amplification, especially when compared with the point-by-point sampling case. The cases when sparser or less-sparse matrices are most appropriate for CS current mapping can eventually depend on the sample to be measured or the background noise of measurements.

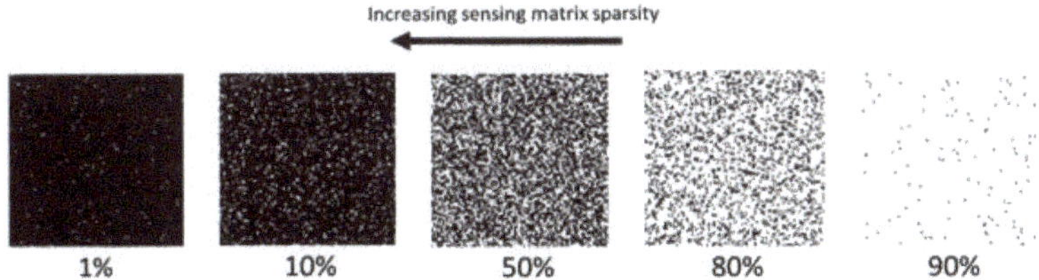

Figure 4. Visualisation of the sparsity of individual patterns of sensing matrices. Sparser patterns have more dark pixels than bright pixels, which is equivalent to a larger proportion of micromirrors being in the "off" state.

3. Results

3.1. Signal Amplification

The photocurrent signal level that a conventional LBIC system has to accurately measure in order to produce the current map can be in the range of nA. In our case, when the optical system is used to implement a point-by-point photocurrent scan, the current values are indeed in the nA range, as can be seen on the right in Figure 5, for the c-Si reference sample. In the same figure, the values of compressive sampling are also presented. All the values are in the range of 0.45–0.50 mA, which means that the current signal is enhanced by at least three orders of magnitude. This is an important feature that can be highly advantageous in cases where the signal level of individual pixel points is very weak to measure with a point-by-point process without a lock-in system.

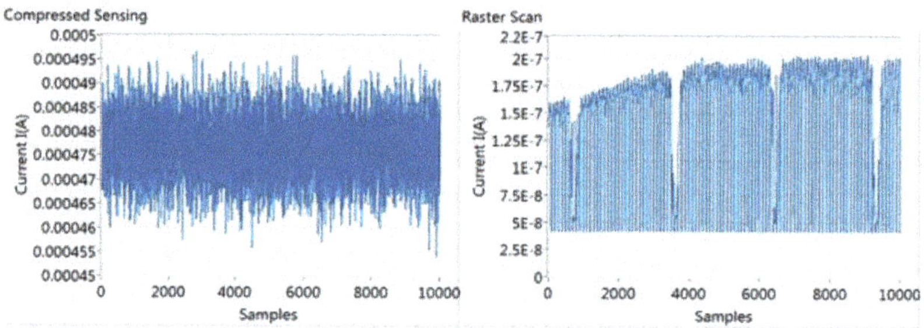

Figure 5. On the left, the current measurements for 10,000 patterns for compressive sampling—all the values are within a very small value range. On the right, the current measurements for a 10,000-pixel point-by-point current map, where the values have a range of 1 order of magnitude.

All values are within a very narrow range (0.45 mA to 0.50 mA), and this never changes during measurements for a specific sample or a sensing matrix sparsity choice. All the necessary information for reconstructing the current map is within the scatter of the measurements. This means that when acquiring measurements, the minimum and maximum instrument reading range can be set easily in a way that provides a very high dynamic range for the sampling procedure, which can increase accuracy of measurements. This specific feature of compressive sampling is utilised in the next section to correct long-term noise during measurements. Additionally, problematic measurements, such as spikes or zero values, will appear as outliers, and can be excluded easily from the reconstruction process, along with their corresponding pattern. Although the signal levels are greatly enhanced with compressive sampling, actual measurements will still contain noise as any measurement, which always influences the reconstruction process.

In practice, while signal levels are significantly enhanced by using compressive sampling, the background measurement noise levels are kept relatively stable, depending on the measurement settings of the instrument. In order to show the influence of measurement SNR on the method's performance, the SNR was calculated for all samples and cases of sensing matrix sparsity. The results are presented in Table 1. The SNR for every projected pattern during compressive sampling is calculated using 30 samples for each measurement (pattern). The measurement SNR is calculated with the Formula (1)

$$\text{SNR (signal-to-noise ratio)} = \frac{Mean\ Value}{Standard\ Deviation} \tag{1}$$

Table 1. Average current values and signal-to-noise ratio (SNR) for different samples and sampling methods.

Sampling Method/Pixels in the "on" State		Raster	CS 1%	CS 50%	CS 99%
Average Current I (A)	Ref cell	1.37×10^{-7}	9.57×10^{-6}	4.77×10^{-4}	9.48×10^{-4}
	OPV	2.27×10^{-8}	1.52×10^{-6}	8.22×10^{-5}	1.25×10^{-4}
	Large cell	1.07×10^{-5}	2.14×10^{-5}	5.34×10^{-4}	1.05×10^{-3}
SNR	Ref cell	54	2637	52,396	44,307
	OPV	19.4	2676	11,963	25,969
	Large cell	1.1	973	7056	9164

SNR is calculated for each projected pattern, and the measurement SNR is the average for all the patterns. The values of average current and measurement SNR for all three samples, and for different cases of sampling procedures, are presented in Table 1. The difference between compressive sampling and the raster scans (point-by-point scans) regarding signal amplitude and SNR is significant. In particular, for the large area mc-Si cell, the dark current present results in high levels of noise for the point-by-point scan. In all cases of different samples, the signal is amplified at least two orders of magnitude compared to the raster scan, as can be observed in Table 1. The sparsity of sensing matrices also has an effect on SNR of measurements. As it can be observed in Table 1 and in Figure 6a, the SNR increases significantly for all of the samples, with decreasing sparsity levels of sensing matrices. Specific "falls" of the SNR trend in Figure 6a (for example, at 30% and 95% for the c-Si reference cell) are due to changes in the measurement range for the photocurrent reading of the instrument as the signal increases. This results in slightly higher background noise levels when a higher value is selected for the range of the instrument, when the measured signal reaches the limit for the previous range. For a specific choice of range, the SNR increases steadily, until it saturates before the range changes. This behaviour can be observed from 30% to 90% of sparsity levels for the c-Si reference cell, for 50% to 90% for the OPV, and from 0% to 90% for the large mc-Si cell. This shows that the choice of sensing matrix sparsity for optimising SNR would also depend on the specific instrument used for measurements.

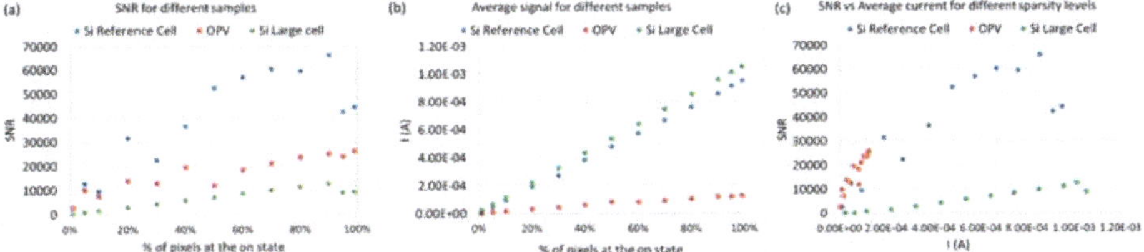

Figure 6. (a) Measurement SNR for the three different samples in the case of compressive sampling, with reducing levels of sensing matrix sparsity. (b) Signal amplitude (current reading) for the three different samples in the case of compressive sampling, with reducing levels of sensing matrix sparsity. (c) SNR as a function of average current of samples, while reducing sparsity levels.

In Figure 6b, the average current measured for each sparsity level is presented. As expected, all the cells demonstrate a linear response, with current increasing while sparsity is decreasing. The silicon devices both demonstrate similar trends, since they have similar efficiencies, while the OPV low-efficiency device produces a much lower current. The falls observed in Figure 6a due to changes in range are not observed in Figure 6b, since what is affected when the range changes is the background noise levels and not the measured signal. The correlation between SNR and measured current is presented in Figure 6c. The same behaviour seen in Figure 6a can be observed for the silicon devices. It is also clear that the OPV device demonstrates the same SNR levels as the c-Si reference device for given measured current values; although the efficiency of the OPV sample is low, and the current gains are not as high as for the silicon devices, its SNR still increases significantly for less sparse sensing matrices.

The influence of measurement SNR can be observed in Figure 7. While a raster scan is possible with this optical system for the two smaller samples, the high noise levels of the large mc-Si sample result in a very noisy current map. On the other hand, with the signal amplification when using compressive sampling, the acquisition of a current map is possible even with such high noise levels. It is clear that the number of pixels in the "on" state of the sensing matrix can be increased in order to amplify the measured signal. This does not affect the reconstruction performance, as will be demonstrated in a following section. Thus, CS current mapping can provide reliable results, even in cases of very weak signals or high noise levels, when a raster scan is not possible. It has to be noted that the SNR analyzed here is the measurement SNR at the sampling stage, and not the final current map SNR. The final current map SNR will also depend on the choice of sensing matrix, transform, reconstruction algorithm, undersampling level, and of course, the initial measurement SNR that is discussed in this work.

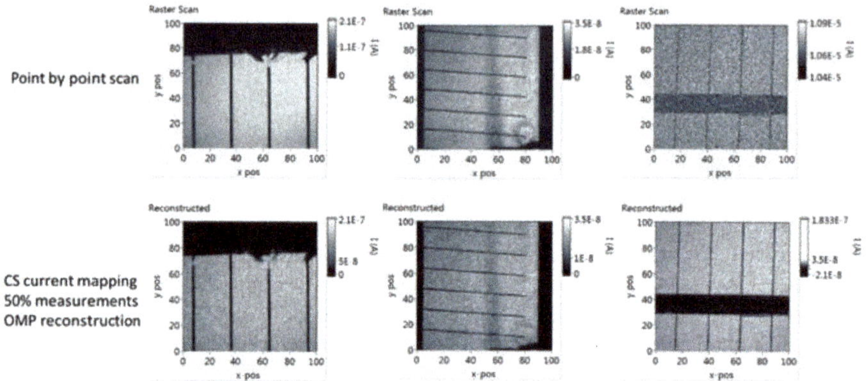

Figure 7. Current maps of the three samples used in this work, with the c-Si reference cell on the left, the OPV in the middle, and the large mc-Si cell on the right. In the top row, point-by-point current maps of the three samples. In the bottom row, CS current maps of the three samples, using the orthogonal matching pursuit (OMP) algorithm for reconstruction and for 50% undersampling.

3.2. Low-Frequency Noise Correction

Although a reference measurement for the laser light source has been implemented into the optical system using a photodiode, there is a more convenient and practical way of removing long-term noise during measurements in the case of compressive sampling. Low frequency noise/drift of signal that is independent of the sample's instantaneous performance can be due to laser instability or temperature changes of the sample. Such changes can be easily filtered out when compressive sampling is applied. As described in the previous section, when compressive sampling is applied, the complete measurement set spans within a very small range of values. This range is constant for a specific current map and sensing matrix, and any changes due to long-term noise will appear as drifts from this range. In addition, any spikes or other significant instantaneous changes of laser power are visible as outliers, and can be removed from the measurement set without losing any information. This is because fewer measurements than the pixels of the current map are applied, removing one more measurement, and the corresponding pattern will have no effect on the reconstruction.

This feature is demonstrated in Figure 8, for a case of compressed sensing current mapping of a small area of the large mc-Si sample with drifting measurement data. The laser source power changed slightly over time, simulating the potential effects of temperature or light source instability. This created a drift of the measured signal, which affected the reconstruction process and resulted in a very noisy current map. The signal was unstable and increased slightly over time. The OMP algorithm is used for reconstruction in this case. As can be observed in Figure 8, this small drift results in a noisy reconstruction of the current map. Nevertheless, since the compressively sampled measurements are expected to always be within a short range of values, this noise can be corrected. Even in this case of more intense deformation of the signal, the actual average signal difference due to this drift is around 2.5%. Still, this affects the reconstruction process if there is no correction of the sampled data. Using a generated curve to normalise the data, the drift is completely removed from the sampled data, after using a polynomial fitting to generate a curve on the corrupted sampled data. Although in most cases it is not necessary, this correction procedure is used for all cases in this work, in order to ensure that any drift of the signal is not affecting reconstruction performance. Such a correction would not be possible with a raster scan, as there would be a chance that real information would be removed.

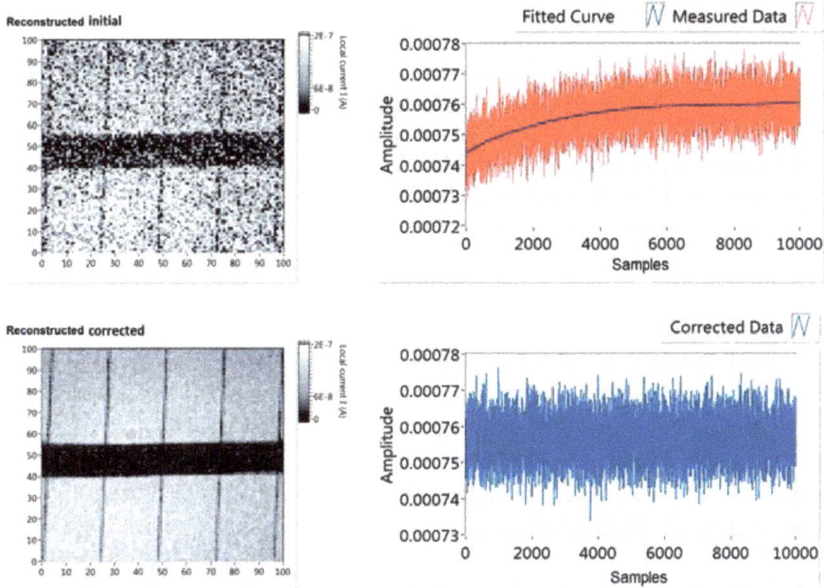

Figure 8. A case of drift correction of sampled data, using 80% pixels "on" patterns. On the top row, the uncorrected map and sampled data, at the bottom, the corrected map and data.

3.3. Reconstruction Performance

For a quantitative evaluation of the performance of the method, depending on sensing matrix sparsity, the point-by-point and reconstructed current maps were compared at a pixel-by-pixel level. This is straightforward to achieve using the DMD optical system, as it includes no moving parts and in both sampling cases, the coordinates of the current maps coincide accurately. Pearson's correlation coefficient $\rho(\hat{x}, x)$ was calculated for different levels of undersampling used for reconstruction, for different levels of sensing matrix sparsity and for the two different algorithms. The correlation coefficient is calculated by dividing the covariance of the point-by-point and reconstructed current map by the product of their standard deviations:

$$\rho(\hat{x}, x) = \frac{\text{cov}(\hat{x}, x)}{\sigma_{\hat{x}} \cdot \sigma_x} \tag{2}$$

where x is the point-by-point current map, and $\hat{x}$ is the CS-reconstructed current map, both in vector form. For the case of the large mc-Si sample, a pixel-by-pixel comparison is not possible with this optical system. Due to the large area of this sample, and since no lock-in is used, the raster-scanned current map is very noisy due to high dark current, and cannot be used as a reference for the reconstructed current maps for a pixel-by-pixel comparison.

In Figure 9, the reconstructed current maps of the c-Si reference cell and the OPV cell are presented along with the raster scan, using the same DMD optical system. By using compressive sampling, the current maps were acquired with half the number of measurements that the raster scan required. A number of sensing matrices with different sparsity levels were used, from 1% of pixels in the "on" state up to 99%. As it can be observed in Figure 9, for this sample, and for a given algorithm, all sensing matrices with different sparsity levels exhibit similar reconstruction performance. In all cases, defects like broken fingers in the silicon device and non-uniformities in the OPV device are clearly imaged. In the case of sensing matrices with 99% of the pixels "on", it is almost as if the whole sample is illuminated, significantly increasing the measured signal and measurement SNR without affecting the accuracy of the reconstructed current map.

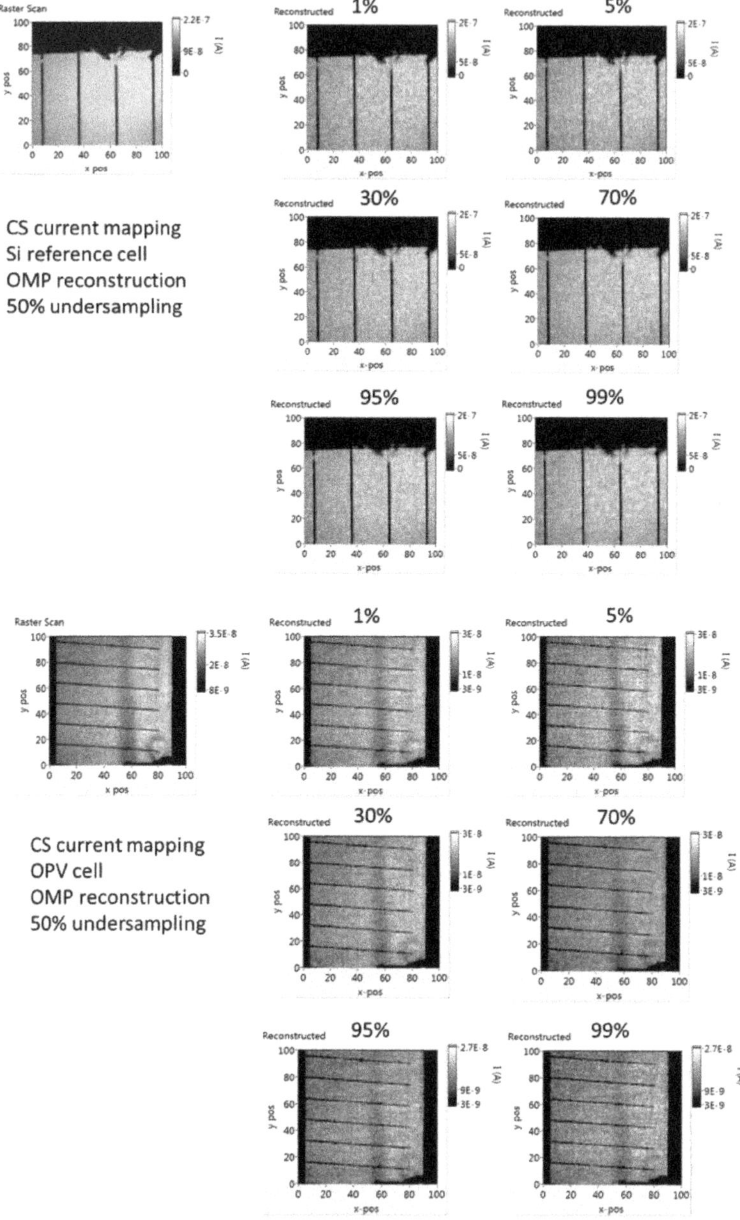

Figure 9. CS current mapping with sensing matrices with different sparsity levels (number of pixels in the "on" state). On the top left, the point-by-point scan is also included for comparison. It can be observed that the differences in reconstruction performance for different sensing matrices are negligible.

On the other hand, when using different reconstruction algorithms, the reconstruction performance can vary. In Figure 10, the correlation coefficient between the point-by-point and the CS current maps for the two samples is presented as a function of measurements used for reconstruction, for sensing matrices with different sparsity levels, and for two different reconstruction algorithms ($\ell 1$, OMP). Although for the c-Si reference cell the differences between different algorithms are not significant, for the OPV sample the reconstruction performance varies significantly between the two algorithms. This shows that some algorithms can have a different performance for different samples, depending on the features of the current map. This has to be taken into consideration when choosing a reconstruction algorithm. Nevertheless, as can be observed in the graphs of Figure 10, sensing matrix sparsity does not affect reconstruction performance for a given reconstruction algorithm. This shows that the right sparsity level of the sensing matrices can be set each time, considering background noise, signal levels, and equipment sensitivity for acquiring the current map of a specific sample.

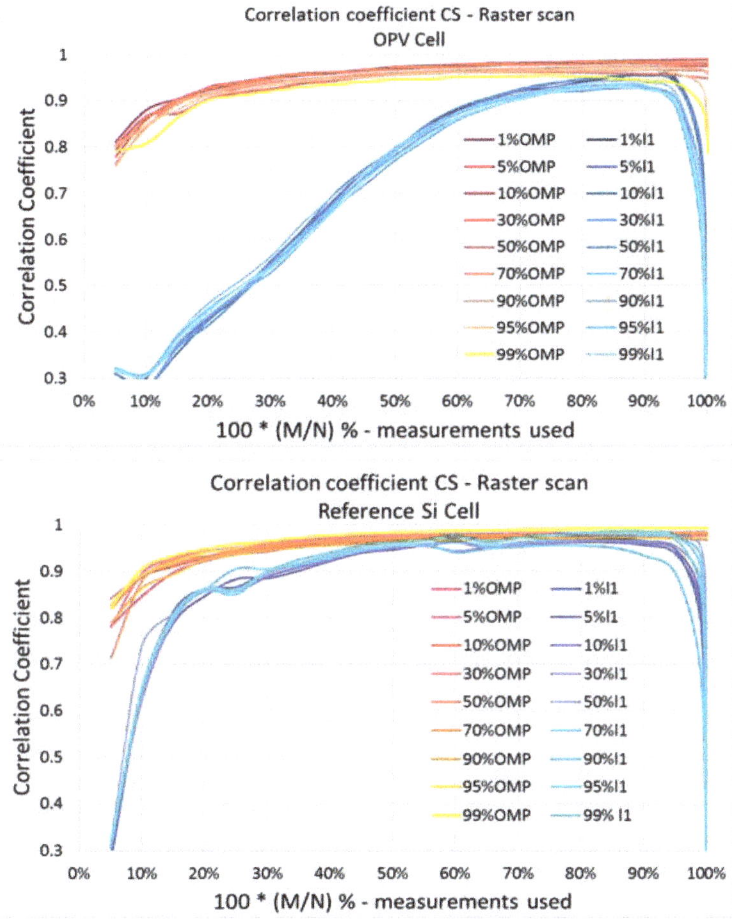

Figure 10. Correlation coefficient as a function of the number of measurements used for reconstruction, for sensing matrices with different sparsity levels, and for the two different algorithms. On the left, the graph for the c-Si reference cell; on the right, the results of the OPV cell.

When approaching 100% of measurements used for reconstruction, the performance of the algorithms, especially that of the $\ell 1$ algorithm, declines. This is because when measurement noise is included, the optimisation algorithm fails to find a solution for 100% of measurements used, as has been previously demonstrated in [22]. This is because the optimisation algorithm is increasingly constrained as we reach 100%, and has fewer degrees of freedom to filter out noise. There are algorithms available in the literature that can expect some noise in the measurements, and would not have such issues when approaching 100%. For the graphs in Figure 10, reconstruction was implemented for 99.0% as well as 99.9% of sampling, in order to accurately draw these curves. In reality, this area of undersampling is meaningless for compressive sampling, and such problems will not arise in real applications.

4. Conclusions

Compressed sensing photocurrent mapping provides fast and reliable measurements with simple experimental layouts. In this work, the signal amplification gains and the ability to optimise CS current mapping by controlling sensing matrix sparsity levels is demonstrated. By setting the right sparsity levels of sensing matrices, a significant increase in the SNR of measurements can be achieved. This provides the means to acquire current maps of samples with very weak currents or high dark currents, where a point-by-point scan would fail. In addition, current mapping systems can be put together without the need of a lock-in amplifier, allowing measurements when the application of lock-in techniques is not possible. The nature of compressive sampling allows long-term noise correction to be applied without the need of a reference measurement of the light source. For this experimental application of CS, the selected sensing matrix sparsity for optimum signal amplification does not affect current map reconstruction performance for a given reconstruction algorithm. As a result, sensing matrix sparsity can be a crucial setting that can be controlled in order to optimise measurement accuracy of CS current mapping.

It is apparent from the results of this work that in CS current mapping, different reconstruction algorithms behave differently for different samples. A future investigation of different algorithms and transforms for this CS application is necessary in order to fully optimise CS current mapping. Since a direct pixel-by-pixel comparison with a raster scan is possible, the DMD-based optical current mapping system of this work offers the opportunity to investigate the performance of different algorithms and transforms for compressive sampling. In this way, tools for this CS application can be evaluated experimentally in a realistic way, including instrument noise and system specific features. Such an evaluation of CS tools can be useful for other optical CS applications, where a comparison with a point-by-point scan is not always possible.

Author Contributions: Conceptualization, G.K.; methodology, G.K.; software, G.K.; formal analysis, G.K.; investigation, G.K.; data curation, G.K.; writing—original draft preparation, G.K.; writing—review and editing, G.K., J.C.B., and F.A.C.; visualization, G.K.; funding acquisition, J.C.B. and F.A.C.

Funding: This research was funded by the European Metrology Programme for Innovation and Research (EMPIR) 16ENG02 PV-Enerate, co-financed by the participating states and from the European Union's Horizon 2020 research and innovation programme.

Conflicts of Interest: The authors declare no conflict of interest.

References

1. Padilla, M.; Michl, B.; Thaidigsmann, B.; Warta, W.; Schubert, M.C. Short-circuit current density mapping for solar cells. *Sol. Energy Mater. Sol. Cells* **2014**, *120*, 282–288. [CrossRef]
2. Redfern, D.A.; Smith, E.P.G.; Musca, C.A.; Dell, J.M.; Faraone, L. Interpretation of current flow in photodiode structures using laser beam-induced current for characterization and diagnostics. *IEEE Trans. Electron Devices* **2006**, *53*, 23–31. [CrossRef]
3. Qiu, W.C.; Hu, W. Da Laser beam induced current microscopy and photocurrent mapping for junction characterization of infrared photodetectors. *Sci. China Phys. Mech. Astron.* **2014**, *58*, 1–13. [CrossRef]
4. Xu, K.; Huang, L.; Zhang, Z.; Zhao, J.; Zhang, Z.; Snyman, L.W.; Swart, J.W. Light emission from a poly-silicon device with carrier injection engineering. *Mater. Sci. Eng. B* **2018**, *231*, 28–31. [CrossRef]

5. Xu, K. Silicon MOS Optoelectronic Micro-Nano Structure Based on Reverse-Biased PN Junction. *Phys. Status Solidi* **2019**, *216*, 1800868. [CrossRef]

6. Bokalič, M.; Jankovec, M.; Topič, M. Solar Cell Efficiency Mapping by LBIC. In Proceedings of the 45th International Conference on Microelectronics, Devices and Materials & The Workshop on Advanced Photovoltaic Devices and Technologies, Postojna, Slovenia, 9–11 September 2009; pp. 269–273.

7. Rinio, M.; Müller, H.J.; Werner, M. LBIC investigations of the lifetime degradation by extended defects in multicrystalline solar silicon. *Solid State Phenom.* **1998**, *63–64*, 115–122. [CrossRef]

8. Carstensen, J.; Schütt, A.; Popkirov, G.; Föll, H. CELLO measurement technique for local identification and characterization of various types of solar cell defects. *Phys. Status Solidi* **2011**, *8*, 1342–1346. [CrossRef]

9. Vorasayan, P.; Betts, T.R.; Gottschalg, R. Limited laser beam induced current measurements: A tool for analysing integrated photovoltaic modules. *Meas. Sci. Technol.* **2011**, *22*, 085702. [CrossRef]

10. Sites, J.R.; Nagle, T.J. LBIC analysis of thin-film polycrystalline solar cells. In Proceedings of the Conference Record of the IEEE Photovoltaic Specialists Conference, Lake Buena Vista, FL, USA, 3–7 January 2005; pp. 199–204.

11. Geisthardt, R.M.; Sites, J.R. Nonuniformity characterization of cdte solar cells using LBIC. *IEEE J. Photovoltaics* **2014**, *4*, 1114–1118. [CrossRef]

12. Carstensen, J.; Popkirov, G.; Bahr, J.; Föll, H. CELLO: An advanced LBIC measurement technique for solar cell local characterization. *Sol. Energy Mater. Sol. Cells* **2003**, *76*, 599–611. [CrossRef]

13. Vorasayan, P.; Betts, T.R.; Tiwari, A.N.; Gottschalg, R. Multi-laser LBIC system for thin film PV module characterisation. *Sol. Energy Mater. Sol. Cells* **2009**, *93*, 917–921. [CrossRef]

14. Seager, C.H. The determination of grain-boundary recombination rates by scanned spot excitation methods. *J. Appl. Phys.* **1982**, *53*, 5968. [CrossRef]

15. Gupta, R.; Breitenstein, O. Digital micromirror device application for inline characterization of solar cells by tomographic light beam-induced current imaging. *Proc. SPIE* **2007**, *6616*, 66160O-1–66160O-9. [CrossRef]

16. Hornbeck, L.J. The DMD^{TM} Projection Display Chip: A MEMS-Based Technology. *MRS Bull.* **2001**, *26*, 325–327. [CrossRef]

17. Riddick, B.C.; Montgomery, E.J.; Fiorito, R.B.; Zhang, H.D.; Shkvarunets, A.G.; Pan, Z.; Khan, S.A. Photocathode quantum efficiency mapping at high resolution using a digital micromirror device. *Phys. Rev. Spec. Top. Accel. Beams* **2013**, *16*, 14–17. [CrossRef]

18. Yoo, J.; Kim, S.; Lee, D.; Park, S. Spatial uniformity inspection apparatus for solar cells using a projection display. *Appl. Opt.* **2012**, *51*, 4563–4568. [CrossRef] [PubMed]

19. Fong, A.Y. Application of digital micromirror devices for spectral-response characterization of solar cells and photovoltaics. In *Emerging Digital Micromirror Device Based Systems and Applications II*; International Society for Optics and Photonics: Leiden, The Netherlands, 2010; Volume 7596, pp. 75960I-1–75960I-8.

20. Missbach, T.; Karcher, C.; Siefer, G. Frequency Division Multiplex Based Quantum Efficiency Determination of Solar Cells. In *2015 IEEE 42nd Photovoltaic Specialist Conference, PVSC 2015*; IEEE: New Orleans, LA, USA, 2015; pp. 1–6.

21. Xu, K. Integrated Silicon Directly Modulated Light Source Using p-Well in Standard CMOS Technology. *IEEE Sens. J.* **2016**, *16*, 6184–6191. [CrossRef]

22. Koutsourakis, G.; Cashmore, M.; Hall, S.R.G.; Bliss, M.; Betts, T.R.; Gottschalg, R. Compressed Sensing Current Mapping Spatial Characterization of Photovoltaic Devices. *IEEE J. Photovolt.* **2017**, *7*, 486–492. [CrossRef]

23. Hall, S.R.G.; Cashmore, M.; Blackburn, J.; Koutsourakis, G.; Gottschalg, R. Compressive Current Response Mapping of Photovoltaic Devices Using MEMS Mirror Arrays. *IEEE Trans. Instrum. Meas.* **2016**, *65*, 1945–1950. [CrossRef]

24. Cashmore, M.T.; Koutsourakis, G.; Gottschalg, R.; Hall, S.R.G. Optical technique for photovoltaic spatial current response measurements using compressive sensing and random binary projections. *J. Photonics Energy* **2016**, *6*, 025508. [CrossRef]

25. Quan, L.; Xie, K.; Xi, R.; Liu, Y. Compressive light beam induced current sensing for fast defect detection in photovoltaic cells. *Sol. Energy* **2017**, *150*, 345–352. [CrossRef]

26. Donoho, D. Compressed sensing. *IEEE Trans. Inf. Theory* **2006**, *52*, 1289–1306. [CrossRef]

27. Candès, E.J.; Romberg, J.K.; Tao, T. Stable signal recovery from incomplete and inaccurate measurements. *Commun. Pure Appl. Math.* **2006**, *59*, 1207–1223. [CrossRef]

28. Lustig, M.; Donoho, D.; Pauly, J.M. Sparse MRI: The application of compressed sensing for rapid MR imaging. *Magn. Reson. Med.* **2007**, *58*, 1182–1195. [CrossRef] [PubMed]
29. Duarte, M.F.; Davenport, M.A.; Takhar, D.; Laska, J.N.; Sun, T.; Kelly, K.F.; Baraniuk, R.G. Single-Pixel Imaging via Compressive Sampling. *IEEE Signal Process. Mag.* **2008**, *25*, 83–91. [CrossRef]
30. Ender, J.H.G. On compressive sensing applied to radar. *Signal Process.* **2010**, *90*, 1402–1414. [CrossRef]
31. Ye, P.; Paredes, J.L.; Arce, G.R.; Wu, Y.; Chen, C.; Prather, D.W. Compressive confocal microscopy. In Proceedings of the 2009 IEEE International Conference on Acoustics, Speech and Signal Processing, Taipei, Taiwan, 19–24 April 2009; Volume 7210, pp. 429–432.
32. Li, Z.; Gao, F.; Greenham, N.C.; McNeill, C.R. Comparison of the Operation of Polymer/Fullerene, Polymer/Polymer, and Polymer/Nanocrystal Solar Cells: A Transient Photocurrent and Photovoltage Study. *Adv. Funct. Mater.* **2011**, *21*, 1419–1431. [CrossRef]
33. Marcia, R.F. Compressed sensing for practical optical imaging systems: A tutorial. *Opt. Eng.* **2011**, *50*, 072601. [CrossRef]
34. Koutsourakis, G.; Cashmore, M.; Bliss, M.; Hall, S.R.G.; Betts, T.R.; Gottschalg, R. Compressed sensing current mapping methods for PV characterisation. In Proceedings of the Conference Record of the IEEE Photovoltaic Specialists Conference, Portland, OR, USA, 5–10 June 2016; Volume 2016–Novem, pp. 1308–1312.
35. Baraniuk, R.; Davenport, M.; DeVore, R.; Wakin, M. A Simple Proof of the Restricted Isometry Property for Random Matrices. *Constr. Approx.* **2008**, *28*, 253–263. [CrossRef]
36. Candes, E.J.; Romberg, J.; Tao, T. Robust uncertainty principles: Exact signal reconstruction from highly incomplete frequency information. *IEEE Trans. Inf. Theory* **2006**, *52*, 489–509. [CrossRef]
37. Tropp, J.A.; Gilbert, A.C. Signal recovery from random measurements via orthogonal matching pursuit. *IEEE Trans. Inf. Theory* **2007**, *53*, 4655–4666. [CrossRef]
38. Ye, P.; Paredes, J.L.; Wu, Y.; Chen, C.; Arce, G.R.; Prather, D.W. Compressive Confocal Microscopy: 3D Reconstruction Algorithms. In *Proceedings of SPIE*; SPIE: Bellingham WA, USA, 2009; Volume 7210, pp. 72100G-1–72100G-12.

Article

The Effect of the Color Filter Array Layout Choice on State-of-the-Art Demosaicing

Ana Stojkovic *,†, **Ivana Shopovska** *,†, **Hiep Luong, Jan Aelterman** * and **Ljubomir Jovanov** and **Wilfried Philips**

TELIN-IPI, Ghent University—imec, 9000 Ghent, Belgium
* Correspondence: Ana.Stojkovic@UGent.be (A.S.); Ivana.Shopovska@UGent.be (I.S.); Jan.Aelterman@UGent.be (J.A.)
† These authors contributed equally to this work.

Received: 20 June 2019; Accepted: 18 July 2019; Published: 21 July 2019

Abstract: Interpolation from a Color Filter Array (CFA) is the most common method for obtaining full color image data. Its success relies on the smart combination of a CFA and a demosaicing algorithm. Demosaicing on the one hand has been extensively studied. Algorithmic development in the past 20 years ranges from simple linear interpolation to modern neural-network-based (NN) approaches that encode the prior knowledge of millions of training images to fill in missing data in an inconspicious way. CFA design, on the other hand, is less well studied, although still recognized to strongly impact demosaicing performance. This is because demosaicing algorithms are typically limited to one particular CFA pattern, impeding straightforward CFA comparison. This is starting to change with newer classes of demosaicing that may be considered generic or CFA-agnostic. In this study, by comparing performance of two state-of-the-art generic algorithms, we evaluate the potential of modern CFA-demosaicing. We test the hypothesis that, with the increasing power of NN-based demosaicing, the influence of optimal CFA design on system performance decreases. This hypothesis is supported with the experimental results. Such a finding would herald the possibility of relaxing CFA requirements, providing more freedom in the CFA design choice and producing high-quality cameras.

Keywords: demosaicing; debayering; color filter array; image interpolation; image reconstruction

---

1. Introduction

Since Bayer's original patent [1], (Bayer) Color Filter Array (CFA) demosaicing has established itself as the de facto standard method of acquiring multi-dimensional color images. General demosaicing in this sense would be defined as the reconstruction of a (multi-dimensional) color signal from an inherently single-dimensional array of (e.g., Charge-Coupled Device (CCD) or Complementary Metal-Oxide Semiconductor (CMOS)) sensors. The mosaiced image is obtained by using a planar sensor that is covered by an interleaved pattern of different color filters, resulting in sensor output that is an interleaved pattern of signal components that represent different parts of the color spectrum. A demosaicing algorithm reconstructs this into a (three-dimensional) full color signal. An optimal demosaicing system design would be constituted of the creation of an optimal interleaving pattern (the color filter array or CFA) and an optimal demosaicing algorithm that achieves the highest color reconstruction quality.

1.1. CFA Pattern Design

A good CFA pattern design satisfies the criteria presented in [2]: cost-effective image reconstruction, robustness to color aliasing, robustness to image sensor imperfections and robustness to optical or electrical influence between the neighboring pixels. The Bayer CFA exploits the fact that the human eye is more sensitive to wavelengths corresponding to the green colors, rather than to the blue and the red colors, i.e., the Bayer CFA consists of a repeating 2×2 pattern of one red, one blue and two green color filters. Given the analysis for the spectra of different CFA patterns, performed by Hirakawa et al. and presented in [3], it can be concluded that the Bayer CFA pattern is susceptible to introducing aliasing artifacts in both vertical and horizontal high spatial frequencies. Therefore, other solutions for CFA design that will overcome the problems with aliasing artifacts introduction for horizontal and/or vertical edges, were proposed. The idea for these designs is either based on the assumption that the physical world is rather horizontally than vertically oriented, or on the opposite assumption. Examples for more oriented patterns are: the Yamanaka CFA pattern [4], the Lukac CFA pattern [2], the Vertical CFA pattern, the Modified Bayer CFA pattern, the Diagonal CFA pattern, etc. With a purpose to increase the quality of the low-light photography, Sony introduced the Quad Bayer CFA sensor. In order to reduce the sensitivity to noise, many camera systems use multi-frame photography, which leads to ghosting artifacts introduction that affects the quality of the captured video. To deal with the problem of ghosting artifacts introduction, Sony introduced the IMX586 smartphone sensor with the Quad Bayer design and 48 MP resolution [5], with which Sony succeeds to obtain high performance and quality gain (as explained in [6]) when used in the high dynamic range (HDR) mode in low-light conditions. According to Sony, in normal light conditions, the camera achieves a similar quality of the captured images, when these are compared with the images captured with a sensor of 12 MP sensor with the Bayer design. Furthermore, to deal with the problem of low sensitivity in low-light conditions, many panchromatic CFA designs were introduced, e.g., Sony 4-Color [7], Kodak Ver.1-3 [8], etc. There also exist works on the multispectral filter array design [9,10] that find application in different fields of multispectral imaging.

In our analysis, we will only test the influence of Bayer-like CFA patterns (with the same sampling ratio as the Bayer CFA), considering them to be sufficient to show that the difference in quality performance between different CFA designs decreases with the increasing power of demosaicing algorithms.

1.2. Demosaicing

Since demosaicing has been an extensively studied research area, it abounds with algorithms that may be classified into different groups. The earliest works are based on using simple interpolation techniques (bilinear, bicubic, spline interpolation, etc.). These reconstruction techniques are prone to artifacts introduction (aliasing) in the regions with high spatial frequencies, i.e., regions with edges and details. Consequently, the research in this field progressed further towards designing algorithms based on more sophisticated reconstruction techniques. For that purpose, numerous survey studies have been proposed [11–13]. In [14], the demosaicing algorithms were roughly classified into five categories, depending on the used techniques and on the reconstruction domain (spatial, frequency, wavelet, etc.). According to this classification, *classical* demosaicing includes: frequency-domain algorithms (good representatives are [15,16]), algorithms based on directional interpolations (among which residual interpolation (RI) algorithms, specifically [14] show superior performance), wavelet-based algorithms (among which good representatives are the algorithms presented in [17–19]), and reconstruction-based (with the algorithms presented in [20,21] as good representatives of this group). Another group of algorithms are the *learning-based* algorithms: dictionary-learning-based [22], regression-based [23], reconstruction-based with machine learning procedures for multispectral demosaicing (with [24] being a good representative algorithm) and neural-network-based algorithms, which we will refer to as *modern learning-based* demosaicing algorithms (with [25] as a representative).

A good overview of the performance of the *classical* demosaicing algorithms (that are not deep-learning-based) is given in the graph from [14], shown in Figure 1. As it can be seen, ARI [14] outperforms all previous algorithms. Its good performance is due to the fact that it is an iterative approach based on the assumption of color consistency along the edges and smooth areas. Additionally, the algorithm considers the color differences in creating the final demosaiced image. However, this method, despite its good performance in the way it was designed, is not generic and is applicable only to Bayer pattern CFA and the multispectral filter array (MSFA) that was presented earlier in [10]. *Classical* demosaicing techniques have an advantage of being applicable to any type of image without requiring training data, while, on the other hand, the *modern learning-based* algorithms show superior performance and can be modified to be applied on different CFA patterns, which makes them generic.

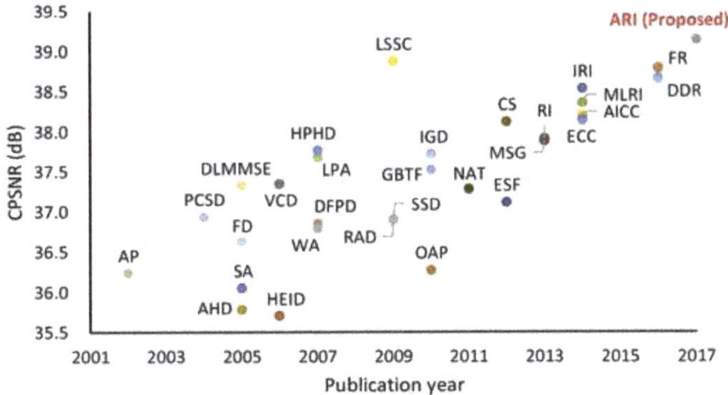

Figure 1. Performance of the demosaicing algorithms over the years (from the analysis presented in [14]). Image source: [14].

1.3. CFA-Demosaicing Co-Design

The co-design of CFA and demosaicing has received little attention. A reason for this is that many *classical* demosaicing techniques are intricate interpolation schemes that are finely tuned to a particular CFA pattern, typically the ubiquitous Bayer pattern. The intricacy of this design precludes the possibility of applying a demosaicing algorithm to a different CFA pattern and therefore inhibits the application of well-performing demosaicing techniques to CFA patterns that they were not designed for. There exist approaches in which a particular CFA design is introduced and a matching demosaicing algorithm is proposed (e.g., Pseudo-randomized CFA pattern [26] design and a demosaicing algorithm presented in [20]). Similarly, in [2], a Bayer-like CFA design is proposed and a generic algorithm is devised [27]. Demosaicing algorithms that are generic, in a sense that may be applied successfully to any CFA, do exist, but are much less common and are typically outperformed by CFA-specific demosaicing. Examples of such algorithms are [20,21], which belong to an earlier generation of demosaicing algorithms. These algorithms were outperformed by ARI (for the Bayer CFA pattern) and by another state-of-the-art algorithm, known as ACUDe [28] that belongs to the group of classical demosaicing methods and is also generic.

Another generic algorithm that belongs to the group of newer generation of *classical demosaicing* algorithms is the algorithm presented in [29], where the authors for reconstruction use the linear minimum mean square error (LMMSE) model. The LMMSE approach was tested on different periodic RGB CFA patterns and the experimental analysis has shown that, for some random CFA patterns, it achieves better performance than for the Bayer CFA pattern. Furthermore, with this analysis, it was shown that, for all analyzed CFA patterns, the reconstruction quality increases as a bigger neighborhood around the pixel to be estimated is taken into consideration.

More advanced generic algorithms are the learning based algorithms, among which neural network based algorithms are superior. Such an algorithm is presented in [30]. Here, authors propose a demosaicing algorithm based on a simple neural-network architecture. In this algorithm, the authors rely on the previously proposed concept of using a superpixel (neighboring area) for estimating the unknown pixel values. This technique was tested on different CFAs and the MSFA presented in [10].

Convolutional neural networks (CNNs) have shown great success in many image processing and computer vision tasks. One of the main strengths of CNNs is that they allow learning features specific for a given domain, compared to earlier approaches where the features were predefined based on domain knowledge. For example, for the problem of demosaicing, the same CNN can be applied for different CFAs by retraining it with different input data, without any structure modifications [25,31–36]. Moreover, some CNN-based algorithms learn an optimal color filter layout jointly with a model for demosaicing of images obtained with the learned pattern [37,38].

1.4. Structure

This paper studies the effect of the CFA design choice on the overall CFA-demosaicing reconstruction quality. Specifically, we test the hypothesis that the impact of the CFA layout choice on the quality performance decreases with the emergence of more sophisticated, *modern learning-based* demosaicing algorithms.

It is structured as follows: in Section 2, we provide a description of demosaicing that is state-of-the-art with respect to quality performance and that lends itself well to adaptation and improvement towards generic CFA. In Section 3, we describe adaptations we made to these methods in an effort to allow evaluation of different CFAs and to achieve better results in terms of reconstruction quality. In Section 4, by starting with the motivation for the performed analysis, we proceed with explanation about the performed experiments, about the used image data-sets and the used CFA patterns. In Section 5, we present the qualitative and the quantitative results from the performed analysis, while in Section 6 we give a summary of the obtained conclusions.

2. State-of-the-Art Demosaicing

In order to demonstrate a trend of state-of-the-art demosaicing becoming less sensitive to the CFA pattern (i.e., to show that the performance of *modern learning-based* demosaicing is less affected by the CFA design), we selected two algorithms, as representatives of two different groups of demosaicing algorithms: in the first group, we consider *classical* demosaicing techniques that are not based on machine learning and in the second group we consider *modern learning-based* techniques. In this section, we describe the two representatives of each group separately. Specifically, the first algorithm, known as ACUDe, belongs to the group of directional interpolation demosaicing algorithms. For reconstructing the full color image, it exploits the color consistency in real images, by using the interchrominance dependency. The second algorithm is a neural network based algorithm called CDMNet [25], chosen as a representative method with a publicly available code.

2.1. Universal Demosaicing of CFA (ACUDe)

The method proposed by Zhang et al. [28], known as ACUDe, is devised to be applied on different designs of CFA. Considered to be a state-of-the-art generic algorithm among the algorithms that are based on directional interpolations, it has similar performance to the state-of-the-art ARI algorithm proposed by Monno et al. [14], which outperforms all demosaicing algorithms that were previously proposed. Furthermore, as ARI was implemented, it is only applicable to the Bayer CFA pattern and the five-band multispectral filter array described in [10,14]. Therefore, we will consider only ACUDe for our analysis. Although the main idea and theory of ACUDe can be applied on multispectral filter arrays, the algorithm was tested and implemented for the three primary color system. Considered to be an adaptive generic method, it exploits the inter-chrominance dependence and the measured CFA response, to estimate the chrominance components. It is implemented in three steps: estimation of the color/demosaicing transform matrix (only dependent from the CFA pattern) and the chrominance direction; distance dependent per-pixel weight generation based on inter-pixel chrominance capture and edge sensitivity; chrominance components estimation by weighting over the CFA image and demosaicing transformation to the three primary colors. The method was evaluated on the KODAK data-set [39] (because it abounds with images with both high and low spatial frequency content) and the IMAX data-set [40] (considered to be more challenging for demosaicing methods because of its high color diversity and also because many images from the data-set contain high spatial frequencies in the chrominance components). In [28], the authors compared their algorithm with two other generic demosaicing algorithms for RGB CFA patterns (Condat's generic method [20] and Menon and Galvagno's regularization approach to demosaicing (RAD) [21]). The obtained results (in terms of measured color peak signal-to-noise ratio (CPSNR) and CIE LAB error) indeed show that ACUDe outperforms the two mentioned algorithms for six different CFA patterns.

2.2. Demosaicing Using a CNN (CDMNet)

The baseline neural-network-based algorithm in this work is CDMNet [25]. In our analysis, we use this algorithm because of its high-quality performance and reconstruction power among the algorithms that belong to the group of *modern learning-based* demosaicing techniques and because our experiments show that it outperforms the state-of-the-art algorithms that belong to the group of *classical* demosaicing. Furthermore, this algorithm with the publicly available code and the retrainable neural-network is suitable for performing modifications towards making it generic (adaptive to any repetitive CFA pattern). In its original design, CDMNet is a three-stage neural network that is among the state-of-the-art, according to evaluations on different datasets. CDMNet relies on the inter-channel correlation for interpolation of the missing values. The problem of RGB demosaicing is split into three stages: (1) reconstruction of the green channel, (2) separate reconstructions of the red and blue channels jointly with the high-quality green channel and (3) joint fine-tuning of all three channels.

3. Modifying State-of-the-Art Demosaicing Algorithms

Here, we present the modifications we made on ACUDe and CDMNet. Since ACUDe is designed to be generic, with our modifications, we achieved slight improvement in the reconstruction quality. The adaptations made on CDMNet apply to the generalisation of this algorithm towards repetitive RGB CFA patterns.

3.1. Modifying ACUDe

The flowchart of the modified ACUDe is presented in Figure 2. The input data that is considered is: the CFA pattern and the CFA filtered image (CFA input). The green arrows represent the CFA pattern dependent data-flow, and the blue arrows represent the CFA filtered image dependent data-flow. For more details about the algorithm and each procedure, we refer to the explanation given in [28]. The modifications that were made consist of inserting the original values of the CFA filtered image to the corresponding locations in the output image. This procedure is applied after obtaining the rough estimate of the demosaiced image (non-adaptive demosaicing). For the main source-code used in our implementation and the obtained results for Bayer pattern, of the originally designed (as explained in [28]), and unmodified ACUDe, we direct the reader to the following web-site [41].

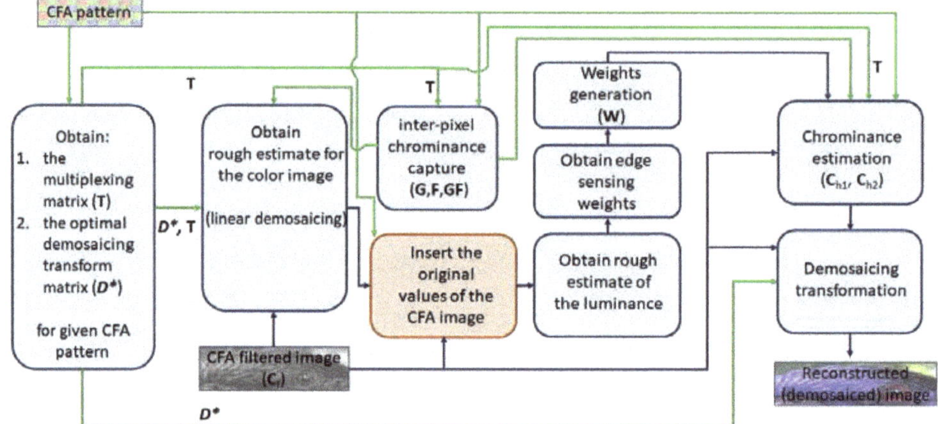

Figure 2. Flowchart of ACUDe and our modifications. Modifications are represented with the procedure in the orange block. Green arrows represent the data-flow between procedures (given with blocks) where only the color filter array (CFA)pattern is used. Blue arrows represent the data-flow between procedures (given with blocks) where also a CFA input image is used.

3.2. Making CDMNet Generic

For this study, we need to compare the performance of the same demosaicing algorithm using different CFA patterns. To make the CDMNet algorithm agnostic of the pattern, we made several modifications of the original version of CDMNet, shown in Figure 3 and explained in more details below:

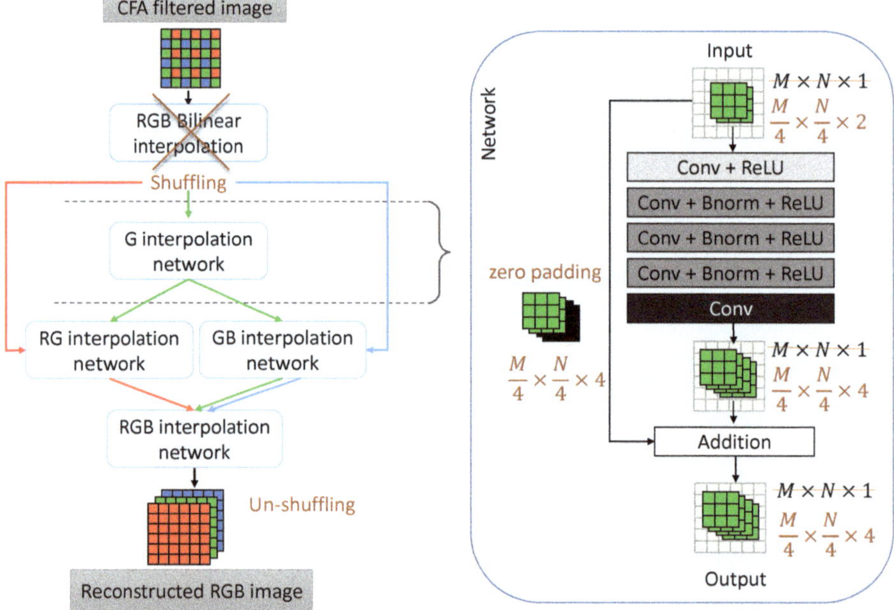

Figure 3. Flowchart of CDMNet and our modifications. The orange color represents locations where we modified the original algorithm to make it agnostic of any CFA pattern: no bilinear demosaicing, pixel shuffling and working in a lower spatial resolution with fixed downscaling factor of 4×4.

1. We do not rely on an initial estimate obtained by bilinear interpolation. Any interpolation technique cannot guarantee equal reconstruction quality when applied on different CFA patterns. Since our goal is to compare the influence of the patterns on the reconstruction quality, it is necessary that all other conditions are equal when training the neural networks. Instead of initial interpolation, the network operates on the original, sub-sampled color channels provided at the input. To compensate for the lower spatial resolution of the input and reconstruct a high-resolution output, the idea of periodic shuffling is applied as explained in the next paragraph.

2. We integrated the idea of periodic shuffling in the CDMNet. Periodic shuffling was proposed for the problem of image super-resolution [42], in order to substitute the deconvolution layer for up-sampling. The effect of this operation is that the network operates on a lower spatial resolution, and interpolates the missing values in the feature channels. The number of channels is proportional to the sub-sampling factor. To obtain the final high-resolution output, the elements of the tensors of low spatial resolution and high dimensionality are re-arranged into a high-resolution RGB image.

3. All patterns used in these experiments were assumed to have the same size. Patterns that are smaller (e.g., 2×2 Bayer pattern or 4×2 Lukac pattern) can be considered as replicated in the appropriate dimension to achieve the largest size of all compared patterns, 4×4. The down-sampling factor r in the periodic shuffling is determined by the size of the pattern, and in this case we have fixed it to 4 in both the horizontal and vertical directions.

These adaptations allow for providing the same conditions when training the neural network for different patterns. Any difference in the reconstruction quality will only be a consequence of the pattern that was used. To train the neural network, we relied on the Waterloo Exploration Dataset (WED) [43] following the same practice as the original CDMNet method. Initially, we randomly select 4644 images to create the training dataset, and the remaining 100 images from WED are used as a validation set. We have then fixed the selected training and validation sets and used the same ones during re-training for all of the patterns.

Approximately 360,000 patches were extracted from the training images. Instead of the originally proposed patch size of 50×50, we extracted patches of size 48×48 to keep the dimensions divisible by the downscaling factor. For each pattern, the neural network was randomly initialized and re-trained for 81 epochs using batches of 64 patches. As in the original version [25], the learning rate was decreased five times every 20 epochs, ranging from 10^{-3} to 10^{-5}. Applying the modified CDMNet model to any camera with a three-color, a repetitive (periodic) CFA pattern would require only one offline re-training and applying a scaling factor proportional to the pattern size.

4. Experimental Analysis

The objective of the performed analysis is to show that, as the performance quality increases with the more sophisticated reconstruction techniques being introduced, it becomes less affected by the CFA design.

4.1. Experiments and Materials

In our study, we tested the two modified generic state-of-the art algorithms (that belong to different classes of algorithms) on Bayer-like patterns (where the sampling ratio between the green, red and blue channel is 2:1:1). We limit our analysis only to Bayer-like patterns, with a purpose of achieving fair comparison and to obtain unbiased (towards more sophisticated and novel CFA patterns) and uninfluenced (by different designs of demosaicing algorithms) conclusions. According to the performed analysis presented in [29], some random patterns perform better than Bayer for the specific LMMSE demosaicing methods. Our assumption is that the worst-performing CFAs (like Quad Bayer) will present the largest differences with respect to output quality as a function of the demosaicing algorithm used. Our experiments show that when neural networks are applied for demosaicing, the choice of the CFA pattern does not significantly influence the quality of the results. Based on our finding and the results of generic algorithms presented in [28] (which are similar for Bayer CFA and for Pseudo Randomized CFA), we expect that this trend will be true for random patterns as well. For this reason, the paper focuses on showing how the performance for the worst-performing CFAs, as these are the ones with the largest differences in performance, changes as a function of chosen demosaicing algorithm. Therefore, we test the hypothesis that if the exact CFA pattern becomes irrelevant or less relevant, this largest difference would diminish with it.

The patterns (see Figure 4) on which the modified state-of-the-art algorithms were tested are: Bayer pattern, Lukac pattern, Yamanaka pattern, Modified Bayer pattern and Quad Bayer pattern.

Following the common practice for quality performance evaluation, in the performed analysis for the two modified state-of-the-art generic algorithms, we used the two well-known image data-sets: KODAK [39] (abounding with highly diverse in content and with plenty of details images) consisted of 24 images in total (18 images with resolution of 768 × 512 pixels and six images with a resolution of 512 × 768 pixels) and IMAX [40] (abounding with highly diverse in content and colors and with plenty of details images) consisted of 18 images with resolution of 500 × 500 pixels. The second image data-set, IMAX, is more challenging for quality performance evaluation of the demosaicing algorithms. These image data-sets do not coincide with WED and are therefore new, previously unseen data for the modified CDMNet. For testing the quality performance of the modified CDMNet, each of the five trained neural network models (one model for every CFA design presented on Figure 4), was applied on the input mosaic images obtained with the appropriate CFA pattern.

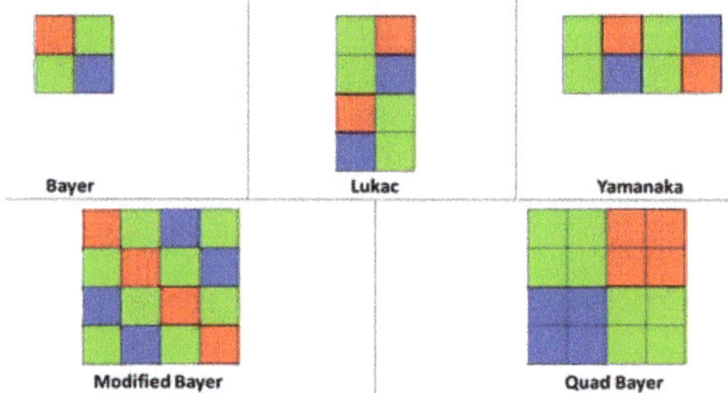

Figure 4. Bayer-like CFA patterns used in the experimental analysis: Bayer, Lukac, Yamanaka, Modified Bayer and Quad Bayer.

In order to test our hypothesis that a generic (in terms of CFA pattern) *modern learning-based* demosaicing technique (such as the modified CDMNet) is more advantageous for reconstruction and less affected by the quality of the CFA pattern than a generic *classical* demosaicing technique that is not learning based (such as the modified ACUDe), we perform three experiments. The first two experiments are part of the quantitative analysis for the quality performance of the two algorithms and these include: objective evaluation (using the average CPSNR and the average PSNR for each color channel, with the standard deviation of the mean, calculated over the reconstructed images from each image data-set) and perceptual evaluation (using the SSIM metric, with the standard deviation of the mean calculated over the reconstructed images from each image data-set). When each quality metric was calculated, 11 pixels were excluded from the two vertical and horizontal image borders. In the third experiment, which is part of the qualitative analysis, we analyze and compare images that were reconstructed with the both algorithms, for the five different Bayer-like CFA patterns.

5. Results

In what follows, we present the results and the conclusions from the experiments of the performed analysis about the quality performance of the two representative state-of-the-art algorithms (the modified ACUDe as representative among the *classical* demosaicing algorithms and the modified CDMNet as representative among the *modern learning-based* demosaicing algorithms).

5.1. Results from the Quantitative Analysis

In Figure 5, we present the averaged CPSNR results with the standard deviation of the mean (over the images from each image data-set) obtained for each analysed Bayer-like CFA pattern, with the representative techniques (the modified ACUDe and the modified CDMNet). In Figures 6–8, we present the average PSNR results with the standard deviation of the mean (over the images from each image data-set) for each channel separately. In the same way, in Figure 9, we present the SSIM results.

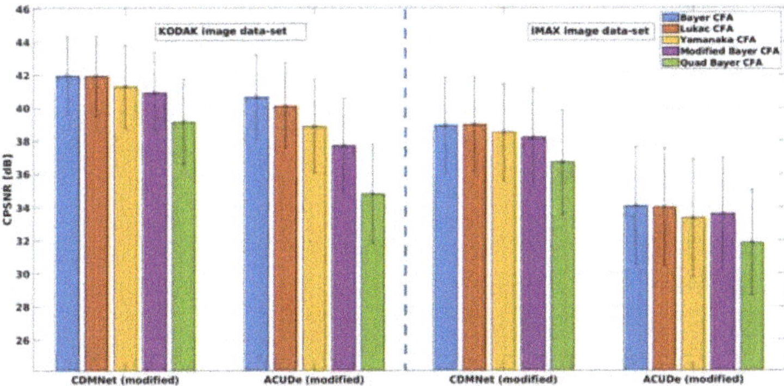

Figure 5. Average color peak signal-to-noise ratio (CPSNR) results with standard deviation of the mean: for the two modified generic algorithms ("CDMNet modified" and "ACUDe modified"), for the two data-sets ("KODAK" and "IMAX"), for five different Bayer-like patterns (Bayer, Lukac, Yamanaka, Modified Bayer and Quad Bayer).

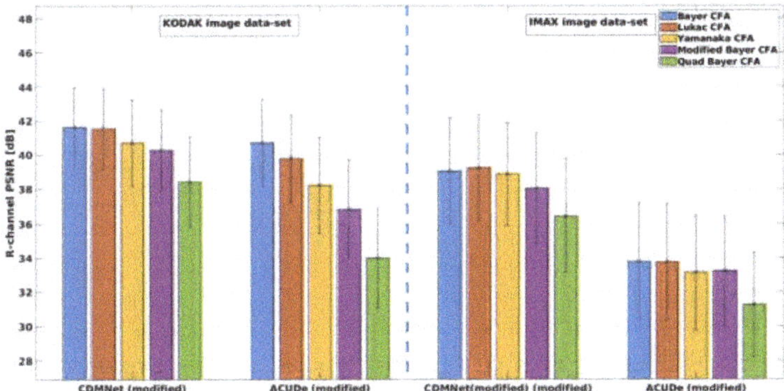

Figure 6. Average PSNR (red channel) results with standard deviation of the mean: for the two modified generic algorithms ("CDMNet modified" and "ACUDe modified"), for the two data-sets ("KODAK" and "IMAX"), for five different Bayer-like patterns (Bayer, Lukac, Yamanaka, Modified Bayer and Quad Bayer).

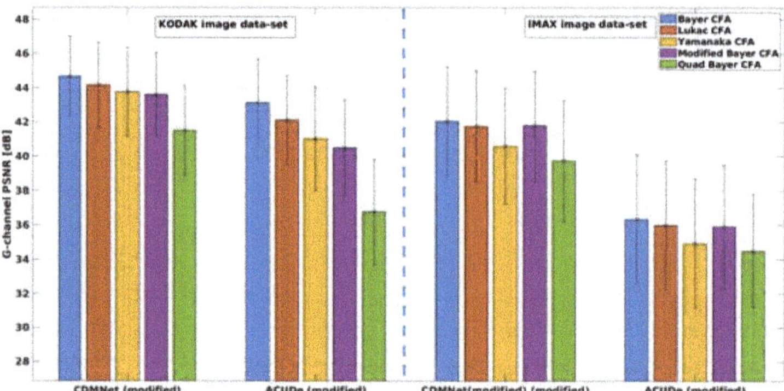

Figure 7. Average PSNR (green channel) results with standard deviation of the mean: for the two modified generic algorithms ("CDMNet modified" and "ACUDe modified"), for the two data-sets ("KODAK" and "IMAX"), for five different Bayer-like patterns (Bayer, Lukac, Yamanaka, Modified Bayer and Quad Bayer).

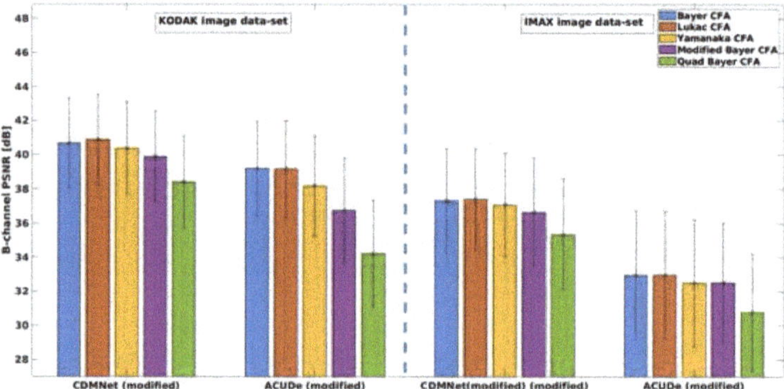

Figure 8. Average PSNR (blue channel) results with standard deviation of the mean: for the two modified generic algorithms ("CDMNet modified" and "ACUDe modified"), for the two data-sets ("KODAK" and "IMAX"), for five different Bayer-like patterns (Bayer, Lukac, Yamanaka, Modified Bayer and Quad Bayer).

From the CPSNR graph (see Figure 5), it can be noticed that the difference in quality, achieved with the modified CDMNet (especially on the KODAK image data-set) across the different Bayer-like CFA patterns, is smaller than the difference in quality of the modified ACUDe. The highest difference in the quality, which may be observed from the both PSNR and SSIM values and as it is expected, is between the Bayer CFA pattern and the Quad Bayer CFA pattern. The color pixels in Bayer CFA are more frequently distributed around the pixel of interest (the pixel whose value is to be estimated) and therefore the reconstruction quality with both algorithms is better for the Bayer CFA pattern. This difference (for the KODAK image data-set) of $\approx$3 dB when the modified CDMNet is applied, is significantly lower, when being compared to the $\approx$6 dB difference when the modified ACUDe is applied. Although not highly pronounced, a similar trend is recognized when the results of the IMAX image data-set are observed. The same conclusion can be derived from the observations on the PSNR results for each color channel separately (see Figures 6–8).

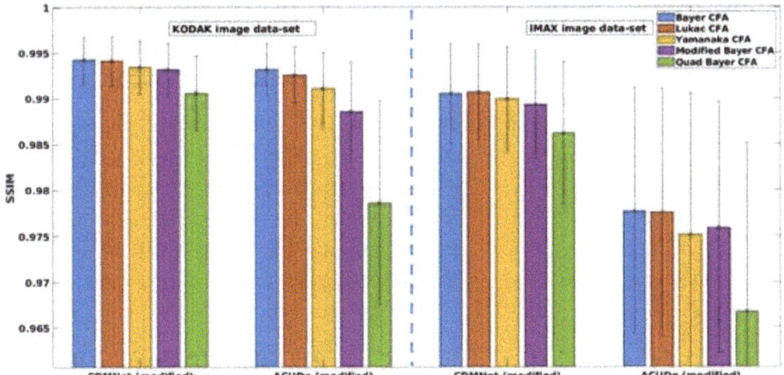

Figure 9. Average SSIM results with standard deviation of the mean: for the two modified generic algorithms ("CDMNet modified" and "ACUDe modified"), for the two data-sets ("KODAK" and "IMAX"), for five different Bayer-like patterns (Bayer, Lukac, Yamanaka, Modified Bayer and Quad Bayer).

To examine if this difference (between the reconstruction quality across the different CFA patterns) becomes more apparent when the modified ACUDe is applied, we proceed with analyzing the SSIM results (see Figure 9). As expected, the derived conclusion of the CPSNR and PSNR results is more notably supported with the SSIM results. This analysis brings us towards a more general conclusion on the improvement of the reconstruction quality, with the emergence of new and more advanced demosaicing techniques. The conclusion is that the reconstruction quality becomes less affected by the CFA pattern, as the demosaicing techniques become more sophisticated and more reliant on powerful deep learning-based approaches.

Although the absolute overall quality performance of both representative demosaicing algorithms (the modified ACUDe and the modified CDMNet) is not the main focus of the performed analysis, in what follows, with a purpose to make the analysis thorough, we will discuss the major differences between the two algorithms for each image data-set. The KODAK image data-set, compared to the IMAX image data-set, consists of natural images that are more color consistent. On the contrary, the IMAX image data-set abounds with images that consider more high spatial color frequencies. Therefore, the better quality performance of the both demosaicing algorithms, on the KODAK image data-set, is expected and justified. On the other hand, the color constancy in natural images is one of the assumptions that the design of many *classical* demosaicing techniques (also including the state-of-the-art algorithms: the generic ACUDe and ARI) is based on. Therefore, our assumption is that the significantly better quality performance of the modified ACUDe on the KODAK image data-set, rather than on the IMAX image data-set, is due to the fact that ACUDe, in the way it was originally designed, is inherently biased towards image content with higher color constancy. If the PSNR results for each color channel (see Figures 6–8) are analyzed, it can be noticed that both algorithms are better in reconstructing the green color channel, rather than the blue and the red color channels. It can also be noticed that the difference in quality reconstruction between the content of the two image data-sets is smaller in the case when the modified CDMNet is applied. Moreover, the standard deviation of the mean (calculated over the reconstructed images from each image data-set and each analyzed CFA pattern) is smaller (especially when SSIM results are observed) in the case when the modified CDMNet is applied.

These results show that the modified CDMNet, as a representative among the *modern learning-based* demosaicing techniques, is more adaptive to different types of scenes and at the same time succeeds with achieving high quality reconstruction for different CFA patterns.

5.2. Results from the Qualitative Analysis

Here, we visually present the results (cropped parts of the reconstructed images) from the selected representative examples of each image data-set (KODAK and IMAX). The ground-truth (GT) images, with the corresponding cropped parts (marked with rectangles), are presented in Figure 10. In Figure 11, we show the results obtained with the both algorithms (the modified ACUDe and the modified CDMNet) for the five Bayer-like CFA patterns presented in Figure 4. The differences between the reconstructed images from the different CFA inputs are visually more pronounced in the case when the modified ACUDe is applied. Note the color aliasing artifacts in the result for the KODAK example and the Yamanaka CFA pattern and the color aliasing artifacts accompanied with zippering artifacts for the KODAK example and the Quad Bayer CFA pattern presented in Figure 11. Some color aliasing artifacts may also be noticed (although negligible) in the reconstructed images with the modified CDMNet (see the result obtained with the modified CDMNet, for the KODAK example and the Yamanaka CFA pattern, presented in Figure 11). If we compare the results for the Quad Bayer CFA pattern and the Bayer CFA pattern, obtained with the modified CDMNet, we notice that there are no demosaicing artifacts present. The only difference between the two results may be perceived as insignificant blurriness in the case of the Quad Bayer CFA pattern. Furthermore, if the results obtained with the modified CDMNet for the IMAX example are observed, almost no differences between the reconstructed images will be noticed.

The consistently higher reconstruction quality across different patterns and data-sets achieved by CDMNet, over ACUDe, can be attributed to the strong representation power of convolutional neural networks. The advantage of CNN models compared to classical methods is the capability of modeling a distribution of natural images, and learning spatial and spectral correlations in the data. The huge number of trainable parameters in CDMNet provides sufficient model complexity for solving the demosaicing problem with similar quality for different input mosaic configurations.

The presented results from the performed experimental analysis, additionally with the results from the experimental evaluation presented in [30], veritably support our initial hypothesis that, when *modern learning-based* techniques are applied for demosaicing, the overall reconstruction quality is less influenced by the choice of the CFA pattern.

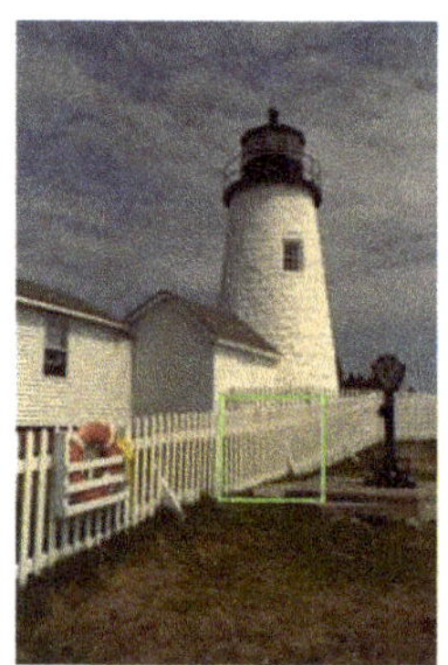

KODAK example: ground truth (GT)

Cropped part: KODAK GT

IMAX example: ground truth (GT)

Cropped part: IMAX GT

Figure 10. Examples of ground truth images from KODAK and IMAX image data-sets and their cropped parts.

Representative example from the KODAK set

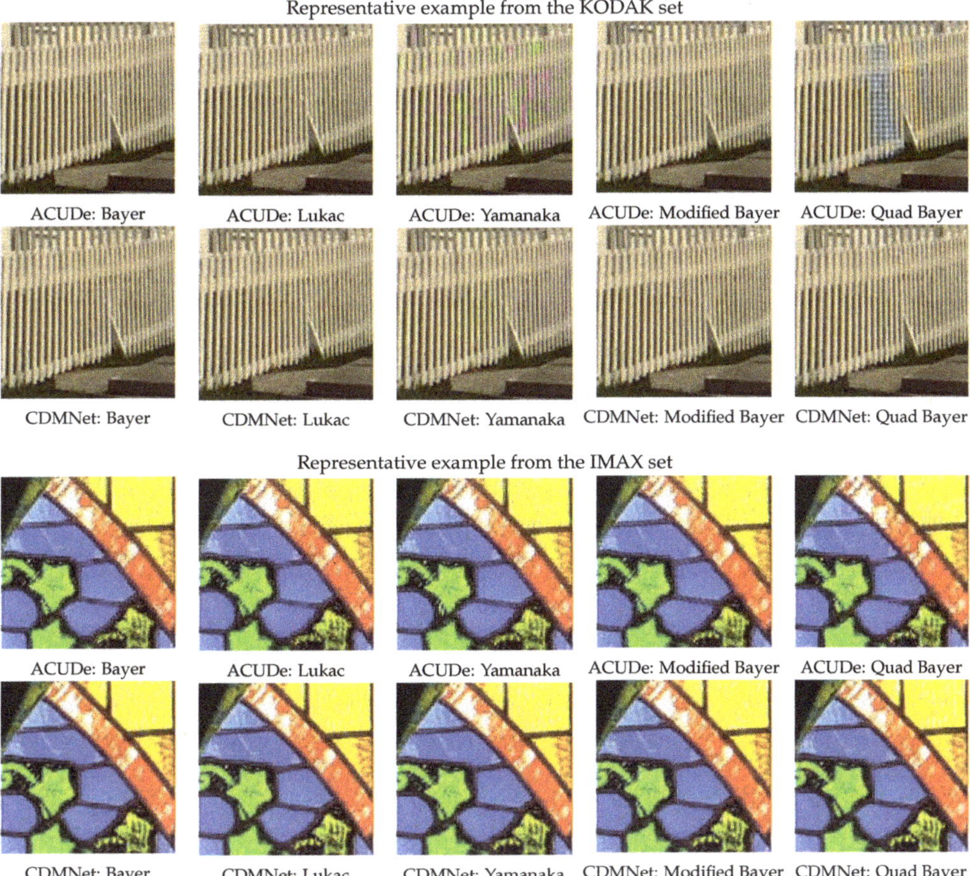

Figure 11. Results obtained for the representative examples from KODAK and IMAX data-sets. Two algorithms were analysed: (1) the modified ACUDe based on the algorithm presented in [28], as a state-of-the-art generic algorithm among the *classical* demosaicing techniques and (2) the modified CDMNet based on the algorithm presented in [25] as a state-of-the-art generic algorithm among the *learning-based* demosaicing techniques. The algorithms were tested on five Bayer-like CFA patterns. There are no big visual differences between the reconstructed images from different CFA inputs when the modified CDMNet (the representative of the *modern learning-based* demosaicing techniques) is applied.

6. Conclusions

Within this study, we analyse the quality performance of two state-of-the-art generic demosaicing algorithms for different Bayer-like CFA patterns. The first one (modified ACUDe) belongs to the group of *classical* demosaicing algorithms, while the second one (modified CDMNet) belongs to the group of *modern learning-based*, i.e., neural-network-based demosaicing algorithms. The aim of the performed analysis is to test the hypothesis that the *modern learning-based* demosaicing techniques (the NN-based approaches) overcome the high difference in quality performance for different CFA patterns and at the same time succeed at achieving quality performance that is higher than the quality performance of the state-of-the-art generic algorithm (here modified ACUDe) that belongs to the group of *classical* demosaicing techniques. The presented results of the analysis for the quality performance indeed show that the hypothesis is true. From this study, we derive a conclusion about the constantly increasing reconstruction power of the *modern learning based* demosaicing algorithms towards adaptiveness to any CFA design without loss in the reconstruction quality (which used to be dependent on the quality of the CFA design). This conclusion leads to a finding regarding the future opportunities for camera manufacturing and image reconstruction, specifically in combining lower hardware requirements with powerful reconstruction techniques. In other words, this means that, with the *modern learning-based* demosaicing methods, camera manufacturers have more freedom in the choice of the CFA pattern layout, without a noticeable loss in the image quality. In that direction, the patterns can be adapted to improve other image properties and facilitate various imaging tasks, such as the Quad Bayer that was designed to improve noise reduction in low-light imaging. Furthermore, this conclusion points towards the advantage of using the easily adaptive and retrainable neural-network based demosaicing techniques in various applications of multispectral imaging.

Author Contributions: A.S. and I.S. made the modifications of the two state-of-the-art demosaicing algorithms and conducted the experimental analysis. A.S., I.S. and J.A. wrote the paper. H.L. and J.A. proposed the hypothesis for which the experimental analysis was conducted. H.L., J.A., L.J. and W.P. supervised the research.

Funding: This work was partially funded by EU Horizon 2020 ECSEL JU research and innovation program under grant agreement 662222 (EXIST).

Acknowledgments: We would like to thank Pengwei Hao for making the implementation of the algorithm presented in [28] publicly available on http://www.eecs.qmul.ac.uk/~phao/CFA/acude/ and for the provided help for our analysis and the derived conclusions about the algorithm presented in [28]. We would also like to thank the authors of the algorithm presented in [25], for making the implementation publicly available. Many thanks to all authors of the publicly available image data-sets that were used in the performed analysis.

Conflicts of Interest: The authors declare no conflict of interest

References

1. Bayer, B.E. Color Imaging Array. U.S. Patent 3,971,065, 1976.
2. Lukac, R.; Plataniotis, K.N. Color filter arrays: Design and performance analysis. *IEEE Trans. Consum. Electron.* **2005**, *51*, 1260–1267. [CrossRef]
3. Hirakawa, K.; Wolfe, P.J. Spatio-spectral color filter array design for optimal image recovery. *IEEE Trans. Image Process.* **2008**, *17*, 1876–1890. [CrossRef] [PubMed]
4. Yamanaka, S. Solid State Color Camera. U.S. Patent 4,054,906, 1977.
5. Available online: https://www.sony.net/SonyInfo/News/Press/201807/18-060E/index.html (accessed on 20 July 2019).
6. Available online: https://www.ubergizmo.com/articles/quad-bayer-camera-sensor/ (accessed on 20 July 2019).
7. Realization of Natural Color Reproduction in Digital Still Cameras, Closer to the Natural Sight Perception of the Human Eye, Sony Corp. 2003. Available online: https://www.sony.net/SonyInfo/News/Press_Archive/200307/03-029E/ (accessed on 20 July 2019).
8. Kijima, T.; Nakamura, H.; Compton, J.; Hamilton, J. Image Sensor with Improved Light Sensitivity. U.S. Patent 20,070,177,236, 2007.
9. Lapray, P.J.; Wang, X.; Thomas, J.B.; Gouton, P. Multispectral filter arrays: Recent advances and practical implementation. *Sensors* **2014**, *14*, 21626–21659. [CrossRef] [PubMed]
10. Monno, Y.; Kikuchi, S.; Tanaka, M.; Okutomi, M. A practical one-shot multispectral imaging system using a single image sensor. *IEEE Trans. Image Process.* **2015**, *24*, 3048–3059. [CrossRef] [PubMed]
11. Gunturk, B.K.; Glotzbach, J.; Altunbasak, Y.; Schafer, R.W.; Mersereau, R.M. Demosaicking: Color filter array interpolation. *IEEE Signal Process. Mag.* **2005**, *22*, 44–54. [CrossRef]
12. Li, X.; Gunturk, B.; Zhang, L. Image demosaicing: A systematic survey. *Proc. SPIE* **2008**, *6822*, 68221J.
13. Menon, D.; Calvagno, G. Color image demosaicking: An overview. *Signal Process. Image Commun.* **2011**, *26*, 518–533. [CrossRef]
14. Monno, Y.; Kiku, D.; Tanaka, M.; Okutomi, M. Adaptive residual interpolation for color and multispectral image demosaicking. *Sensors* **2017**, *17*, 2787. [CrossRef]
15. Alleysson, D.; Susstrunk, S.; Hérault, J. Linear demosaicing inspired by the human visual system. *IEEE Trans. Image Process.* **2005**, *14*, 439–449.

16. Dubois, E. Frequency-domain methods for demosaicking of Bayer-sampled color images. *IEEE Signal Process. Lett.* **2005**, *12*, 847–850. [CrossRef]

17. Lu, Y.M.; Karzand, M.; Vetterli, M. Demosaicking by alternating projections: Theory and fast one-step implementation. *IEEE Trans. Image Process.* **2010**, *19*, 2085–2098. [CrossRef] [PubMed]

18. Menon, D.; Calvagno, G. Demosaicing based on wavelet analysis of the luminance component. In Proceedings of the 2007 IEEE International Conference on Image Processing, San Antonio, TX, USA, 16–19 September 2007; Volume 2; pp. II-181–II-184.

19. Aelterman, J.; Goossens, B.; De Vylder, J.; Pižurica, A.; Philips, W. Computationally efficient locally adaptive demosaicing of color filter array images using the dual-tree complex wavelet packet transform. *PLoS ONE* **2013**, *8*, e61846. [CrossRef] [PubMed]

20. Condat, L. A generic variational approach for demosaicking from an arbitrary color filter array. In Proceedings of the 2009 16th IEEE International Conference on Image Processing (ICIP), Cairo, Egypt, 7–10 November 2009; pp. 1605–1608.

21. Menon, D.; Calvagno, G. Regularization approaches to demosaicking. *IEEE Trans. Image Process.* **2009**, *18*, 2209–2220. [CrossRef] [PubMed]

22. Mairal, J.; Bach, F.R.; Ponce, J.; Sapiro, G.; Zisserman, A. Non-local sparse models for image restoration. *ICCV. Citeseer* **2009**, *29*, 54–62.

23. Wu, J.; Timofte, R.; Van Gool, L. Demosaicing based on directional difference regression and efficient regression priors. *IEEE Trans. Image Process.* **2016**, *25*, 3862–3874. [CrossRef] [PubMed]

24. Amba, P.; Thomas, J.B.; Alleysson, D. N-LMMSE demosaicing for spectral filter arrays. *J. Imaging Sci. Technol.* **2017**, *61*, 40407:1–40407:11. [CrossRef]

25. Cui, K.; Jin, Z.; Steinbach, E. Color Image Demosaicking Using a 3-Stage Convolutional Neural Network Structure. In Proceedings of the 2018 25th IEEE International Conference on Image Processing (ICIP), Athens, Greece, 7–10 October 2018; pp. 2177–2181.

26. Condat, L. A new random color filter array with good spectral properties. In Proceedings of the 2009 16th IEEE International Conference on Image Processing (ICIP), Cairo, Egypt, 7–10 November 2009; pp. 1613–1616.

27. Lukac, R.; Plataniotis, K.N. Universal demosaicking for imaging pipelines with an RGB color filter array. *Pattern Recognit.* **2005**, *38*, 2208–2212. [CrossRef]

28. Zhang, C.; Li, Y.; Wang, J.; Hao, P. Universal demosaicking of color filter arrays. *IEEE Trans. Image Process.* **2016**, *25*, 5173–5186. [CrossRef]

29. Amba, P.; Dias, J.; Alleysson, D. Random Color Filter Arrays are Better than Regular Ones. *J. Imaging Sci. Technol.* **2016**, *60*, 50406:1–50406:6. [CrossRef]

30. Amba, P.; Alleysson, D.; Mermillod, M. Demosaicing using Dual Layer Feedforward Neural Network. In *Color and Imaging Conference*; No.1; Society for Imaging Science and Technology: Springfield, VA, USA; Volume 2018, pp. 211–218.

31. Gharbi, M.; Chaurasia, G.; Paris, S.; Durand, F. Deep joint demosaicking and denoising. *ACM Trans. Graph. (TOG)* **2016**, *35*, 191. [CrossRef]

32. Tan, R.; Zhang, K.; Zuo, W.; Zhang, L. Color image demosaicking via deep residual learning. In Proceedings of the 2017 IEEE International Conference on Multimedia and Expo (ICME), Hong Kong, China, 10–14 July 2017; pp. 793–798

33. Kokkinos, F.; Lefkimmiatis, S. Deep image demosaicking using a cascade of convolutional residual denoising networks. In Proceedings of the European Conference on Computer Vision (ECCV), Munich, Germany, 8–14 September 2018; pp. 303–319.

34. Kokkinos, F.; Lefkimmiatis, S. Iterative Residual Network for Deep Joint Image Demosaicking and Denoising. *arXiv* **2018**, arXiv:1807.06403.

35. Syu, N.S.; Chen, Y.S.; Chuang, Y.Y. Learning deep convolutional networks for demosaicing. *arXiv Preprint* **2018**, arXiv:1802.03769.

36. Tan, D.S.; Chen, W.Y.; Hua, K.L. DeepDemosaicking: Adaptive image demosaicking via multiple deep fully convolutional networks. *IEEE Trans. Image Process.* **2018**, *27*, 2408–2419. [CrossRef]

37. Chakrabarti, A. Learning sensor multiplexing design through back-propagation. *Adv. Neural Inf. Process. Syst.* **2016**, 3081–3089.

38. Henz, B.; Gastal, E.S.; Oliveira, M.M. Deep joint design of color filter arrays and demosaicing. *Comput. Graph. Forum* **2018**, *37*, 389–399. [CrossRef]

39. Available online: http://r0k.us/graphics/kodak/ (accessed on 20 July 2019).
40. Available online: https://www4.comp.polyu.edu.hk/~cslzhang/DATA/McM.zip (accessed on 20 July 2019).
41. Available online: http://www.eecs.qmul.ac.uk/~phao/CFA/acude/ (accessed on 20 July 2019).
42. Shi, W.; Caballero, J.; Huszár, F.; Totz, J.; Aitken, A.P.; Bishop, R.; Rueckert, D.; Wang, Z. Real-time single image and video super-resolution using an efficient sub-pixel convolutional neural network. In Proceedings of the IEEE Conference on Computer Vision and Pattern Recognition, Las Vegas, NV, USA, 27–30 June 2016; pp. 1874–1883.
43. Ma, K.; Duanmu, Z.; Wu, Q.; Wang, Z.; Yong, H.; Li, H.; Zhang, L. Waterloo exploration database: New challenges for image quality assessment models. *IEEE Trans. Image Process.* **2016**, *26*, 1004–1016. [CrossRef]

Article

A Novel Method for Early Gear Pitting Fault Diagnosis Using Stacked SAE and GBRBM

Jialin Li [1], **Xueyi Li** [1], **David He** [1,2,*] **and Yongzhi Qu** [3]

1   School of Mechanical Engineering and Automation, Northeastern University, Shenyang 110000, China;
    jialinli_neu@163.com (J.L.); lixueyineu@gmail.com (X.L.)
2   Department of Mechanical and Industrial Engineering, The University of Illinois at Chicago, Chicago,
    IL 60607, USA
3   School of Mechanical and Electronic Engineering, Wuhan University of Technology, Wuhan 430000, China;
    yongzhiqu@hotmail.com
*   Correspondence: davidhe@uic.edu; Tel.: +86-132-6253-3830

Received: 16 January 2019; Accepted: 9 February 2019; Published: 13 February 2019

Abstract: Research on data-driven fault diagnosis methods has received much attention in recent years. The deep belief network (DBN) is a commonly used deep learning method for fault diagnosis. In the past, when people used DBN to diagnose gear pitting faults, it was found that the diagnosis result was not good with continuous time domain vibration signals as direct inputs into DBN. Therefore, most researchers extracted features from time domain vibration signals as inputs into DBN. However, it is desirable to use raw vibration signals as direct inputs to achieve good fault diagnosis results. Therefore, this paper proposes a novel method by stacking spare autoencoder (SAE) and Gauss-Binary restricted Boltzmann machine (GBRBM) for early gear pitting faults diagnosis with raw vibration signals as direct inputs. The SAE layer is used to compress the raw vibration data and the GBRBM layer is used to effectively process continuous time domain vibration signals. Vibration signals of seven early gear pitting faults collected from a gear test rig are used to validate the proposed method. The validation results show that the proposed method maintains a good diagnosis performance under different working conditions and gives higher diagnosis accuracy compared to other traditional methods.

Keywords: early gear pitting fault diagnosis; vibration signals; SAE; GBRBM

---

1. Introduction

Gears play an important role in mechanical transmission systems. It is necessary to diagnose gear faults to ensure stable and reliable operation of the systems. The methods of fault diagnosis can be roughly divided into two categories: model-driven methods and data-driven methods [1]. Model-based diagnostic methods require a deep understanding of the systems, and many parameter adjustments need to be performed to build the model. Therefore, this paper applies data-driven methods to diagnose gear faults. The data-driven diagnostic process involves two steps: (1) establish a data model based on known state data, (2) use the established model to diagnose mechanical faults. The fault diagnosis process can be regarded as the process of applying the model for pattern recognition. When building a fault diagnosis model, there are generally two processes: feature extraction and pattern recognition [2]. The purpose of feature extraction is to convert high-dimensional data into low-dimensional features, which can better perform pattern recognition. There are many methods for feature extraction such as statistical analysis methods, fast Fourier transform (FFT), Hilbert–Huang transform (HHT) [3], empirical mode decomposition (EMD) [4], wavelet transform (WT) [5], principal components analysis (PCA) [6], and so on. There are many traditional pattern recognition methods including Bayesian classifier [7], K-nearest neighbor (KNN) algorithms [8], artificial neural network

(ANN), support vector machine (SVM) [9], etc. Traditional ANN can only distinguish less complex features, and the diagnosis results are greatly affected by process of feature extraction and feature selection. In recent years, research on deep learning is getting popular. Deep learning can improve the shortcomings of traditional neural network. There are many types of deep learning methods applied in fault diagnosis, which can be divided into supervised methods such as deep neural network (DNN), convolutional neural network (CNN) [10] and unsupervised methods such as deep belief network (DBN), and autoencoder (AE).

Zhang et al. [11] used DNN to diagnose bearing faults and directly used the collected vibration signals as the inputs of neural network, removing the error caused by the feature extraction process. Tested by two publicly available data from University of Cincinnati Center for Intelligent Maintenance System (IMS) and Case Western Reserve University (CWRU), their proposed method was shown to be able to effectively diagnose bearing faults. Chen et al. [12] used a CNN to diagnose gearbox faults. FFT was performed on the vibration signals. Statistical methods were used to extract features from the time-domain and the frequency domain signals as inputs to the CNN. Chen et al. [13] applied ensemble empirical mode decomposition (EEMD) to extract features, and then applied DBN to classify the gear faults. Wang et al. [14] applied unsupervised continuous sparse autoencoder (CSAE) for feature learning and connected a layer of back propagation (BP) networks behind the CSAE. The training process first applied CSAE for unsupervised classification, and then used BP for supervised learning.

DBN has been used for fault diagnosis. The research using DBN for fault diagnosis is reviewed next. Tran et al. [15] used Teager–Kaiser energy operator (TKEO) and DBN to diagnose reciprocating compressor valves faults. In their paper, TKEO was proposed to estimate the amplitude envelopes. The collected vibration signal was processed by WT denoising, and then the time domain signal was converted into a frequency domain. Finally, the statistical methods were used to extract the feature as inputs of the DBN. The diagnostic method used by Han et al. [16] was similar to the method in reference [15], except that it adds a particle swarm optimization-support vector machine (PSO-SVM) to classify extracted parameters. Shao et al. [17] applied the dual-tree complex wavelet packet for feature extraction, and then used statistical methods for feature selection. Finally, an adaptive DBN was used for fault classification. Wang et al. [18] also applied statistical methods to process time-frequency domain signals, and then used DBN to detect multiple faults in axial piston pumps. Lee et al. [19] used DBN to diagnose the air handling unit (AHU). Ahmed et al. [20] combined DBN and softmax classifiers to diagnose rolling bearing faults. Tao et al. [21], He et al. [22], and Chen et al. [23] also applied statistical methods for feature extraction, and then applied DBN for fault classification.

Deutsch et al. [24] integrates DBN and a particle filter for bearing remaining useful life (RUL) prediction. Geng et al. [25] were inspired by the glial chains to improve the structure of the restricted Boltzmann machines (RBMs). An improved greedy layer-wise learning algorithm was used to improve the diagnostic accuracy. Ren et al. [26] combined deep belief networks and multiple models (DBN-MMs) to diagnose complex systems faults. Shao et al. [27] combined the CNN with the DBN to process the compressed sensing (CS). In addition, exponential moving average (EMA) technique was used to improve diagnostic accuracy of the constructed deep model. Jiang et al. [28] proposed a feature fusion DBN method to diagnose rotating machinery fault. Moreover, the locality preserving projection (LPP) was used to fusion deep features to further improve the quality of the deep features.

SAE has been used for fault diagnosis recently. The research using SAE to diagnose faults is reviewed next. Shao et al. [29] proposed ensemble deep auto-encoders (EDAEs) to diagnose bearing faults. The effects of different activation functions and various AEs on diagnostic results were discussed in the article. Maurya et al. [30] used stacked autoencoder to fuse the low-level feature. And then a multi-class SVM was used as classifier. Shao et al. [31] applied deep autoencoder to diagnose rotating machinery faults. The maximum cross entropy was used as the loss function and artificial fish swarm (AFS) algorithm was applied to optimize the key parameters of the deep autoencoder. Meng et al. [32] used denoising autoencoder to diagnose bearing faults. They improved the fault diagnosis rate by reusing the data points between the adjacent samples. The hyper parameter was adjusted by changing the number of units per layer to adapt to the different resilience of the layer.

This paper proposed integrates SAE with GBRBM to diagnose early gear pitting faults. The SAE is used to convert high-dimensional data into low-dimensional data, and GBRBM is used to accommodate the continuous distribution of the inputs. The rest of this paper is organized as follows. In Section 2, the proposed method based the stacked SAE and GBRBM is explained in details. In Section 3, the description of the experimental test rig used for collecting the vibration data for the seven gear pitting faults is provided. In Section 4, the validation results and the discussion of the validation results are presented. Finally, Section 5 concludes the paper.

2. The Proposed Method

2.1. Framework of Proposed Method

Most of the data-driven diagnosis methods involve separate manual feature extraction process. Manual feature extraction mostly relies on human expertise, and the manual feature extraction process is time-consuming and labor intensive. Moreover, the diagnostic results are greatly affected by the feature extraction method. Therefore, diagnostic methods that do not include separate manual feature process are more desirable. Inspired by the unsupervised learning process, this paper proposes a diagnostic method that combines supervised learning with unsupervised learning. The framework of the diagnostic method is shown in Figure 1.

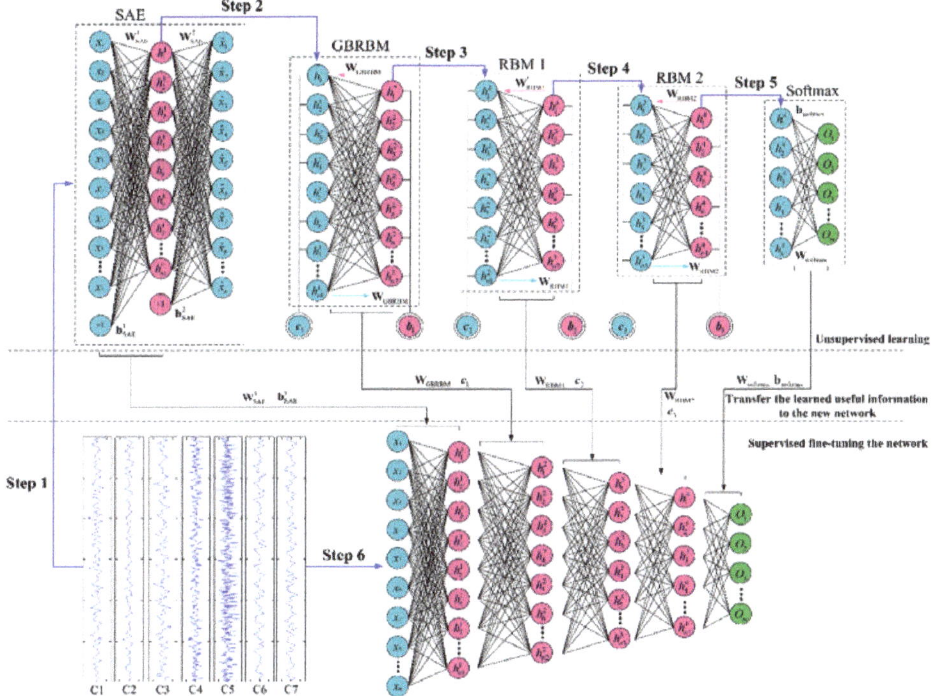

Figure 1. The framework of the proposed method.

As shown in Figure 1, the framework of the proposed method includes three parts: (1) unsupervised feature learning, (2) transfer the learned useful information to the new network, (3) supervised fine-tuning the restructured network. Stacked SAE, GBRBM and RBMs are combined to work as a simultaneous signal processing and unsupervised feature extraction process. The blue circles in the figure represent the input layer neurons, the red circles represent the hidden layer neurons, and the green circles represent the output layer neurons. The entire diagnostic model has a total of 6 layers of neurons. The specific training process contains 6 steps as shown in Figure 1, raw vibration signals are first used for feature extraction through unsupervised learning, and the data is forwarded through the SAE, GBRBM, two-layer RBM and softmax layers, then fine-tune the weights and biases from unsupervised learning process of each layer according to the cross entropy error function.

The network training in Figure 1 consists of 6 steps. Table 1 shows the detailed calculation principle for the 6 steps, and also includes input values, output values, and parameters transferred for each layer. Figure 1 and Table 1 in combination gives a general understanding of the training procedure. First, the unsupervised learning is performed layer by layer. Then, the learned useful information is transferred into the new network. Finally, supervised fine-tune is performed to adjust the entire network. The detailed equations are shown in Table 1.

Table 1. Detailed process of proposed method.

Overall process:
(1) Unsupervised: SAE→GBRBM→RBM$^{(1,2)}$→Softmax→(2) Supervised: Back propagation

Step 1: SAE training
Input: training data $\mathbf{x}$, $\mathbf{W}$ and $\mathbf{b}$, λ, β, η_1, max-epochs$^{(1)}$
for i to max-epochs$^{(1)}$

- $\mathbf{h} = sigm(\mathbf{W}_1\mathbf{x} + \mathbf{b}_1)$
- $\hat{\mathbf{x}} = sigm(\mathbf{W}_2\mathbf{x} + \mathbf{b}_2)$
- $J_{SAE} = J_{MSE} + \lambda \cdot J_{weight} + \beta \cdot J_{sparse}$
- $\Delta w_{ij}^l = \Delta_1 \partial J_{sparse}/\partial w_{ij}^l$, $\Delta b_i^l = \Delta_1 \partial J_{sparse}/\partial b_i^l$

  end
Output: $\mathbf{h}_{SAE}$, $\mathbf{W}_1^{SAE}$, $\mathbf{b}_1^{SAE}$
Step 2: GBRBM training
Input: $\mathbf{h}_{SAE}$, max-epochs$^{(2)}$, w_{ij}^1, c_i^1, b_j^1, σ_i^2, η_2, α_1
for i to max-epochs$^{(2)}$

- $p(v_i = v|\mathbf{h}) = N(v, c_i^1 + \sum_j w_{ij}^1 \cdot h_j, \sigma_i^2)$
- $p(h_j = 1|\mathbf{v}) = sigm(b_j^1 + \sum_i \frac{v_i}{\sigma_i^2} w_{ij}^1)$
- $\Delta w_{ij}^{new} = \alpha_1 \Delta w_{ij}^1 + \eta_2(< v_i^{(0)} h_j^{(0)} - v_i^{(1)} h_j^{(1)} >)$
- $\Delta b_j^{new} = \alpha_1 \Delta b_j^1 + \eta_2(< h_j^{(0)} - h_j^{(1)} >)$
- $\Delta c_i^{new} = \alpha_1 \Delta c_i^1 + \eta_2(< v_i^{(0)} - v_i^{(1)} >)$

  end
Output: $\mathbf{h}_{GBRBM}$, $\mathbf{W}_{GBRBM}$, $\mathbf{b}_{GBRBM}$
Step 3: RBM1 training
Input: $\mathbf{h}_{GBRBM}$, max-epochs$^{(3)}$, w_{ij}^2, c_i^2, b_j^2, η_3, α_2
for i to max-epochs$^{(3)}$

- $p(v_i = 1|\mathbf{h}) = sigm(c_i^2 + \sum_j w_{ij}^2 h_j)$

- $p(h_j = 1|\mathbf{v}) = sigm(b_j^2 + \sum_i v_i w_{ij}^2)$

- The update process of Δw_{ij}^{new}, Δb_j^{new} and Δc_i^{new} is similar to GBRBM

  end
Output: $\mathbf{h}_{RBM1}$, $\mathbf{W}_{RBM1}$, $\mathbf{b}_{RBM1}$

Step 4: RBM2 training
Input: $\mathbf{h}_{RBM1}$, max-epochs$^{(4)}$, w_{ij}^3, c_i^3, b_j^3, η_4, α_3

- The training process is similar to step 3.

Output: $\mathbf{h}_{RBM2}$, $\mathbf{W}_{RBM2}$, $\mathbf{b}_{RBM2}$
Step 5: Softmax layer
Input: $\mathbf{h}_{RBM2}$, w_{ij}^4, d_j

- $y_j = softmax(\sum_{i=1}^p (h_i w_{ij}^4 + d_j))$
- $softmax(z_i) = (e^{z_j} / \sum_{j=1}^q e^{z_j})$

Output: $\mathbf{y}$, $\mathbf{W}_{softmax}$, $\mathbf{d}$
Step 6: Back propagation
Input: $\mathbf{y}$, max-epochs$^{(5)}$, η_5
for i to max-epochs$^{(5)}$

- $\mathbf{W}_1^{SAE}$, $\mathbf{b}_1^{SAE}$, $\mathbf{W}_{GBRBM}$, $\mathbf{b}_{GBRBM}$, $\mathbf{W}_{RBM1}$, $\mathbf{b}_{RBM1}$, $\mathbf{W}_{RBM2}$, $\mathbf{b}_{RBM2}$, $\mathbf{W}_{softmax}$, $\mathbf{d}$ as the weight and bias of fully connected DNN.
- $E_{cross-entropy} = -((o \log y) + (1 - o) \log (1 - y))$
- $\Delta w = \frac{\partial E_{cross-entropy}}{\partial w}$, $\Delta b = \frac{\partial E_{cross-entropy}}{\partial b}$

  end all
Output: the trained network
Step 7: Test the trained network with test sample

2.2. Spare Autoencoder

Sparse autoencoder (SAE) [33,34] is an unsupervised learning network mainly used for data dimensionality reduction and feature extraction. The SAE includes three layers: input layer ($n + 1$ neuron), hidden layer ($m + 1$ neuron, $m < n$), and output layer (n neurons). Figure 1 show the structure of SAE, which can be seen to contain two processes of encoding and decoding.

The encoding process of SAE can be implemented by Equation (1), and the decoding process can be implemented by Equation (2).

$$\mathbf{h} = sigm(\mathbf{W}_1\mathbf{x} + \mathbf{b}_1) \tag{1}$$

$$\hat{\mathbf{x}} = sigm(\mathbf{W}_2\mathbf{h} + \mathbf{b}_2) \tag{2}$$

where $\mathbf{x}$ is the input matrix, $\mathbf{W}_1$ and $\mathbf{b}_1$ are the weight matrix and bias vector between input layer and hidden layer, $\mathbf{h}$ is the hidden matrix, $\mathbf{W}_2$ and $\mathbf{b}_2$ are the weight matrix and bias vector between hidden layer and output layer, and $\hat{\mathbf{x}}$ is the output matrix; function $sigm\,(\cdot)=1/(1 + e^{-z})$.

When the mean square error (MSE) is used as the loss function of SAE, the expected processing results usually cannot be achieved. In order to make SAE perform better, a new loss function is designed as Equation (3), which consists of three parts: J_{MSE}, J_{weight}, and J_{sparse} [35]. The purpose of using J_{weight} is to control the value of the connected weights to avoid overfitting [36]. The added J_{sparse} is a sparsity penalty term, which can make SAE learn more features from the input by forcing SAE to maintain a degree of sparsity [37,38].

$$J_{SAE} = J_{MSE} + \lambda \cdot J_{weight} + \beta \cdot J_{sparse} \tag{3}$$

where J_{MSE} is the mean square error term as show in Equation (4), J_{weight} is the weight penalty item as show in Equation (5), J_{sparse} is the sparsity penalty term as show in Equation (6), λ is the regularization parameter of weight term, and β is the coefficient of sparsity penalty term.

$$J_{MSE} = \frac{1}{2s}\sum_{i=1}^{s}\|x_i - \hat{x}_i\| \tag{4}$$

$$J_{weight} = \frac{1}{2}\sum_{l=1}^{k-1}\sum_{j=1}^{n_l}\sum_{i=1}^{n_{l-1}}W_{ij}^l \tag{5}$$

$$J_{spare} = \sum_{j=1}^{m}\mathrm{KL}(\rho\|\hat{\rho}_j) = \sum_{j=1}^{m}\left(\rho\log\left(\frac{\rho}{\hat{\rho}_j}\right) + (1-\rho)\log\left(\frac{1-\rho}{1-\hat{\rho}_j}\right)\right) \tag{6}$$

$$\hat{\rho}_j = \frac{1}{s}\sum_{i=1}^{s}\mathbf{h}_{ij} \tag{7}$$

where s is the sample size of training set, k is the number of layers in the network, n_l is the neurons in layer l, ρ is the set neuron sparsity parameter, and $\hat{\rho}_j$ is the sparsity of the j-th neuron as show in Equation (7).

2.3. Develop the GBRBM based on RBM

Restricted Boltzmann Machine (RBM) is the basic component of the deep belief network (DBN) [39,40]. Similar to the SAE, it is also an unsupervised learning network that can be used for feature extraction. The RBM contains two layers: visible layer (contains n visible units) and hidden layer (contains m hidden units). The neurons in the same layer are not connected, and neurons in different layer are connected in each other. The weight matrix connecting the two layers is denoted by $\mathbf{W}$, the bias vector of the visible layer is denoted by $\mathbf{c}$, and the bias vector of the hidden layer is denoted by $\mathbf{b}$.

Inspired by statistical physics, it can be found that any probability distribution can be transformed into an energy-based model. The joint probability distribution of the visible layer and the hidden layer is proportional to the energy equation [41], as shown in Equation (8). And the joint probability distribution of v and h can be obtained as shown in Equation (9).

$$-\log P(\mathbf{v},\mathbf{h}) \propto E(\mathbf{v},\mathbf{h}|\theta) = -\sum_{i=1}^{n} c_i v_i - \sum_{j=1}^{m} b_j h_j - \sum_{i=1}^{n}\sum_{j=1}^{m} v_i w_{ij} h_j \tag{8}$$

$$P(\mathbf{v},\mathbf{h}|\theta) = \frac{1}{Z(\theta)} \exp(-E(\mathbf{v},\mathbf{h}|\theta)) \tag{9}$$

where v_i is the visible layer unit, h_j is the hidden layer unit, w_{ij} is the weights between visible layer and hidden layer, c_i and b_j are the bias of two layers; m hidden units in hidden layer, n visible units in visible layers, $\theta = \{w_{ij}, c_i, b_j\}$ are the parameters of RBM, and $Z(\theta) = \sum_n \sum_m \exp(-E(\mathbf{v},\mathbf{h}|\theta))$ is a partition function.

The probability function of the visible layer is given by Equation (10).

$$\begin{aligned} P(\mathbf{v};\theta) \;&= \sum_h P(v,h;\theta) \\ &= \frac{1}{Z(\theta)} \sum_h \exp\left(\sum_{i=1}^{n} c_i v_i + \sum_{j=1}^{m} b_j h_j + \sum_{i=1}^{n}\sum_{j=1}^{m} v_i w_{ij} h_j \right) \\ &= \frac{1}{Z(\theta)} \exp\left(\sum_{i=1}^{n} c_i v_i \right) \times \prod_{j=1}^{m} \sum_h \exp\left(b_j h_j + \sum_{i=1}^{n} v_i w_{ij} h_j \right) \end{aligned} \tag{10}$$

Combining Equations (9) and (10), the conditional probability of the hidden layer can be obtained as shown in Equation (11).

$$\begin{aligned} P(\mathbf{h}|\mathbf{v};\theta) \;&= \frac{P(\mathbf{v},\mathbf{h};\theta)}{P(\mathbf{v};\theta)} \\ &= \prod_j \frac{\exp\left(b_j h_j + \sum_{i=1}^{n} v_i w_{ij} h_j \right)}{\sum_h \left(b_j h_j + \sum_{i=1}^{n} v_i w_{ij} h_j \right)} = \prod_j P(h_j|v) \end{aligned} \tag{11}$$

Similarly, the conditional probability of the visible layer can be based on the joint probability of **v** and **h** divided by independent probability of hidden layer, as show in Equation (12).

$$P(\mathbf{v}|\mathbf{h};\theta) = \frac{P(\mathbf{v},\mathbf{h};\theta)}{P(\mathbf{h};\theta)} = \prod_i P(v_i|h) \tag{12}$$

The neurons in the same layer are not connected, meaning that the units are conditionally independent. So the conditional probability of the visible layer and hidden layer can be calculated by Equations (13) and (14).

$$P(v_i = 1|\mathbf{h}) = sigm\left(c_i + \sum_j w_{ij} h_j \right) \tag{13}$$

$$P(h_j = 1|\mathbf{v}) = sigm\left(b_j + \sum_i v_i w_{ij} \right) \tag{14}$$

where $sigm(x) = 1/(1 + \exp(-x))$ is the sigmoid function.

The parameter update of the RBM can be obtained by performing a stochastic gradient descent on the negative log-likelihood probability of the training data. The gradient of the negative log probability visible layer to the network parameters can be calculated by Equations (15)–(17). The value of <·>data is easy to get, but the value of $<\cdot>_{model}$ is difficult to get. Therefore, the contrastive divergence (CD) algorithm was proposed by Hinton [42].

$$\frac{\partial \log p(\mathbf{v};\theta)}{\partial w_{ij}} = (<v_i h_j>_{data} - <v_i h_j>_{model}) \tag{15}$$

$$\frac{\partial \log p(\mathbf{v};\theta)}{\partial \mathbf{b}} = (<h_j>_{data} - <h_j>_{model}) \tag{16}$$

$$\frac{\partial \log p(\mathbf{v};\theta)}{\partial \mathbf{c}} = (<v_i>_{data} - <v_i>_{model}) \tag{17}$$

where $<\cdot>_{data}$ indicates expectations for data distribution and $<\cdot>_{model}$ is the expectation of the distribution of the model definition.

Both the visible layer and the hidden layer of RBM are binary layers. It is not appropriate to construct the RBM with the binary visible layer when the input is a continuous valued data. So this paper is to develop the Gauss-Binary RBM (GBRBM) [43–45] instead of standard RBM, and the energy function of the standard RBM in Equation (8) is changed to Equation (18).

$$E(\mathbf{v},\mathbf{h}|\theta) = -\sum_{i=1}^{n} \frac{(v_i - c_i)^2}{2\sigma_i^2} - \sum_{j=1}^{m} b_j h_j - \sum_{i=1}^{n}\sum_{j=1}^{m} \frac{v_i}{\sigma_i^2} w_{ij} h_j \tag{18}$$

where σ_i^2 is the variance of Gaussian distribution.

With the energy equation in Equation (18), the conditional probability between the visible layer and the hidden layer can be obtained according to the derivation process in Section 2.2.

$$P(v_i = v|\mathbf{h}) = N\left(v; c_i + \sum_j w_{ij} h_j, \sigma_i^2\right) \tag{19}$$

$$P(h_j = 1|\mathbf{v}) = sigm\left(b_j + \sum_i \frac{v_i}{\sigma_i^2} w_{ij}\right) \tag{20}$$

where $N(\cdot, \mu, \sigma_i^2)$ is Gaussian distribution, also called normal distribution, μ is the mean, and σ_i^2 is the variance.

The softmax classification layer is commonly used in the last layer of the neural network, and its working principle is shown in Equations (21) and (22).

$$y_j = softmax(\sum_{i=1}^{p} (h_i w_{ij} + d_j)) \tag{21}$$

$$softmax(z_i) = e^{z_j} / \sum_{j=1}^{q} e^{z_j} \tag{22}$$

where w_{ij} and d_j are weights and bias of softmax layer, h_i is the input of softmax layer, p is the number of neurons in input layer, and q is the number of neurons in output layer.

3. Experiment Setup and Data Acquisition

In this paper, vibration data collected from experiments of seven gears with early gear pitting faults on a gear test rig were used to validate the proposed method. Figure 2 shows the gear test rig and the seven gears with the early gear pitting faults. The gearbox in the test rig consists of a pair of spur gears. The pinion gear is the driving gear (including 40 teeth, module 3 mm), and the large gear is the driven gear (including 72 teeth, module 3 mm). The gearbox is powered by two Siemens servo motors with a power of 45 kW. Motor 1 is the driving motor and motor 2 is the loading motor.

The gearbox is equipped with a lubrication and cooling system. The tri-axial acceleration sensor was mounted on the gearbox housing (the red box in the figure) with a sampling rate of 10240 Hz, and the vibration signals in the three directions of X, Y and Z were collected.

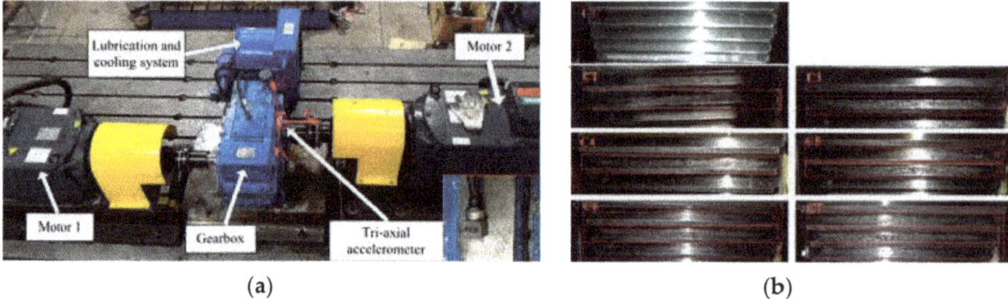

(a)   (b)

Figure 2. (a) Experimental test rig (b) gear pitting type.

The gear pitting faults were artificially manufactured by the drill on the driven gear surface. The specific conditions of the gear pitting faults are shown in Table 2. The fault degree is gradually increased and the latter one fault includes all of the previous fault conditions.

Table 2. Driven gear pitting type.

Label	Gear Pitting Type		
	72th Tooth	**First Tooth**	**Second Tooth**
C1	healthy	healthy	healthy
C2	healthy	10% in middle	healthy
C3	healthy	30% in middle	healthy
C4	healthy	50% in middle	healthy
C5	10% in middle	50% in middle	healthy
C6	10% in middle	50% in middle	10% in middle
C7	30% in middle	50% in middle	10% in middle

The vibration signals were collected under 25 working conditions. The 25 working conditions included combinations of five speeds (100–500 rpm) and five torque levels (100–500 Nm). Taking the working condition of 500 rpm–500 Nm as an example, each of seven gear types performed five independent data acquisitions and resulted in a total of 35 sets (120,000 data points per set) of data. 80% of all the data was used for training and the remaining data was used for testing. Hence, a training data matrix of 120,000 × 28 and testing data matrix of 120,000 × 7 were generated.

If the data matrix is directly used as the inputs, the network will be complex and the training will be slow. Therefore each data set was divided into several segmentations. For the sampling rate of 10240 Hz and a rotation speed of 500 RPM, approximately 1200 data points per gear rotation can be computed. In each segment, 300 data points (quarter of the collected data per gear rotation) were included [46]. In this case, the training data matrix dimension was 300 × 11200 and test data matrix dimension was 300 × 2800. Figure 3a shows sample vibration signals of the seven gears in Z-axis under 500 rpm–500 Nm working condition and Figure 3b represents one segment of the corresponding sample vibration signals.

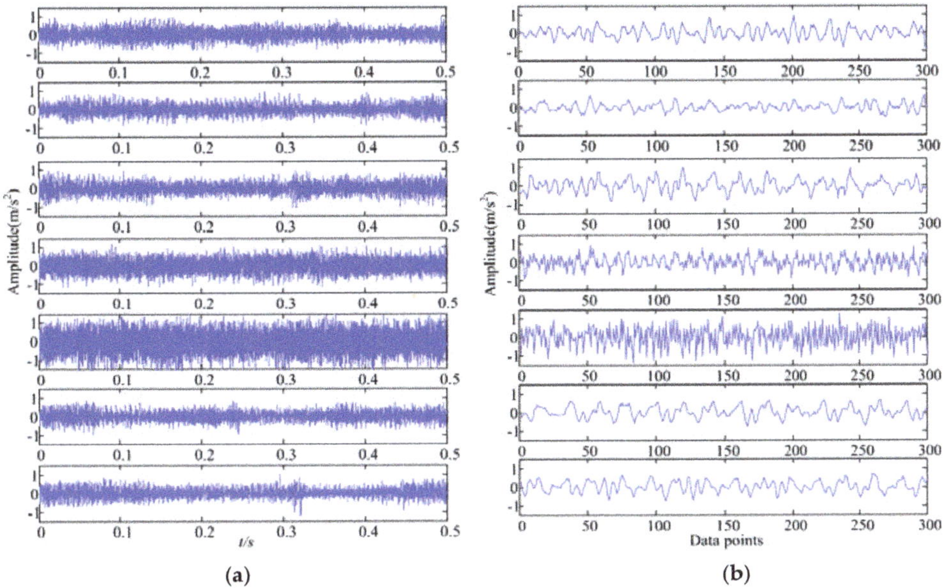

Figure 3. Sample vibration signals of seven gear types in Z-axis under 500 rpm–500 Nm: (**a**) signal with length of 0.5 s, (**b**) signal segment containing 300 data points.

4. Results and Discussion

4.1. PCA Data Visualization During the Training Process

To show the effectiveness by stacking SAE and GBRBM for extracting useful gear pitting fault information from the raw vibration signals, the network was trained with data from working condition 500 rpm–500 Nm. A total of six layers of neurons constitute the proposed diagnostic model, as shown in Figure 1. The structure of the proposed diagnostic model had the following structure: SAE: 300 × 300 (300 neurons in the input layer and 300 neurons in the hidden layer), GRRBM: 300 × 200 (300 neurons in the visible layer and 200 neurons in the hidden layer), RBM 1: 200 × 100 (200 neurons in the visible layer and 100 neurons in the hidden layer), RBM 2: 100 × 50 (100 neurons in the visible layer and 50 neurons in the hidden layer), Softmax: 50 × 7 (50 neurons in the input layer and seven neurons in the output layer). The size of the weight matrix and the biases were determined by the structure of the proposed model. The initial weights (W1 and W2) of SAE layer were randomly generated between 0 and 1. The initial weights of the softmax layer were randomly generated between 0 and 0.5. The remaining initial weights and biases were set to 0. The proposed diagnostic model was trained layer by layer. Steps 1, 2, 3, 4, and 6 were trained in 300 epochs, respectively. The parameter λ of SAE layer was set to be 0.005, β set to 1.5, and ρ set to 0.1. The learning rate of GBRBM was set to 0.005, the learning rate of RBM set to 0.5, and the learning rate of the back propagation process set to 0.05. The minimum training error of the back propagation process was set to 0.05. The entire network was calculated on a mini-batch with the batch size set to 100. There are many related parameters affecting the performance of the diagnostic model. The key parameters such as learning rate, structure of the network, and training epochs that have a great impact on the diagnostic results will be discussed in Section 4.3 below.

The outputs of each layer in the network structure were obtained and these outputs were further processed by PCA. The first two principal components of the PCA results are used to draw a scatter plot in Figure 4 to show the changes of data. The effectiveness of each layer of the network can be judged by observing the changes in the data through each layer of the neural network. In the experiment, the training and testing of the diagnostic model were performed using MATLAB 2014a software. The PCA results shown in Figure 4 were also obtained using the MATLAB codes. All the computational experiments were carried out on a PC with Windows 7 system and a CPU of Intel(R) Core i5-6500 @ 3.2GHz.

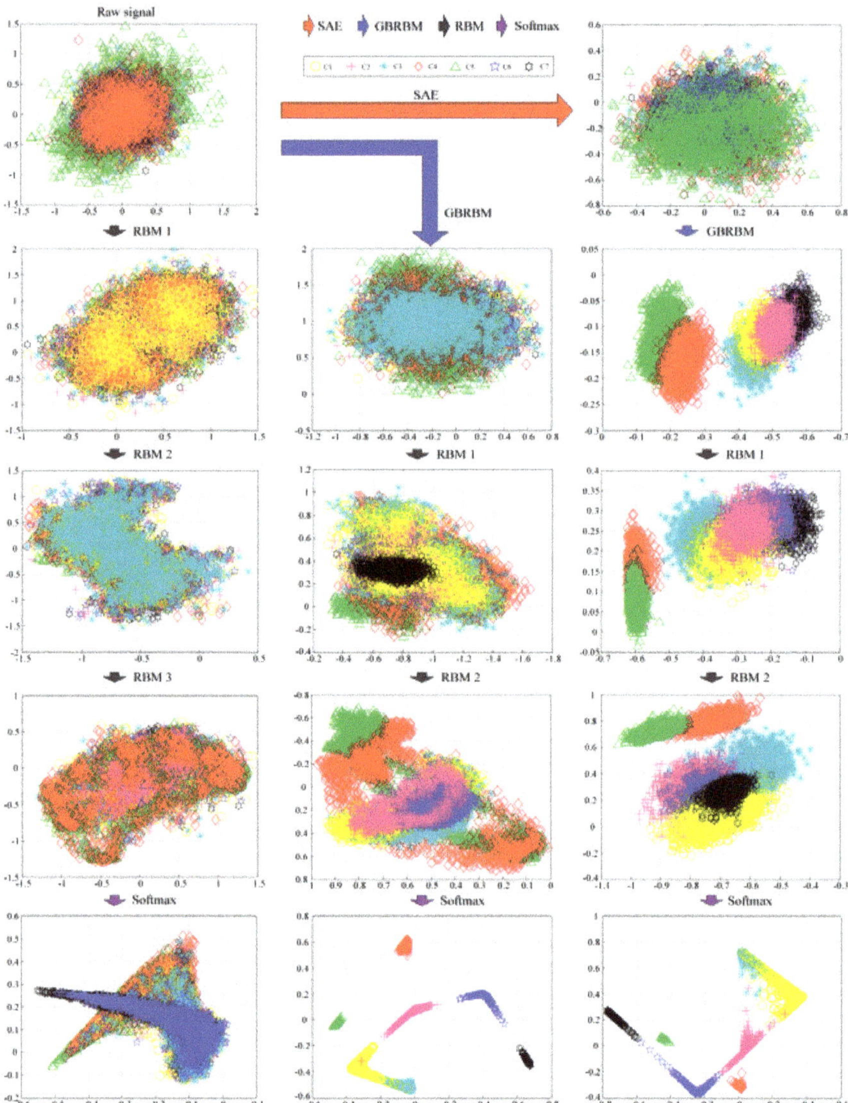

Figure 4. Each layer PCA result of three methods.

In Figure 4, three methods are shown. The first column in Figure 4 represents a standard DBN. The middle column represents the method with the first RBM layer of the standard DBN replaced with a GBRBM. The third column represents the proposed method by adding the SAE layer. As can be seen in Figure 4, the proposed method has the best fault separation result, and the separation result of the middle method is better than the standard DBN. Also seen from Figure 4, as the data moves from top down, the level of the fault separation is getting better. Figure 5 shows the confusion matrix of the gear pitting fault diagnosis results of the three methods. Again, as shown in Figure 5, the proposed method has the best diagnosis accuracy of 0.9346, the method with the first RBM layer of the standard DBN replaced with a GBRBM has a diagnosis accuracy of 0.8939, and the standard DBN has the worst accuracy of 0.3954. Even though the confusion matrix shown in Figure 5b looks similar to that in Figure 5c obtained by the proposed method, the diagnostic accuracy for the confusion matrix shown in Figure 5b is 0.8939 while the diagnostic accuracy for the confusion matrix shown in Figure 5c is 0.9346. Therefore, the proposed method gives more accurate diagnosis results. As shown in Figure 5, the graph located at the 2nd row in the middle column represents the PCA result without going through the SAE layer, while the graph located at the 2nd row in the 3rd column represents the PCA result after being processed by the SAE layer. By comparing these two graphs in Figure 4, one should note that the PCA result obtained by the SAE layer in the proposed method gives a better pitting fault separation. The results have shown the effectiveness of SAE layer in the proposed method for extracting useful fault features when it is used for processing the vibration signals.

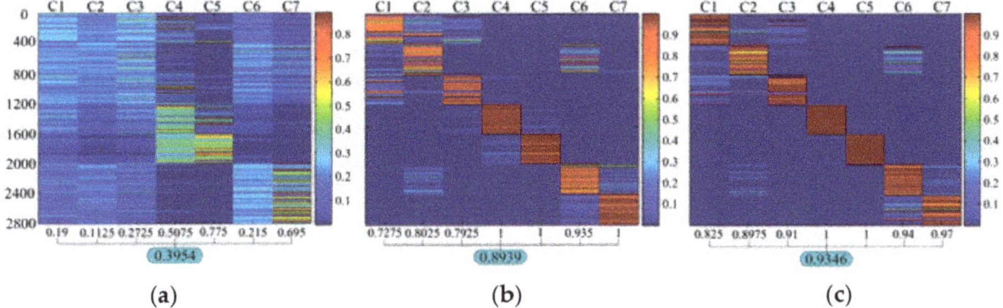

Figure 5. Confusion matrix of the fault diagnosis results: (**a**) standard DBN, (**b**) the first RBM layer of DBN replaced by GBRBM, (**c**) the proposed method.

4.2. Diagnostic Results of Proposed Method

Figure 6 shows the diagnostic accuracy of the proposed method and the other seven traditional methods. The 7 traditional methods include: (1) The first RBM layer of DBN replaced by GBRBM, (2) standard DBN, (3) standard DNN, (4) ANN with time domain vibration features, (5) ANN with frequency domain vibration features, (6) SVM with time domain vibration features, and (7) SVM with frequency domain vibration features. The results include the diagnostic accuracy for each gear pitting fault condition under 500 rpm–500 Nm working condition and the averaged accuracy over seven gear pitting fault conditions. From Figure 6, in comparison with other methods, the performance of the proposed method is significantly better than other methods. It can also be seen that the diagnostic accuracy for gear pitting conditions C4 and C5 is maintained at a high level in various methods, indicating that they are easier to diagnose than other fault conditions. This can be explained by observing the vibration signal in Figure 3b. It can be found that the vibration signal of C4 and C5 are clearly distinguished from the other gear pitting fault signals.

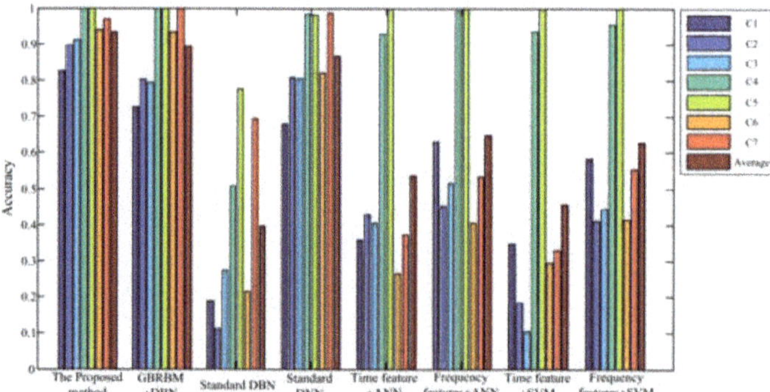

Figure 6. Diagnosis accuracy for all seven gear pitting conditions and the averaged accuracy under 500 rpm–500 Nm working condition.

Figure 7 shows the averaged diagnostic accuracy over all seven gear pitting conditions under 500 rpm–500 Nm working condition in ten trials with eight different methods. It can be seen that the proposed method has the highest diagnostic accuracy. In comparison with the proposed method, the accuracy of the method with the first RBM layer of the standard DBN replaced with a GBRBM is slightly lower. The standard DNN methods also have more prominent diagnosis results.

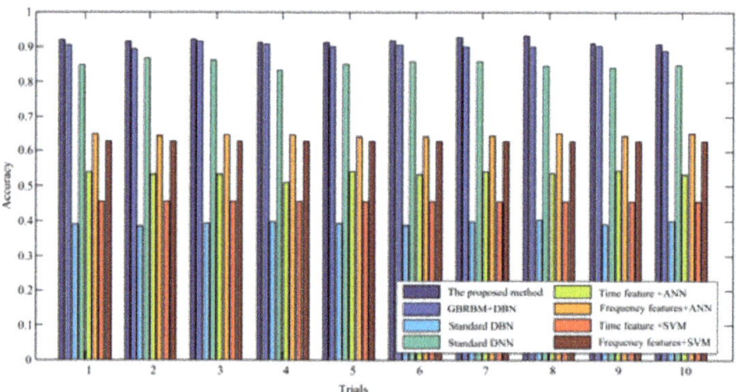

Figure 7. Averaged diagnosis accuracy of the ten trails under 500 rpm–500 Nm working condition.

As shown in Figures 6 and 7, among the methods compared with the proposed method, standard DNN has shown a competitive performance under the 500 rpm–500 Nm working condition. To show the performance of the proposed method in comparison with DNN for all the working conditions, the vibration signals under 25 working conditions were used compute the diagnostic accuracy for both the proposed method and the standard DNN. The results are provided in Figure 8, Tables 3 and 4.

In Figure 8, the averaged diagnosis accuracy over seven gear pitting conditions under 25 working conditions is provided for both the proposed method and the standard DNN. Further, the average accuracy over all five torque levels for each speed in Figure 8 is computed and provided in Table 3. The average accuracy over all five speeds for each torque level in Figure 8 is computed and provided in Table 4.

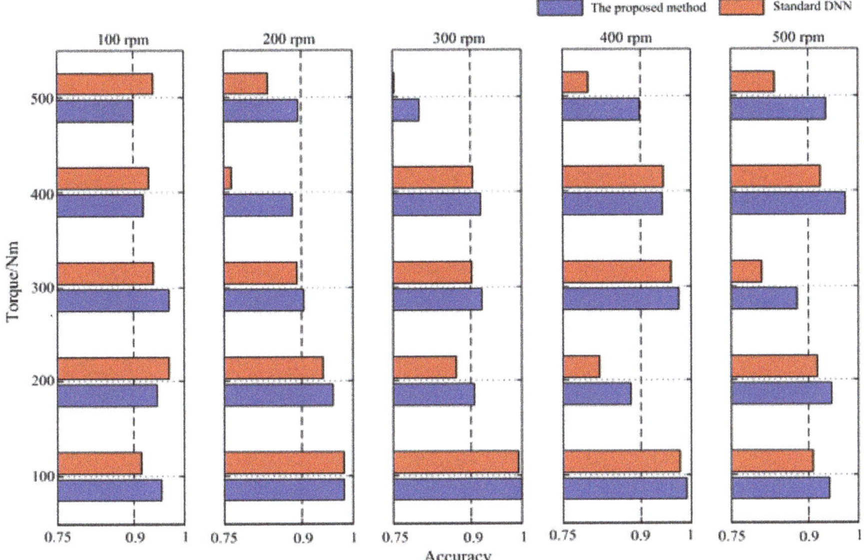

Figure 8. Diagnostic accuracy under 25 working condition.

Table 3. Averaged accuracy under 5 speeds.

Speed	Proposed Method	Standard DNN
100 rpm	**0.9374**	0.9372
200 rpm	**0.9245**	0.8824
300 rpm	**0.9091**	0.8831
400 rpm	**0.9372**	0.9003
500 rpm	**0.9344**	0.8791

Table 4. Averaged accuracy under 5 torques.

Torque	Proposed Method	Standard DNN
100 Nm	**0.9729**	0.9546
200 Nm	**0.9279**	0.9036
300 Nm	**0.9294**	0.8997
400 Nm	**0.9275**	0.8935
500 Nm	**0.8848**	0.8307

It can be seen from Figure 8, Tables 3 and 4 that the average diagnostic accuracy of the proposed method is higher than that of the standard DNN under various speeds and torque conditions.

It can be seen from Tables 3 and 4 that the diagnostic accuracy under 100 Nm working condition can reach to 0.9729. In order to prove the repeatability of the diagnosis results, five consecutive diagnoses were performed for five working conditions under 100 Nm. The diagnostic results are shown in Table 5. The averaged diagnostic accuracy of the five diagnosis results under 100Nm working condition is 0.9744, indicating that the proposed diagnostic method has high diagnostic reliability.

Table 5. Diagnosis accuracy of 5 trials under 5 working conditions.

Working Condition		Trial 1	Trial 2	Trial 3	Trial 4	Trial 5	Row Average
	100 rpm-100 Nm	0.9546	0.9554	0.9621	0.9354	0.9843	0.9584
	200 rpm-100 Nm	0.9861	0.9636	0.9736	0.9236	0.9857	0.9665
100 Nm	300 rpm-100 Nm	1	0.9986	0.9975	0.9989	0.9961	0.9982
	400 rpm-100 Nm	0.9954	0.9968	0.9950	0.9921	0.9961	0.9951
	500 rpm-100 Nm	0.9582	0.9557	0.9446	0.9550	0.9557	0.9539
Column Average		0.9789	0.9740	0.9746	0.9610	0.9836	**0.9744**

4.3. The Effect of the Parameters on the Diagnostic Accuracy

To investigate effect of the parameters of the proposed method on the performance of the gear pitting fault diagnosis, experiments were performed. In the first experiment, diagnostic accuracy results with epochs increased from 30 to 300 in an increment of 5 were obtained. In the network structure of the proposed method, the number of neurons in the input layer and the output layer were 300 and 7.

In order to investigate the impact of the network structure on the performance of the proposed method, a structure parameter $N\lambda$ was designed to represent the middle layer. Let $N\lambda$ be an integer coefficient between 1 and 10. In this case, the network structure of the proposed method can be represented as: $300\text{-}N\lambda\times(30\text{-}20\text{-}10\text{-}5)\text{-}7$. In the second experiment, diagnostic accuracy results with $N\lambda$ increased from 1 to 10 in an increment of 1 were obtained. The results of the first and second experiments are provided in Figure 9. From Figure 9a, the average accuracy of ten trials gradually increases when the training epochs increased from 30 to 120, and reached to constant level after 120 epochs. Figure 9b shows the effect of the parameter N_λ on the diagnostic accuracy of the network structure. When N_λ is increased from 1 to 4, the diagnostic accuracy is greatly improved. However, as N_λ reaches over 4, the improvement becomes insignificant.

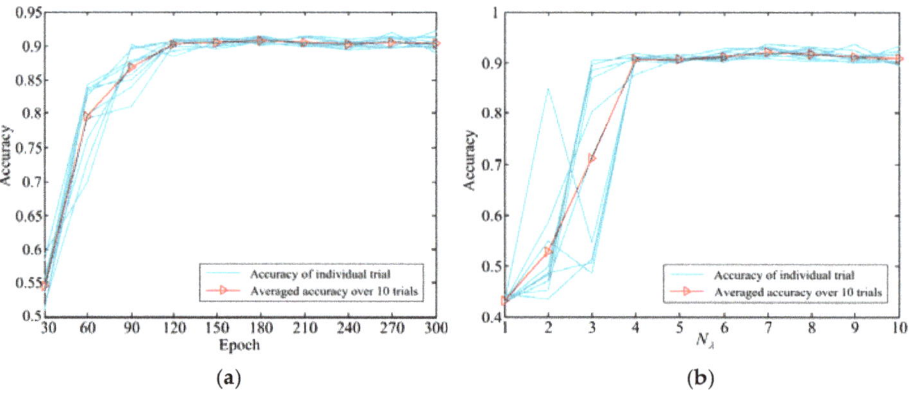

Figure 9. Parameters affecting the diagnosis accuracy: (a) epochs, (b) N_λ.

To investigate the impact of the learning rate on the performance of the proposed method, in the third experiment, diagnostic accuracy results with the different learning rates (lr) in RBM and GBRBM were obtained. The results are provided in Figure 10. As seen from Figure 10, the learning rate of GBRBM has a greater impact on the diagnostic accuracy. When the learning rate of GBRBM is greater than 0.03, the accuracy decreased rapidly.

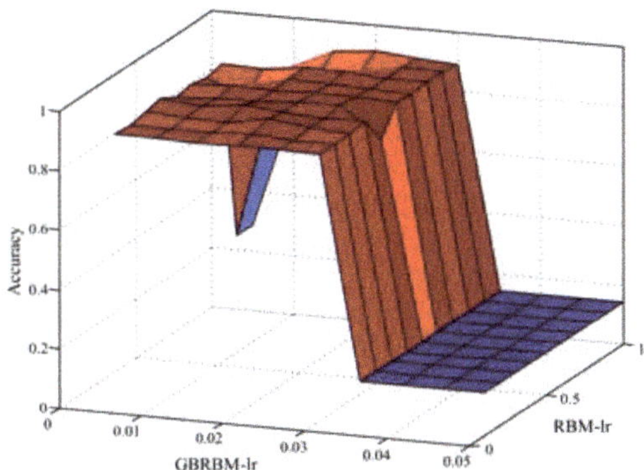

Figure 10. The influence of learning rate on the diagnosis accuracy.

5. Conclusions

In this paper, a novel method for early gear pitting fault diagnosis with raw vibration signals as direct inputs was presented. The method was developed by stacking a spare autoencoder (SAE) and a Gauss-Binary restricted Boltzmann machine (GBRBM). The vibration data collected from the gear test rig was used to validate the diagnostic capability of the proposed method. The validation results have shown that the proposed method is capable of gear pitting fault diagnosis with high accuracy. The performance of the proposed method was also compared with other 7 methods including: (1) The first RBM layer of DBN replaced by GBRBM, (2) standard DBN, (3) standard DNN, (4) ANN with time domain vibration features, (5) ANN with frequency domain vibration features, (6) SVM with time domain vibration features, and (7) SVM with frequency domain vibration features. The results of the comparison have shown that the proposed method outperform the other methods in terms of the gear pitting fault diagnostic accuracy. The effect of parameters of the proposed method on the diagnostic performance of the proposed method was investigated and discussed in the paper.

Author Contributions: Conceptualization, D.H. and J.L.; methodology, J.L.; software, J.L.; validation, D.H., Y.Q., J.L. and X.L.; formal analysis, J.L.; investigation, D.H. and Y.Q.; resources, D.H. and Y.Q.; data curation, J.L. and X.L.; writing—original draft preparation, J.L.; writing—review and editing, D.H.; visualization, J.L.; supervision, D.H.; project administration, J.L. and X.L.; funding acquisition, D.H. and Y.Q.

Funding: This work was funded by the National Natural Science Foundation of China (No. 51675089 and No. 51505353).

Conflicts of Interest: The authors declare no conflict of interest.

References

1. Khan, S.; Yairi, T. A review on the application of deep learning in system health management. *Mech. Syst. Signal Process.* **2018**, *107*, 241–265. [CrossRef]
2. Liu, R.; Yang, B.; Zio, E.; Chen, X. Artificial intelligence for fault diagnosis of rotating machinery: A review. *Mech. Syst. Signal Process.* **2018**, *108*, 33–47. [CrossRef]
3. Soualhi, A.; Medjaher, K.; Zerhouni, N. Bearing Health Monitoring Based on Hilbert-Huang Transform, Support Vector Machine, and Regression. *IEEE Trans. Instrum. Meas.* **2015**, *64*, 52–62. [CrossRef]
4. Mejia-Barron, A.; Valtierra-Rodriguez, M.; Granados-Lieberman, D.; Olivares-Galvan, J.C.; Escarela-Perez, R. The application of EMD-based methods for diagnosis of winding faults in a transformer using transient and steady state currents. *Measurement* **2018**, *117*, 371–379. [CrossRef]
5. Bhattacharyya, A.; Pachori, R.; Upadhyay, A.; Acharya, U. Tunable-Q Wavelet Transform Based Multiscale Entropy Measure for Automated Classification of Epileptic EEG Signals. *Appl. Sci.* **2017**, *7*, 385. [CrossRef]
6. Gajjar, S.; Kulahci, M.; Palazoglu, A. Real-time fault detection and diagnosis using sparse principal component analysis. *J. Process Control* **2018**, *67*, 112–128. [CrossRef]

7. Bennacer, L.; Amirat, Y.; Chibani, A.; Mellouk, A.; Ciavaglia, L. Self-Diagnosis Technique for Virtual Private Networks Combining Bayesian Networks and Case-Based Reasoning. *IEEE Trans. Autom. Sci. Eng.* **2015**, *12*, 354–366. [CrossRef]

8. Denœux, T.; Kanjanatarakul, O.; Sriboonchitta, S. EK-NNclus: A clustering procedure based on the evidential K-nearest neighbor rule. *Knowl.-Based Syst.* **2015**, *88*, 57–69. [CrossRef]

9. Ziegier, J.; Gattringer, H.; Mueller, A. Classification of Gait Phases Based on Bilateral EMG Data Using Support Vector Machines. In *Proceedings of the 2018 7th IEEE International Conference on Biomedical Robotics and Biomechatronics (Biorob)*; IEEE: Enschede, The Netherlands, 2018; pp. 978–983.

10. Ince, T.; Kiranyaz, S.; Eren, L.; Askar, M.; Gabbouj, M. Real-Time Motor Fault Detection by 1-D Convolutional Neural Networks. *IEEE Trans. Ind. Electron.* **2016**, *63*, 7067–7075. [CrossRef]

11. Zhang, R.; Peng, Z.; Wu, L.; Yao, B.; Guan, Y. Fault Diagnosis from Raw Sensor Data Using Deep Neural Networks Considering Temporal Coherence. *Sensors* **2017**, *17*, 549. [CrossRef]

12. Chen, Z.; Li, C.; Sanchez, R.-V. Gearbox Fault Identification and Classification with Convolutional Neural Networks. *Shock Vib.* **2015**, *2015*, 1–10. [CrossRef]

13. Chen, K.; Zhou, X.-C.; Fang, J.-Q.; Zheng, P.; Wang, J. Fault Feature Extraction and Diagnosis of Gearbox Based on EEMD and Deep Briefs Network. *Int. J. Rotating Mach.* **2017**, *2017*, 1–10. [CrossRef]

14. Wang, L.; Zhao, X.; Pei, J.; Tang, G. Transformer fault diagnosis using continuous sparse autoencoder. *SpringerPlus* **2016**, *5*. [CrossRef] [PubMed]

15. Tran, V.T.; AlThobiani, F.; Ball, A. An approach to fault diagnosis of reciprocating compressor valves using Teager–Kaiser energy operator and deep belief networks. *Expert Syst. Appl.* **2014**, *41*, 4113–4122. [CrossRef]

16. Han, D.; Zhao, N.; Shi, P. A new fault diagnosis method based on deep belief network and support vector machine with Teager–Kaiser energy operator for bearings. *Adv. Mech. Eng.* **2017**, *9*, 168781401774311. [CrossRef]

17. Shao, H.; Jiang, H.; Wang, F.; Wang, Y. Rolling bearing fault diagnosis using adaptive deep belief network with dual-tree complex wavelet packet. *ISA Trans.* **2017**, *69*, 187–201. [CrossRef] [PubMed]

18. Wang, S.; Xiang, J.; Zhong, Y.; Tang, H. A data indicator-based deep belief networks to detect multiple faults in axial piston pumps. *Mech. Syst. Signal Process.* **2018**, *112*, 154–170. [CrossRef]

19. Lee, D.; Lee, B.; Woo Shin, J. Fault Detection and Diagnosis with Modelica Language using Deep Belief Network. In Proceedings of the 11th International Modelica Conference, Versailles, France, 21–23 September 2015; pp. 615–623.

20. Ahmed, H.O.A.; Dennis Wong, M.L.; Nandi, A.K. Effects of deep neural network parameters on classification of bearing faults. In Proceedings of the IECON—42nd Annual Conference of the IEEE Industrial Electronics Society, Florence, Italy, 23–26 October 2016; pp. 6329–6334.

21. Tao, J.; Liu, Y.; Yang, D. Bearing Fault Diagnosis Based on Deep Belief Network and Multisensor Information Fusion. *Shock Vib.* **2016**, *2016*, 1–9. [CrossRef]

22. He, J.; Yang, S.; Gan, C. Unsupervised Fault Diagnosis of a Gear Transmission Chain Using a Deep Belief Network. *Sensors* **2017**, *17*, 1564. [CrossRef]

23. Chen, Z.; Deng, S.; Chen, X.; Li, C.; Sanchez, R.-V.; Qin, H. Deep neural networks-based rolling bearing fault diagnosis. *Microelectron. Reliab.* **2017**, *75*, 327–333. [CrossRef]

24. Deutsch, J.; He, M.; He, D. Remaining Useful Life Prediction of Hybrid Ceramic Bearings Using an Integrated Deep Learning and Particle Filter Approach. *Appl. Sci.* **2017**, *7*, 649. [CrossRef]

25. Geng, Z.; Li, Z.; Han, Y. A new deep belief network based on RBM with glial chains. *Inf. Sci.* **2018**, *463–464*, 294–306. [CrossRef]

26. Ren, H.; Chai, Y.; Qu, J.; Ye, X.; Tang, Q. A novel adaptive fault detection methodology for complex system using deep belief networks and multiple models: A case study on cryogenic propellant loading system. *Neurocomputing* **2018**, *275*, 2111–2125. [CrossRef]

27. Shao, H.; Jiang, H.; Zhang, H.; Duan, W.; Liang, T.; Wu, S. Rolling bearing fault feature learning using improved convolutional deep belief network with compressed sensing. *Mech. Syst. Signal Process.* **2018**, *100*, 743–765. [CrossRef]

28. Jiang, H.; Shao, H.; Chen, X.; Huang, J. A feature fusion deep belief network method for intelligent fault diagnosis of rotating machinery. *J. Intell. Fuzzy Syst.* **2018**, *34*, 3513–3521. [CrossRef]

29. Shao, H.; Jiang, H.; Lin, Y.; Li, X. A novel method for intelligent fault diagnosis of rolling bearings using ensemble deep auto-encoders. *Mech. Syst. Signal Process.* **2018**, *102*, 278–297. [CrossRef]

30. Maurya, S.; Singh, V.; Dixit, S.; Verma, N.K.; Salour, A.; Liu, J. Fusion of Low-level Features with Stacked Autoencoder for Condition based Monitoring of Machines. In Proceedings of the 2018 IEEE International Conference on Prognostics and Health Management (ICPHM), Seattle, WA, USA, 11–13 June 2018; pp. 1–8.
31. Shao, H.; Jiang, H.; Zhao, H.; Wang, F. A novel deep autoencoder feature learning method for rotating machinery fault diagnosis. *Mech. Syst. Signal Process.* **2017**, *95*, 187–204. [CrossRef]
32. Meng, Z.; Zhan, X.; Li, J.; Pan, Z. An enhancement denoising autoencoder for rolling bearing fault diagnosis. *Measurement* **2018**, *130*, 448–454. [CrossRef]
33. Sohaib, M.; Kim, J.-M. Reliable Fault Diagnosis of Rotary Machine Bearings Using a Stacked Sparse Autoencoder-Based Deep Neural Network. *Shock Vib.* **2018**, *2018*, 1–11. [CrossRef]
34. Saufi, S.R.; bin Ahmad, Z.A.; Leong, M.S.; Lim, M.H. Differential evolution optimization for resilient stacked sparse autoencoder and its applications on bearing fault diagnosis. *Meas. Sci. Technol.* **2018**, *29*, 125002. [CrossRef]
35. Gao, X.; Wang, H.; Gao, H.; Wang, X.; Xu, Z. Fault diagnosis of batch process based on denoising sparse auto encoder. In Proceedings of the 2018 33rd Youth Academic Annual Conference of Chinese Association of Automation (YAC), Nanjing, China, 18–20 May 2018; pp. 764–769.
36. Mahdi, M.; Genc, V.M.I. Post-fault prediction of transient instabilities using stacked sparse autoencoder. *Electr. Power Syst. Res.* **2018**, *164*, 243–252. [CrossRef]
37. Xu, L.; Cao, M.; Song, B.; Zhang, J.; Liu, Y.; Alsaadi, F.E. Open-circuit fault diagnosis of power rectifier using sparse autoencoder based deep neural network. *Neurocomputing* **2018**, *311*, 1–10. [CrossRef]
38. Amini, S.; Ghaernmaghami, S. Sparse Autoencoders Using Non-smooth Regularization. In Proceedings of the 2018 26th European Signal Processing Conference (EUSIPCO), Rome, Italy, 3–7 September 2018; pp. 2000–2004.
39. Shao, H.; Jiang, H.; Zhang, X.; Niu, M. Rolling bearing fault diagnosis using an optimization deep belief network. *Meas. Sci. Technol.* **2015**, *26*, 115002. [CrossRef]
40. Jiang, H.; Shao, H.; Chen, X.; Huang, J. Aircraft Fault Diagnosis Based on Deep Belief Network. In Proceedings of the 2017 International Conference on Sensing, Diagnostics, Prognostics, and Control (SDPC), Shanghai, China, 16–18 August 2017; pp. 123–127.
41. Qin, X.; Zhang, Y.; Mei, W.; Dong, G.; Gao, J.; Wang, P.; Deng, J.; Pan, H. A cable fault recognition method based on a deep belief network. *Comput. Electr. Eng.* **2018**, *71*, 452–464. [CrossRef]
42. Hinton, G.E.; Osindero, S.; Teh, Y.-W. A Fast Learning Algorithm for Deep Belief Nets. *Neural Comput.* **2006**, *18*, 1527–1554. [CrossRef] [PubMed]
43. Li, Z.; Cai, X.; Liu, Y.; Zhu, B. A Novel Gaussian–Bernoulli Based Convolutional Deep Belief Networks for Image Feature Extraction. *Neural Process. Lett.* **2018**. [CrossRef]
44. Cho, K.H.; Raiko, T.; Ilin, A. Gaussian-Bernoulli deep Boltzmann machine. In Proceedings of the 2013 International Joint Conference on Neural Networks (IJCNN), Dallas, TX, USA, 4–9 August 2013; pp. 1–7.
45. Keronen, S.; Cho, K.; Raiko, T.; Ilin, A.; Palomaki, K. Gaussian-Bernoulli restricted Boltzmann machines and automatic feature extraction for noise robust missing data mask estimation. In Proceedings of the 2013 IEEE International Conference on Acoustics, Speech and Signal Processing, Vancouver, BC, Canada, 26–31 May 2013; pp. 6729–6733.
46. Lu, W.; Wang, X.; Yang, C.; Zhang, T. A novel feature extraction method using deep neural network for rolling bearing fault diagnosis. In Proceedings of the 27th Chinese Control and Decision Conference (2015 CCDC), Qingdao, China, 23–25 May 2015; pp. 2427–2431.

Article

Self-Adaptive Spectrum Analysis Based Bearing Fault Diagnosis

Jie Wu, Tang Tang *, Ming Chen and Tianhao Hu

School of Mechanical Engineering, Tongji University, Shanghai 201804, China; 1710020@tongji.edu.cn (J.W.);
chen.ming@tongji.edu.cn (M.C.); sirius.hu@tongji.edu.cn (T.H.)
* Correspondence: tang.tang@tongji.edu.cn; Tel.: +86-139-1645-8293

Received: 16 September 2018; Accepted: 1 October 2018; Published: 2 October 2018

Abstract: Bearings are critical parts of rotating machines, making bearing fault diagnosis based on signals a research hotspot through the ages. In real application scenarios, bearing signals are normally non-linear and unstable, and thus difficult to analyze in the time or frequency domain only. Meanwhile, fault feature vectors extracted conventionally with fixed dimensions may cause insufficiency or redundancy of diagnostic information and result in poor diagnostic performance. In this paper, Self-adaptive Spectrum Analysis (SSA) and a SSA-based diagnosis framework are proposed to solve these problems. Firstly, signals are decomposed into components with better analyzability. Then, SSA is developed to extract fault features adaptively and construct non-fixed dimension feature vectors. Finally, Support Vector Machine (SVM) is applied to classify different fault features. Data collected under different working conditions are selected for experiments. Results show that the diagnosis method based on the proposed diagnostic framework has better performance. In conclusion, combined with signal decomposition methods, the SSA method proposed in this paper achieves higher reliability and robustness than other tested feature extraction methods. Simultaneously, the diagnosis methods based on SSA achieve higher accuracy and stability under different working conditions with different sample division schemes.

Keywords: fault diagnosis; feature extraction; self-adaptive spectrum analysis; bearing

---

1. Introduction

Bearings are critical parts in rotating machines and their health condition has a great impact on production. However, because of non-linear factors such as frictions, clearance and stiffness, vibration signals of bearings acquired in real application scenarios are characterized by non-linearity and instability which make bearing fault diagnosis difficult [1].

The general fault diagnosis process involves three main steps, namely signal acquisition and processing, fault feature extraction and fault feature classification [2]. Sensors are utilized to acquire signals with noises, and signal processing techniques are applied subsequently to improve the signal-to-noise ratio [3]. Particularly, ideal fault feature extraction can express the feature information of filtered signals comprehensively and efficiently, and it is the basis to produce an accurate fault feature classification. Therefore, a reasonable and efficient fault feature extraction plays an important role in fault diagnosis. Current fault extraction methods mainly include time domain, frequency domain and time-frequency domain analysis [4].

Time domain analysis is one of the earliest methods studied and applied. It calculates various statistical parameters in the time domain, for instance peak amplitude, kurtosis and skewness [5–7] to construct feature vectors. Frequency domain analysis transforms signals from the time domain into the frequency domain first, mainly focusing on Fourier Transform (FT) [8], then the periodical features, frequency features and distribution features of signals are extracted with methods such as cepstrum analysis and envelope spectrum analysis to construct feature vectors [9,10].

However, time domain or frequency domain analysis only extracts the information in the corresponding domain, resulting in the loss of information in the other domain. With in-depth study, time-frequency domain fault feature extraction methods were developed accordingly. They can extract both time and frequency information. They also have shown superiority for analyzing nonlinear and unstable signals.

As a typical time and frequency domain analysis method, Short Time Fourier Transform (STFT) [11] improves the analysis capability for unstable signals by introducing a fixed-width time window function. However, a fixed-width time window function in STFT cannot guarantee optimal time and frequency resolution simultaneously. The Wavelet Transform (WT) [12] introduces time and frequency scale coefficients to overcome the drawbacks of STFT. WT is based on the theory of inner product mapping and a reasonable basis function is the key to guarantee the effectiveness of WT. However, it is difficult to select a proper basis function. Therefore, to improve the adaptive analysis capability to signals, Empirical Mode Decomposition (EMD) [13] and Local Mean Decomposition (LMD) [14] methods were successively studied and applied. According to the local characters of signals themselves, EMD and LMD adaptively decompose a signal into various components which have better statistical characters for later analysis. Compared with each other, EMD is a mature tool for long-term study and usage, while LMD has an improved decomposition process and better decomposition results with physical explanations [15].

In recent years, EMD and LMD have been extensively studied and implemented. Mejia-Barron et al. [16] developed a method based on EMD to decompose signals and extract features, completing the fault diagnosis of winding faults. Saidi et al. [17] introduced a synthetical application of bi-spectrum and EMD to detect bearing faults. Cheng et al. [18] combined EEMD and entropy fusion to extract fault features for planetary gearboxes, and furthermore implemented fault diagnosis successfully. Yi et al. [19] also utilized EEMD to pre-process signals for further fault diagnosis for bearings. Liu and Han et al. [20] applied LMD and multi-scale entropy methods to extract fault features and analyzed faults successfully. Yang et al. [21] proposed an ensemble local mean decomposition method and applied it in rub-impact fault diagnosis for rotor systems. Han and Pan et al. [22] integrated LMD, sample entropy and energy ratio to process vibration signals and realized the fault feature extraction and fault diagnosis in rolling element bearings. Yasir and Koh et al. [23] adopted LMD and multi-scale permutation entropy and realized bearing fault diagnosis. Guo et al. [24] studied an improved fault diagnosis method for gearbox combining LMD and a synchrosqueezing transform.

Fault feature classification is implemented after fault feature extraction. Nowadays, shallow machine learning methods are extensively utilized to solve the classification problem. Support Vector Machine, Artificial Neural Network and Fuzzy Logical System are widely applied in condition monitoring and fault diagnosis [25]. Particularly, SVM is based on statistics and minimum theory of structured risk, and it has better classification performance when dealing with the practical problems of a small amount of and non-linear samples. To solve the multi-class classification problems, based on SVM, Cherkassky [26] proposed a one-against-all (oaa) strategy in his studies, transforming a N-class classification problem into N binary classification problems. Also, Kressel [27] used a method to transform a N-class classification problem into $N(N-1)/2$ binary classification problems, namely the one-against-one (oao) strategy. Wu et al. [28] adopted SVM to diagnosis via analyzing the full-spectrum to extract fault features. Saimurugan et al. [29] improved the diagnosis performance by integrating SVM and avdecision tree. Santos et al. [30] selected SVM for classification in wind turbine fault diagnosis with several trails of different kernels.

Currently, researchers all over the world have carried out extensive studies on bearing fault diagnosis. To our best knowledge, fault diagnosis methods still need further study, although various solutions have been investigated from different aspects. The main problems to be solved in this paper are summarized as follows:

(1) Vibration signals acquired in real application scenarios are non-linear and unstable and their statistical characters are time-varying. Hence, it is difficult to extract effective and comprehensive fault features only in the time-domain or in frequency-domain.

(2) Conventional fault feature extraction methods take the overall characteristics of signals into account via calculating statistical parameters to construct feature vectors with fixed dimensions, however, local detailed characteristics are neglected. Therefore, fault information contained in vectors may be insufficient or redundant in different working conditions because vectors have a fixed dimension, consequently leading to lower reliability and robustness of fault feature extraction. Meanwhile, data-driven classifiers are sensitive to classification features and minor changes in classification features may result in performance reduction [31].

In order to improve the fault diagnosis performance, in this paper, SSA is proposed to adaptively extract fault features and construct unfixed-dimension feature vectors according to local characters of signals. Then, SSA is implemented under the designed framework. Signals are decomposed firstly to obtain components with better analyzability, LMD and EEMD are both utilized to decompose signals into different components from different analysis aspects. SSA is utilized to extract fault features adaptively and feature vectors with non-fixed dimensions are constructed subsequently. Finally, SVM is selected to classify the fault features considering its inherent advantages to small amount train samples.

2. Methodology

2.1. Self-Adaptive Spectrum Analysis

Aiming at solving the problem that conventional feature extraction methods neglect local details of signal and fault information may be redundant or insufficient because of fixed-dimension feature vectors, Self-adaptive Spectrum Analysis (SSA) is proposed. With the SSA method, unfixed-dimension feature vectors are constructed by extracting the local characteristics of signals adaptively.

At first, a number of signals corresponding to different categories of fault types are selected. To implement SSA method efficiently, Fast Fourier Transform (FFT) is used to transform the signals into frequency domain to get corresponding spectrums for better readability. Then an overall frequency-window is set to all spectrums according to the fluctuation in spectrums, and local feature information inside the frequency-window is extracted to construct feature vectors.

In order to implement the proposed SSA, some definitions are given:

Definition 1. *Differential frequency f_z.*

f_z is the minimum frequency unit in SSA. Normally, feature information is extracted at points corresponding to nf_z ($n = 1, 2, 3, \dots$), where f_z is calculated as follows:

Firstly, in each spectrum, the maximum amplitude and corresponding frequency value are found. All the frequency values are denoted as $f_1, f_2, f_3, \dots , f_m$, where m means the sequence number of signals. More than two fault categories must be included within the selected signals.

Secondly, the frequency values are arranged into different vectors according to the categories of samples; vectors are denoted as:

$$v_i = \left[f_{(i-1)m/k+1}, f_{(i-1)m/k+2}, \cdots, f_{(i-1)m/k+m/k} \right] \tag{1}$$

where k means k kinds of faults, $i = [1, 2, \dots k]$. Here we assume that different categories have the same amount of signals. Then, the average values of all elements in each vector are figured out and denoted as $\bar{v}_1, \bar{v}_2, \dots , \bar{v}_k$, respectively, then a vector $f = [\bar{v}_1, \bar{v}_2 \dots \bar{v}_k]$ is constructed.

Thirdly, minimum frequency value $f_{\min}$ and the maximum frequency value $f_{\max}$ are selected in vector f. Then, two neighboring frequency values are also selected in f, between which there is the maximum value among the differences between every two neighboring frequencies, the lower frequency is denoted as f_{low} and the higher one is denoted as f_{high}.

Finally, f_{min}, f_{max}, f_{low}, f_{high} are arranged in ascending order, and absolute values f_{diff} of differences between every neighboring two frequencies are calculated. The minimum non-zero f_{diff} value is picked to be the value of f_z:

$$f_z = min(f_{diff}) \tag{2}$$

Definition 2. *Frequency Window* $W = [f_l, f_r]$.

The frequency window is a specific frequency section for extracting feature information, f_l is the left boundary while f_r is the right boundary. Frequency window is determined with fixed boundaries, and feature information is extracted inner the window. Boundaries are calculated as follows:

$$f_l = floor\left(\frac{f_{min}}{f_z}\right) * f_z \tag{3}$$

$$f_r = ceil\left(\frac{f_{max}}{f_z}\right) * f_z \tag{4}$$

where $floor(*)$ is a round down function, $ceil(*)$ is a round up function.

Definition 3. *Tolerance* μ.

Tolerance μ denotes that in a section which is centered with nf_z, μ is taken as the semidiameter to determine the searching section $(nf_z - \mu, nf_z + \mu]$, and the maximum amplitude value corresponding to a frequency within this section can be regarded as the amplitude value to nf_z. μ is calculated as follows:

$$\mu = floor\left(\frac{f_z}{2}\right) \tag{5}$$

Definition 4. *Peak value ratio coefficient h.*

h denotes the degree of peak amplitude value. It is utilized to judge whether the amplitude value is normal or not and all h construct fault feature vectors. h is calculated as follows:

Firstly, average value of all the amplitude values in frequency window $[f_l, f_r]$ is calculated, denoted as A_{ave}, also the maximum amplitude value in section $(nf_z - \mu, nf_z + \mu]$ is selected and denoted as A_{max}. Finrfally, h can be calculated as follows:

$$h = \frac{A_{max}}{A_{ave}} \tag{6}$$

Figure 1 gives a description of the definitions mentioned above.

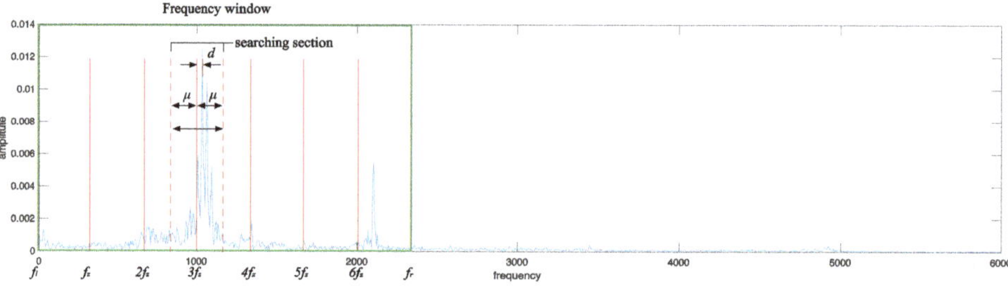

Figure 1. Sketch map of parameters in adaptive spectrum analysis.

Combined with Figure 1, SSA is implemented on each spectrum as follows:

(1)   Calculating values of differential frequency f_z and boundaries f_l, f_r, frequency window W is determined;

(2)   Calculating all the nf_z values, taking μ as side intervals to determine different searching sections;

(3)   Selecting the maximum amplitude in each searching section and corresponding frequency value, calculating the absolute frequency interval d between this frequency value and section center nf_z, also, h are calculated, frequency interval vector $D = [d_1, d_2, d_3 \ldots d_n]$, Peak value ratio coefficient vector $H = [h_1, h_2, h_3 \ldots h_n]$;

(4)   Setting a threshold value h_t for h, and h_t could be optimized automatically by the overall accuracy. h_t is used to judge if an anomaly exists in sections. When $h > h_t$, the corresponding section is regarded as an abnormal one;

(5)   If an anomaly is found, figuring out whether all the frequency values corresponding to abnormal sections are on the same side of nf_z (n = 1, 2, 3, ...) along the frequency axis simultaneously. If they are on the same side, selecting a minimum d in D, and shifting the spectrum to the opposite direction by d. Subsequently, repeating steps 1 to 3. While, if they are not on the same side, skip steps 5 and 6;

(6) H is taken as the fault feature vector extracted from the spectrum.

2.2. Framework Construction of Fault Diagnosis

The overall framework construction of the proposed fault diagnosis method based on SSA in our research is shown in Figure 2.

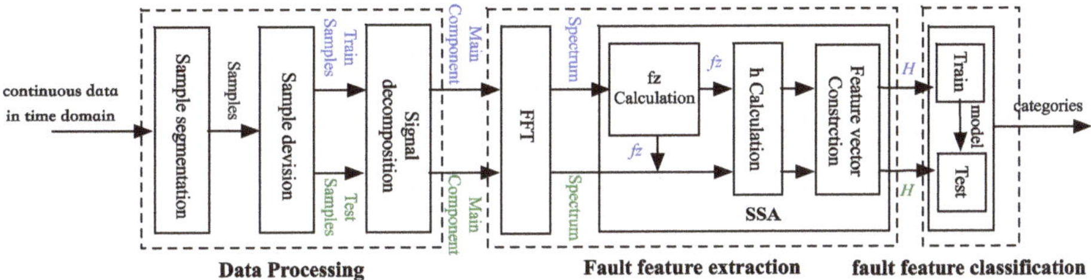

Figure 2. SSA-based diagnostic framework.

The proposed fault diagnosis method includes three parts, namely data processing, fault feature extraction and fault feature classification.

2.2.1. Data Processing

As shown in Figure 3, a signal segment containing 120,000 points is selected, then it is segmented into 100 parts with a same length. In total, 100 samples are extracted from one signal segment. Therewith, 100 samples are separated into a training sample set and a test sample set.

Each sample is decomposed into a set of components with better analyzability with a time-frequency analysis method, LMD and EEMD are two commonly used ones. The very first component in each set of components is chosen to extract fault features because they accumulate the main part of the energy.

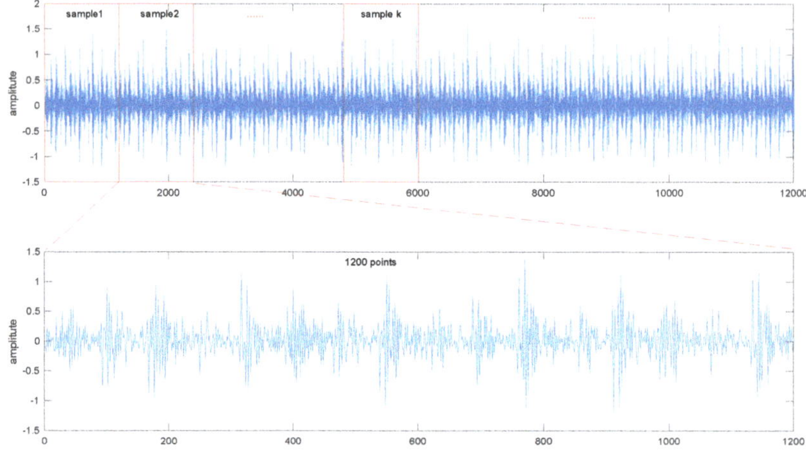

Figure 3. Segmentation of samples.

2.2.2. Fault Feature Extraction

FFT is utilized to transform the decomposed component into the frequency domain, and then SSA is implemented to extract fault features. First the components of the training samples are selected to calculate f_z, and f_z is utilized for both the training samples and test samples to extract fault features.

2.2.3. Fault Feature Classification

Fault feature vectors are classified into different fault patterns. Vectors extracted from the training samples are utilized to train the classification model and parameters are tuned to optimize the model. Here, SVM is selected because of its better performance in classification with small samples. Eventually, categories are output with the well-trained model.

2.3. Experiment Preparation

2.3.1. Data Selection and Processing

Vibration signals acquired from bearings are utilized for validation. In this paper, selected bearing data published by Case Western Reverse University were used [32]. Single point faults are introduced to the test bearings on different parts (ball, inner race and outer race) to simulate different kinds of faults. Vibration signals of different kinds of faults with different failure degrees are collected under different loads to construct the experimental data set.

The data set consisted of vibration data collected on SKF bearings, and the sampling frequency is 12 kHz. Twelve kinds of combinations under four kinds of loads (0, 1, 2 and 3 hp) and three kinds of failure degrees (0.007, 0.014 and 0.021 inch) form 12 different working conditions.

Under each working condition, four kinds of fault mode (normal, ball fault, inner race fault and outer race fault) are simulated, and four time-varying signals corresponding to the faults are collected, respectively. Each signal is processed with the proposed method given in Figure 3 to extract 100 samples, and 100 feature vectors are subsequently constructed. Eventually, 400 feature vectors are determined under every working condition.

2.3.2. Parameter Determination

Parameters corresponding to decomposition methods, fault feature extraction process and fault feature classification modeling process are determined as follows:

Parameters to be determined in signal decomposition methods:

(1)   In LMD, parameters are determined according to reference [33];

(2)   In EEMD, parameters are determined according to reference [34];

   Parameters to be determined in SSA method:

(1) f_z, differential frequency value is calculated according to Equation (2);

(2) f_l, left boundary value is calculated according to Equation (3);

(3) f_r, right boundary value is calculated according to Equation (4);

(4) μ, tolerance value is calculated according to Equation (5);

(5) h, peak value coefficient ratio is calculated according to Equation (6);

(6) h_t, the minimum value in vector H is selected as the threshold value of h;

   Parameters to be determined in pattern recognition method:

(1)   In SVM, cost c is a basic parameter while g is a specific one in RBF kernel. In this paper, Grid search [35] is applied and overall accuracy is taken into consideration to tune the two parameters.

3. Experiments and Results

Experiment Results and Analysis

   In this subsection, a simulated signal $x(t)$ is utilized to evaluate the effectivity of decomposition methods [33]. $x(t)$ consists of two superimposed component signals: $x(t) = (1 + 0.5cos(9\pi t))cos(200\pi t + 2cos(10\pi t)) + 3cos(20\pi t^2 + 6\pi t)\ t \in [0, 1]$

   The LMD and EEMD methods are used to decompose the signal. Figure 4 illustrates the results of the decomposition.

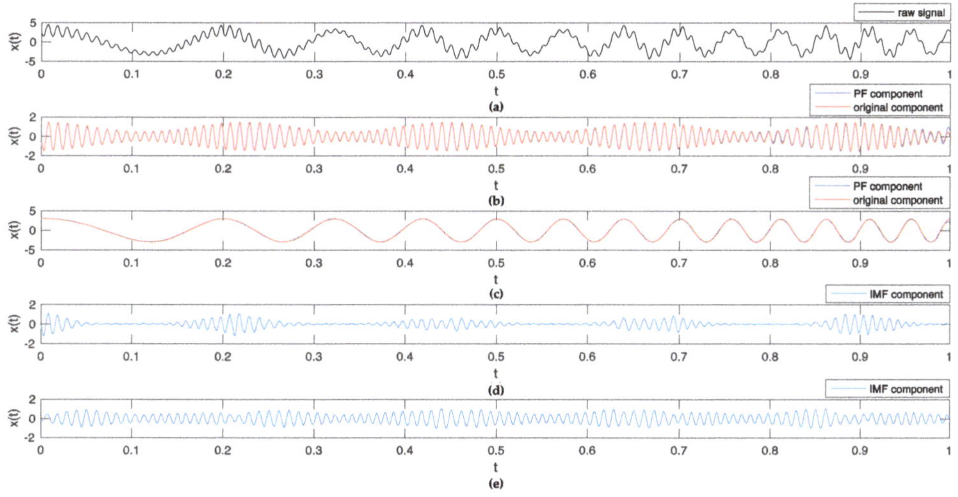

Figure 4. Results of decomposition for simulated signal.

   Figure 4a shows the oscillograph of the simulated signal. In Figure 4b,c, the oscillographs in red are two original components of the raw simulated signal, and the ones in blue are the Product Function (PF) components extracted with the LMD method. Obviously, the original components and extracted PF components have a high similarity except for several end points on the right. In Figure 4d,e, the oscillographs in blue are the first two Intrinsic Mode Function (IMF) components extracted with the EEMD method; both of them have less similarity with the original ones. These results prove that LMD adopted in the research can effectively decompose the raw signal into PF components which have physical significance, and EEMD can decompose the raw signal into IMF components by another mechanism [18].

Four experimental sets are designed combining two different signal decomposition methods: two different feature extraction methods and a fault feature classification method. The four experimental sets are arranged as shown in Table 1. LMD and EEMD are utilized to decompose the signals. The fault feature extraction methods include the proposed SSA and the combination of Sample Entropy (SE) and Energy Ratio (ER) [36], and the LIBSVM [37] software package is selected to implement the pattern classification.

Table 1. Arrangement of experiments.

Experiment Set	Signal Decomposition	Fault Feature Extraction	Fault Feature Classification
Set 1	LMD	SSA	SVM
Set 2	LMD	SE&ER	SVM
Set 3	EEMD	SSA	SVM
Set 4	EEMD	SE&ER	SVM

In each experiment set, 12 kinds of working conditions (a working condition is denoted as a load-fault five kinds of sample division scheme are tested (a sample division scheme is denoted as: number of training samples in 100 samples to every fault/number of test samples in 100 samples to every fault, for example 5/95, 10/90, 20/80, 40/60, 60/40), with each scheme, 10 independent experiments are repeated. Ultimately, 2400 experiments are carried out in total within the four experiment sets. Table 2 shows the f_z values and dimensions of feature vectors under 12 working conditions with sample division schemes of 5/95 and 60/40, respectively, in experimental set 1.

Table 2. Values of differential frequency and dimensions of character vectors.

Division Scheme	Working Condition	0–0.007	0–0.014	0–0.021	1–0.007	1–0.014	1–0.021
5/95	f_z (Hz)	94	234	369	533	217	486
	Dimension	37	15	9	7	16	7
60/40	f_z (Hz)	229	334	375	451	176	504
	Dimension	16	11	9	8	20	7

Division Scheme	Working Condition	2–0.007	2–0.014	2–0.021	3–0.007	3–0.014	3–0.021
5/95	f_z (Hz)	563	234	656	586	580	574
	Dimension	7	15	6	6	6	6
60/40	f_z (Hz)	463	240	598	580	440	568
	Dimension	8	15	6	6	8	6

The results illustrate that when the working condition or division scheme changes, the differential frequency f_z value and dimension value of the feature vectors change accordingly.

Without considering sample division schemes, the overall diagnostic capability of proposed model is evaluated. The average values and variance of accuracy values to all independent experiments (50 times) under each working condition are listed in Tables 3 and 4.

Table 3. Average diagnostic accuracy of all independent experiments corresponding to 12 different working conditions respectively in 1st–4th experiment sets.

Working Condition	Set 1	Set 2	Set 3	Set 4
0–0.007	98.14	94.15	97.65	98.12
0–0.014	98.75	81.83	94.41	74.59
0–0.021	96.48	88.61	99.36	65.52
1–0.007	97.91	96.85	99.80	97.51
1–0.014	99.02	84.93	98.75	63.61
1–0.021	95.07	97.69	99.53	71.55
2–0.007	99.03	96.90	99.97	98.42
2–0.014	97.89	85.03	97.36	67.33
2–0.021	97.18	97.70	99.54	74.77
3–0.007	97.62	97.74	99.42	99.46
3–0.014	94.67	87.44	94.20	74.16
3–0.021	97.26	97.71	99.26	79.23
Average	97.42	92.21	98.27	80.36

Table 4. Variances of diagnostic accuracy of all independent experiments corresponding to 12 different working conditions respectively in 1st–4th experiment sets.

Working Condition	Set 1	Set 2	Set 3	Set 4
0–0.007	1.94	25.00	4.85	1.93
0–0.014	1.50	19.08	9.39	49.01
0–0.021	13.75	42.70	0.63	8.15
1–0.007	3.90	9.32	0.13	3.32
1–0.014	1.36	40.30	1.27	58.80
1–0.021	6.32	10.01	0.10	19.94
2–0.007	0.49	8.23	0.01	1.74
2–0.014	2.82	32.84	6.12	43.05
2–0.021	4.77	6.04	1.55	13.25
3–0.007	6.55	2.12	0.61	0.71
3–0.014	15.34	20.66	12.60	35.93
3–0.021	10.95	1.16	0.82	10.36
Average	5.81	18.12	3.17	20.52

Figure 5a,b transforms Tables 3 and 4 in graphic ways, respectively.

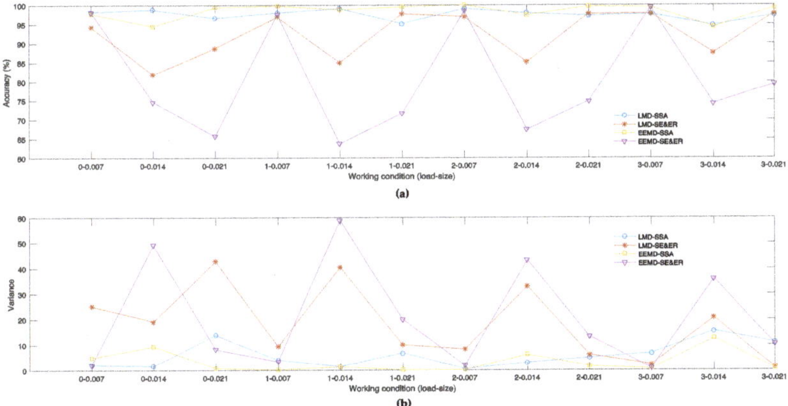

Figure 5. (**a**) Average diagnostic accuracy of all independent experiments corresponding to 12 different working conditions respectively in 1st–4th experimental sets. (**b**) Variance of diagnostic accuracy of all independent experiments corresponding to 12 different working conditions respectively in 1st–4th experimental set.

Table 3 and Figure 5a show that Set 3 achieves the best average accuracies under six kinds of working conditions and Set 1 achieves the best under five kinds of working conditions, while Set 4 only ranks the first place under one kind of working conditions. Overall, the average accuracies of Set 1 and Set 3 are 97.42% and 98.27%, maintaining a higher level, yet the average accuracies of Set 2 and Set 4 are only 92.21% and 80.36%, being especially worse in severe failure situations. Meanwhile, the variance of average accuracies under different working conditions of Set 1 is 1.99, and the numbers of Set 2, Set 3 and Set 4 are 37.84, 4.08 and 195.72, respectively. Obviously, the statistics of Set 1 and Set 3 indicate better performances than Set 2 and Set 4.

Table 4 and Figure 5b show that Set 3 obtains the smallest variances under nine working conditions and Set 1 is the least under to kinds of working conditions, while Set 4 only performs best under one working condition. Clearly, the average value of variance values under 12 different working conditions for Set 1, Set 2, Set 3 and Set 4 are 5.81, 18.12, 3.17 and 20.52, respectively. Set 1 and Set 3 outperform Set 2 and Set 4. Meanwhile the values of the variances of Set 2 and Set 4 show obvious fluctuation under severe failure conditions.

As a matter of fact, to simulate a situation that labeled data are rare in real application scenarios, a small amount of samples division scheme is tested. Average values and variances of the accuracy values of all independent experiments (10 times) under each working condition with a (5/95) sample division scheme are shown in Tables 5 and 6.

Table 5. Average diagnostic accuracy to all independent experiments corresponding to 12 different working conditions respectively with the scheme (5/95) in 1st–4th experiment sets.

Working Condition	Set 1	Set 2	Set 3	Set 4
0–0.007	97.26	86.82	96.84	97.37
0–0.014	97.74	77.68	93.53	64.89
0–0.021	93.92	79.13	98.84	64.50
1–0.007	96.39	93.16	99.68	95.68
1–0.014	97.95	75.68	98.00	52.34
1–0.021	93.45	95.03	99.61	68.68
2–0.007	98.97	93.26	99.97	97.45
2–0.014	96.32	75.47	95.53	59.37
2–0.021	96.03	94.32	99.74	72.00
3–0.007	96.39	96.18	99.45	99.34
3–0.014	92.05	83.37	91.13	66.00
3–0.021	94.82	96.50	98.79	76.16
Average	95.94	87.22	97.59	76.15

Table 6. Variances of diagnostic accuracy to all independent experiments corresponding to 12 different working conditions respectively with the scheme of (5/95) in 1st–4th experiment sets.

Working Condition	Set 1	Set 2	Set 3	Set 4
0–0.007	2.80	14.43	2.74	4.45
0–0.014	3.43	20.42	15.94	47.56
0–0.021	17.47	59.47	1.40	10.03
1–0.007	8.50	20.01	0.21	5.60
1–0.014	2.67	40.65	1.28	42.14
1–0.021	12.41	25.03	0.07	14.24
2–0.007	0.16	13.75	0.01	2.22
2–0.014	5.08	12.28	8.14	18.92
2–0.021	8.26	12.88	0.15	10.14
3–0.007	3.40	6.16	0.90	0.87
3–0.014	41.64	18.63	15.50	11.85
3–0.021	22.03	2.46	1.68	6.45
Average	10.65	20.51	4.00	14.54

Figure 6a,b also illustrates the results of Tables 5 and 6 in graphic ways, respectively.

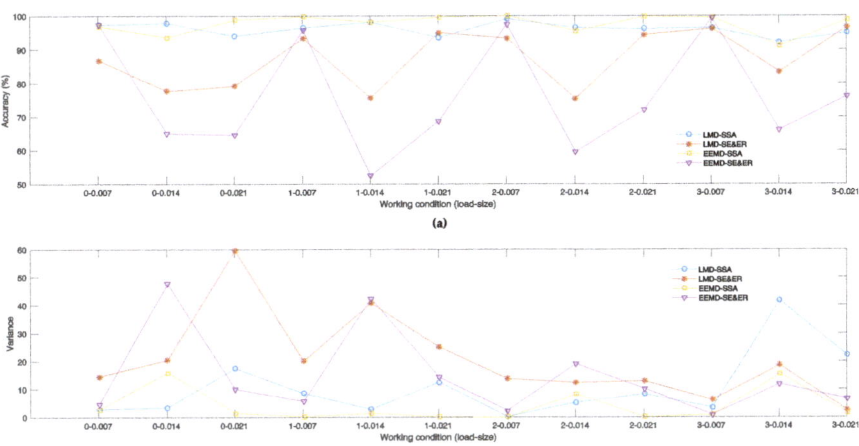

Figure 6. (**a**) Average diagnostic accuracy to all independent experiments corresponding to 12 different working conditions respectively with the scheme of (5/95) in the 1st–4th experimental sets. (**b**) Variance of diagnostic accuracy to all independent experiments corresponding to 12 different working conditions, respectively, with the scheme of (5/95) in the 1st–4th experimental sets.

Table 5 and Figure 6a show that Set 3 achieves the best average accuracies under eight kinds of working conditions and Set 1 achieves the best under two kinds of working conditions, while Set 4 only ranks the first place under two kinds of working conditions. Overall, the average accuracies of Set 1 and Set 3 are 95.94% and 97.59%, still maintaining a high level with slight decreases compared to the overall average accuracies, yet the average accuracies of Set 2 and Set 4 are only 87.22% and 76.15%, being especially worse in severe failure situations, and showing sharp decreases compared to overall average accuracies. Meanwhile, the variance of average accuracies under different working conditions of Set 1 and Set 3 are 10.65 and 4.00, and the numbers for Set 2 and Set 4 are 20.51 and 15.4, respectively. Obviously, the results in Set 1 and Set 3 are better than the results in Set 2 and Set 4.

Table 6 and Figure 6b show that the variance values of 50 independent experiments in Set 3 under each working condition maintain a steady low level and the mean value of 12 values is 4.00. While in Set 1, variance values appear obvious fluctuation under 3–0.014 only, and the mean value is 10.65; values in Set 2 and Set 4 fluctuate wildly and the mean values are 20.51 and 14.54, respectively.

The convergence performance with the increase of the amount of training samples is also a key indicator to evaluate a model. Experiments are conducted under different working conditions with different sample division schemes, also, mean value and variance of accuracy values of all independent experiments (10 times) are calculated and listed in Table 7. Figure 7 transforms Table 7 into diagrams.

In Table 7, 60 comparisons are conducted under different working conditions with different sample division schemes, and the results show that Set 1 and Set 3 have better performance in average accuracy with 56 comparisons out of 60, while Set 2 or Set 4 only get higher accuracies under the working condition of 0–0.007 with four kinds of schemes. As shown in Figure 7, with the increase of the number of training samples, the average accuracies of Set 1 and Set 3 obviously converge toward the highest value faster than Set 2 and Set 4.

Table 7. Average diagnostic accuracy to all independent experiments corresponding to 12 different working conditions respectively with different schemes in 1st–4th experiment sets.

	0–0.007					0–0.014				
	5/95	10/90	20/80	40/60	60/40	5/95	10/90	20/80	40/60	60/40
Set 1	97.26	97.31	98.28	98.96	98.88	97.74	98.33	99	99.13	99.56
Set 2	86.82	92.36	96.25	97.08	98.25	77.68	79.22	82.41	84.71	85.13
Set 3	96.84	95.92	97.22	98.79	99.5	93.53	93.72	95.34	94.63	94.81
Set 4	97.37	97.83	97.97	98.79	98.63	64.89	71.78	76.19	78.71	81.38

	0–0.021					1–0.007				
	5/95	10/90	20/80	40/60	60/40	5/95	10/90	20/80	40/60	60/40
Set 1	93.92	94.97	97.06	98.79	97.69	96.39	98.03	98.19	98.5	98.44
Set 2	79.13	87.81	90.78	92.83	92.5	93.16	96.39	97.16	98.63	98.94
Set 3	98.84	99.64	99.66	99.33	99.31	99.68	99.81	99.84	99.92	99.75
Set 4	64.5	64.36	65.28	66.63	66.81	95.68	97.19	97.63	98.38	98.69

	1–0.014					1–0.021				
	5/95	10/90	20/80	40/60	60/40	5/95	10/90	20/80	40/60	60/40
Set 1	97.95	99.17	98.91	99.58	99.5	93.45	94.81	96	95.17	95.94
Set 2	75.68	82.83	86.91	89.88	89.38	95.03	96.81	98.41	98.92	99.31
Set 3	98	98.08	98.94	99.17	99.56	99.61	99.53	99.44	99.38	99.69
Set 4	52.34	60.03	66.44	68.75	70.5	68.68	68.86	69.34	74.88	76

	2–0.007					2–0.014				
	5/95	10/90	20/80	40/60	60/40	5/95	10/90	20/80	40/60	60/40
Set 1	98.97	99.25	98.47	99.46	99	96.32	97.42	98.34	98.79	98.56
Set 2	93.26	96.78	96.63	98.58	99.25	75.47	84.67	87.06	89	88.94
Set 3	99.97	99.97	99.97	99.96	100	95.53	96.72	97.25	98.88	98.44
Set 4	97.45	97.36	98.91	99.08	99.31	59.37	60.86	71.88	71.33	73.19

	2–0.021					3–0.007				
	5/95	10/90	20/80	40/60	60/40	5/95	10/90	20/80	40/60	60/40
Set 1	96.03	96.86	97.25	97.13	98.63	96.39	95.5	98.41	98.79	99
Set 2	94.32	98.03	98.22	98.33	99.63	96.18	97.89	98.13	98.46	98.06
Set 3	99.74	98.53	99.66	99.96	99.81	99.45	99.31	98.97	99.54	99.81
Set 4	72	72.86	74.41	76.83	77.75	99.34	99.08	99.22	99.71	99.94

	3–0.014					3–0.021				
	5/95	10/90	20/80	40/60	60/40	5/95	10/90	20/80	40/60	60/40
Set 1	92.05	93.44	94.75	96.71	96.38	94.82	96.03	97.53	99.17	98.75
Set 2	83.37	83.92	88.59	90.96	90.38	96.5	97.97	97.91	98	98.19
Set 3	91.13	93.14	95.06	94.79	96.88	98.79	99.25	98.94	99.63	99.69
Set 4	66	72.78	74.28	77.67	80.06	76.16	76.72	79.78	81.42	82.06

Considering all the results comprehensively, further analysis is carried out. LMD and EEMD can decompose nonlinear and unstable signals into a set of components in the time domain, and these components have better analyzability. The proposed SSA method can adaptively extract feature information according to local characteristics, and construct unfixed-dimension fault feature vectors, and it is proved to have better efficiency and robustness. SSA-based fault diagnosis methods can obtain higher accuracies under different working conditions with different sample division schemes in most comparisons (56/60), and the accuracies show less fluctuation between different conditions. With the increasing number of samples, the accuracies achieved with the SSA-based method converge towards the highest values faster. Especially with a small sample division scheme (5/95), the results have shown that methods based on SSA still maintain high accuracy and stability and they are proved specially suitable for practical application in scenarios with small amounts of training samples.

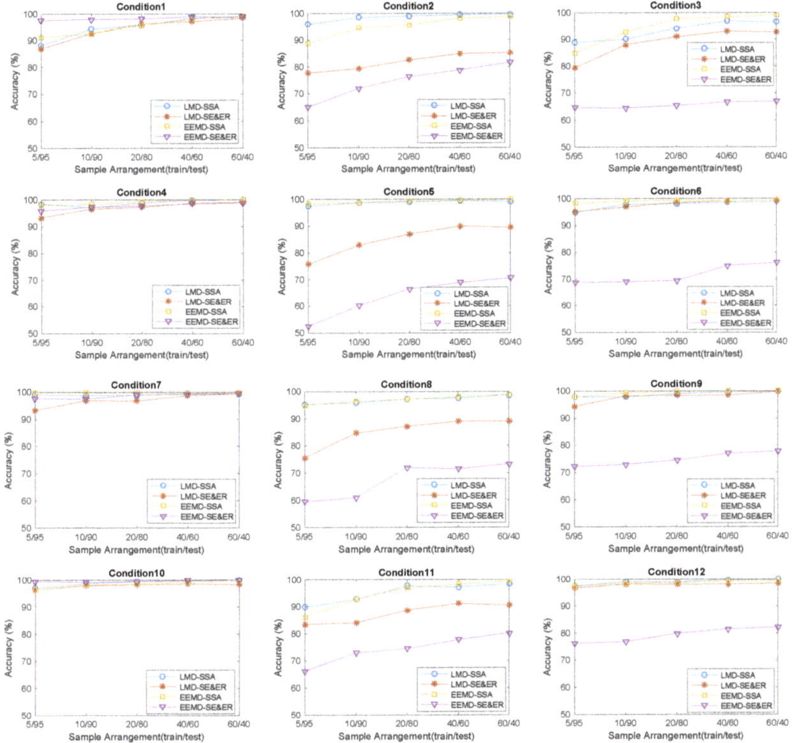

Figure 7. Average diagnostic accuracy to all independent experiments corresponding to 12 different working conditions, respectively, with different schemes in the 1st–4th experimental sets.

4. Conclusions

To improve the fault extraction performance, SSA is proposed in this paper. Combined with signal decomposition methods, SSA extracts fault features from non-linear and unstable signals effectively, then fault features are classified with SVM. Bearing data under 12 different working conditions obtained from CWRU are utilized to evaluate the diagnosis methods. The conclusions may be summarized as follows:

1. SSA extracts fault features and constructs unfixed-dimension vectors adaptively, it has reduced the side effects caused by information insufficiency and redundancy. Moreover, SSA has higher efficiency and robustness in fault extraction.
2. Fault diagnosis methods based on SSA can achieve higher accuracies and stability than other methods under the same proposed framework and with an increased number of training samples, the accuracies achieved with the SSA-based method converge to the highest value faster.
3. Especially, with a small amount training samples, the SSA-based method still provides high accuracy with more obvious superiority in accuracy and stability, therefore they have the potential to be implemented in real application scenarios.

5. Future Lines of Work

In recent years, deep learning has been adopted gradually in fault diagnosis. It can extract fault features automatically because of its multi-layer structure, this characteristic can improve the feature extraction further. At the same time, transfer learning [38] has achieved great success in many fields.

Its generalization capability can be also utilized in fault diagnosis to promote the diagnostic theories to applications. Therefore, our future work will be focused on the study of implementation of the combination of deep learning and transfer learning in fault diagnosis.

Author Contributions: J.W. conceived the original ideas, then designed and conducted all the experiments, subsequently drafted the manuscript. T.T. and T.H. and L.W. contributed to writing-review and editing. M.C. provided supervision to the project.

Funding: This research was funded by the Ministry of Industry and Information Technology of the People's Republic of China (MIIT; Projects: 2017 Prefabricated Building New Material Intelligent Manufacturing Construction Project, 2018 Microwave Component Digital Factory New Mode Application, 2018 Remote Operation and Maintenance Standards and Experimental Verification of Key Equipment for Integrated Circuit Packaging), and the Young Excellent Talent Cultivation Project at Tongji University (2016KJ020).

Conflicts of Interest: The authors declare no conflict of interest.

References

1. Yan, X.A.; Jia, M.P. A novel optimized SVM classification algorithm with multi-domain feature and its application to fault diagnosis of rolling bearing. *Neurocomputing* **2018**, *313*, 47–64. [CrossRef]
2. Hoang, D.T.; Kang, H.J. Rolling element bearing fault diagnosis using convolutional neural network and vibration Image. *Cogn. Syst. Res.* **2018**, in press. [CrossRef]
3. Lu, S.L.; He, Q.B.; Zhang, H.B.; Kong, F.R. Rotating machine fault diagnosis through enhanced stochastic resonance by full-wave signal construction. *Mech. Syst. Signal Process.* **2017**, *85*, 82–97. [CrossRef]
4. Ma, J.X.; Xu, F.Y.; Huang, K.; Huang, R. GNAR-GARCH model and its application in feature extraction for rolling bearing fault diagnosis. *Mech. Syst. Signal Process.* **2017**, *93*, 175–203. [CrossRef]
5. Lin, B.; Chang, P. Fault diagnosis of rolling element bearing using more robust spectral kurtosis and intrinsic time-scale decomposition. *J. Vib. Control.* **2014**, *22*, 2921–2937.
6. Jia, F.; Lei, Y.G.; Shan, H.K.; Lin, J. Early Fault Diagnosis of Bearings Using an Improved Spectral Kurtosis by Maximum Correlated Kurtosis Deconvolution. *Sensors* **2015**, *15*, 29363–29377. [CrossRef] [PubMed]
7. Fu, S.; Liu, K.; Xu, Y.G.; Liu, X. Rolling bearing diagnosing method based on time-domain analysis and adaptive fuzzy C-means clustering. *Shock. Vib.* **2016**, *2016*, 9412787.
8. Rai, V.K.; Mohanty, A.R. Bearing fault diagnosis using FFT of intrinsic mode functions in Hilbert–Huang transform. *Mech. Syst. Signal Process.* **2007**, *21*, 2607–2615. [CrossRef]
9. Borghesani, P.; Pennacchi, P.; Randall, R.B.; Sawalhi, R.B.; Ricci, R. Application of cepstrum pre-whitening for the diagnosis of bearing faults under variable speed conditions. *Mech. Syst. Signal Process.* **2013**, *36*, 370–384. [CrossRef]
10. Kang, M.; Kim, J.; Wills, L.M.; Kim, J.M. Time-varying and multiresolution envelope analysis and discriminative feature analysis for bearing fault diagnosis. *IEEE Trans. Ind. Electron.* **2015**, *62*, 7749–7761. [CrossRef]
11. Mateo, C.; Talavera, J.A. Short-Time Fourier Transform with the Window Size Fixed in the Frequency Domain. *Digit. Signal Process.* **2018**, *77*, 13–21. [CrossRef]
12. Yan, R.Q.; Gao, R.X.; Chen, X.F. Wavelets for fault diagnosis of rotary machines: A review with applications. *Signal Process.* **2014**, *96*, 1–15. [CrossRef]
13. Huang, N.E.; Shen, Z.; Long, S.R.; Wu, M.C.; Shih, H.H.; Zheng, Q.; Yen, N.-C.; Tung, C.C.; Liu, H.H. The empirical mode decomposition and Hilbert spectrum for nonlinear and non-stationary time series analysis. *Math. Phys. Eng. Sci.* **1998**, *454*, 903–995. [CrossRef]
14. Smith, J.S. The local mean decomposition and its application to EEG perception data. *J. R. Soc. Interface* **2005**, *2*, 443–454. [CrossRef] [PubMed]
15. Wang, Y.X.; He, Z.J.; Zi, Y.Y. A comparative study on the Local mean decomposition and empirical mode decomposition and their applications to rotating machinery health diagnosis. *J. Vib. Acoust.* **2010**, *132*, 021010. [CrossRef]
16. Mejia-Barron, A.; Valtierra-Rodriguez, M.; Granados-Lieberman, D.; Olivares-Galvan, J.C.; Escarela-Perez, R.E. The application of EMD-based methods for diagnosis of winding faults in a transformer using transient and steady state currents. *Measurement* **2017**, *117*, 371–379. [CrossRef]

17. Saidi, L.; Ali, J.B.; Fnaiech, F. Bi-spectrum based-EMD applied to the non-stationary vibration signals for bearing faults diagnosis. *ISA Trans.* **2014**, *53*, 1650–1660. [CrossRef] [PubMed]
18. Cheng, G.; Chen, X.H.; Li, H.Y.; Li, P.; Liu, H.G. Study on planetary gear fault diagnosis based on entropy feature fusion of ensemble empirical mode decomposition. *Measurement* **2016**, *91*, 140–154. [CrossRef]
19. Yi, C.; Wang, D.; Fan, W.; Tsui, K.L.; Lin, J.H. EEMD-Based Steady-State Indexes and Their Applications to Condition Monitoring and Fault Diagnosis of Railway Axle Bearings. *Sensors* **2018**, *18*, 704. [CrossRef] [PubMed]
20. Liu, H.H.; Han, M.H. A fault diagnosis method based on local mean decomposition and multi-scale entropy for roller bearings. *Mechani. Mach. Theory* **2014**, *75*, 67–78. [CrossRef]
21. Yang, Y.; Cheng, J.S.; Zhang, K. An ensemble local means decomposition method and its application to local rub-impact fault diagnosis of the rotor systems. *Measurement* **2012**, *45*, 561–570. [CrossRef]
22. Han, M.H.; Pan, J.L. A fault diagnosis method combined with LMD sample entropy and energy ratio for roller bearings. *Measurement* **2015**, *76*, 7–19. [CrossRef]
23. Yasir, M.N.; Koh, B.H. Data Decomposition Techniques with Multi-Scale Permutation Entropy Calculations for Bearing Fault Diagnosis. *Sensors* **2018**, *18*, 1278. [CrossRef] [PubMed]
24. Guo, Y.J.; Chen, X.F.; Wang, S.B.; Sun, R.B.; Zhao, Z.B. Wind Turbine Diagnosis under Variable Speed Conditions Using a Single Sensor Based on the Synchrosqueezing Transform Method. *Sensors* **2017**, *17*, 1149.
25. Bordoloi, D.J.; Tiwari, R. Optimum multi-fault classification of gears with integration of evolutionary and SVM algorithms. *Mechani. Mach. Theory* **2014**, *73*, 49–60. [CrossRef]
26. Cherkassky, V. The nature of statistical learning theory. *IEEE Trans. Neural Netw.* **2002**, *38*, 409–409. [CrossRef] [PubMed]
27. Kressel, B.U. Pairwise classification and support vector machines. In *Advances in Kernel Methods: Support Vector Learning*; MIT Press: Cambridge, MA, USA, 1999; pp. 255–268.
28. Wu, F.Q.; Meng, G. Compound rub malfunctions feature extraction based on full-spectrum cascade analysis and SVM. *Mechani. Syst. Signal Process.* **2006**, *20*, 2007–2021.
29. Saimurugan, M.; Ramachandran, K.I.; Sugumaran, V.; Sakthivel, N.R. Multi component fault diagnosis of rotational mechanical system based on decision tree and support vector machine. *Expert Syst. Appl.* **2011**, *38*, 3819–3826. [CrossRef]
30. Santos, P.; Villa, L.F.; Reãones, A.; Maudes, J. An SVM-based solution for fault detection in wind turbines. *Sensors* **2015**, *15*, 5627–5648. [CrossRef] [PubMed]
31. Lu, W.N.; Liang, B.; Cheng, Y.; Meng, D.S.; Yang, J.; Zhang, T. Deep model-based domain adaptation for fault diagnosis. *IEEE Trans. Ind. Electron.* **2017**, *64*, 2296–2305. [CrossRef]
32. Case Western Reverse University. Available online: http://csegroups.case.edu/bearingdatacenter/pages/apparatus-procedures (accessed on 7 May 2018).
33. Zhang, K. Research on Local Mean Decomposition Method and Its Application to Rotating Machinery Fault Diagnosis. Ph.D. Thesis, Hunan University, Changsha, China, 2012.
34. Yeh, J.R.; Shieh, J.S.; Huang, N.E. Complementary ensemble empirical mode decomposition: A novel noise enhanced data analysis method. *Adv. Adapt. Data Anal.* **2010**, *2*, 135–156. [CrossRef]
35. Liu, Bo.; Xiao, Y.S.; Cao, L.B. SVM-based multi-state-mapping approach for multi-class classification. *Knowl.-Based Syst.* **2017**, *129*, 79–86. [CrossRef]
36. Ju, B.; Zhang, H.J.; Liu, Y.B.; Dai, Z.J. A feature extraction method using improved multi-scale entropy for rolling bearing fault diagnosis. *Entropy* **2018**, *20*, 212. [CrossRef]
37. Chang, C.C.; Lin, C.J. LIBSVM: A library for support vector machines. *ACM Trans. on Intell. Syst. Technol.* **2011**, *2*, 27. [CrossRef]
38. Weiss, K.; Khoshgoftaar, T.M.; Wang, D.D. A survey of transfer learning. *J. Big Data* **2016**, *3*, 9. [CrossRef]

Article

Ultrasonic Flaw Echo Enhancement Based on Empirical Mode Decomposition

Wei Feng [1], Xiaojun Zhou [1], Xiang Zeng [2] and Chenlong Yang [1,*]

[1] State Key Lab of Fluid Power and Mechatronic Systems, Zhejiang University, Hangzhou 310027, China; fengweizju@126.com (W.F.); cmeesky@163.com (X.Z.)

[2] CRRC Zhuzhou Institute Co. Ltd., Zhuzhou 412001, China; zzjjuu0104@163.com

* Correspondence: yangchenlong@zju.edu.cn; Tel.: +86-135-8874-5549

Received: 3 December 2018; Accepted: 6 January 2019; Published: 9 January 2019

Abstract: The detection of flaw echoes in backscattered signals in ultrasonic nondestructive testing can be challenging due to the existence of backscattering noise and electronic noise. In this article, an empirical mode decomposition (EMD) methodology is proposed for flaw echo enhancement. The backscattered signal was first decomposed into several intrinsic mode functions (IMFs) using EMD or ensemble EMD (EEMD). The sample entropies (SampEn) of all IMFs were used to select the relevant modes. Otsu's method was used for interval thresholding of the first relevant mode, and a window was used to separate the flaw echoes in the relevant modes. The flaw echo was reconstructed by adding the residue and the separated flaw echoes. The established methodology was successfully employed for simulated signal and experimental signal processing. For the simulated signals, an improvement of 9.42 dB in the signal-to-noise ratio (SNR) and an improvement of 0.0099 in the modified correlation coefficient (MCC) were achieved. For experimental signals obtained from two cracks at different depths, the flaw echoes were also significantly enhanced.

Keywords: ultrasonic flaw echo enhancement; empirical mode decomposition; sample entropy; Otsu's method for thresholding; flaw echo separation

---

1. Introduction

The ultrasonic technique has been widely used in nondestructive testing. Usually, the backscattered signal is complex due to the existence of electronic noise and backscattering noise. Consequently, flaw echo detection may be challenging. Numerous methods have been proposed to enhance flaw echoes, such as split spectrum processing [1–4], wavelet transforms [5–10], the Stockwell transform [11–14], and empirical mode decomposition (EMD) [15–22] (including the so-called ensemble EMD, i.e., EEMD [23]).

Split spectrum processing has significant advantages in processing ultrasonic signals with scattered noise. Split-spectrum analysis separates the spectrum of the signals to obtain several sub-bands, and uses some nonlinear de-noising criteria (such as thresholding method, etc.) to process the signals in each sub-band to achieve the purpose of de-noising. The difficulty of split spectrum processing is how to determine the filter type, central frequency, bandwidth and other parameters. In addition, split spectrum processing lacks the capability of multiresolution analysis.

Wavelet transform is a classical multiresolution analysis method. The difficulty of wavelet transform is how to choose the appropriate wavelet base function and decomposition layer. The disadvantage of conventional wavelet transform is that the phase information of signals is lost.

S transform is the development of short-time Fourier transform and wavelet transform. S transform combines the multiresolution analysis ability of wavelet transform and the phase retention ability of short-time Fourier transform. Meanwhile, the S transform adopts Gaussian window function, which satisfies the normalization characteristic, so the S transform is invertible, that is, the original signal can be obtained from the converted time spectrum. However, due to the fact that the standard deviation of Gaussian window function in S transform is inversely proportional to the frequency and lacks flexibility, S transform may output the result of poor time-frequency resolution. At present, to compensate for the limitations of S transform, researchers introduce additional parameters to control the window function morphology, so that the generalized S transform has the ability to flexibly adjust the time-frequency resolution.

For a single component signal, Hilbert transform can be applied to obtain its analytical signal, and then the envelope spectrum and instantaneous frequency of the signal can be obtained. Detection of a flaw signal from the envelope spectrum is a common method of ultrasonic nondestructive testing. For a multi-component signal, it is necessary to decompose them into single component signals, and then obtain their analytical signals separately. Empirical mode decomposition (EMD) is an adaptive decomposition method. The original signal is decomposed into a series of intrinsic mode functions. The main disadvantages of EMD are the possibility of endpoint effect and mode mixing. Ensemble empirical mode decomposition (EEMD) can solve the mode mixing issue. EEMD decomposes the original signal by adding Gaussian white noise, and takes the result of the lumped average as the mode function. The main difficulty of EEMD is to select the intensity of Gaussian white noise and the times of lumped average.

In this article, an EMD-based methodology for ultrasonic flaw echo enhancement was established. The proposed methodology enhanced the flaw echo through six steps. First, the backscattered signal was adaptively decomposed into several intrinsic mode functions (IMFs) by EMD or EEMD. Second, the sample entropies (SampEn) [24,25] of all IMFs were calculated, and the differences in consecutive SampEn values were studied. Third, those IMFs with a large SampEn were considered to be irrelevant modes and were discarded, significantly suppressing the electronic noise. Fourth, the intervals containing the flaw echo were determined based on IMF interval thresholding and mode cell merging. Otsu's method [26] was used to search the threshold in IMF interval thresholding. Fifth, a Turkey-Hanning window was used to separate the flaw echo for each relevant mode. Finally, the denoised signal was reconstructed by combining the separated flaw echoes and the residue.

The remainder of this article is organized as follows. In Section 2, reviews of the required tools, including EMD and EEMD, EMD-based denoising methods, SampEn, and Otsu's method for thresholding, are presented. An analysis of the modes extracted from the backscattered signal, including the mixing of noise and flaw echoes, and the SampEn is given in Section 3. The proposed EMD-based methodology for ultrasonic flaw echo enhancement is given in Section 4 and tested using a simulated signal in Section 5. In Section 6, experimental validations of the proposed EEMD-based methodology are presented. Finally, conclusions are drawn in Section 7.

2. Required Tools

2.1. EMD and EEMD

In EMD, a signal is adaptively decomposed into a collection of IMFs. The EMD results can be presented as

$$x(t) = \sum_{i=1}^{L} h^{(i)}(t) + r(t) \tag{1}$$

where $x(t)$ is the observed signal, $h^{(i)}(t)(i \leq L)$ are the extracted IMFs, and $r(t)$ is the residue.

Unfortunately, EMD is susceptible to mode-mixing. EEMD is an effective technique for alleviating mode-mixing in EMD by repeatedly adding Gaussian white noise and finding the mean of the individual ensemble IMFs as the final IMF.

2.2. Denoising Strategies Based on EMD

Partial reconstruction, direct thresholding, and interval thresholding are three typical strategies adopted in EMD-based denoising.

Partial reconstruction removes noise from the observed signal by discarding irrelevant modes, which can be expressed as

$$\hat{x}(t) = \sum_{i=M_1}^{L} h^{(i)}(t) + r(t) = x(t) - \sum_{i=1}^{M_1-1} h^{(i)}(t) \tag{2}$$

where $h^{(i)}(t)\,(i < M_1)$ are the irrelevant modes.

Direct thresholding is a direct application of wavelet thresholding in the EMD case. For the hard thresholding case, the denoised IMF is given by

$$\widetilde{h}^{(i)}(t) = \begin{cases} h^{(i)}(t), & \left| h^{(i)}(t) \right| > T_i \\ 0, & \left| h^{(i)}(t) \right| \leq T_i \end{cases} \tag{3}$$

where T_i is the threshold of $h^{(i)}(t)$. The denoised signal can be given by

$$\widetilde{x}(t) = \sum_{i=M_1}^{M_2} \widetilde{h}^{(i)}(t) + \sum_{i=M_2+1}^{L} h^{(i)}(t) + r(t) \tag{4}$$

The interval thresholding divides an IMF into several mode cells and treats each mode cell as a whole to perform thresholding. Generally, a mode cell is defined as the signal between two adjacent zero-crossings. For the interval $\mathbf{z}_j^{(i)} = [z_j^{(i)} z_{j+1}^{(i)}]$ defined by two zero-crossings $z_j^{(i)}$ and $z_{j+1}^{(i)}$, the denoised IMF in the hard thresholding case is given as

$$\widetilde{h}^{(i)}(\mathbf{z}_j^{(i)}) = \begin{cases} h^{(i)}(\mathbf{z}_j^{(i)}), & \left| h^{(i)}(r_j^{(i)}) \right| > T_i \\ 0, & \left| h^{(i)}(r_j^{(i)}) \right| \leq T_i \end{cases} \tag{5}$$

where $h^{(i)}(\mathbf{z}_j^{(i)})$ are all of the samples from $z_j^{(i)}$ to $z_{j+1}^{(i)}$, and $h^{(i)}(r_j^{(i)})$ is the single extremum of $h^{(i)}(t)$ in the interval $\mathbf{z}_j^{(i)}$.

Interval thresholding generally outperforms direct thresholding as it avoids catastrophic consequences for the continuity of the reconstructed signal, which are inevitable in direct thresholding.

2.3. Sample Entropy

Approximate entropy [27] and SampEn are two popular metrics for signal complexity measurement. Entropy values increase with increased signal complexity. It has been reported that SampEn outperforms approximate entropy in many aspects, such as reduced bias, independence from the signal, and relative consistency. SampEn was used here for signal complexity assessment.

The SampEn of a specified time series $\{u(i), i = 1, 2, \cdots, N\}$ can be obtained through the following steps.

Step 1: Form $m-$dimensional vectors as

$$U_m(i) = [u(i), u(i+1), \ldots, u(i+m-1)] \tag{6}$$

where $i = 1, 2, \ldots, N - m + 1$.

Step 2: Define the distance of two such vectors:

$$d[U_m(i), U_m(j)] = \max_{k=0-(m-1)} |u(i+k) - u(j+k)| \tag{7}$$

Step 3: Consider the first $N - m$ vectors of length m so that for $i = 1, \cdots, N - m$, both $U_m(i)$ and $U_{m+1}(i)$ can be defined.

Step 4: Given a threshold $r > 0$, define

$$\begin{aligned} B_i^m(r) &= \tfrac{1}{N-m-1}\sum \Theta(r - d[U_m(i), U_m(j)]) \\ A_i^m(r) &= \tfrac{1}{N-m-1}\sum \Theta(r - d[U_{m+1}(i), U_{m+1}(j)]) \end{aligned} \tag{8}$$

where $j = 1, 2, \ldots, N - m, j \neq i$, and $\Theta(\cdot)$ is defined as

$$\Theta(x) = \begin{cases} 1, & x \geq 0 \\ 0, & x < 0 \end{cases} \tag{9}$$

Step 5: Calculate $B^m(r)$ and $A^m(r)$:

$$\begin{aligned} B^m(r) &= \tfrac{1}{N-m}\sum_{i=1}^{N-m} B_i^m(r) \\ A^m(r) &= \tfrac{1}{N-m}\sum_{i=1}^{N-m} A_i^m(r) \end{aligned} \tag{10}$$

Step 6: For a limited series, the SampEn (denoted by s) is estimated as

$$s = SampEn(m, r) = -\ln\frac{A^m(r)}{B^m(r)} \tag{11}$$

In general, the dimension and threshold are often set to $m = 2$ and $r = (0.1 \sim 0.25)SD_u$, where SD_u is the standard deviation of the time series $\{u(i)\}$.

2.4. Otsu's Method for Thresholding

Otsu's method determines a threshold by maximizing the between-class variance σ_B^2. For a histogram with H levels (i.e., bins), the probability at each level can be first obtained:

$$p_i = \frac{n_i}{N}, i = 1, 2, \cdots, H \tag{12}$$

where n_i is the number of elements in the i_{th} level, and N is the number of elements in the histogram. Obviously, $p_i \geq 0$ and $\sum p_i = 1$ are satisfied.

The histogram can be divided into two classes, C_1 and C_2, with a threshold. σ_B^2 is defined as

$$\sigma_B^2 = w_1(\mu_1 - \mu_0)^2 + w_2(\mu_2 - \mu_0)^2 \tag{13}$$

where

$$w_1 = \sum_{C_1} p_i, \quad w_2 = \sum_{C_2} p_i, \quad w_1 + w_2 = 1 \tag{14}$$

μ_0, μ_1, μ_2 are the means of the histogram, class C_1 and class C_2, respectively. Therefore, Equation (15) can be obtained:

$$\mu_0 = w_1\mu_1 + w_2\mu_2 \tag{15}$$

According to Equations (13)–(15), σ_B^2 can also be expressed as

$$\sigma_B^2 = w_1 w_2(\mu_1 - \mu_2)^2 \tag{16}$$

3. Analysis of Modes from Ultrasonic Signals

3.1. The Clutter Model

For metallic materials, when an incident ultrasonic wave propagates into the specimen, the backscattered signal will be primarily composed of three components: (1) the flaw echo signal $s(t)$, (2) backscattering noise $v(t)$ due to the grains, and (3) electronic noise $n(t)$ due to the instruments and the environment. $n(t)$ can be approximated as Gaussian white noise. The frequency spectra of $s(t)$ and $v(t)$ can be expressed as [2]

$$V(\omega) = H_t^2(\omega) \sum_{k=1}^{K} \beta_k \frac{\omega^2}{x_k} e^{-\alpha_s 2 x_k \omega^4} e^{-i\omega \frac{2x_k}{c_0}} \tag{17}$$

$$S(\omega) = H_t^2(\omega) \exp(-\alpha_s 2 d_{\text{flaw}} \omega) \exp(-i\frac{2d_{\text{flaw}}}{c_0}) \tag{18}$$

where $H_t(\omega)$ is the frequency response of the ultrasonic transducer, and d_{flaw} is the location of the flaw. α_s is the material attenuation coefficient, c_0 is the velocity of the longitudinal waves, and K is the total number of scatterers. β_k and x_k are the scattering coefficient and the position of the k_{th} scatterer, respectively. The amplitudes of $s(t)$ and $v(t)$ are often normalized for brevity.

The similarity function obtained by deconvolution can be used to distinguish the flaw signals in ultrasonic inspection from other no flaw signals, such as specimen geometric reflection. It was found that the deconvolution patterns of the geometric reflection were impulse-like patterns, whereas those of flaws were bipolar patterns [28]. Therefore, geometric reflection was not considered in the model.

The observed backscattered signal $x(t)$ can be expressed as

$$x(t) = s(t) + \mu v(t) + \sigma_n n_0(t) \tag{19}$$

where $n_0(t)$ is standard Gaussian white noise. μ and σ_n are scale factors.

Typical simulated results are depicted in Figure 1. The centre frequency of the ultrasonic transducer was 5 MHz, and the sampling frequency was 100 MHz. The scale factors were set to $\mu = 0.3$ and $\sigma_n = 0.2$. It can be seen in Figure 1a that the flaw echo had been polluted by intense noise. In addition, frequency aliasing of the flaw echo, backscattering noise, and wide-band Gaussian white noise can be found in Figure 1b.

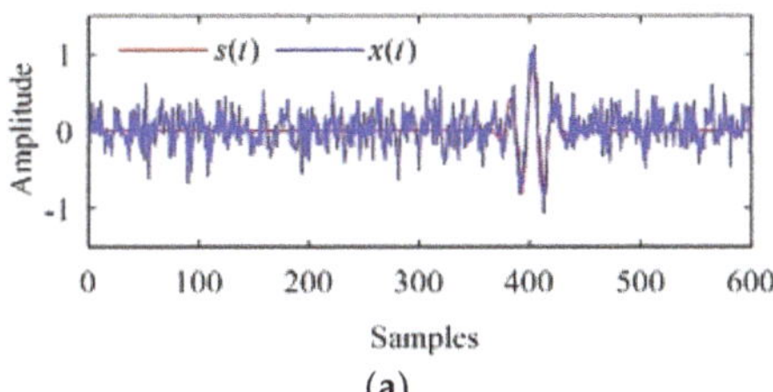

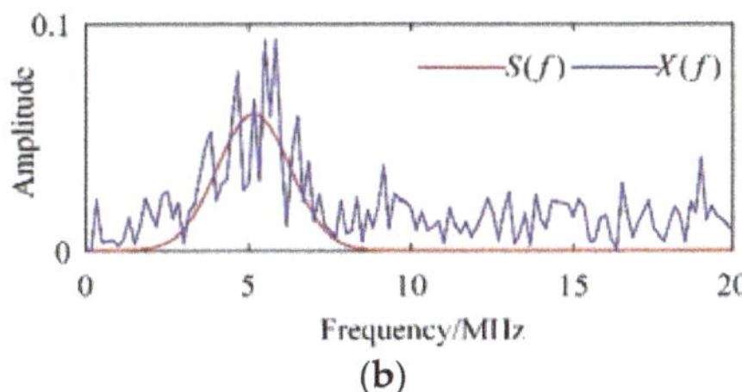

Figure 1. Simulated results. (**a**) Waveforms and (**b**) frequency spectra.

3.2. Signal Decomposition

The EMD results for the observed signal $x(t)$ are shown in Figure 2.

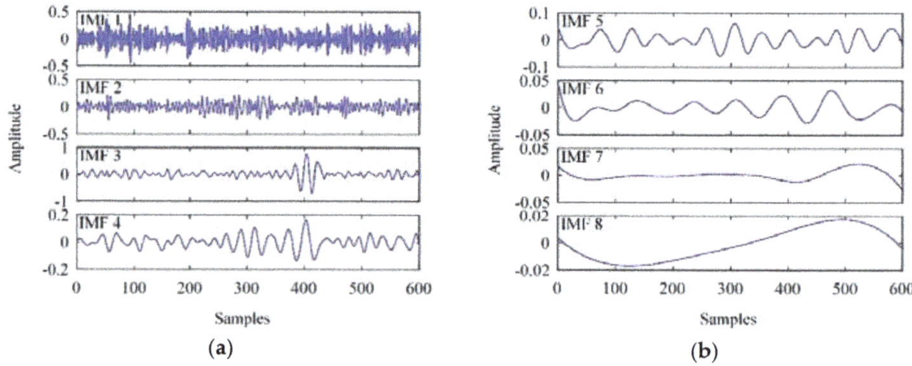

Figure 2. Intrinsic mode functions (IMFs) extracted. (**a**) IMF 1 ~ IMF 4; (**b**) IMF 5 ~ IMF 8.

As shown in Figure 2, intense white noise was found in the low-order IMFs; in particular, IMF 1 resembled pure noise. The flaw echo was clearly detected in IMF 3. Consequently, the flaw echo was significantly enhanced by discarding IMF 1 and IMF 2 from the observed signal. However, partial reconstruction is often inadequate; further processing is required. Here, we take IMF 3 as an example.

The flaw echo is an instant signal. Ideally, oscillations can only be detected in the interval in which the flaw echo is located in IMF 3, in contrast to Figure 2. The difference mainly arises from frequency aliasing of the flaw echo, backscattering noise and wide-band Gaussian white noise. In addition, even in the noiseless case, the IMFs still contain false oscillations, as they resemble AM-FM modulated sinusoids. Consequently, further denoising is required to suppress the mixed noise lying outside the interval in which the flaw echo is located.

3.3. SampEn Values of IMFs

The SampEn values of all of the IMFs were calculated and are listed in Table 1, where the threshold r in the SampEn calculations was set to $(0.1, 0.15, 0.2)SD_u$.

Table 1. SampEn of the IMFs.

r/SD_u	0.1	0.15	0.2
s_1	2.0121	1.7943	1.535
s_2	1.8823	1.4872	1.213
s_3	0.7224	0.6242	0.5682
s_4	0.646	0.5984	0.5616
s_5	0.5501	0.4921	0.4472
s_6	0.4717	0.3589	0.2708
s_7	0.1095	0.0821	0.065
s_8	0.0526	0.036	0.0276

It can be noted that the SampEn tends to decrease with increasing IMF order, which indicates that the noise intensity in each IMF decreases as IMF order increases. It is noteworthy that the SampEn values of the first two IMFs, i.e., s_1 and s_2, were much higher than the others. In addition, a sharp drop between s_2 and s_3 was detected.

4. Proposed Methodology

According to Section 3.2, flaw echo enhancement can be achieved by discarding irrelevant modes and suppressing the mixed noise in the remaining relevant modes. Specifically, the relevant modes are determined by the SampEn of all IMFs or their differences, and the mixed noise is suppressed by separating the flaw echo from those relevant modes by windowing.

4.1. Relevant Mode Selection

The IMFs with intense noise were of much higher complexity than other IMFs. Consequently, those relevant modes were determined according to the SampEn of the IMFs. Specifically, the parameter M_1, which determined the first relevant mode, was determined by s_i or the difference of s_i.

Given a predefined SampEn threshold T_s, M_1 can be determined as

$$M_1 = (\max i) + 1, \quad \text{s.t. } s_i > T_s \tag{20}$$

Using the difference of s_i, i.e., ds_i, M_1 can also be determined:

$$M_1 = \operatorname*{argmin}_i\{ds_i\} = \operatorname*{argmin}_i\{s_i - s_{i-1}\} \tag{21}$$

To determine T_s, we should note that the SampEn is dependent on the threshold r. For example, Table 1 shows that s_1 and s_2 decreased rapidly with increasing r. Consequently, the selection of T_s was dependent on r. For this article, $r = 0.15SD_u$ and $T_s = 1$ were selected.

4.2. Mixed Noise Suppression

Two steps are required to suppress mixed noise: determine the location of the flaw echo in the first relevant mode and separate the flaw echo from all of the relevant modes.

For the first relevant mode, i.e., $h^{(M_1)}(t)$, a collection of mode cells is built according to the zero-crossings. The set of absolute values of the extrema in $h^{(M_1)}(t)$, which is denoted $\left|\mathbf{r}_j^{(i)}\right| = [\left|r_1^{(i)}\right|, \left|r_2^{(i)}\right|, \cdots]$, can be determined. Otsu's method is used for searching the threshold of $\left|\mathbf{r}_j^{(i)}\right|$, i.e., T_i. It is performed on the first relevant mode, and all of the mode cells in which the flaw echoes are located can be detected.

All of the adjacent mode cells in which the flaw echo is located are further merged, yielding an interval in which the flaw echo is located. For example, three adjacent mode cells, $\mathbf{z}_j^{(i)} = [z_j^{(i)} z_{j+1}^{(i)}]$, $\mathbf{z}_{j+1}^{(i)} = [z_{j+1}^{(i)} z_{j+2}^{(i)}]$ and $\mathbf{z}_{j+2}^{(i)} = [z_{j+2}^{(i)} z_{j+3}^{(i)}]$, can be merged into one interval $[z_j^{(i)} z_{j+3}^{(i)}]$. In other words, the interval in which the flaw echo is located is often composed of several mode cells containing the flaw echo.

Thus far, the location of the flaw echo in the first relevant mode has been detected. Next, the flaw echoes in all relevant modes will be separated.

A window was determined and used for separating the flaw echoes in the relevant modes. Herein, the Turkey-Hanning window was used. The window was defined as

$$w(m) = \begin{cases} \frac{(1-\cos 2\pi \frac{m}{M+1})}{2}, & 1 \le m \le \frac{M}{2} \\ 1, & \frac{M}{2} < m \le \frac{M}{2} + W \\ \frac{1-\cos 2\pi \frac{m-W}{M+1}}{2}, & \frac{M}{2} + W < m \le M + W \end{cases} \tag{22}$$

where W determined the width of the pass zone and M determined the width of the transition zone. As an example, Figure 3 shows the waveform of a Turkey-Hanning window.

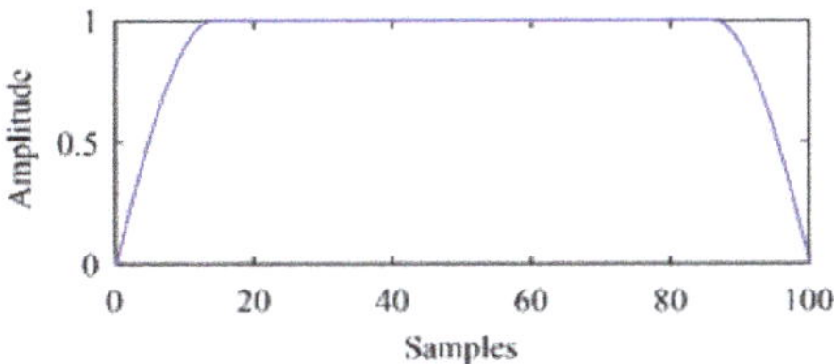

Figure 3. Waveform of a Turkey-Hanning window.

The parameters W and M were determined according to the width of the interval in which the flaw echo was located. To preserve the flaw echo, W was set to the width of this interval. M was selected more flexibly. In this article, $M = [W/4]$ was adopted, where the operator "$[]$" indicated rounding.

4.3. Summarization of the Proposed Methodology

In this methodology, the flaw echo was enhanced in six steps.

Step 1. Signal decomposition. Decompose the observed signal $x(t)$ into a collection of the residue $r(t)$ and the IMFs' $h^{(i)}(t)(i = 1, 2, \cdots, L)$ using EMD or EEMD.

Step 2. SampEn calculation. Obtain the SampEn of all of the IMFs.

Step 3. Relevant mode selection. Determine the first relevant mode using the SampEn or the differences between them.

Step 4. Determine the interval in which the flaw echo is located in the first relevant mode. Otsu's method is used for threshold selection. Any two adjacent mode cells in which the flaw echo is located are merged into one interval.

Step 5. Separate the flaw echo using the Turkey-Hanning window in each relevant mode, which yields a collection of denoised modes $\tilde{h}^{(i)}(t)$ $(M_1 \leq i \leq L)$.

Step 6. Reconstruct the denoised signal:

$$\tilde{x}(t) = \sum_{i=M_1}^{L} \tilde{h}^{(i)}(t) + r(t) \tag{23}$$

5. Simulated Signal Processing

5.1. Performance Assessment

To assess the performance of the EMD-based methodology, two metrics are introduced.

The signal-to-noise ratio (SNR) is the first metric. Suppose that $s(t)$ is a noiseless signal; the SNR of signal $x(t)$ is then defined as

$$SNR = 10\lg \frac{\sum\limits_{i=1}^{N} s^2(i)}{\sum\limits_{i=1}^{N} (x(i) - s(i))^2} \tag{24}$$

Well-preserved flaw echoes are expected in practice. Specifically, the amplitudes and shapes of flaw echoes are expected to be unchanged. This can be assessed by the modified correlation coefficient (MCC), which is defined as

$$MCC = \left| \frac{A_x - A_s}{A_s} \right| (1 - \frac{\sum\limits_{i=m}^{n} (s(i) - \bar{s})(x(i) - \bar{x})}{\sqrt{\sum\limits_{i=m}^{n} (s(i) - \bar{s})^2 \sum\limits_{i=m}^{n} (x(i) - \bar{x})^2}}) \tag{25}$$

where m and n are instants defining the interval in which the flaw echo is located. A_x and A_s are the amplitudes of the flaw echoes in $x(t)$ and $s(t)$, respectively.

A high SNR and low MCC are expected.

5.2. Signal Processing Results

The relevant modes were first determined. On one hand, with the predefined parameters $r = 0.15SD_u$ and $T_s = 1$, it can be seen in Table 1 that $s_1, s_2 > T_s$ and $s_3 < T_s$ and were satisfied. Consequently, $M_1 = 3$ was determined according to Equation (20). On the other hand, it could be inferred from Table 1 that $ds_3 = s_3 - s_2 < ds_k(k \neq 3)$ was satisfied, which also indicates $M_1 = 3$ according to Equation (21). Thus, these two methods yielded the same relevant mode selection results.

Consequently, the electronic noise could be significantly suppressed if we reconstructed the flaw echo via partial reconstruction by discarding $h^{(1)}(t)$ and $h^{(2)}(t)$. The corresponding reconstructed signal, $\hat{x}(t)$, is depicted in Figure 4a.

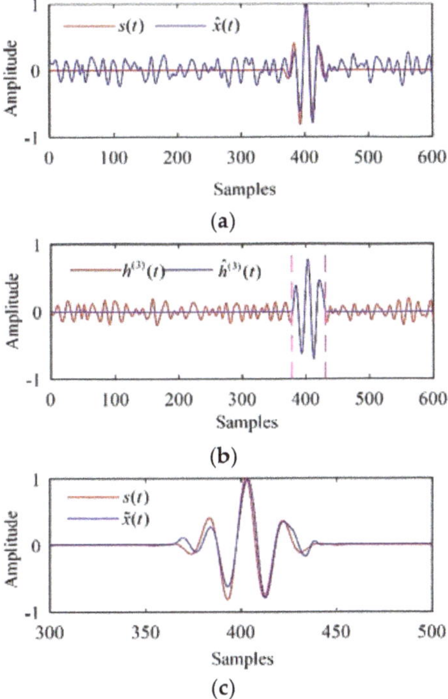

Figure 4. Processing results of the simulated signal. (**a**) The reconstructed signal given by partial reconstruction; (**b**) the detection interval in which the flaw echo was located; (**c**) the denoised signal.

$h^{(3)}(t)$ was used to determine the interval in which the flaw echo was located. The corresponding result is depicted in Figure 4b, for which the threshold given by Otsu's method was 0.2902. The flaw echoes in all relevant modes $h^{(i)}(t)$ ($3 \leq i \leq 8$) were separated by the Turkey-Hanning window, and the denoised signal was reconstructed. The denoised signal is depicted in Figure 4c, which indicated that the electronic noise and backscattering noise were significantly suppressed, and the flaw echo were preserved well.

The performance of the EMD-based methodology was assessed using the SNR and the MCC. The results are provided in Table 2, which shows that the methodology achieved a high SNR and low MCC as expected. The improvements in SNR and MCC were 12.89 dB and 0.0099, respectively.

Table 2. Performance assessment.

	SNR/dB	MCC/ $\times 10^{-2}$
observed	−3.47	1.05
partial reconstruction denoised	1.81	0.06
	9.42	0.06

6. Experimental Study

The ultrasonic testing system was primarily composed of an ADVANTECH industrial personal computer (IPC), an Olympus ultrasonic probe with a centre frequency of 5 MHz, and a PCIUT3100 ultrasonic acquisition card installed on the IPC. A 6061 aluminium alloy specimen with two artificial cracks was used for ultrasonic testing. The specimen material density was 2.7×10^3 kg/m^3 and the sound wave velocity was 6300 m/s. The two cracks, denoted F1 and F2, were machined using wire electrical discharge machining. The buried depths of F1 and F2 were 60 mm and 35 mm, respectively, and the sampling frequency was 100 MHz. The ultrasonic testing system and the specimen are shown in Figure 5.

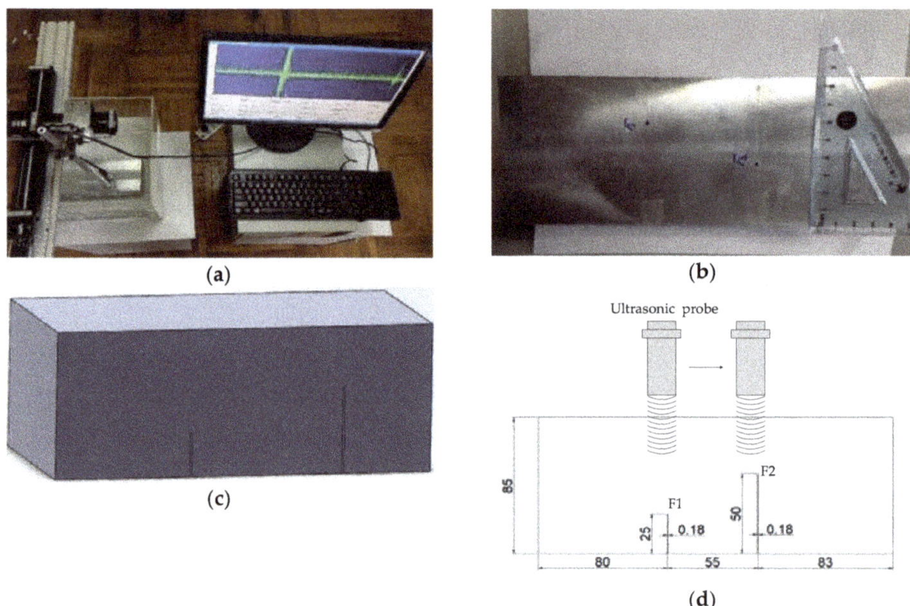

Figure 5. The ultrasonic test system and specimen. (**a**) The ultrasonic testing system. (**b**) The specimen. (**c**) Geometry of the specimen. (**d**) Schematic diagram of flaws location and size, with dimensions in millimeters.

The ultrasonic signals acquired from cracks F1 and F2 are shown in Figure 6. The backscattered signals from instant 1001 to instant 3000 were used for further analysis.

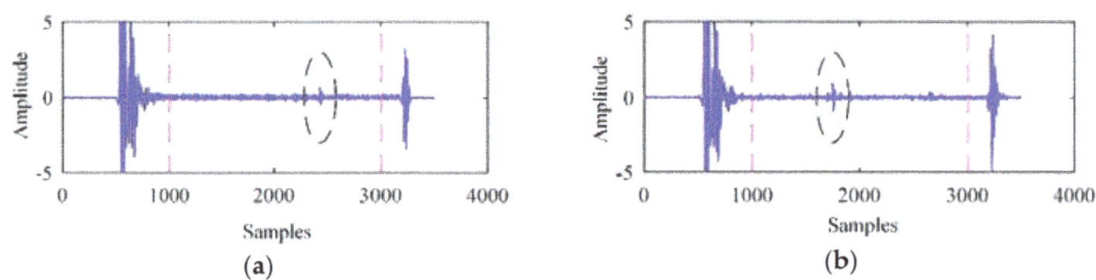

Figure 6. The ultrasonic signals. (**a**) Acquired from F1; (**b**) acquired from F2.

The processing results of the backscattered signal from crack F1 are shown in Figure 7. The backscattered signal was first decomposed into a collection of IMFs, as shown in Figure 7a. For brevity, only the first five IMFs are shown. Correspondingly, the SampEn values of the first five IMFs were calculated and are listed in Table 3. As the first relevant mode, the third IMF was used to determine the interval in which the flaw echo was located based on Otsu's method for thresholding and merging of adjacent mode cells, as shown in Figure 7b. The denoised signal is shown in Figure 7c, indicating that both the backscattering noise and the electronic noise were significantly suppressed.

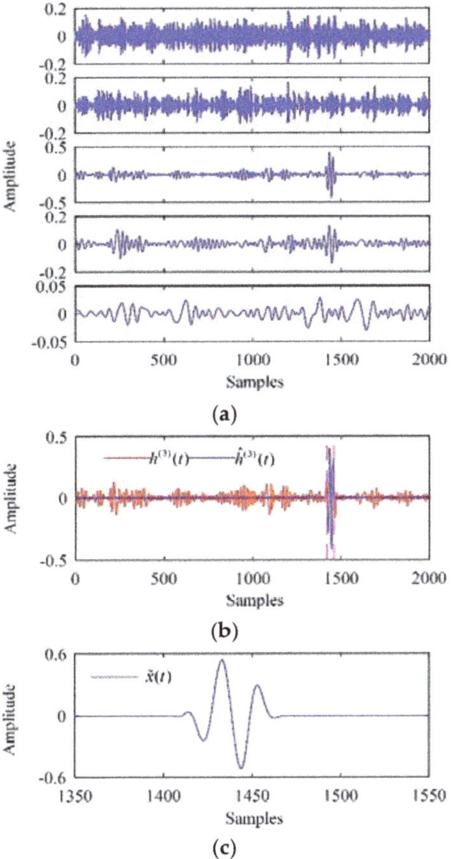

Figure 7. Processing results of the backscattered signal from F1. (**a**) The IMFs extracted; (**b**) the interval in which the flaw echo was located; (**c**) the denoised signal.

Table 3. SampEn of the first five IMFs.

s_1	s_2	s_3	s_4	s_5
1.4902	1.1739	0.5684	0.5508	0.4764

As supplementary information, Figure 8 shows the denoised signal of the backscattered signal from F2, with a significant reduction of the noise.

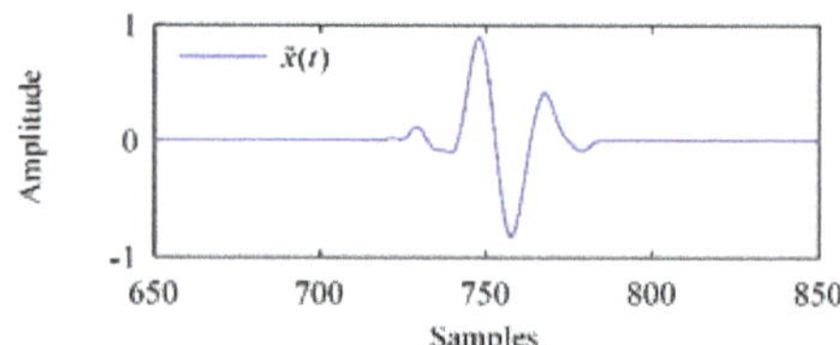

Figure 8. The denoised signal of the ultrasonic signal from F2.

7. Conclusions

An EMD-based methodology was proposed for ultrasonic flaw echo enhancement. The observed signal was decomposed into IMFs by EMD or EEMD. The relevant modes were determined according to the SampEn of the IMFs. Otsu's method was used for interval thresholding of the first relevant mode, obtaining the interval in which the flaw echo was located. The flaw echoes in all IMFs were separated by the Turkey-Hanning window. The separated flaw echo and the residue were added together, yielding the denoised signal. Simulation results demonstrated that the EMD-based methodology achieved a high SNR and low MCC. Applications of the EMD-based methodology in flaw echo enhancement in experimental signals was also presented.

Author Contributions: Conceptualization, W.F., X.Z. (Xiaojun Zhou) and C.Y.; formal analysis, W.F.; funding acquisition, C.Y.; investigation, W.F., X.Z. (Xiang Zeng) and C.Y.; methodology, W.F. and X.Z. (Xiang Zeng); resources, X.Z. (Xiaojun Zhou) and C.Y.; supervision, X.Z. (Xiaojun Zhou); writing—original draft, W.F.; writing—review and editing, W.F., X.Z. (Xiaojun Zhou), X.Z. (Xiang Zeng) and C.Y.

Funding: This research was funded by 1: The Fundamental Research Funds for the Central Universities under Grant No.2018QNA4001; 2: Zhejiang Provincial Natural Science Foundation of China under Grant No.LY18E050002.

Conflicts of Interest: The authors declare no conflict of interest.

References

1. Shankar, P.M.; Karpur, P.; Newhouse, V.L.; Rose, J.L. Split-spectrum processing: Analysis of polarity threshold algorithm for improvement of signal-to-noise ratio and detectablity in ultrasonic signals. *IEEE Trans. Ultrason. Ferroelectr. Freq. Control* **1989**, *36*, 101–108. [CrossRef] [PubMed]

2. Gustafsson, M.G.; Stepinski, T. Studies of split spectrum processing, optimal detection, and maximum likelihood amplitude estimation using a simple clutter model. *Ultrasonics* **1997**, *35*, 31–52. [CrossRef]

3. Rodriguez, A.; Miralles, R.; Bosch, I.; Vergara, L. New analysis and extensions of split-spectrum processing algorithms. *NDT E Int.* **2012**, *45*, 141–147. [CrossRef]

4. Jafar, S.; Daniel, T.N.; Kevin, D.D. Analysis of order statistic filters applied to ultrasonic flaw detection using split-spectrum processing. *IEEE Trans. Ultrason. Ferroelectr. Freq. Control* **1991**, *38*, 133–140. [CrossRef]

5. Lazaro, J.C.; San Emeterio, J.L.; Ramos, A.; Fernandez-Marron, J.L. Influence of thresholding procedures in ultrasonic grain noise reduction using wavelets. *Ultrasonics* **2002**, *40*, 263–267. [CrossRef]

6. Matz, V.; Smid, R.; Starman, S.; Kreidl, M. Signal-to-noise ratio enhancement based on wavelet filtering in ultrasonic testing. *Ultrasonics* **2009**, *49*, 752–759. [CrossRef]

7. Peng, C.Y.; Gao, X.R.; Wang, A. Novel wavelet self-optimisation threshold denoising method in axle press-fit ultrasonic defect detection. *Insight* **2016**, *58*, 145–151. [CrossRef]

8. Liu, Y.; Li, Z.; Zhang, W. Crack detection of fibre reinforced composite beams based on continuous wavelet transform. *Nondestruct. Test. Eval.* **2010**, *25*, 25–44. [CrossRef]

9. Chen, Y.C.; Zhou, X.J.; Yang, C.L.; Li, Z. The ultrasonic evaluation method for the porosity of variable-thickness curved CFRP workpiece: Using a numerical wavelet transform. *Nondestruct. Test. Eval.* **2014**, *29*, 195–207. [CrossRef]

10. Praveen, A.; Vijayarekha, K.; Abraham, S.T.; Venkatraman, B. Signal quality enhancement using higher order wavelets for ultrasonic TOFD signals from austenitic stainless steel welds. *Ultrasonics* **2013**, *53*, 1288–1292. [CrossRef]

11. Stockwell, R.G.; Mansinha, L.; Lowe, R.P. Localization of the complex spectrum: The S transform. *IEEE Trans. Signal Process.* **1996**, *44*, 998–1001. [CrossRef]

12. Ari, S.; Das, M.K.; Chacko, A. ECG signal enhancement using S-Transform. *Comput. Biol. Med.* **2013**, *43*, 649–660. [CrossRef] [PubMed]

13. Muhammad, A.M.; Jafar, S. S-transform applied to ultrasonic nondestructive testing. In Proceedings of the IEEE Ultrasonics Symposium, Beijing, China, 2–5 November 2008; IEEE: Piscataway, NJ, USA, 2008; pp. 184–187. [CrossRef]

14. Benammar, A.; Drai, R.; Guessoum, A. Ultrasonic flaw detection using threshold modified S-transform. *Ultrasonics* **2014**, *54*, 676–683. [CrossRef] [PubMed]

15. Huang, N.E.; Shen, Z.; Long, S.R.; Wu, M.C.; Shih, H.H.; Zheng, Q.N.; Yen, N.C.; Tung, C.C.; Liu, H.H. The empirical mode decomposition and the Hilbert spectrum for nonlinear and non-stationary time series analysis. *Proc. R. Soc. A-Math. Phys. Eng. Sci.* **1998**, *454*, 903–995. [CrossRef]

16. Gabriel, R.; Patrick, F. One or Two Frequencies? The empirical mode decomposition answers. *IEEE Trans. Signal Process.* **2008**, *56*, 85–95. [CrossRef]

17. Kazys, R.; Tumsys, O.; Pagodinas, D. Ultrasonic detection of defects in strongly attenuating structures using the Hilbert–Huang transform. *NDT E Int.* **2008**, *41*, 457–466. [CrossRef]

18. Abdel, O.B.; Jean, C.C. EMD-based signal filtering. *IEEE Trans. Instrum. Meas.* **2007**, *56*, 2196–2202. [CrossRef]

19. Zhang, Q.; Que, P.W.; Liang, W. Applying sub-band energy extraction to noise cancellation of ultrasonic NDT signal. *J. Zhejiang Univ. Sci. A* **2008**, *9*, 1134–1140. [CrossRef]

20. Yannis, K.; Stephen, M. Development of EMD-based denoising methods inspired by wavelet thresholding. *IEEE Trans. Signal Process.* **2009**, *57*, 1351–1362. [CrossRef]

21. Yang, G.L.; Liu, Y.Y.; Wang, Y.Y.; Zhu, Z.L. EMD interval thresholding denoising based on similarity measure to select relevant modes. *Signal Process.* **2015**, *109*, 95–109. [CrossRef]

22. Bouden, T.; Djerfi, F.; Nibouche, M. Adaptive split spectrum processing for ultrasonic signal in the pulse echo test. *Russ. J. Nondestruct. Test.* **2015**, *51*, 245–257. [CrossRef]

23. Sharma, G.K.; Kumar, A.; Jayakumar, T.; Rao, B.P.; Mariyappa, N. Ensemble empirical mode decomposition based methodology for ultrasonic testing of coarse grain austenitic stainless steels. *Ultrasonics* **2015**, *57*, 167–178. [CrossRef] [PubMed]

24. Richman, J.S.; Moorman, J.R. Physiological time-series analysis using approximate entropy and sample entropy. *Am. J. Physiol. Heart Circ. Physiol.* **2000**, *278*, H2039–H2049. [CrossRef]

25. Hu, X.S.; Jiang, J.C.; Cao, D.P.; Bo, E. Battery Health Prognosis for Electric Vehicles Using Sample Entropy and Sparse Bayesian Predictive Modeling. *IEEE Trans. Ind. Electron.* **2016**, *63*, 2645–2656. [CrossRef]

26. Otsu, N. A threshold selection method from gray-level histograms. *IEEE Trans. Syst. Man Cybern.* **1979**, *9*, 62–66. [CrossRef]

27. Steven, M.P. Approximate entropy as a measure of system complexity. *Proc. Natl. Acad. Sci. USA* **1991**, *88*, 2297–2301. [CrossRef]

28. Jung, H.J.; Kim, H.J.; Song, S.J.; Kim, Y.H. Model-based enhancement of the TIFD for flaw signal identification in ultrasonic testing of welded joints. In Proceedings of the 29th Annual Review of Progress in Quantitative Nondestructive Evaluation, Western Washington University, Bellingham, WA, USA, 14–19 July 2002; Amer Inst Physics: College Park, MD, USA, 2002; pp. 628–634. [CrossRef]

Article

An Intelligent Fault Diagnosis Method for Bearings with Variable Rotating Speed Based on Pythagorean Spatial Pyramid Pooling CNN

Sheng Guo, Tao Yang *, Wei Gao, Chen Zhang and Yanping Zhang

School of Energy and Power Engineering, Huazhong University of Science & Technology, Wuhan 430074, China; levykwok@hust.edu.cn (S.G.); gw@hust.edu.cn (W.G.); zhangchen710@yeah.net (C.Z.); zyp2817@hust.edu.cn (Y.Z.)
* Correspondence: hust_yt@hust.edu.cn; Tel.: +86-027-87542817

Received: 16 October 2018; Accepted: 7 November 2018; Published: 9 November 2018

Abstract: Deep learning methods have been introduced for fault diagnosis of rotating machinery. Most methods have good performance when processing bearing data at a certain rotating speed. However, most rotating machinery in industrial practice has variable working speed. When processing the bearing data with variable rotating speed, the existing methods have low accuracies, or need complex parameter adjustments. To solve this problem, a fault diagnosis method based on continuous wavelet transform scalogram (CWTS) and Pythagorean spatial pyramid pooling convolutional neural network (PSPP-CNN) is proposed in this paper. In this method, continuous wavelet transform is used to decompose vibration signals into CWTSs with different scale ranges according to the rotating speed. By adding a PSPP layer, CNN can process CWTSs in different sizes. Then the fault diagnosis of variable rotating speed bearing can be carried out by a single CNN model without complex parameter adjustment. Compared with a spatial pyramid pooling (SPP) layer that has been used in CNN, a PSPP layer locates as front layer of CNN. Thus, the features obtained by PSPP layer can be delivered to convolutional layers for further feature extraction. According to experiment results, this method has higher diagnosis accuracy for variable rotating speed bearing than other methods. In addition, the PSPP-CNN model trained by data at some rotating speeds can be used to diagnose bearing fault at full working speed.

Keywords: convolutional neural network; spatial pyramid pooling; fault diagnosis; bearing; wavelet transform

---

1. Introduction

As most rotating machinery are the key equipment in production and work in a high-speed rotating environment, their failures will cause major economic losses and safety accidents. Therefore, it is important to detect equipment failure as soon as possible. An intelligent fault diagnosis method has the motivation that it can detect the fault of operating machinery in real time without manual operation. With the popularization of online vibration monitoring systems, manufacturers have accumulated a large amount of data that can support intelligent fault diagnosis methods. Many intelligent fault diagnosis methods have been proposed to diagnose bearing faults [1–5]. However, most of them use a simple classifier and focus on fault feature extraction algorithms. When these methods are used to diagnose bearings with complex operating conditions, the simple classifier cannot process large amounts of monitoring data and can easily cause over fitting.

As the advanced representation of intelligent algorithms, deep learning methods have greatly changed our daily life. They are successfully applied in many different areas such as computer vision, object detection, natural language processing, and even disease diagnosis [6–10]. Deep learning methods can recognize high-dimensional complex input, get rid of the reliance on signal processing techniques and hand-engineered feature extraction algorithms. Unlike most previous artificial intelligent methods that can only process one-dimensional hand-engineered features [11], they can process the two-dimensional results of some basic signal processing methods directly. With these advantages, deep learning methods have been introduced to the fault diagnosis of rotating machinery, such as convolutional neural network (CNN), deep belief nets (DBN), and recurrent neural networks (RNN) [12–17]. Most methods perform well when dealing with bearing data at a certain rotating speed. However, in practical use, most machinery, such as wind turbines, pumps, and fans, have varying rotating speed. The performance of the methods has not been verified when processing data with varying rotating speed. Liu et al. [18] propose a dislocated time series CNN method for bearing diagnosis and apply it to varying rotating speed data by testing different network parameters to achieve the best result. However, the number of parameters that can be tested is limited, and the method is difficult to realize in practical applications. Meanwhile, in practical use, faults may not happen fully in all working speeds. Then there will not be enough fault data that covers full working speeds. Therefore, there needs to be a new deep learning fault diagnosis framework which can deal with data with varying rotating speed without complex parameters attempt. In addition, it needs to be able to realize diagnosis in all working speed based on limited fault data.

The continuous wavelet transform (CWT) has been proved to be a useful method to analyze vibration signals [19–21]. As a time-frequency analysis method, the result contains the complete time-frequency domain information of the vibration signal and avoids information loss of the original signal. In addition, it is suitable for detecting bearing faults which is usually presented as shock signals. When used in fault diagnosis of bearings, it has advantages compared with some other time-frequency methods, such as Short-time Fourier Transform (STFT), Discrete Wavelet Transform (DWT), Wavelet Package Decomposition (WPD) and Empirical Mode Decomposition (EMD).

Short-time Fourier Transform is the time-frequency transform based on Fourier transform. Because the window size is fixed, it only applies to stationary signals with small frequency fluctuations. In addition, the result is susceptible to noise interference. DWT is the discretization of the scale and displacement of CWT according to the power of 2. It retains less time-frequency information than CWT and may lose the key information near fault characteristic frequencies of bearing. WPD is an improved method of CWT. It provides a more detailed decomposition of high-frequency components. The fault characteristic frequencies of a bearing are usually less than 12 multiples of the rotating frequency. With the high sampling frequency, all the information near fault characteristic frequencies can be got by CWT. There is no need to use WPD, which is a more time-consuming method than CWT. Empirical Mode Decomposition decomposes the signal based on the time scale characteristics of the signal itself. Without a preset basis function, the location of the fault feature is uncertain in EMD when it applies to signals of different sensors. Therefore, when used for deep learning method, CWT, which uses a fixed wavelet basis function to decompose all the signals, is a better choice.

However, in most cases, a feature exaction method is used to exact a one-dimensional vector from the two-dimensional continuous wavelet transform coefficients. It may result in the loss of key fault information. Continuous wavelet transform scalogram (CWTS) contains all the continuous wavelet transform coefficients. It is a two-dimensional matrix which contains the complete time-frequency domain information of the vibration signal. With its powerful image recognition ability, CNN is the most suitable deep learning method to deal with CWTS. When applied to data with varying rotating speed, the CWTSs will have different size if they have the same frequency multiplication range of the rotating frequency. Without a cropping or warping operation, ordinary CNN can only process the input in the same size. Therefore, a CNN with new structure is needed to process CWTSs in different sizes.

To overcome the problems and challenges above, this paper proposes a fault diagnosis method based on continuous wavelet transform and Pythagorean spatial pyramid pooling (PSPP) CNN. This method uses a continuous wavelet transform to decompose vibration signals into CWTS in different scales according to the rotating speed. Using the PSPP strategy, CNN could then process the different size scalograms. Therefore, the fault diagnosis of data at variable rotating speed can be carried out by a single CNN model. The PSPP strategy is an improvement on spatial pyramid pooling (SPP). A PSPP layer can locate as front layer of CNN. Thus, the features obtained by the PSPP layer can be delivered to the convolutional layers for further feature extraction. Experiments are carried out on data from two different testbeds, constant rotating speed data and variable rotating speed data, respectively. The results demonstrate the effectiveness of the proposed approach. The contributions of the proposed approach are as follows:

(1) Compared with features extraction method used before when dealing with continuous wavelet transform coefficients, using a two-dimensional CWTS for fault diagnosis directly can retain the complete time-frequency domain information of signal and avoid the loss of fault information.

(2) A PSPP layer is proposed based on the SPP layer. In contrast with SPP-CNN, PSPP-CNN can place convolutional layers after the PSPP layer for further feature extraction. A PSPP layer can also retain position information of input feature maps. Experiment results show that PSPP-CNN performs better than SPP-CNN.

(3) A CWTS cropping method is presented to crop CWTSs to different sizes according to rotating speed and sample frequency. The objects recognition using CNN is concerned with the shape of the object. However, in signal processing area, the location of the signal features should also be paid attention to. The cropped CWTSs have the same frequency and time domain range. It helps the PSPP-CNN to achieve a more accurate and faster convergence.

(4) The proposed method can process data in different rotating speeds using a single CNN without complex parameter selection. PSPP-CNN trained by data at some rotating speeds can be used to diagnose bearing fault in full working speed. The experiments provide a good result.

The paper is organized as follows. Section 2 presents the fault diagnosis method that combines CWTS and PSPP-CNN for fault diagnosis, with a detailed procedure of the proposed method and SPP, the proposed PSPP layer, and the structure of PSPP-CNN used in this paper. Experimental verification of the method which includes constant rotating speed data and variable rotating speed data is described in Section 3. Finally, the concluding remarks are given in Section 4.

2. Proposed Method

As described above, CNN has been successfully applied to fault diagnosis. However, the proposed fault diagnosis models lack the diagnosis of variable rotating speed data, which has practical engineering value for online diagnosis of variable speed equipment such as the wind turbine. Therefore, this paper proposes a fault diagnosis framework based on CWTS and PSPP-CNN. Figure 1 illustrates the procedure of the proposed method. First, accelerators are used to collect the vibration signals of bearing. Second, continuous wavelet transform is used to decompose vibration signals into CWTSs. Next, as fault characteristic frequencies of bearings are related to rotating speed, the CWTSs are cropped into different sizes according to rotating speed. Then using PSPP strategy, CNN can deal with the input of different sizes. Therefore, the CWTSs in different sizes can be trained using a single PSPP-CNN. Finally, the test signals of bearing with variable rotating speed need to be decomposed into CWTSs with different scale ranges according to rotating speed. Using the CWTSs as the input of the trained PSPP-CNN, fault diagnosis of the signals can be achieved. Details of the main steps of the proposed method are described as follows:

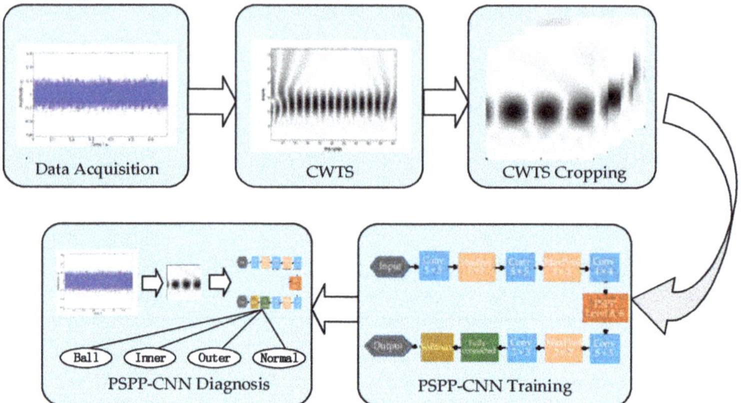

Figure 1. Flow chart of proposed fault diagnosis method. CWTS represents Continuous Wavelet Transform Scalogram, and PSPP-CNN represents Pythagorean Spatial Pyramid Pooling Convolutional Neural Network.

2.1. Continuous Wavelet Transform Scalogram

The continuous wavelet transform decomposes a signal in the time-frequency domain by using a family of wavelet functions to obtain feature values. Next, by analyzing the continuous wavelet coefficients or using the classification algorithm, we gain insight about the fault condition of the equipment. The process of continuous wavelet transformation can be described as:

$$\Psi_{a,b}(t) = |a|^{-\frac{1}{2}} \Psi \left(\frac{t-b}{a} \right) \quad a,b \in R \, a \neq 0 \tag{1}$$

$$C_a(k) = \int x(t) \overline{\Psi}_{a,b}(t) dt \tag{2}$$

where $\Psi_{a,b}(t)$ is a wavelet function whose shape and displacement are determined by a, the scale parameter, and b, the translation parameter. x is a signal with m data points. The wavelet coefficient of $x(t)$ at the a-th scale is C_a ($a = 1,2,3,\cdots,l$). k is time order ($k = 1,2,3,\cdots,m$). $\overline{\Psi}_{a,b}(t)$ is the complex conjugate of the wavelet function at scale a and translation b.

To show the change of wavelet coefficients intuitively, a CWTS is proposed [22]. The CWTS expresses continuous wavelet coefficients by a two-dimensional image in the time-frequency domain. Put all wavelet coefficients in a matrix $P = [C_1, C_2, \cdots, C_l]$. The graph of wavelet coefficients matrix P is called a CWTS.

Figure 2 shows the CWTS of a ball fault bearing signal with a 2400 rpm rotating speed sampled at 12 kHz. It is decomposed by the Morlet wavelet from 1 to 300-scale and has 300 data points in time series. The horizontal axis represents the position along the time direction, and the vertical axis represents the scale. Morlet wavelet is chosen as the wavelet used in this paper. Because it has the similar shape with the shock signal caused by bearing faults [23]. In addition, the signal extracted by the Morlet wavelet has the higher energy-to-Shannon entropy ratio than the other common wavelet types. Energy-to-Shannon entropy ratio is an important indicator to measure the fitness of wavelet functions [24].

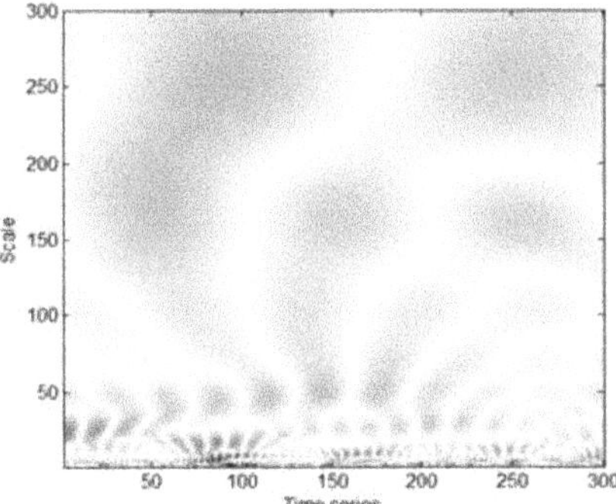

Figure 2. Continuous Wavelet Transform Scalogram (CWTS) of a ball fault bearing signal. The darker pixels correspond to larger wavelet coefficients.

2.2. Continuous Wavelet Transform Scalogram Cropping

The object recognition using CNN is concerned with the shape of the object. If the shape appears in the image, the existence of the object is detected. However, in signal processing area, the axes of images constructed usually have clearly defined meanings. The appearance of the same shape in different location may indicate different fault modes. Therefore, the location of the features should also be paid attention to.

As the vertical axis of CWTS represents the scale, different positions on the vertical axis relate to different frequencies. As we know, the fault characteristic frequencies of bearings are related to rotating speed. If the input CWTSs of PSPP-CNN can be ensued to have the same time domain range and frequency multiplication of the rotating frequency, the fault characteristic of the same fault could appear at the similar position in CWTSs. Thus, the classification of CWTSs will achieve a comparatively accurate result. Therefore, a CWTS cropping step is proposed in this paper.

Suppose a vibration signal $x(i)$ ($i = 1,2, \ldots m$) is collected at a sampling frequency f (Hz) with m sampling data points. The rotating speed is n (rpm), corresponding to a machine rotating frequency $f_m = n/60$. The integer multiple f to f_m is

$$q = \frac{f}{f_m} + \frac{1}{2} = \frac{60f}{n} + \frac{1}{2} \tag{3}$$

From Equation (1), we can calculate that the central frequency of a wavelet function is inversely proportional to scale a. Suppose $f_j = k/j$, where f_j is the central frequency at scale j, and k is the proportionality coefficient. According to the Morlet wavelet function,

$$k = f_0 \times f \tag{4}$$

where f_0 is the center frequency of the wavelet function, and the range is from 0.796 to 0.955. In this paper, 0.955 is chosen as f_0. Therefore, we choose scale 1 as the starting scale of cropping which corresponds to a high frequency. To make the end scale the same multiple of the rotating frequency, we choose q as the end scale. Scale q corresponds to the frequency:

$$f_q = k/q = k/f \times f_m = f_0 \times f_m \tag{5}$$

Thus, f_q is the f_0 multiplication of the rotating frequency. As the fault characteristic frequencies of bearings are larger than the rotating frequency, the cropping from scale 1 to q at the scale axis is sufficient for bearing fault diagnosis. The cropped CWTS will contain all the fault characteristic frequencies of bearings needed for analysis.

For the time domain axis, the time of q length data points is $t = q/f \approx 60/n$, which is about the time duration of a rotor rotation cycle.

Therefore, in this paper, the original CWTS is cropped from scale 1 to q, and q length in the time domain axis. Thus, the cropped CWTSs with different rotating speeds have the same time domain range and frequency multiplication relative to the rotating speed.

2.3. Pythagorean Spatial Pyramid Pooling Convolutional Neural Network Training

2.3.1. Pythagorean Spatial Pyramid Pooling Convolutional Neural Network

A CNN comprises convolutional layers, pooling layers, and fully connected layers. Most CNNs require a fixed input size. So before being sent into the first CNN layer, images need a cropping or warping operation. Convolutional layers and pooling layers do not need a fixed input size. However, the fully connected layers require a fixed input and output size to maintain constant number of the full connections. SPP can pool the mixed-size images into fixed-length outputs, thus meeting the need for fixed inputs in the fully connected layers.

Spatial pyramid pooling (or spatial pyramid matching) was first used in computer vision. Used together with feature extraction and classification algorithms, it has shown good results in image classification [25–27], object recognition [28–30], semantic concept detection [31], and image memorability [32]. Next, the SPP layer was introduced to CNN to remove the fixed-size input constraint of CNN [33]. The SPP-CNN method has been used in remote sensing hyperspectral image classification [34], handwritten word image categorization [35], and action recognition [36]. According to these applications, SPP is useful in CNN. It can reduce the cropping and warping operations used to fit a fixed-size CNN input, and avoid information loss in the operations.

The ordinary pooling layer in CNN is used to compress the input feature maps. It helps to reduce the feature maps and simplify the computational complexity of the network. It also extracts the main features from the original maps. There are two general types of pooling operations: average pooling and max pooling. Figure 3 shows an example of max pooling process in CNN, where *filter* is the filter size that indicates the range of pooling operation. *stride* is the space between pooling operations. It is clear that if the input image size changes more than the stride, the output size will change. This will make the classification algorithm impossible to continue.

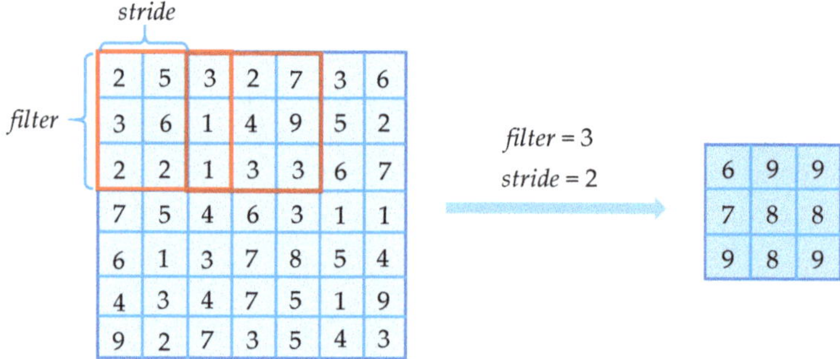

Figure 3. Max pooling process. Get the maximal value in the range of each pooling operation.

To resolve the requirement for a fixed input size, SPP is introduced to CNN as the last pooling layer. As shown in Figure 4, the feature images are pooled to different levels in the SPP layer. We will get an $l \times l$ size image in level l. Thereafter, a fixed-length output can be obtained by n level pooling. Feature values $\sum_{i=1}^{n} i^2$ will be sent into the fully connected layer for classification.

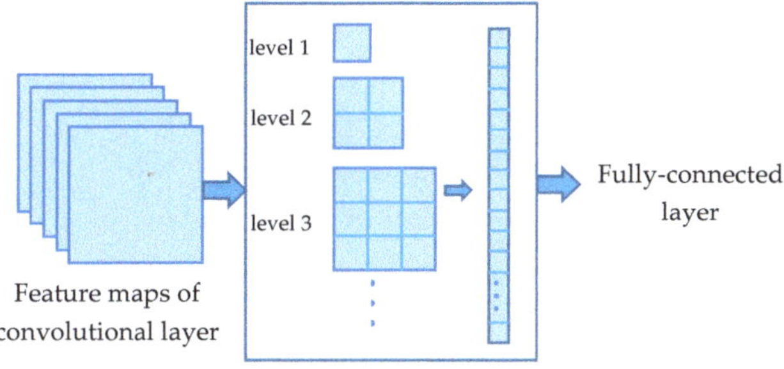

Figure 4. Convolutional neural network (CNN) with a spatial pyramid layer. Each input image is pooled to several levels. The results are transformed into one-dimension vectors to form the spatial pyramid pooling (SPP) output.

To get an $l \times l$ size image in level l, the filter size and stride should change by level. The *filter* and *stride* can be computed by

$$filter = [m/l] \tag{6}$$

$$stride = [m/l] \tag{7}$$

where $m \times m$ is the size of the feature maps from the last layer. Table 1 shows an example of 4-level SPP. Two input images with different size 15×15 and 20×20 get the same 30 length output by a 4-level SPP layer. If the input size changes, the pooling parameters will change to ensure the outputs have the same length.

Table 1. Parameters of a 4-level SPP.

Input Size	Level	Filter	Stride	Output Size	Output Length
	1	15	15	1×1	
	2	8	7	2×2	
15×15	3	5	5	3×3	30
	4	4	3	4×4	
	1	20	20	1×1	
	2	10	10	2×2	
20×20	3	7	6	3×3	30
	4	5	5	4×4	

Although SPP-CNN has shown good performance in image classification, there are still some problems. As the SPP layer lies at the last, before the fully connected layer in CNN, the outputs of the SPP layer are sent directly into the fully connected layer for classification. However, with no convolutional layer, the features obtained by the SPP layer may not be fully used. In addition, some of the location information will be lost. Meanwhile, the fully connected layer will have a large input matrix. It will greatly increase the connections in fully connected layer. Then there will be much more parameters to be trained. Therefore, a PSPP layer is proposed to make full use of the features and reduce the parameters.

The structure of the PSPP layer is shown in Figure 5. SPP is used to pool the input images into two different levels. Next, the output of the two levels will be used to compose new feature maps rather than a feature vector in the SPP layer. Thus, the output feature maps can be delivered to the convolutional layer for another round of feature extraction.

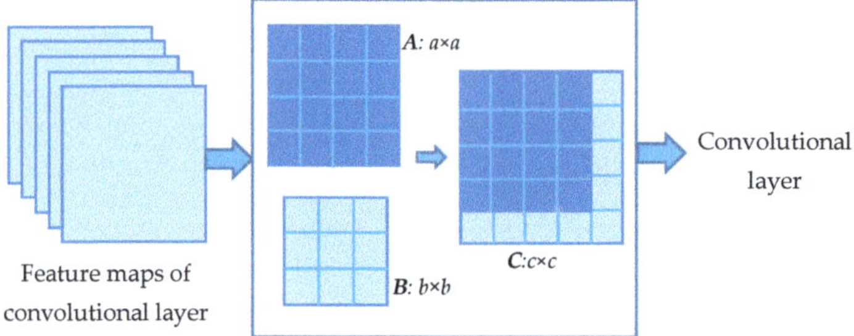

Figure 5. Structure of Pythagorean spatial pyramid pooling (PSPP) layer. It constructs the output using the results of two pooling levels. The position information of the higher-level pooling result A are retained.

To facilitate the composition of two SPP outputs, the pooling levels a, b ($a > b$) are chosen from the smaller two numbers of the Pythagorean triple. Hence, the output feature maps will have the size of the largest number c in the Pythagorean triple. To retain some position information of the feature maps, the composition is processed in the following way. The output matrix of the higher pooling level A will be retained. Next, the smaller output matrix B is reshaped as $(c + a) \times (c - a)$. The reshaped matrix is used to expand A to C on the right side and down side.

Using the PSPP layer in CNN, the fixed input problem can be solved. In addition, the output of the layer are square matrices which can be extracted in the following steps. The structure of the PSPP-CNN used in multi-size training of this paper is shown in Figure 6. Two convolutional layers are added after the PSPP layer for further feature extraction. The convolutional layers will also reduce the size of feature maps. Then the connections in fully connected layer is reduced.

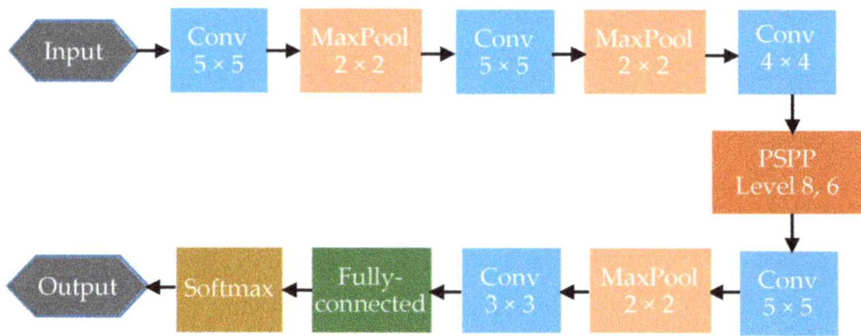

Figure 6. Structure of PSPP-CNN used in this paper. It is the PSPP-CNN used in all the following experiments, as PSPP-CNN has ability to process multi-size input.

It is recommended that the PSPP layer be in the middle layers of PSPP-CNN. As the PSPP layer has a larger feature reduction than normal 2×2 or 3×3 max pooling, the ahead position PSPP layer will lead to the premature loss of features. The PSPP layer at the back position is more like an SPP layer without enough convolutional layers to use the features obtained.

The size of convolutional and pooling kernels is changeable according to the input image size. However, big kernel size may result in information loss and increase computational complexity. Therefore, we choose to add more convolutional layers when processing large input images.

2.3.2. Pythagorean Spatial Pyramid Pooling Convolutional Neural Network Training Method

According to the previous description, the forward process is easy to realize. The filter size and stride can be pre-computed before the pooling. However, the back-propagation process in PSPP-CNN training requires some strategy.

When a back-propagation result is received from the last layer, the result is first divided to levels in the same order as the forward process. Next, the result in each PSPP level is restructured as a square matrix. The square matrices apply back-propagation separately. Thus, 2 back-propagation matrices are obtained. Hence, there are two ways to calculate the back-propagation matrix of an PSPP layer: (1) the mean of 2 back-propagation matrices, and (2) the weighted mean of 2 back-propagation matrices according to the level. The calculations can be presented by (8) and (9)

$$d(i) = \frac{1}{2} \sum_{k=1}^{2} d_k(i+1) \tag{8}$$

$$d(i) = \frac{1}{\sum_{k=1}^{2} k^2} \sum_{k=1}^{2} k^2 d_k(i+1) \tag{9}$$

where $d(i)$ is the back-propagation matrix of layer i in CNN, $d_k(i+1)$ is the level k back-propagation matrix of layer $i+1$ which is an PSPP layer. The two methods are tested using the same CNN structure, samples, and learning rate. The samples are part of the data used in Section 3.1. The results are listed in Table 2. Training error rate less than 0.05% is consider as achieving convergence.

Table 2. Convergence time and accuracy using two back-propagation methods.

Method	Training Steps	Convergence Time/Min	Time of Each Step/Min	Accuracy/%
1	63	324	5.14	92.43
2	51	263	5.16	92.52

As shown in Table 2, the two methods have similar accuracy and during time of each training step. However, the back-propagation using the second method has a faster convergence rate. This is because that the high-level pooling in PSPP reserves more features from feature maps. Therefore, the high weight of high-level pooling will lead the training to the right direction.

When PSPP-CNN is applied to multi-size images training, an important problem is the training order of the multi-size samples. To prevent the network from fitting a single image size, the multi-size samples in our work will be trained by turns. After all the samples of one size are trained, we will switch to another size. When the training error rates of samples in each size are less than 0.1%, the PSPP-CNN is considered to be achieving convergence.

3. Experiment

To verify the validity of the proposed method, two series of experiments are presented in this paper. Fault diagnosis of constant and variable rotating speed data are conducted using the proposed method.

3.1. Constant Rotating Speed Data

The bearing fault data from the Case Western Reserve University (CWRU) Bearing Data Center [37] is selected to verify the validity of the method in a constant rotating speed environment. The bearing test stand used in the experiment is shown in Figure 7. There are four bearing states: normal, ball fault, inner race fault, and outer race fault. In each bearing fault state, the bearings have fault diameters of 0.007 inches, 0.014 inches, and 0.021 inches. There are also 0.028 inches fault data of ball fault and inner race fault. Thus, there are 12 conditions in total. The fault bearings are installed on the drive end. Three accelerometers are installed on the fan end, drive end, and the base, respectively.

The rotating speed of the shaft is about 1800 rpm with the motor load ranging from 0 to 3 HP. All the data selected were sampled at a frequency of 12 kHz.

Figure 7. Bearing test stand used by Case Western Reserve University (CWRU).

In our fault diagnosis experiment, all the data are divided into 12 conditions according to the fault states and fault diameters. The influences of fault bearing location, accelerometer location, and motor load are ignored. Because the data were stored as a long array with more than 250,000 data points, the data were divided into several samples. Each sample contains 1024 data points. The size of each condition used for training set and test set are listed in Table 3. The selection of training samples is random. The ratio of training samples to test samples is 2 to 1.

Table 3. Sizes of training and test sets in 12 conditions.

Fault	None (NO)	Ball (BA)				Inner Race (IR)				Outer Race (OR)			Total
Diameters/in	0	0.007	0.014	0.021	0.028	0.007	0.014	0.021	0.028	0.007	0.014	0.021	
Training set size	24720	10080	10080	10080	3360	10080	10080	10080	3360	30240	10080	30240	162480
Test set size	12360	5040	5040	5040	1680	5040	5040	5040	1680	15120	5040	15120	81240

Because the sampling frequency of all data is the same, and the change in rotating speed is very small, it can be considered that the sampling frequency of all data is equal to the same multiple of the rotating frequency, namely a 400 multiple. At the same time, because the characteristic frequency of these bearing faults is higher than a two multiple of the rotating frequency, in this experiment, the continuous wavelet transform of the bearing data is carried out from 1 to 200 scales. Hence, 200×200 CWTSs are obtained as the CNN input.

To compare the diagnosis effectiveness of the three networks, CNN, SPP-CNN, and PSPP-CNN, on constant rotating speed data, three models are built to diagnose the samples. The structures of the CNNs are listed in Table 4:

As shown in Table 4, the first five layers of the CNNs are set as the same. Because the SPP layer and PSPP layer can reduce more image size than the max pooling layer, the original CNN has more layers and more connections. PSPP-CNN has two more convolutional layers than SPP-CNN for further feature exaction. The selection of CNNs parameters is problem dependent and obtained by trial and error. The selection of the parameters of the first 5 layers was based on the principles proposed in [38]. Then a validation set was built to optimize the parameters of the layers after layer 5 in three networks. The parameters that have best performance on validation set were chosen as the final CNN parameters. The training rate of all three CNNs is set to 0.002 and changed to 0.0005 when the training error is reduced to 1%.

Table 4. Structure of three Convolutional neural networks.

Layer	1	2	3	4	5	6	7	8	9	10	11
CNN	Conv $5 \times 5 \times 1$ 50	MaxPool 2×2	Conv $5 \times 5 \times 50$ 50	MaxPool 2×2	Conv $4 \times 4 \times 50$ 100	MaxPool 2×2	Conv $5 \times 5 \times 100$ 100	MaxPool 2×2	Conv $4 \times 4 \times 100$ 200	MaxPool 3×3	FC
SPP-CNN	Conv $5 \times 5 \times 1$ 50	MaxPool 2×2	Conv $5 \times 5 \times 50$ 50	MaxPool 2×2	Conv $4 \times 4 \times 50$ 100	MaxPool 2×2	Conv $5 \times 5 \times 100$ 100	SPP 5	FC		
PSPP-CNN	Conv $5 \times 5 \times 1$ 50	MaxPool 2×2	Conv $5 \times 5 \times 50$ 50	MaxPool 2×2	Conv $4 \times 4 \times 50$ 100	PSPP (8,6)	Conv $5 \times 5 \times 100$ 100	MaxPool 2×2	Conv $3 \times 3 \times 100$ 200	FC	

MATLAB is used to implement the training on the computer with two E5-2667 v3 CPUs, a GTX1080Ti GPU, 32 GB memory, and a 1 TB driver. Based on Matconvnet toolkit [39], the SPP-CNN and PSPP-CNN layer are implemented by adding new layer types to it. After the convergence of the CNNs, the test samples are sent into CNNs for fault diagnosis. The convergence and accuracy of CNNs are listed in Table 5:

Table 5. Convergence and accuracy of Convolutional neural networks.

Model	Number of Parameters	Training Steps	Convergence Time/Min	Accuracy/%
CNN	1.1e6	38	208	97.86%
SPP-CNN	1.5e6	48	281	97.23%
PSPP-CNN	5.8e5	44	211	97.79%

From Table 5, it is clear that the accuracy of SPP-CNN is a bit lower than that of CNN in Constant rotating speed data classification. SPP-CNN may lose some important information during SPP layer which is a big feature reduction. However, the convolutional layer following a PSPP layer can re-extract fault features. PSPP-CNN has a diagnosis accuracy similar with the CNN method, and better than SPP-CNN. It has a much shorter training time of each steps, for it has much less parameters to be trained. This shows that the PSPP layer retains the main fault features while reducing the total number of features. The fault diagnosis accuracy of PSPP-CNN in 12 conditions is shown in Table 6. All the conditions have an accuracy greater than 91.03%. Compared with diagnosis accuracies listed in [37], PSPP-CNN has equivalent accuracy, lower proportion of training samples and more conditions. This shows that the method we proposed has good performance in a constant rotating speed data diagnosis.

Table 6. Accuracy of test samples in 12 conditions using Pythagorean Spatial Pyramid Pooling Convolutional Neural Network (PSPP-CNN).

Fault	None	Ball				Inner Race				Outer Race			Total
Diameters/in	0	0.007	0.014	0.021	0.028	0.007	0.014	0.021	0.028	0.007	0.014	0.021	
Accuracy/%	99.98	95.32	95.99	91.03	99.64	99.94	94.17	99.01	99.88	98.09	96.96	99.31	97.79

3.2. Variable Rotating Speed Data

The fault diagnosis of equipment with variable rotating speed is an important issue that has not been solved satisfactorily. The proposed method is capable of handling variable rotating speed data. Therefore, a variable rotating speed experiment is carried out to show its advantages.

The test bed used in this experiment, the Machinery Fault Simulator-Rotor Dynamics Simulator (MFS-RDS), is shown in Figure 8. Bearing fault experiments were conducted on this test bed. The bearing used is ER-16K LINK-BELT (LBX Company LLC, Lexington, KY, USA). There are four bearing fault modes: normal (NO), ball fault (BA), inner race fault (IR), and outer race fault (OR). The fault bearings can be installed at the drive or non-drive end, and each experiment has at most one fault bearing. Two accelerators are installed on the vertical direction of the two bearing housings. The load is constant in the experiment. Under each bearing fault condition, three sets of data are collected at the rotating speed of 1800 rpm, 2400 rpm, and 2900 rpm, respectively, and the sampling time of each set is approximately 10 min. The sampling frequency of all data is 12 kHz.

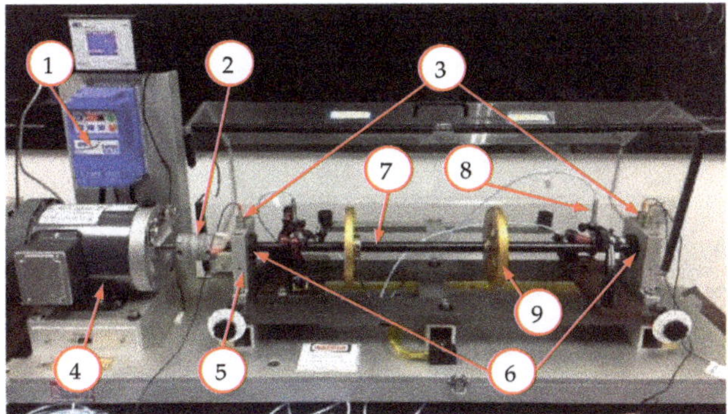

Figure 8. Machinery Fault Simulator-Rotor Dynamics Simulator (MFS-RDS) test bed. It can be used to simulate shaft, motor and bearing faults. The eddy current sensors are used to monitor the state of shaft. The data of them are not used for bearing diagnosis experiment. (1) speed controller, (2) rigid coupling, (3) accelerometer, (4) electromotor, (5) bearing base, (6) bearing, (7) shaft, (8) eddy current sensor, (9) rotary table.

In the fault diagnosis experiment, the data are divided into 12 cases according to the fault status and rotating speeds. The influence of fault bearing position and sensor position is neglected. Because the data are stored as continuous time series, the data are divided into several samples. Each sample contains 1200 data points. Next, 5376 samples are obtained in each case; half of them, 2688 samples, are used for training, and the remaining 2688 samples are for test. The ratio of training samples to test samples is 1:1. Therefore, we have 32,256 training samples and 32,256 test samples totally. The fault diagnosis aims to classify the data into four categories based on the fault status.

Because all data samples have the same sampling frequency of 12 kHz, this corresponds to 400, 300, and 248 multiples of the rotating speed at 1800 rpm, 2400 rpm, and 2900 rpm, respectively. Thus, the continuous wavelet transform of three sets of bearing data is carried out from 1 to 400 scales, 300 scales, and 248 scales, respectively. In the time axis, the middle 400, 300, and 248 coordinates are chosen, because they have data points of one rotating period and can neglect the first and last few points of each sample. Hence, the CWTSs are chopped to square CWTSs that have a size of 400 × 400, 300 × 300, and 248 × 248, respectively.

The PSPP-CNN structure used in this experiment is shown in Figure 4 and Table 4. There are five convolutional layers, three max pooling layers, one PSPP layer, and one fully connected layer. The training rate is initially set to 0.002 and changed to 0.0005 when the training error is reduced to 1%. The training environment is the same as the constant rotating speed data training. It takes 317 min to achieve convergence after 44 training steps which means the error of training samples is less than 0.1%.

The confusion matrix of fault diagnosis result is shown in Table 7. The first row represents the rotating speed and labels of the test data. The first shows the diagnosis result labels. The method has a high diagnosis accuracy of 99.11%, and an accuracy of more than 97.61% is obtained for the data of each fault condition. This indicates that the PSPP-CNN method is suitable for bearing fault diagnosis with variable rotating speed.

To compare the diagnosis effect with CNN and SPP-CNN, two other models are built using the CNN and SPP-CNN structures, as listed in Table 4. As CNN can only accept fixed-size images, all the data to 400 × 400, 300 × 300, and 248 × 248 CWTSs are translated to train the CNN separately. Accordingly, the CNN structure changes by adding a convolutional layer of a different size before the fully connected layer. SPP-CNN uses the same input images as those used by PSPP-CNN. 400 × 400, 300 × 300, and 248 × 248 CWTSs are also used to train SPP-CNN and PSPP-CNN. The diagnosis accuracy of seven different CNN models are listed in Table 8 and Figure 9.

Table 7. Accuracy of test samples at different rotating speeds using PSPP-CNN.

Labels	NO 1800 rpm	IR 1800 rpm	OR 1800 rpm	BA 1800 rpm	NO 2400 rpm	IR 2400 rpm	OR 2400 rpm	BA 2400 rpm	NO 2900 rpm	IR 2900 rpm	OR 2900 rpm	BA 2900 rpm	Accuracy %
NO	2631	0	0	48	2686	1	0	17	2688	3	0	31	99.27
IR	0	2687	7	0	0	2685	5	3	0	2678	8	25	99.83
OR	0	1	2681	0	0	2	2683	0	0	7	2680	69	99.75
BA	57	0	0	2640	2	0	0	2668	0	0	0	2563	97.61
Accuracy%	97.88	99.96	99.74	98.21	99.93	99.89	99.81	99.26	100	99.63	99.70	95.35	99.11

Table 8. Accuracy of test samples using different CNN models.

Input Size	CNN	SPP-CNN	PSPP-CNN
248 × 248	94.74	94.62	94.70
300 × 300	95.31	95.26	95.76
400 × 400	95.89	96.34	96.55
400 × 400, 300 × 300, 248 × 248		96.79	99.11

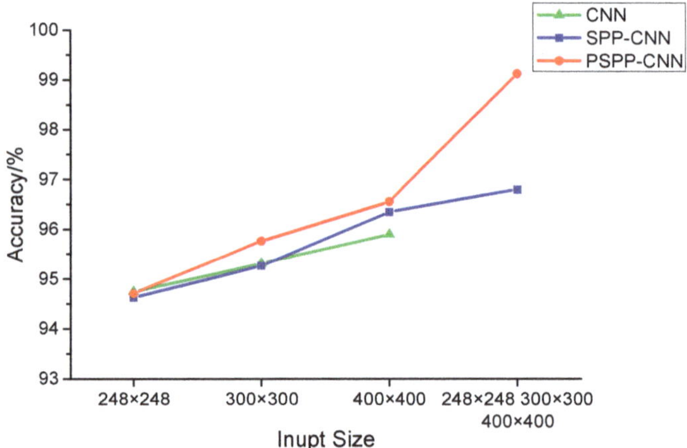

Figure 9. Accuracy of test samples using different input size and CNN models.

As shown in Table 8, PSPP-CNN has a better diagnosis accuracy than other CNN and SPP-CNN models with single-size or multi-size input. Figure 9 shows that the accuracy increases along with the size and diversity of input samples. The method we propose using PSPP-CNN and multi-size input has the best diagnosis accuracy 96.58%. It shows a nearly two-percent improvement over SPP-CNN. The SPP-CNN model with multi-size does not get a big accuracy improvement. It is because that different from image recognition tasks, fault diagnosis needs precisely feature positioning in CWTS. A SPP layer before fully connect layer will lose more position information in CWTS than a PSPP layer which can locate as front layer of PSPP-CNN. In addition, using the normal CNN, the diagnosis accuracy increases with the increase of input image size. According to our analysis, this is mainly owing to the increase of 1800 rpm data accuracy which are 92.05%, 94.22%, 95.53% separately. The reason may be that the 300 × 300 and 248 × 248 cropping of CWTSs will lose some of the fault features of 1800 rpm data. An input size larger than 400 × 400 will not increase the accuracy.

To compare this PSPP-CNN method with other CNN-based fault diagnosis methods, several proposed methods are applied using the same dataset. Deep Convolution Neural Network with Wide first-layer kernels (WDCNN) [12] is a bearing diagnosis method that uses raw vibration signals as input, and wide kernels in first layer for feature extraction and high-frequency noise suppression. Dislocated Time Series Convolutional Neural Network (DTS-CNN) [18] is proposed for mechanical signals by adding a dislocate layer to CNN. DTS-CNN can extract the relationship between signals with different intervals in periodic mechanical signals. Resample-CNN [38] uses a resample method to normalize the data to make sampling frequencies are the same multiples of the rotating frequency. The three proposed methods are used to compare the fault diagnosis accuracy using the same dataset, for they are all proposed for the diagnosis of machinery using vibration signals based on CNN. Because the data used in the papers have similar testbed structure, sampling frequency and rotating speed, WDCNN and DTS-CNN use the same network parameters as those in the papers. Resample-CNN uses the 400 × 400 CNN structure for they have the same input size. As all the three methods use the ordinary CNN and have the similar CNN structures, the changes of CNN parameters will not change the diagnosis result greatly. What we need to focus on is the construction of deep learning input. As the accuracies shown in Table 9, the method proposed in this paper has the best performance in the diagnosis of variable rotating speed data.

Table 9. Fault diagnosis result using other proposed CNN-based methods.

Model	Deep Convolution Neural Network with Wide first-layer kernels	Dislocated Time Series Convolutional Neural Network	Resample-CNN	PSPP-CNN
Accuracy/%	97.76	96.20	98.15	99.11

The reason for the comparison made on this set of data is that the accuracies on this case can reflect the effectives of the methods both on constant speed and variable speed data. In addition, in practical use, most variable speed machines work like this case at a certain speed range or some preset optimal speed points.

To further study the effectiveness of the method at full working speed range, an experiment to diagnosis the full working speed data is carried on using the PSPP-CNN trained above. The data of four bearing conditions were collected with 12 kHz on MFS-RDS. The fault bearing was installed on drive end. The rotation speed ranges from 300 to 3000 r/min. The rotation frequency increases at 0.15 Hz/s. Figure 10 shows the vibration signals of drive end accelerator in four fault conditions.

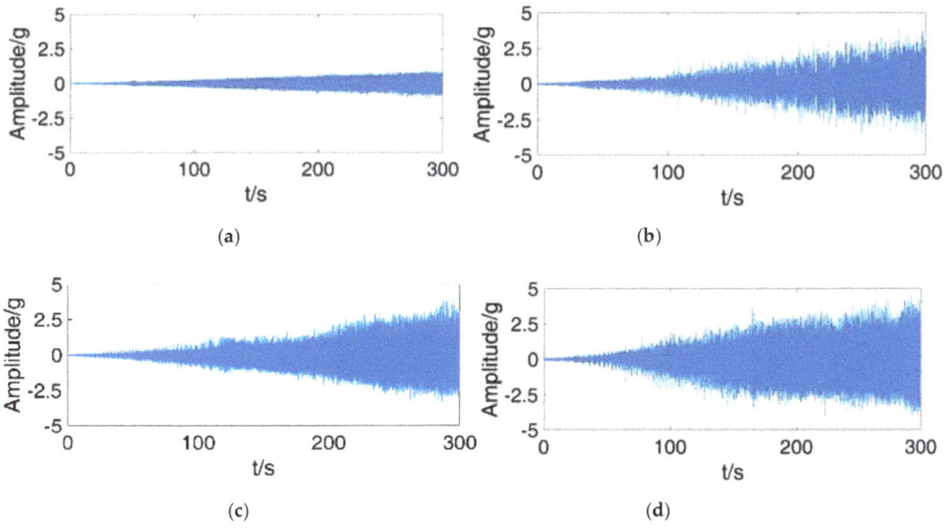

Figure 10. Vibration signals of four fault conditions. (**a**) normal (**b**) ball (**c**) inner race (**d**) outer race.

To test the PSPP-CNN trained, the data of drive end accelerator in each condition are divided into 440 samples. Each sample contains 8192 data points. We get the cropped CWTS of each sample according to rotating speed. Therefore, the size of cropped CWTSs are ranges from 240×240 to 2400×2400. Then the cropped CWTSs are sent into PSPP-CNN for fault diagnosis. The accuracy of each fault condition is listed in Table 10.

Table 10. Accuracy of four fault conditions at full rotating speeds using PSPP-CNN.

Fault	None	Ball	Inner race	Outer race	Total
Accuracy/%	90.23	91.82	92.05	92.95	91.76

As shown in Table 10, PSPP-CNN has diagnosis accuracies more than 90% for each fault conditions. It means that the PSPP-CNN trained by data at some rotating speed can be used to diagnosis bearing fault in full working speed. Through analysis, the accuracy of data under 1200 rpm is a little lower. Adding an incremental training using low speed data will increase the accuracy. It shows that the PSPP-CNN trained using data of few certain rotating speeds can be used to diagnose bearing fault in full working speed.

Through the fault diagnosis experiments of constant and variable rotating speed data, we can know that the PSPP-CNN method proposed is an effective solution for fault diagnosis of bearing. When applied to intelligent diagnosis system, the method has some advantages. First, the PSPP-CNN proposed in this paper can be easily implemented by adding a PSPP layer to the ordinary CNN code based on max pooling layer. There are some mature CNN frameworks based on MATLAB, Net Framework or Python. All of them can be easily built. Second, the fault diagnosis process has high power efficiency. As shown in Table 5, PSPP-CNN has less parameters than an ordinary CNN with the same front layers. It reduces the computation of each training and test epoch. Although the training of PSPP-CNN is still time-consuming in the computer without GPU accelerated computing, the diagnosis process using trained PSPP-CNN model can be completed quickly even at laptop computer. Third, with the de-noising ability of wavelet transform, the diagnosis system has good robustness. Diagnosis result will not be affected by the background noise in signal of practical equipment.

In practical applications, we can gather the experiment data or online monitoring data from the varying working speed bearing and use the data to train PSPP-CNN for fault diagnosis. The intelligent fault diagnosis method proposed in this paper has been used for online fault diagnosis of wind turbine bearings in a wind farm. The fault diagnosis software is exploited by C#& MATLAB combined programming. The signal is transmitted to MATLAB for CWTS calculation and classification. The diagnosis result can be obtained in 5 s. The training data was collected from the online vibration monitoring system installed on wind turbines. The fault data was picked out by referencing fault records.

4. Conclusions

In this paper, we propose an intelligent fault diagnosis method for a variable rotating speed bearing. The proposed approach is built upon CWTS and PSPP-CNN. This method decomposes vibration signals of bearing into CWTSs of different scales according to the rotating speed. In addition, the different size CWTSs are sent into PSPP-CNN for fault diagnosis. The PSPP-CNN that we proposed is an improvement of the SPP-CNN. The PSPP layer can fully use the pooling result for further feature extraction than SPP layer as they all can pool the input of different sizes to a fixed size. A series of experiments are carried out using constant rotating speed data and variable rotating speed data. The results show that the proposed approach is an effective solution. The method has been used in practical applications.

Although many fault diagnosis methods based on deep learning have been proposed, most methods are totally data-driven and focus on the small improvement of deep learning algorithm.

The domain knowledge that has been used for fault diagnosis in recent decades is barely used. In addition, the working condition information is not considered in the input of deep learning algorithm. This paper takes the fault characteristics frequency of bearing and working speed into consideration. However, more domain knowledge and working condition information, such as load and output power, can be combined with deep learning. It may improve the accuracy and robustness, and the features extracted will be more interpretable.

Author Contributions: T.Y. and W.G. conceived this study. S.G. and T.Y. designed the experiments. S.G. and C.Z. performed the experiments. S.G. analyzed the data and wrote the paper. T.Y., W.G. and Y. Z. reviewed the paper.

Funding: This research was supported by the National Program of International Science and Technology Cooperation of China (Project No. 2014DFA60990).

Conflicts of Interest: The authors declare no conflict of interest.

References

1. Ali, J.B.; Fnaiech, N.; Saidi, L.; Chebel-Morello, B.; Fnaiech, F. Application of empirical mode decomposition and artificial neural network for automatic bearing fault diagnosis based on vibration signals. *Appl. Acoust.* **2015**, *89*, 16–27.
2. Zhang, B.; Sconyers, C.; Byington, C.; Patrick, R.; Orchard, M.E.; Vachtsevanos, G. A Probabilistic Fault Detection Approach: Application to Bearing Fault Detection. *IEEE Trans. Ind. Electron.* **2011**, *58*, 2011–2018. [CrossRef]
3. Tian, Y.; Ma, J.; Lu, C.; Wang, Z. Rolling bearing fault diagnosis under variable conditions using LMD-SVD and extreme learning machine. *Mech. Mach. Theory* **2015**, *90*, 175–186. [CrossRef]
4. Li, Y.; Xu, M.; Wei, Y.; Huang, W. A new rolling bearing fault diagnosis method based on multiscale permutation entropy and improved support vector machine based binary tree. *Measurement* **2016**, *77*, 80–94. [CrossRef]
5. Zhang, B.; Sconyers, C.; Orchard, M.; Patrick, R.; Vachtsevanos, G. Fault progression modeling: An application to bearing diagnosis and prognosis. In Proceedings of the 2010 American Control Conference, Baltimore, MD, USA, 30 June–2 July 2010; pp. 6993–6998.
6. Hoochang, S.; Roth, H.R.; Gao, M.; Lu, L.; Xu, Z.; Nogues, I.; Summers, R.M. Deep Convolutional Neural Networks for Computer-Aided Detection: CNN Architectures, Dataset Characteristics and Transfer Learning. *IEEE Trans. Med. Imaging* **2016**, *35*, 1285.
7. Abdulnabi, A.H.; Wang, G.; Lu, J.; Jia, K. Multi-Task CNN Model for Attribute Prediction. *IEEE Trans. Multimed.* **2015**, *17*, 1949–1959. [CrossRef]
8. Ren, S.; He, K.; Girshick, R.; Sun, J. Faster R-CNN: Towards Real-Time Object Detection with Region Proposal Networks. *IEEE Trans. Pattern Anal. Mach. Intell.* **2017**, *39*, 1137–1149. [CrossRef] [PubMed]
9. Kalchbrenner, N.; Grefenstette, E.; Blunsom, P. A Convolutional Neural Network for Modelling Sentences. *arxiv* **2014**, arXiv:1404.2188.
10. Kamnitsas, K.; Ledig, C.; Newcombe, V.F.; Simpson, J.P.; Kane, A.D.; Menon, D.K.; Rueckert, D.; Glocker, B. Efficient multi-scale 3D CNN with fully connected CRF for accurate brain lesion segmentation. *Med. Image Anal.* **2016**, *36*, 61. [CrossRef] [PubMed]
11. Zhang, B.; Georgoulas, G.; Orchard, M.; Saxena, A.; Brown, D.; Vachtsevanos, G.; Liang, S. Rolling element bearing feature extraction and anomaly detection based on vibration monitoring. In Proceedings of the 2008 16th Mediterranean Conference on Control and Automation, Ajaccio, France, 25–27 June 2008; pp. 1792–1797.
12. Zhang, W.; Peng, G.; Li, C.; Chen, Y.; Zhang, Z. A New Deep Learning Model for Fault Diagnosis with Good Anti-Noise and Domain Adaptation Ability on Raw Vibration Signals. *Sensors* **2017**, *17*, 425. [CrossRef] [PubMed]
13. Sun, W.; Yao, B.; Zeng, N.; Chen, B.; He, Y.; Cao, X.; He, W. An Intelligent Gear Fault Diagnosis Methodology Using a Complex Wavelet Enhanced Convolutional Neural Network. *Materials* **2017**, *10*, 790. [CrossRef] [PubMed]
14. Hu, J.; Cai, Z.; Cai, Z.; Wang, Y. An intelligent fault diagnosis system for process plant using a functional HAZOP and DBN integrated methodology. *Eng. Appl. Artif. Intell.* **2015**, *45*, 119–135. [CrossRef]

15. Zhao, G.; Liu, X.; Zhang, B.; Zhang, G.; Niu, G.; Hu, C. Bearing Health Condition Prediction Using Deep Belief Network. In Proceedings of the Annual Conference of Prognostics and Health Management Society, Orlando, FL, USA, 2–5 October 2017.

16. Bruin, T.D.; Verbert, K.; Babuška, R. Railway Track Circuit Fault Diagnosis Using Recurrent Neural Networks. *IEEE Trans. Neural Netw. Learn. Syst.* **2017**, *28*, 523–533. [CrossRef] [PubMed]

17. Xia, M.; Li, T.; Xu, L.; Liu, L.; de Silva, C.W. Fault Diagnosis for Rotating Machinery Using Multiple Sensors and Convolutional Neural Networks. *IEEE/ASME Trans. Mechatron.* **2018**, *23*, 101–110. [CrossRef]

18. Liu, R.; Meng, G.; Yang, B.; Sun, C.; Chen, X. Dislocated Time Series Convolutional Neural Architecture: An Intelligent Fault Diagnosis Approach for Electric Machine. *IEEE Trans. Ind. Inform.* **2017**, *13*, 1310–1320. [CrossRef]

19. Wang, Y.; He, Z.; Zi, Y. Enhancement of signal denoising and multiple fault signatures detecting in rotating machinery using dual-tree complex wavelet transform. *Mech. Syst. Signal Process.* **2010**, *24*, 119–137. [CrossRef]

20. Shen, C.; Wang, D.; Kong, F.; Peter, W.T. Fault diagnosis of rotating machinery based on the statistical parameters of wavelet packet paving and a generic support vector regressive classifier. *Measurement* **2013**, *46*, 1551–1564. [CrossRef]

21. Hu, Q.; He, Z.; Zhang, Z.; Zi, Y. Fault diagnosis of rotating machinery based on improved wavelet package transform and SVMs ensemble. *Mech. Syst. Signal Process.* **2007**, *21*, 688–705. [CrossRef]

22. YanPing, Z.; ShuHong, H.; JingHong, H.; Tao, S.; Wei, L. Continuous wavelet grey moment approach for vibration analysis of rotating machinery. *Mech. Syst. Signal Process.* **2006**, *20*, 1202–1220. [CrossRef]

23. Zhang, D.; Sui, W.T.; Zhang, Y. Bearing Fault Diagnosis Based on Optimal Morlet Wavelet. *Bearing* **2009**, *10*, 48–51.

24. Yan, R.; Gao, R.X. Base wavelet selection for bearing vibration signal analysis. *Int. J. Wavel. Multiresolut. Inf. Process.* **2009**, *7*, 411–426. [CrossRef]

25. Han, H.; Han, Q.; Li, X.; Gu, J. Hierarchical spatial pyramid max pooling based on SIFT features and sparse coding for image classification. *Iet Comput. Vis.* **2013**, *7*, 144–150. [CrossRef]

26. Malinowski, M.; Fritz, M. Learnable Pooling Regions for Image Classification. *arXiv* **2013**, arXiv:1301.3516.

27. Russakovsky, O.; Kai, Y.; Kai, Y.; Fei-Fei, L. Object-Centric spatial pooling for image classification. In *European Conference on Computer Vision*; Springer: Berlin, Germany, 2012; pp. 1–15.

28. Wang, G.; Fan, B.; Pan, C. Ordinal pyramid pooling for rotation invariant object recognition. In Proceedings of the IEEE International Conference on Acoustics, Speech and Signal Processing, Brisbane, QLD, Australia, 19–24 April 2015; pp. 1349–1353.

29. Chen, L.; Zhou, Q.; Fang, W. Spatial Pyramid Pooling in Structured Sparse Representation for Flame Detection. In Proceedings of the International Conference on Internet Multimedia Computing and Service, Xi'an, China, 19–21 August 2016; pp. 310–313.

30. Ou, Y.; Zheng, H.; Chen, S.; Chen, J. Vehicle logo recognition based on a weighted spatial pyramid framework. In Proceedings of the 17th International IEEE Conference on Intelligent Transportation Systems (ITSC), Qingdao, China, 8–11 October 2014; pp. 1238–1244.

31. Kawai, Y.; Fujii, M. Semantic Concept Detection based on Spatial Pyramid Matching and Semi-supervised Learning. *Ite Trans. Media Technol. Appl.* **2013**, *1*, 190–198. [CrossRef]

32. Celikkale, B.; Erdem, A.; Erdem, E. Visual Attention-Driven Spatial Pooling for Image Memorability. In Proceedings of the Computer Vision and Pattern Recognition Workshops, Portland, OR, USA, 23–28 June 2013; pp. 976–983.

33. He, K.; Zhang, X.; Ren, S.; Sun, J. Spatial Pyramid Pooling in Deep Convolutional Networks for Visual Recognition. *IEEE Trans. Pattern Anal. Mach. Intell.* **2014**, *37*, 1904–1916. [CrossRef] [PubMed]

34. Yue, J.; Mao, S.; Li, M. A deep learning framework for hyperspectral image classification using spatial pyramid pooling. *Remote Sens. Lett.* **2016**, *7*, 875–884. [CrossRef]

35. Toledo, J.I.; Sudholt, S.; Fornés, A.; Cucurull, J.; Fink, G.A.; Lladós, J. Handwritten Word Image Categorization with Convolutional Neural Networks and Spatial Pyramid Pooling. In *Joint Iapr International Workshops on Statistical Techniques in Pattern Recognition*; Springer International Publishing: Cham, Switzerland, 2016; pp. 543–552.

36. Wang, P.; Cao, Y.; Shen, C.; Liu, L.; Shen, H.T. Temporal Pyramid Pooling Based Convolutional Neural Network for Action Recognition. *IEEE Trans. Circuits Syst. Video Technol.* **2017**, *27*, 2613–2622. [CrossRef]

37. Xiong, Q.; Zhang, W.; Lu, T.; Mei, G.; Liang, S. A Fault Diagnosis Method for Rolling Bearings Based on Feature Fusion of Multifractal Detrended Fluctuation Analysis and Alpha Stable Distribution. *Shock Vib.* **2015**, *2016*, 1–12. [CrossRef]
38. Guo, S.; Yang, T.; Gao, W.; Zhang, C. A Novel Fault Diagnosis Method for Rotating Machinery Based on a Convolutional Neural Network. *Sensors* **2018**, *18*, 1429. [CrossRef] [PubMed]
39. Vedaldi, A.; Lenc, K. Matconvnet: Convolutional neural networks for Matlab. In Proceedings of the 23rd ACM International Conference on Multimedia, Brisbane, Australia, 26–30 October 2015; pp. 689–692.

Article

Collision Detection and Identification on Robot Manipulators Based on Vibration Analysis

Feiyan Min [1,2], Gao Wang [1,2] and Ning Liu [1,2,*]

[1] Department of Electronic Engineering, College of Information Science and Technology, Jinan University, Guangzhou 510632, China; minfeiyan@aliyun.com (F.M.); twangg@jnu.edu.cn (G.W.)
[2] Robotics Research Institue of Jinan University, Guangzhou 510632, China
* Correspondence: tliuning@jnu.edu.cn; Tel.: +86-020-8522-3063

Received: 25 December 2018; Accepted: 27 February 2019; Published: 3 March 2019

Abstract: Robot manipulators should be able to quickly detect collisions to limit damage due to physical contact. Traditional model-based detection methods in robotics are mainly concentrated on the difference between the estimated and actual applied torque. In this paper, a model independent collision detection method is presented, based on the vibration features generated by collisions. Firstly, the natural frequencies and vibration modal features of the manipulator under collisions are extracted with illustrative examples. Then, a peak frequency based method is developed for the estimation of the vibration modal along the manipulator structure. The vibration modal features are utilized for the construction and training of the artificial neural network for the collision detection task. Furthermore, the proposed networks also generate the location and direction information about contact. The experimental results show the validity of the collision detection and identification scheme, and that it can achieve considerable accuracy.

Keywords: manipulator; model independent method; collision detection; collision identification; vibration analysis; artificial neural network

---

1. Introduction

Industrial robots play an important role in the modern manufacturing industry, and human-friendly robots will soon become flexible and versatile coworkers in the industrial setting [1]. One of the core problems in human–robot interaction is the detection of collisions between robots and the industrial environment, including humans and other manufacturing structures. Indeed, industrial robots should be able to operate in very dynamic, unstructured, and partially unknown environments, sharing the workspace with the human user, and preventing upcoming and undesired collisions [2].

Furthermore, researchers are getting more interested in gathering the maximum amount of information from the impact event, such as the contact position, direction and intensity, in order to let the robot react in the most appropriate fashion. Related concepts include collision avoidance [3], and collision isolation, identification, classification and reaction [2]. Particularly, the concept collision isolation aims at localizing the contact point, or at least which link out of the n-body robot collided. Furthermore, collision identification is to determine the directional information of the generalized collision force [2].

Different approaches for detection of robot collisions have been presented in the literature. A first intuitive approach is to monitor the current transient in robot electric drives, looking for shock changes within currents caused by collisions [4–7]. A second approach is based on the tactile sensors laying inside robot skins. The more common approach is the so-called model-based method (the state observer or Kalman filter method), and the detection algorithms are mainly based on the evaluation of monitoring signals (motor currents, difference between actual and predicted torques, etc.), which should be below some setting values, otherwise collision alarms are generated.

A major practical problem in these methods is the selection of thresholds for the monitoring signals, since the modeling error and sensor noise affect the monitoring signal in the same way as collision disturbance. A good detection algorithm therefore must distinguish the effect of modeling error and sensor noise on monitoring signal from that of a real collision. For this reason, it usually leads to a tradeoff between sensitivity and false alarm rate, with a risk of excessively conservative threshold [2,5]. To overcome this problem, some different methods are proposed. A dynamic threshold is defined in [4] to represent the residual dependence on the state of the robot (position, velocity, acceleration) using fuzzy logic rules. Furthermore, in [6], authors propose an adaptive detection algorithm based on a state-dependent dynamic threshold. In recent years, some extended state observer [8] and sliding mode observer [9] methods are proposed to obtain more effective detection performance.

Most of these evaluation algorithms mainly concentrate on the time domain information of motor torque deviation (The motor currents can be considered as another form of torque). A considerable alternative is to use frequency domain features for detection purposes.

This paper describes a novel detection method based on vibration modal feature generated by collision. The natural frequency and vibration modal features of collision experiments are extracted for the constructing and training of several structures of Back Propagation (BP) neural network. The research result shows that this method can be used for the detection of collision with considerable accuracy, not only for the detection of collision occurrence, but also for the positioning and direction determination.

A remarkable contribution of this paper is to introduce frequency and vibration information for the detection of robot collision. The vibration information is independent of the dynamic model and can be easily acquired with acceleration sensor, which has greatly developed in recent years. Since the occurrence of collision can happen anywhere along the robot structure, the acceleration sensor is easy to stall and suitable for the detection use. Furthermore, the industrial robots are not usually equipped with torque sensor because of cost and structural constraints, and this research provides an opportunity for MEMS accelerometer.

Accurate analysis of natural frequency and modal shape is fundamental for mechanical design, dynamic identification and control of high-speed manipulator [10,11]. In recent years, the acceleration and vibration analysis method has been gradually used for the status monitoring of manipulator robot. In [12], the authors propose a fault detection method for industrial welding robot. In their study, joint acceleration of robot is considered as evaluation criteria and their evaluation algorithm is based on neural network. The article [13] presents a method for processing and analyzing the measurement signals used in the problem of diagnosing the state of a manipulator's tool. The analysis algorithm is performed within the time and frequency domain. The signal utilized in the research is the mechanical vibrations and the rotation speed of the tool. In the work [14], frequency domain analysis method is researched for the event classification in robot assisted deburring. Power spectrum density (PSD) of sensor data acquired between certain sample rates is calculated, and it is then used for classifying vibration signal generated by the spindle from the vibration signal acquired during the deburring process. The paper [15] presents a signal fusion method based on accelerometer and encoder in serial robots. Besides, a number of studies have presented the data-based method for mechanical system monitoring based on vibration signals [16–20]. In most of this research, specific forms of artificial neural network are proposed and implemented [21–25].

The rest of this paper is organized as follows. Section 2 describes the frequency domain features and parameters of vibration of manipulator in case of collisions. Section 3 presents the architecture of our detection neural network, together with vibration modal analysis method; Section 4 presents some results and experiments obtained with the proposed method; and finally, Section 5 addresses the main conclusion and future work.

2. Vibration Modeling and Feature Extraction

This section address the vibration presentation and its characteristics. First, the modeling method of vibration response is proposed with dynamic equations and transfer functions. Then, the typical features are analyzed based on the mathematical model, together with illustrative examples.

2.1. Vibration Response under Collision and Its Mathematical Modeling

In this paper, our research focuses on the vibration response of multiple test points along manipulator structure under several experiments. An effective method for dealing with vibration of kinematic chain mechanism is the elastodynamic modeling and analysis [10,11]. In this method, each critical mechanical structure and component is simplified with stiffness, viscous and mass parameter. As for a $n-$dof manipulator, it is composed of a series of motors, gears, links and joints, some of which will generate deformation and vibration under collision. For this reason, the dominant vibration structures can be considered as elastic bodies with certain stiffness and viscous coefficients.

We consider a $n-$dof manipulator with m dominant vibration structures. Its axis displacement vector can be defined as:

$$q = \begin{bmatrix} q_D \\ q_M \end{bmatrix} \in \mathbf{R}^N \tag{1}$$

where $q_D = [q_1, q_2, \cdots, q_m]^T$ is the vibration deviation, and $q_M = [q_{m+1}, q_{m+2}, \cdots, q_{m+n}]^T$ denotes the joint displacements. Furthermore, we assume the equilibrium points of q_D is $\bar{q}_D = [\bar{q}_1, \cdots, \bar{q}_m]^T$. The dynamic equation of manipulator can be written as:

$$M(q)\ddot{q} + C(q,\dot{q})\dot{q} + G(q) = \begin{bmatrix} \tau_D \\ \tau_M + \tau_f \end{bmatrix} \tag{2}$$

the variable τ_M, τ_f and τ_D denotes the joint torque generated by motor, friction and structure deformation respectively. We denote $\tau_D = [\tau_1, \cdots, \tau_m]^T$, and $\tau_M = [\tau_{m+1}, \cdots, \tau_{m+n}]^T$. The subscript D denotes the dominant vibration structures, and M denotes the drive motor of manipulator.

On the other hand, the torque generated by the deformation of vibration structures can be given by:

$$\tau_D = K_p(\overline{q_D} - q_D) - K_v q_D \tag{3}$$

where $K_p = diag(k_{p1}, k_{p2}, \cdots, k_{pm})$, $K_v = diag(k_{v1}, k_{v2}, \cdots, k_{vm})$ is the vibration dynamic coefficient matrix. Furthermore, k_{pi}, k_{vi} denotes the stiffness and vicious coefficient of ith dominant vibration structure respectively.

We assume that the gravity and friction is compensated by feedback control loop, and the Coriolis and centrifugal effect generated by structure deformation is relatively small. In most cases, the displacement of q_D is much smaller than q_M, and the inertia matrix $M(q)$ is mainly determined by joint displacement. Furthermore, the torque vector generated by collision is assumed as f_{ext}. Then we get the dynamic function of manipulator as follows

$$\begin{bmatrix} 0 \\ \tau_M \end{bmatrix} = M(q_M)\begin{bmatrix} \ddot{q}_D \\ \ddot{q}_M \end{bmatrix} + \begin{bmatrix} K_v & C_D(q_M, \dot{q}_M) \\ 0 & C_M(q_M, \dot{q}_M) \end{bmatrix}\begin{bmatrix} \dot{q}_D \\ \dot{q}_M \end{bmatrix} + \begin{bmatrix} K_p \\ 0 \end{bmatrix}(q_D - \overline{q_D}) + \begin{bmatrix} J_D{}^T \\ J_M{}^T \end{bmatrix}f_{ext} \tag{4}$$

where $J_D{}^T$ and $J_M{}^T$ is the associated geometric contact Jacobian matrix to m vibration structures and n robot motors, respectively. However, the collision torque vector f_{ext} and contact Jacobian $J_D{}^T$, $J_M{}^T$ is typically unknown.

Denoting $x_1 = \begin{bmatrix} \dot{q}_D \\ q_D - \overline{q_D} \end{bmatrix}$, $y_1 = q_D - \overline{q_D}$, and $x_2 = \begin{bmatrix} \dot{q}_M \\ q_M \end{bmatrix}$, $y_2 = \dot{q}_M$, an n inputs and $m + n$ outputs state-space equation is obtained

$$\begin{cases} \dot{x}_1 = A_1 x_1 + B_{11} y_2 + B_{12} f_{ext} \\ \dot{x}_2 = A_2 x_2 + B_{21} u + B_{22} f_{ext} \\ y_1 = C_1 x_1 \\ y_2 = C_2 x_2 \end{cases} \tag{5}$$

where u is the active motor torque, and f_{ext} is torque vector generated by collision, and the related parameter matrix is given by:

$$A_1 = \begin{bmatrix} -M(q_M)^{-1} K_v & -M(q_M)^{-1} K_p \\ I & 0 \end{bmatrix},$$

$$B_{11} = \begin{bmatrix} -M(q_M)^{-1} C_D(q_M, \dot{q}_M) \\ 0 \end{bmatrix}, \quad B_{12} = \begin{bmatrix} J_D^T \\ 0 \end{bmatrix}, \quad C_1 = \begin{bmatrix} 0 \\ I \end{bmatrix}^T$$

$$A_2 = \begin{bmatrix} -M(q_M)^{-1} C_M(q_M, \dot{q}_M) & 0 \\ I & 0 \end{bmatrix},$$

$$B_{21} = \begin{bmatrix} -M(q_M)^{-1} \\ 0 \end{bmatrix}, \quad B_{22} = \begin{bmatrix} J_M^T \\ 0 \end{bmatrix}, \quad C_2 = \begin{bmatrix} I \\ 0 \end{bmatrix}^T$$

Then we get the modal analyzed transfer function from collision torque f_{ext} to vibration deformation y_1 as follows

$$P(s) = \frac{Y_1(s)}{F_{ext}(s)} = C_1 (sI - A_1)^{-1} [B_{12} + B_{11} C_2 (sI - A_2)^{-1} B_{22}] \tag{6}$$

We considered the structure of system matrix A_1 and A_2, and we got that $rank(A_1) = 2m$ and $rank(A_2) = n$. Let $\lambda_{Di}, \overline{\lambda}_{Di}(i = 1, 2, \cdots, m)$ be $2m$ eigenvalues of A_1, and $\lambda_0(=0), \lambda_{Mi}(i = 1, 2, \cdots, n)$ be the $n + 1$ eigenvalues of A_2.

From the partial fraction expansion, the modal analyzed transfer function $P(s)$ becomes

$$P(s) = \sum_{k=1}^{m} \frac{\Phi_k}{s^2 + 2\xi_k \omega_{Dk} s + \omega_{Dk}^2} + P_1(s) \tag{7}$$

By modal analysis, $P(s)$ can be expressed as a linear sum of $m + n$ vibration modes, m modes of which are generated from the m dominant vibration structures. Furthermore, the other n poles which come from A_2, correspond to the n-dof active dynamic of manipulator.

Then the natural frequency of vibration along manipulator can be obtained by $\omega_{Di} = |\lambda_{Di}| (i = 1, 2, \cdots, m)$ and $\omega_{Mj} = |\lambda_{Mj}| (j = 1, 2, \cdots, n)$.

The matrix Φ_k corresponds to the rigid body vibration mode and is positive semidefinite, and it has the following structure

$$\Phi_{..i} = \begin{bmatrix} \phi_{11i} & \phi_{12i} & \cdots & \phi_{1ni} \\ \phi_{21i} & \phi_{22i} & \cdots & \phi_{2ni} \\ \cdots & \cdots & \cdots & \cdots \\ \phi_{l1i} & \phi_{l2i} & \cdots & \phi_{lni} \end{bmatrix} \tag{8}$$

For any $\phi_{kji} \in \Phi_{..i}$, the subscript $k \in (1, 2, \cdots, l)$ denotes the serial number of vibration test position, $j \in (1, 2, \cdots, n)$ denotes the serial number of collision torque in vector f_{ext}, and $i \in (1, 2, \cdots, m)$ is the number of vibration modes.

We considered the structure of system matrix $A_1 = \begin{bmatrix} -M(q_M)^{-1}K_v & -M(q_M)^{-1}K_p \\ I & 0 \end{bmatrix}$, it is mainly determined by inertia matrix $M(q_M)$, vibration dynamic coefficient matrix K_p and K_v, of which the latter two remain nearly unchanged. That means the natural frequencies $\omega_{Di}(i = 1, 2, \cdots, m)$ will vary with joint displacement q_M.

As the stiffness coefficients $k_{pi}(i = 1, 2, \cdots, m)$ of manipulator structure are usually considerably large, there exist several eigenvalues of A_1 relatively larger than that of A_2, which comes from the active dynamic of robot and reflects the normal working frequency of manipulator. For this reason, there exist several natural frequencies of vibration that are obviously bigger than the frequency band of active dynamic of robot, i.e., $\exists \omega_{Di}(i = 1, 2, \cdots, m)$, and $\omega_{Di} > \omega_{Mj}(j = 1, 2, \cdots, n)$ hold.

So we get an important conclusion: the dominant natural frequency of vibration under collisions is independent of the dynamic property of robot. When the inertia matrix is given, the natural frequency under collision should remain the same during different dynamic processes or static statuses.

Furthermore, the modal matrix Φ_k depends on the value of geometric contact Jacobian matrix $J_D{}^T$ and $J_M{}^T$. For this reason, the contact position information can be induced by the value of vibration mode matrix Φ_k.

2.2. Vibration Features Analysis and Illustrative Example

This section illustrates the characteristics of vibration modal shape of manipulator under collisions with example. The test data were collected on the STR6-05 robot arm (see Figure 1), a 6-DOF heavy-load manipulator. Four consecutive collision experiments were conducted while the end-effector moved in line; the collision conditions are listed in Table 1.

Table 1. Experiment conditions for vibration modal test.

Experiment Index	Contact Position	Contact Direction	Contact Material
1	Near end-effector	X	Aluminum impact hammer
2	Near end-effector	Z	Aluminum impact hammer
3	Near base	Y	Aluminum impact hammer
4	Near end-effector	X	Human hand

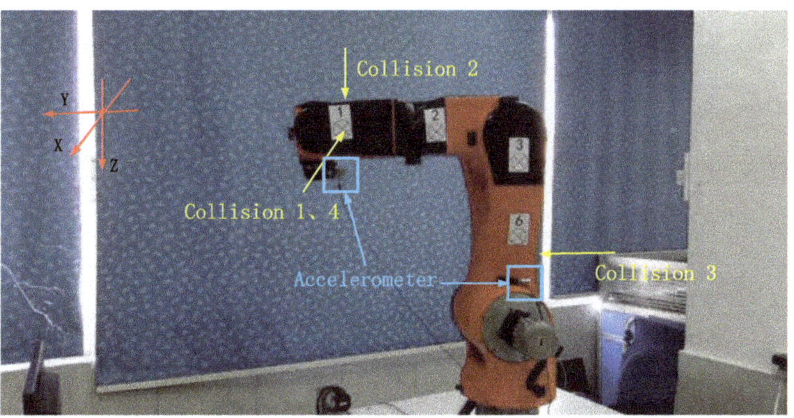

Figure 1. Experiment setup.

The joint displacement and motor current signals of joint 1, 2 and 3 are shown in Figure 2. We label the contact time of corresponding collision events on the figure. We find that there are some shock changes in the motor currents but little change in joint displacements. However, it is hard to detect collision events from current signals directly because of dynamic change and the noise involved in signals, particular for heavy-load manipulator. As shown in the figure, we cannot find some obvious features for collision 1.

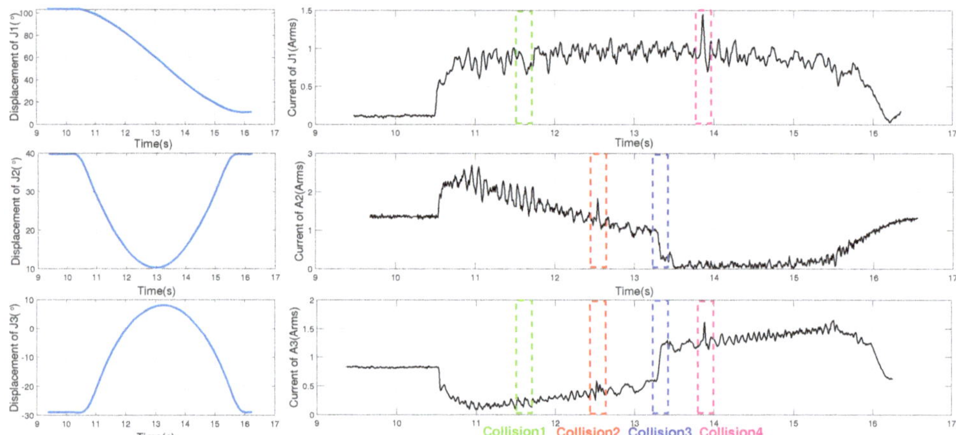

Figure 2. Joint displacement and current during the experiment.

For vibration signal measuring, the 1A113E and 1A114A industrial accelerometer (made by Donghua Testing Technology) with NI-9232 data acquisition module is used. The 1A113E accelerometer (uniaxial) is mounted beside joint 2, perpendicular to the direction of joint 1 and joint 2. The 1A114A accelerometer (triaxial) is equipped beside the end-effector of manipulator.

The vibration acceleration and vibration modal is shown in Figure 3. The first signal comes from 1A113A located beside joint 2, and the latter comes from 1A114A beside end-effector, corresponding to two perpendicular directions.

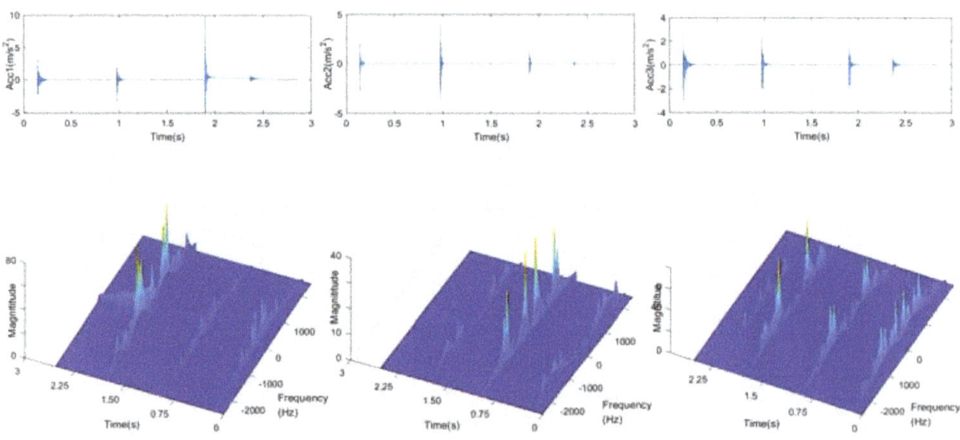

Figure 3. Vibration modes in experiment.

The frequency characteristic of each accelerometer within sliding windows is shown in Figure 3. There are four obvious frequency charts corresponding to four collisions in each vibration signal. The peak frequencies, i.e., the natural frequencies of three accelerations corresponding to one collision are approximately the same, with different energy densities. The magnitudes of all the peak frequencies constitute the modal matrix Φ in Equation (8).

Figure 3 shows that the vibration modes of different parts along robot structure have the following characteristics:

1. The contact position information can be analyzed with vibration modes of different sensors. The main vibration frequency of nearer sensors is usually higher than sensors far from contact position. That is because manipulator structure can be seen as a low-pass filter, the higher frequency vibration signal is reduced during the propagation process. For example, the magnitude of the 3rd collision in the 1st sensor is comparatively higher than that in the 2nd sensor signals. That is because the 3rd collision is conducted near the base, which is much nearer to the 1st signal.
2. Similarly, the contact direction can also be determined with vibration modes. The relative energy density value of different directions in the triaxial accelerometer can be used to detect the contact direction. Generally speaking, the vibration of collision direction may have the comparatively higher energy density. For example, the 1st and 2nd collision took place at the same part with different directions, the 2nd magnitude is higher than the 1st one in the 2nd signal while the 2nd magnitude is smaller than the 1st one in the 3rd signal, as is shown in the figure.
3. Furthermore, contact material information can also be reflected by vibration modal. The band width of the 4th collision is much lower than the other 3. That is because the frequency band of human hand contact force is comparatively narrow.

Limiting the range of magnitude below 1 dB, we got the frequency characteristic of normal dynamic comparative to collisions in Figure 4. It shows that the natural frequency of active dynamic appears at low frequency segment (always below 50 Hz), while the vibration frequency by collision appears at intermediate segment. That means the eigenvalues of A_2 is much smaller than that of A_1 in Equation (5), and the dynamics of normal operation do not affect the natural frequencies of collisions.

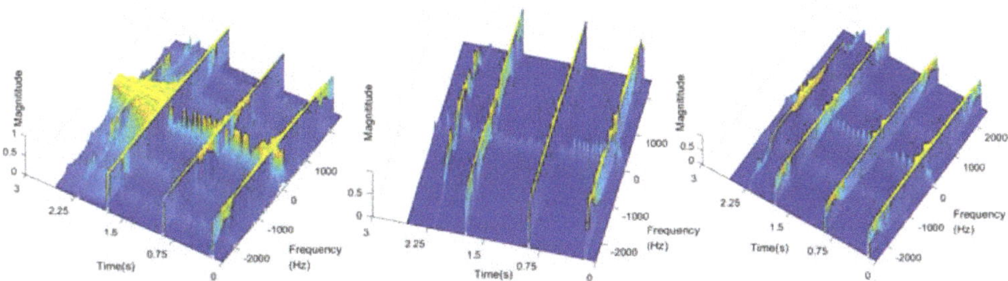

Figure 4. Low magnitude segment of vibration modes in experiment.

Obviously, the natural frequencies and modal is mainly dependent on the inertia matrix, and independent of the robot dynamic. This means that the detection algorithm can be designed without considering the dynamic property of the robot, and the training and test samples for the model independent method can be collected in some simple or static scenarios.

Since collision will generally cause shock vibration, and its natural frequencies are mainly contained in the high-frequency domain, several symptom parameters in the frequency domain can be selected to represent the collision event. Extracting collision information from frequency domain signal requires proper understanding of the process. For frequency features, such as natural frequencies, the vibration modal shown in the spectrum often has direct or indirect connection to certain dynamic events.

3. Collision Detection and Identification Method Based on Vibration Features

Based on the vibration features discussed above, we proposed a learning-based algorithm for the detection, isolation and identification of collisions.

The proposed method mainly contains three parts: vibration mode analysis, collision detection, and collision identification. As shown in Figure 5, the vibration signals are first recorded by accelerometer sensors. The vibration mode related features are then extracted from the vibration signals. Any change in the vibration mode from normal condition can indicate the occurrence of collision. Three different Back Propagation neural networks are developed for the detection of collisions (BP1), the isolation of contact part (BP2), and identification of direction (BP3). BP2 and BP3 should be activated only if some collisions are detected by BP1, and the input of BP3 varies with the output of BP2.

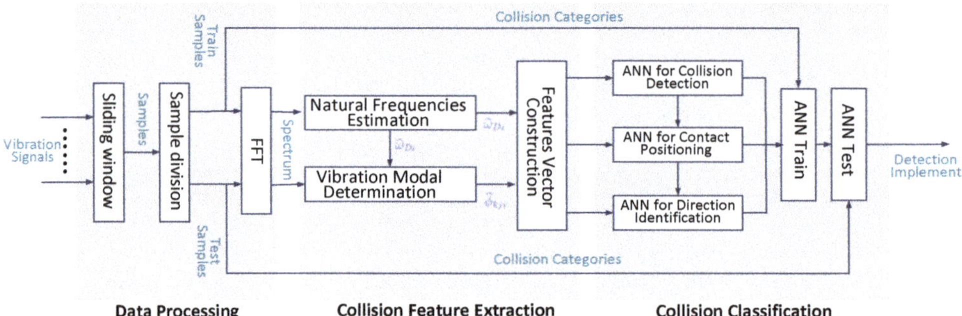

Figure 5. Vibration signal based detection framework.

3.1. Vibration Mode Estimation

The main objective of vibration mode estimation is to determine the main natural frequencies $\omega_{Di}(i = 1, 2, \cdots, m)$, and the values of vibration mode corresponding to each natural frequency $\phi_{kji} \in \Phi_{..j}$, where k denotes the serial number of sensors, j denotes the serial number of collision force, and $i \in (1, 2, \cdots, m)$ denotes the number of vibration modes.

In the collision experiment, the result of measurement is strings of acceleration values in discrete moments. The acceleration signal should be properly processed in order to build the collision classifier. For the discrete string $a1(k)$, $a2(k)$, and $a3(k)$, fast Fourier transform (FFT) is used to determine vibration spectrum within a sliding window which has lower computational cost and better accuracy performance. In our research, the sampling rate is 3.2 k Hz, and the width of sliding window is 320 samples with 50% overlap. Furthermore, the cycle time of detection algorithm is 0.1 s.

For the spectrum of each sampling window, we proposed a peak frequency-based method to determine the natural frequency $\widehat{\omega}_{Di}(i = 1, 2, \cdots, m)$, the estimation of ω_{Di}. For each, the proposed approach consists of the following steps:

Step 1: For acceleration signal $k \in (1, 2, \cdots, l)$, set a tolerance error Δ and δ for spectrum analysis;

Step 2: Find all the local maximum power and local minimum power from start frequency f_0 to cut-off frequency f_e, such that there is no other power value larger than current maximum value between the two adjacent local minimum powers, and there is no other power value less than current minimum value between the two adjacent local minimum power densities; The difference between adjacent maximum and minimum power density should be larger than Δ;

Step 3: Collect all the local maximum power values and corresponding frequencies;

Step 4: Calculate Num, the number of local maximum power density;

Step 5: If Num $> m$, then $\Delta = \Delta - \delta$, repeat step 2 to step 4;

Step 6: Record current local maximum power values $P_{ki}(i = 1, 2, \cdots, m)$ and corresponding frequencies $f_{ki}(i = 1, 2, \cdots, m)$;

Step 7: For other acceleration signal $k \in (1, 2, \cdots, l)$, repeat step 1 to step 6;

Step 8: Rank all the frequency values $f_{ki}(k = 1, 2, \cdots, l; i = 1, 2, \cdots, m)$, and divide them into m groups with maximum intervals.

Step 9: Calculate the average frequencies within each group, these values are the estimation of modal frequency, $\widehat{\omega}_{Di}(i = 1, 2, \cdots, m)$.

With the estimation of modal frequencies, the vibration modal matrix $\widehat{\Phi}$ can be determined by extracting the magnitude of corresponding frequency in the spectrum chart, as it shows in Figure 6. In this figure, we get 6 natural frequencies $\widehat{\omega}_{D1}, \widehat{\omega}_{D2}, \cdots, \widehat{\omega}_{D6}$ from the measurements of three acceleration signals in collision experiment j. The magnitude values at each characteristic frequency are the estimation of vibration modal. Obviously, there is little error of estimation, e.g., the estimation $\widehat{\phi}_{2j5}$. However, this error has limited influence on the final detection result because this detection is based on the synthesis of multiple vibration modal. In this way, we get the estimation of vibration modal matrix of experiment j:

$$\widehat{\Phi}_{\cdot j \cdot} = \begin{bmatrix} \widehat{\phi}_{1j1} & \widehat{\phi}_{1j2} & \cdots & \widehat{\phi}_{1j6} \\ \widehat{\phi}_{2j1} & \widehat{\phi}_{2j2} & \cdots & \widehat{\phi}_{2j6} \\ \widehat{\phi}_{3j1} & \widehat{\phi}_{3j2} & \cdots & \widehat{\phi}_{3j3} \end{bmatrix} \tag{9}$$

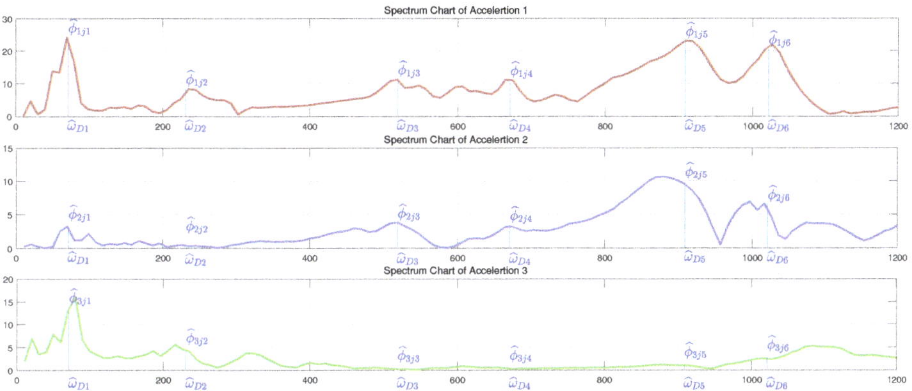

Figure 6. Joint displacement and current during the experiment.

3.2. Proposed Artificial Neural Network

In our method, 3 BP networks are implemented for the collision detection, positioning and direction identification respectively. The collision detection artificial neural network together with the modal analysis algorithm should be executed within each sliding window. Once a collision event is detected, the collision positioning artificial neural network is launched with the current vibration modal data. In addition, the input of the 3rd network is dependent on the output of collision position information. The operation process of relevant algorithms is displayed in Figure 7.

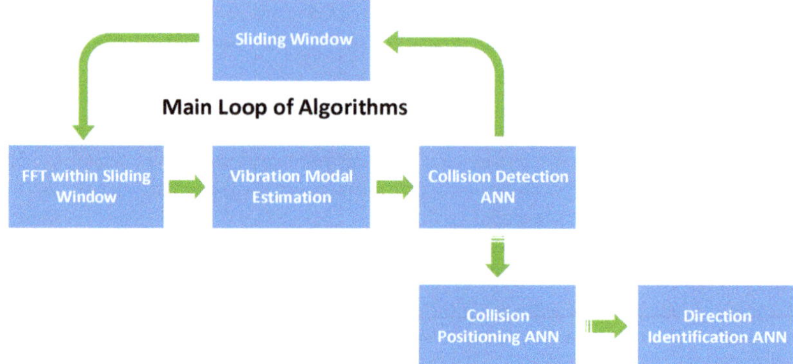

Figure 7. Procedure of detection algorithm.

B-P ANN consists of an input layer, hidden layers, and an output layer of neurons. A neuron serves as a processing unit in which output is a linear or nonlinear transformation of its inputs. The neurons, as a group, serve to map the input vibration modal features to the desired collision patterns. The structure of BP network is shown in Figure 8.

The output signal of hidden layer and output layer can be described in the following equations:

$$m_j(t) = f(\sum_{j=1}^{10} w_{ij}(t)t + b_j) \tag{10}$$

where $m_j(t)$ is the output of current neurons, w_{ij} is the weight of the connection between current layer neurons and its input layer neurons, b_j is the bias of the jth neuron of current layer. The activation function of output layer is linear function, while the hidden layer uses sigmoid function

$$f(t) = \frac{1}{1 + e^{-t}} \tag{11}$$

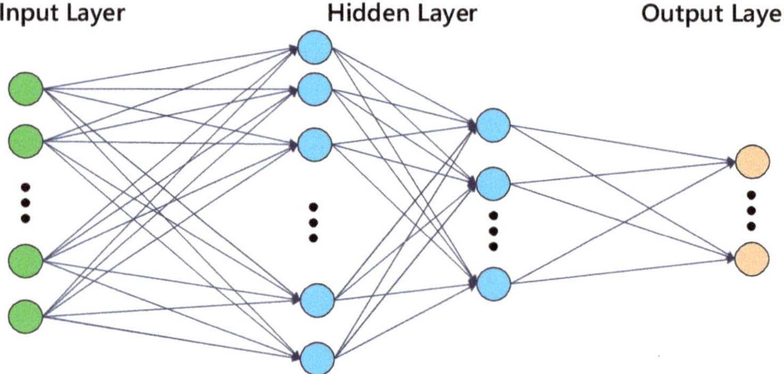

Figure 8. Back Propagation (BP) artificial neural network structure.

Considering the characteristic of vibration modes discussed in Section 2, we selected the most important features for each kind of detection task, as shown in Tables 2–4.

As the natural frequencies are mainly dependent on inertia matrix M_{qM}, we selected the displacement of joint 2 and joint 3 as input features of detection. The displacements of other joints have little influence on inertia matrix.

Table 2. The features used for the detection of collisions.

Features	Function Equation
Geometric Appearance	q_{M2}, q_{M3}
vibration Frequencies	$\hat{\omega}_{Di} \quad (i = 1, 2, \cdots, m)$
vibration modal	$\hat{\phi}_{kji} \quad (k = 1, 2, \cdots, l; j = 1, 2, \cdots, n; i = 1, 2, \cdots, m)$

The contact position and direction of collision can affect the comparative vibration magnitude of sensor on different positions and directions. We select relative modal between test position for the positioning of contact, and relative modal between test direction beside contact position for direction identification.

Table 3. The features used for the isolation of contact position.

Features	Function Equation
Geometric Appearance	q_{M2}, q_{M3}
Vibration Frequencies	$\hat{\omega}_{Di} \quad (i = 1, 2, \cdots, m)$
Vibration modal of end-effector	$\hat{\phi}_{1ji}, \quad \hat{\phi}_{1ji}(i = 1, 2, \cdots, m)$
Relative modal between test position	$\hat{\phi}_{k,j,i} / \hat{\phi}_{k-1,j,i} \quad (k = 2, 3, \cdots, k; i = 1, 2, \cdots, m)$

Table 4. The features used for the identification of collision direction.

Features	Function Equation
Geometric Appearance	q_{M2}, q_{M3}
Vibration Frequencies	$\hat{\omega}_{Di} \quad (i = 1, 2, \cdots, m)$
Relative modal between test direction of contact position	$\hat{\phi}_{cx,j,i} / \hat{\phi}_{cy,j,i}, \quad \hat{\phi}_{cx,j,i} / \hat{\phi}_{cz,j,i}, (i = 1, 2, \cdots, m)$

4. Experiment and Discussion

4.1. Experiment Dataset Preparation and Training Procedure

The experiment dataset is collected on the platform in Figure 1. The robot is controlled with PD control law during the experiment. An uniaxial accelerometer is mounted beside joint 2, perpendicular to the direction of joint 1 and joint 2. In addition, a triaxial accelerometer is equipped beside the end-effector of manipulator. All acceleration data is recorded by NI-9232 data acquisition module. Our experiments are conducted with an aluminum impact hammer, which is also connected to NI-9232 data acquisition module. In this way, the collision time and force data is recorded, as shown in Figure 9.

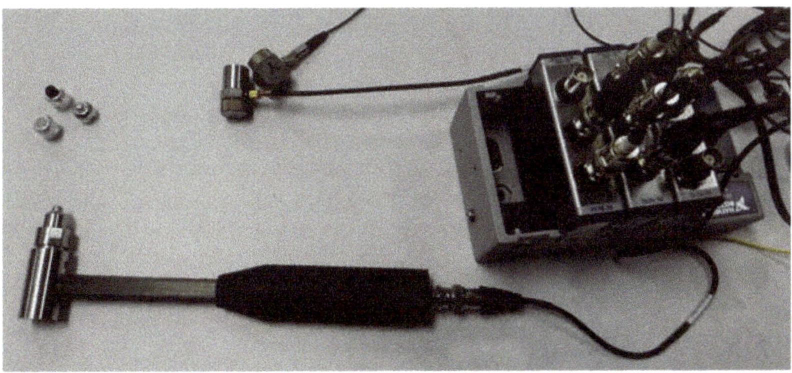

Figure 9. Data acquisition module for collision experiment.

The collision experiments are performed with eight selected contact points, with different direction and force intensity. As the natural frequencies of vibration are mainly dependent on inertial matrix and joint displacement (see Section 2.1), we select five typical working patterns (see Figure 10) as testing standard. Our collision experiments are performed while the robot is transforming randomly from one pattern to another.

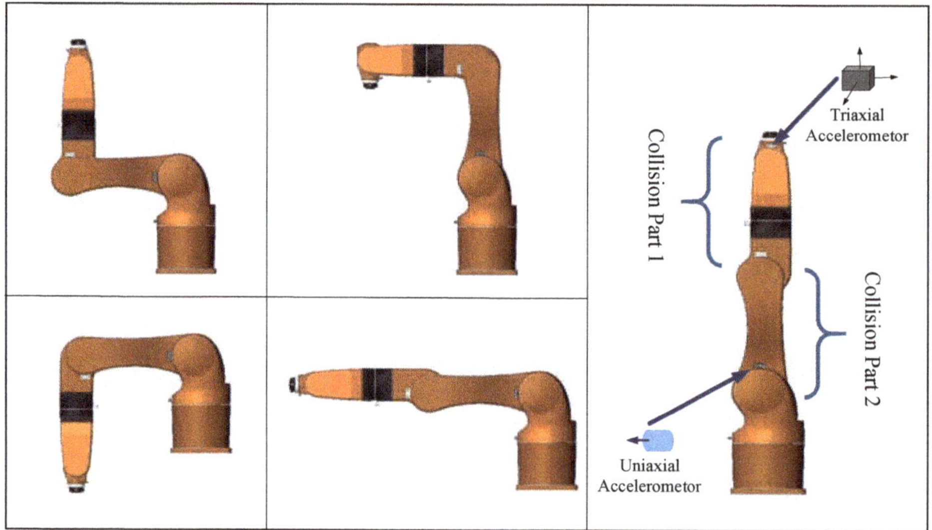

Figure 10. Five testing patterns of manipulator.

Hundreds of collision experiments are performed on the platform. The vibration modal data of acceleration signals during these experiments are analyzed with the sliding-window method introduced above. Typical vibration modal data sets with corresponding experiment conditions are collected for classification research. Meanwhile, we also collect the vibration modal data during collision-free operations. We get in total 800 experiment samples, of which 50 percent are collision samples performed on 8 different contact positions along robot structure. Eighty-five percent of these samples are selected randomly for training and the remaining are used for testing.

4.2. Detection and Identification Results

In this subsection, several test divisions of experiment data are utilized to evaluate the efficiency of the proposed method. Our BP-ANN algorithms are taken from the Neural Network toolbox in Matlab. At the training stage, we optimize the weights and bias parameters by minimizing mean squared error, according to Levenberg-Marquardt optimization.

For the detection of collisions, BP networks with different architecture are constructed and trained. Table 5 lists the accuracy of detection ANN of different architectures (The structure $i - j - k - o$ means a network with i input neurons and o output neurons, and j and k stands for the number of neurons in hidden layers). It is clear to see from the table that the proposed method has considerable accuracy for collision detection, and an average accuracy of 0.95 can be obtained with 1 or 2 hidden layers.

Table 5. Accuracy of collision detection.

Network Architecture	Actual Status	Number of Samples	Detection Result		
			Collision	Non-Collision	Accuracy
27-5-1	Collision	53	51	2	0.962
	Non-collision	59	3	56	0.949
27-10-1	Collision	53	51	2	0.962
	Non-collision	59	4	55	0.932
27-6-4-1	Collision	53	50	3	0.943
	Non-collision	59	3	56	0.946

All the vibration modal data of actual collision are used for the training and testing of positioning neural networks. Considering the geometric layout of the STR6-05 robot, the moveable structure is divided as two parts, i.e., one part is link 3, and the other contains link 4, link 5 and link 6, as shown in Figure 10. Table 6 lists the accuracies of different layers of positioning networks. In addition, the BP networks with 2 hidden layers are suitable for the positioning task.

Table 6. Accuracy of collision positioning.

Network Architecture	Actual Position	Number of Samples	Positioning Result		
			Part 1	Part 2	Accuracy
27-10-1	Part 1	23	18	5	0.783
	Part 2	31	5	26	0.838
27-5-5-1	Part 1	23	20	3	0.870
	Part 2	31	4	27	0.871
27-10-5-1	Part 1	23	21	2	0.913
	Part 2	31	4	27	0.871
27-15-10-3-1	Part 1	23	20	3	0.870
	Part 2	31	2	29	0.936

As the vibration modal features of collision direction are mainly reflected by the acceleration signals nearby the collision point, the input of direction identification network is determined by the output of positioning network. Table 7 lists the accuracy of network of 3 architectures. The training and testing samples come from the vibration data of collisions on link 4, link 5 and link 6, which is near to the acceleration sensors on end-effector. It is obvious that the BP network with 2 hidden layers can obtain comparative stabilization accuracy.

Table 7. Accuracy of direction identification.

Network Architecture	Actual Direction	Number of Samples	Identification Result			
			X-Direction	Z-Direction	Y-Direction	Accuracy
33-8-2	X-direction	18	14	3	1	0.778
	Z-direction	10	3	7	0	0.700
	Y-direction	3	1	0	2	0.667
33-10-6-2	X-direction	18	15	2	1	0.833
	Z-direction	10	1	9	0	0.900
	Y-direction	3	1	0	2	0.667
33-15-10-2	X-direction	18	15	3	0	0.833
	Z-direction	10	2	8	0	0.800
	Y-direction	3	0	0	3	1.000

Considering all the results comprehensively, the proposed method can be utilised for the detection, positioning and identification of collisions with considerable accuracy. By analyzing, we find that the positioning error and identification error is mainly derived from boundary samples, that is the collisions near joint 3. One more accelerometer located beside joint 3 may be used for the enhancement of accuracy.

By analysing the training and test procedure, we can find that an experiment of 300 collision samples is enough for the artificial network training for any $6 - dof$ manipulator with different sensor placement scheme, and it can be accomplished in half or one hour with well designed scenarios and test procedures.

4.3. Rapid Prototyping System Design

In order to realize the online test of the proposed method, the computation complexity and detection time of the related algorithms should be estimated. As discussed above, the main calculation consumption includes three parts, namely, the fast Fourier transform of vibration acceleration data, the estimation of natural frequency and modal, and the node outputs update of neural network.

It can be seen from Section 2.2 that the collision vibration frequency of the robot is mainly between 50 Hz and 1500 Hz, and we set the sampling rate to 3200 samples/second. On the other hand, in order to ensure the spectral characteristics have a sufficient resolution, we select a sliding window of 0.1 s for each cycle, and the number of sampling points $N = 320$ for Fourier transform, as shown in Figure 11. The overlap amount of the sliding window is 50 percent, that is, the calculation cycle of the proposed algorithm is 0.05 s. Within each computing period, Butterfly fast Fourier transform is adopted, and the total number of real multiplications required by FFT of N points is $2N * log_2(N)$, and the total number of real numbers added is $2N * log_2(N)$ for each vibration signal.

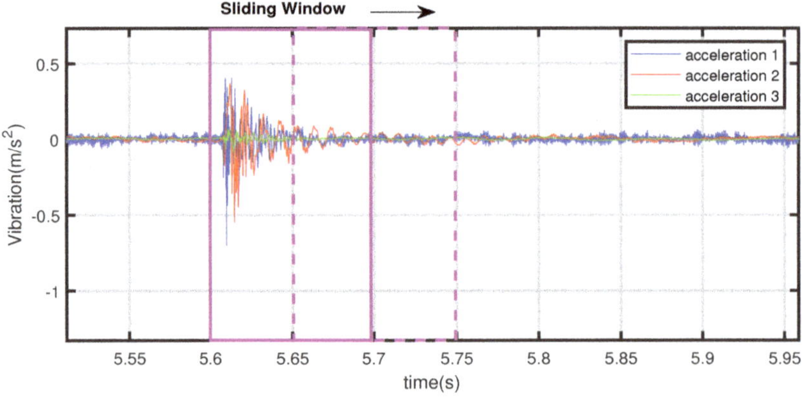

Figure 11. Sliding window for online vibration detection test.

For the natural frequency and modal estimation algorithm, the main computational cost of the algorithm is the sorting algorithm of spectrum amplitude, and the computational complexity of the algorithm is $O(N * log_2(N))$ [26], that means, the computational cost of the frequency and modal estimation algorithm can be ignored relative to the fast Fourier transform.

For the status update of the neural network, each neuron includes multiple addition operations, multiplication operations and exponential operations, and the exponential operation can be converted into a number of addition and multiplication operations. For the collision detection neural network with typical structure of 27-6-4-1, the addition calculation times of a single update is 278, and the total number of multiplication is 98. For the positioning and direction identification neural network with one or two hidden layers, the calculation cost is of the same magnitude.

Therefore, the main computational cost of this method is derived from the fast Fourier transform. In addition, within a single update cycle (50 ms), the addition and multiplication times of the algorithm is about 22,000 and 21,000 respectively. The total computation consumption of this method is about the same size as the traditional state observer method with Newton-Euler Function, as discussed in [27].

Based on the above analysis, we use Simulink/Data Acquisition Toolbox to develop a rapid prototyping system, as shown in Figure 12. The vibration data within each sliding window is buffered, and then used for FFT, modal estimated, and neural network status updates. The system is a slower-than-real-time system due to data caching and operating system. It shows that the collision detection time is about 0.1 s, and the isolation and identification time will be a little longer.

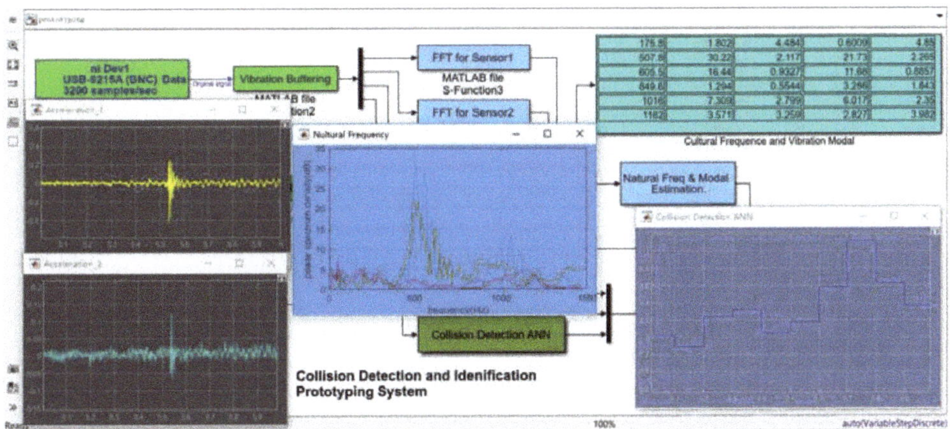

Figure 12. A Rapid Prototyping System.

5. Conclusions and Discussion

In this work, we present a model independent method for robot collision detection, positioning and identification. The vibration signals are analyzed and used for the construction and training of BP neural network. The test results of the experiment confirm that it is possible to build a monitoring algorithm with considerable accuracy. The conclusions may be summarized as follows:

1. With a small amount of training samples (about 300–500 samples), the proposed method can provide considerably high accuracy, and it can be conducted in half an hour or one hour with any kind of manipulators. Therefore, this method has the potential to be implemented in real application scenarios.
2. The detection and identification method is mainly dependent on the frequency domain features of collision, the time-domain features can also be added to improve detection accuracy and computational efficiency in further research.
3. The bandwidth of collisions on heavy load manipulator is mainly below 1500 Hz, and the bandwidth of light robot should be much less than that value. This means some high-performance MEMS accelerometers may be utilised on some occasions.
4. The main calculation consumption of the proposed method comes from the FFT of vibration signal, some dedicated FFT chips can be utilized to improve detection performance.

Author Contributions: F.M. conceived the original ideas, designed and conducted all the experiments, and subsequently drafted the manuscript. G.W. contributed to the construction of the experiment platform and writing-review. N.L. provided supervision to the project.

Funding: This research was funded by the Fundamental Research Fund for the Central University (No. 11618403) and the Guangdong Natural Science Foundation (No. 2018030310482).

Conflicts of Interest: The authors declare no conflict of interest.

References

1. Vorndamme, J.; Schappler, M.; Haddadin, S. Collision detection, isolation and identification for humanoids. In Proceedings of the 2017 IEEE International Conference on Robotics and Automation, Singapore, 29 May–3 June 2017; pp. 4754–4761.
2. Haddadin, S.; Luca, A.D.; Albu-Schaffer, A.; Tjahjowidodo, T. Robot Collisions: A Survey on Detection, Isolation, and Identification. *IEEE Trans. Robot.* **2017**, *6*, 1292–1312. [CrossRef]
3. Ennen, P.; Ewert, D.; Schilberg, D.; Jeschke, S. Efficient collision avoidance for industrial manipulators with overlapping workspaces. *Procedia CIRP* **2014**, *20*, 62–66. [CrossRef]
4. Makarov, M.; Caldas, A.; Grossard, M. Adaptive Filtering for Robust Proprioceptive Robot Impact Detection under Model Uncertainties. *IEEE/ASME Trans. Mech.* **2014**, *6*, 1917–1928. [CrossRef]
5. Luca, A.D.; Albu-Schaffer, A.A.; Haddadin, S.; Hirzinger, G. Collision Detection and Safe Reaction with the DLR-III Lightweight Manipulator Arm. In Proceedings of the 2006 IEEE/RSJ International Conference on Intelligent Robots and Systems, Beijing, China, 9–15 October 2006; pp. 1623–1630.
6. Caldas, A.; Makarov, M.; Grossard, M. Adaptive Residual Filtering for Safe Human-Robot Collision Detection under Modeling Uncertainties. In Proceedings of the 2013 IEEE/ASME International Conference on Advanced Intelligent Mechatronics, Wollongong, Australia, 9–12 July 2013; pp. 722–727.
7. Yamada, Y.; Hirasawa, Y.; Huang, S. Human-robot contact in the safeguarding space. *IEEE/ASME Trans. Mech.* **2002**, *4*, 230–236. [CrossRef]
8. Ren, T.; Dong, Y.; Wu, D.; Chen, K. Collision detection and identification for robot manipulators based on extended state observer. *Control Eng. Pract.* **2018**, *79*, 144–153. [CrossRef]
9. Freddi, A.; Longhi, S.; Monteriù, A.; Ortenzi, D. Fault Tolerant Control Scheme for Robotic Manipulators Affected by Torque Faults. *IFAC-PapersOnLine* **2018**, *24*, 886–893. [CrossRef]
10. Ou, M.; Jiegao, W. Model order reduction for impact-contact dynamics simulations of flexible manipulators. *Robotica* **2007**, *25*, 397–407.
11. Ueda, J.; Yoshikawa, T. Mode-shape compensator for improving robustness of manipulator mounted on flexible base. *IEEE Trans. Robot. Autom.* **2010**, *2*, 256–268.
12. Caccavale, F.; Marino, A.; Muscio, G.; Pierri, F. Discrete-Time Framework for Fault Diagnosis in Robotic Manipulators. *IEEE Trans. Control Syst. Technol.* **2013**, *21*, 1858–1873. [CrossRef]
13. Gierlak, P.; Burghardt, A.; Szybicki, D. On-line manipulator tool condition monitoring based on vibration analysis. *Mech. Syst. Signal. Process.* **2017**, *89*, 14–26. [CrossRef]
14. Pappachan, B.K.; Caesarendra, W.; Tjahjowidodo, T. Frequency Domain Analysis of Sensor Data for Event Classification in Real-Time Robot Assisted Deburring. *Sensors* **2017**, *6*, 1247. [CrossRef] [PubMed]
15. Munoz-Barron, B.; Rivera-Guillen, J.R.; Osornio-Rios, R.A.; Romero-Troncoso, R.J. Sensor Fusion for Joint Kinematic Estimation in Serial Robots Using Encoder, Accelerometer and Gyroscope. *J. Intell. Robot. Syst.* **2015**, *2*, 529–540.
16. Cheng, F.; Peng, Y.; Qu, L. Current-Based Fault Detection and Identification for Wind Turbine Drivetrain Gearboxes. *IEEE Trans. Ind. Appl.* **2017**, *2*, 878–887. [CrossRef]
17. Putra, I.; Brusey, J.; Gaura, E. An Event-Triggered Machine Learning Approach for Accelerometer-Based Fall Detection. *Sensors* **2018**, *18*, 20. [CrossRef] [PubMed]
18. Wu, J.; Tang, T.; Chen, M.; Hu, T.H. Self-Adaptive Spectrum Analysis Based Bearing Fault Diagnosis. *Sensors* **2018**, *18*, 3312. [CrossRef] [PubMed]
19. El-Zahab, S.; Mohammed, A.E.; Zayed, T. An accelerometer-based leak detection system. *Mech. Syst. Signal Process.* **2018**, *108*, 276–291. [CrossRef]
20. Liu, J.T.; Yang, X.X. Learning to See the Vibration: A Neural Network for Vibration Frequency Prediction. *Sensors* **2018**, *18*, 2530. [CrossRef] [PubMed]
21. Lu, S.; Chung, J.H.; Velinsky, S.A. Human-Robot Collision Detection and Identification Based on Wrist and Base Force/Torque Sensors. In Proceedings of the IEEE International Conference on Robotics & Automation, Barcelona, Spain, 18–22 April 2005; pp. 3796–3801.
22. Eski, I.; Erkaya, S.; Savas, S. Fault detection on robot manipulators using artificial neural networks. *Robot. Comput.-Int. Manuf.* **2011**, *1*, 115–123. [CrossRef]

23. Putra, I.P.E.S.; Brusey, J.; Gaura, E. A Cascade-Classifier Approach for Fall Detection. In Proceedings of the International Conference on Wireless Mobile Communication and Healthcare, London, UK, 14–16 October 2015.
24. Vemuri, A.T.; Polycarpou, M.M.; Diakourtis, S.A. Neural network based fault detection in robotic manipulators. *IEEE Trans. Robot. Autom.* **1998**, *2*, 342–348. [CrossRef]
25. Dimeas, F.; Avendao-Valencia, L.D.; Aspragathos, N. Human-robot collision detection and identification based on fuzzy and time series modelling. *Robotica* **2015**, *9*, 1886–1898. [CrossRef]
26. Baase, S. *Computer Algorithms, Introduction to Design and Analysis*; Pearson Education: Delhi, India, 1988.
27. Siciliano, B.; Sciavicco, L.; Villani, L. *Robotics: Modelling, Planning and Control*; Springer Publishing Company, Incorporated: New York, NY, USA, 2010.

Article

Integration of Terrestrial Laser Scanning and NURBS Modeling for the Deformation Monitoring of an Earth-Rock Dam

Hao Xu [1], Haibo Li [2], Xingguo Yang [2], Shunchao Qi [1] and Jiawen Zhou [1,*]

[1]   State Key Laboratory of Hydraulics and Mountain River Engineering, Sichuan University, Chengdu 610065, China; 2017223060078@stu.scu.edu.cn (H.X.); shunchaoqi@scu.edu.cn (S.Q.)
[2]   College of Water Resource and Hydropower, Sichuan University, Chengdu 610065, China; hbli@stu.scu.edu.cn (H.L.); 89022251@163.com (X.Y.)
*   Correspondence: jwzhou@scu.edu.cn; Tel.: +86-28-8546-5055

Received: 9 November 2018; Accepted: 19 December 2018; Published: 21 December 2018

Abstract: A complete picture of the deformation characteristics (distribution and evolution) of the geotechnical infrastructures serves as superior information for understanding their potential instability mechanism. How to monitor more completely and accurately the deformation of these infrastructures (either artificial or natural) in the field expediently and roundly remains a scientific topic. The conventional deformation monitoring methods are mostly carried out at a limited number of discrete points and cannot acquire the deformation data of the whole structure. In this paper, a new monitoring methodology of dam deformation and associated results interpretation is presented by taking the advantages of the terrestrial laser scanning (TLS), which, in contrast with most of the conventional methods, is capable of capturing the geometric information at a huge amount of points over an object in a relatively fast manner. By employing the non-uniform rational B-splines (NURBS) technology, the high spatial resolution models of the monitored geotechnical objects can be created with sufficient accuracy based on these point cloud data obtained from application of the TLS. Finally, the characteristics of deformation, to which the geotechnical infrastructures have been subjected, are interpreted more completely according to the models created based on a series of consecutive monitoring exercises at different times. The present methodology is applied to the Changheba earth-rock dam, which allows the visualization of deformation over the entire dam during different periods. Results from analysis of the surface deformation distribution show that the surface deformations in the middle are generally larger than those on both sides near the bank, and the deformations increase with the increase of the elevations. The results from the present application highlight that the adhibition of the TLS and NURBS technology permits a better understanding of deformation behavior of geotechnical objects of large size in the field.

Keywords: earth-rock dam; 3D visualization; deformation monitoring; terrestrial laser scanning (TLS); NURBS

---

1. Introduction

The deformation distribution and evolution are present as important indications of the instability of large artificial and natural structures such as tunnels, bridges, and landslide [1–4]. Characterizing the in-situ deformation behavior of huge/important structures helps to understand the underlying mechanism of sliding and to predict the possibilities of catastrophic collapse. Thus, periodic monitoring of deformation of the structure has been an important task over the whole lifecycle of infrastructures of great importance, e.g., for ensuring their safety during construction as well as for their post-construction maintenances. Take the example of a dam: many monitoring instruments have been adopted with proper methodologies to appraise the condition and safety of the dam over the past decades [5–7]. Using traditional methods are possible for monitoring over a relatively large area. However, one of the common features of all these approaches e.g. general geodesy method and GPS static method, is that they can only provide point-wise information. For the dams with large scale, the number of the monitoring points is usually rather limited in terms of sufficiently characterizing the deformation characteristics. Synthetic Aperture Radar (SAR) is a relatively new technology, whose variant, GB-SAR, is used for the deformation measurement and monitoring of dam [8–10]. It needs to be recognized that SAR offers high sensitivity to small displacements. The most advanced GB-SAR is even capable of providing millimeter precision. Conversely, the main shortcoming of SAR is that it can only allow for deformation detection along the sensors-target line of sight. Furthermore, the data processing and analysis tools of SAR are quite complex, which makes it difficult to be applied into practice.

In contrast, the terrestrial laser scanning has a definite advantage of providing point clouds composed of millions of points with high accuracy and high spatial resolution, and the point clouds can be used to detect vertical and horizontal deformation through relatively simple process. Nevertheless, the point cloud data acquired from the laser scanner cannot be processed in the same way as that for the data from the traditional methods, since the laser pulse emitted from the machine is not necessarily aimed at the same place of the target object in different scanning operations [11]. Additionally, the point density of the target object varies in different measurement campaigns, which making direct comparison between points cloud obtained sequentially is not favorable.

Hence, how to extract information of the dam deformation from the terrestrial laser scanning (TLS) data is a challenging and important task, which is the focus of many recent works carried out on the deformation detection by the application of TSL [12–16]. A simple means of extracting the deformation is to compare the point-cloud gathered in different epochs [17–19]. This, however, can only show the general deformation tendency exhibited by the target object, which implies that it only allows for a qualitative analysis of the deformation rather than a quantitative analysis. The traditional method for surface reconstruction is point cloud gridding. A typical type of gridding algorithm, called the Delaunay triangulation algorithm, is widely used in modern programs, like Rapid Form and Riscan Pro. Due to its flexibility and adaptability, the Delaunay triangulation algorithm is widely accepted for reconstructing the surface of the objects with irregular shape, such as land relief and mechanical parts [20–22]. Unfortunately, the accuracy of the triangulated irregular network model is not satisfactory, making the subsequent deformation detection difficult to accomplish. Besides, the triangulation algorithm is rather disappointing in the aspect of noise reduction.

In this paper, a detailed deformation measurement method, based on combination of TLS and NURBS technologies, is presented. The method is established on the basis of the quadrangular surface domain parameters for deformation monitoring. By specifying the quadrangular surface domain parameters, the four corner points of the NURBS fitting surface are exactly positioned at the four vertexes of the control mesh. Thus, the precision of the NURBS fitting surface reaches up to a couple of millimeters and is higher than the traditional NURBS surface. The performance of this present method in the field is illustrated by an application to a selected hydraulic structure, the Changheba Dam, located in Southwest China. To acquire the detailed and accurate 3D data rapidly and efficiently, the TLS technology is first applied to take inventories. To fully take advantage of the point cloud data, the digital surface model of the Changheba Dam with high accuracy and high spatial resolution is established by using the NURBS surface modeling technology. Through the comparison between multi-temporal models, the deformation characteristics of the Changheba Dam in different stages are analyzed. The results presented in this study illustrate the applicability of the present methodology to the precise deformation monitoring over large regions.

2. Background

2.1. The Changheba Dam

Located about 360 km southwest of Chengdu, the capital city of Sichuan province in Southwest China, the Changheba Hydropower Station retains the Dadu River over a basin area of 56,648 km^2 (Figure 1a,b). The hydro-technical system, with the primary purpose of exploiting the hydropower potential of the Dadu River, consists of an earth-rock dam, a spillway system and a water diversion and power generation facilities. The construction of the whole project was completed in April, 2018. With the maximum height of 240.0 m, the Changheba dam is the highest earth-rock dam built with gravel and soil core wall in China at present (Figure 1c). The length and width of the dam crest are 502.8 m and 16.6 m, respectively. Its gross filling volume is about 3.42×10^7 m^3, the largest volume of core wall earth-rock dam constructed in China. The dam can be divided into eight different zones according to the filling materials. The distinct mechanical properties of the filling materials make the deformation of the dam very intricate.

Figure 1. Location and layout of the dam at Changheba Hydropower Station: (**a**) and (**b**) location of the Changheba Hydropower Station, (**c**) layout of the dam at Changheba Hydropower Station.

2.2. Geodetic Network

A geodetic network in the nearby of the earth-rock dam has been constructed since the commencement of the project in the year of 2011, which consists of nine control points materialized by observation monuments with a forced centering device. Of the nine control points, five are located on the left bank of the Dadu River, with other four on the right bank, and Figure 2 shows the layout of the geodetic network. The maximum and minimal side lengths of the network are about 1200 m and 210 m, respectively. Datum points TN03, TN04 and TN06 are equipped with inverted plumb line system. Thanks to the good conservation of these observation monuments, the geodetic network has been working well since the beginning. Two measurements of the geodetic network have been performed in October 2016, April 2017, respectively, by means of a Leica TM30 GeoRobot (provided by Leica Geosystems AG, Heerbrugg, Switzeland). The observation results of the TN04 and TN06 based on the inverted plumb line system shows the two points of the control net are stable. Thus, they are used as starting point for the classic free network adjustment calculation. The mean square error of points is less than +2 mm using the classic free network adjustment. Of the two measurements, the mean square error of the weakest point is 1.61 mm and 1.76 mm, respectively. And the average relative mean square error of the weakest side is 1/283,000 and 1/258,000, respectively. The results of the measurements and adjustment calculation indicate that the geodetic network is fine enough for the coordinate measurement of the scanner.

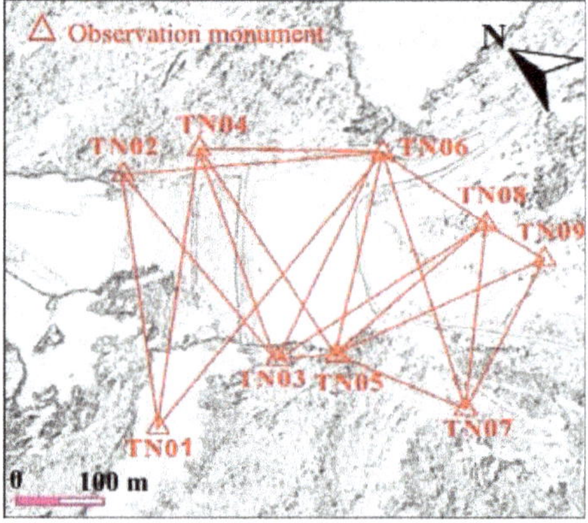

Figure 2. Layout of the geodetic network.

3. Methods

3.1. Terrestrial Laser Scanning

Terrestrial laser scanning is a quite new surveying method in geodesy. The scanner illuminates the target object with pulsed laser and records the returned pulse of the laser. It calculates the distance from itself to the target object surface automatically by timing the round-trip time of one pulse of the laser. Utilizing the time-of-flight (TOF) technology, the TLS is able to record dense point-clouds over an extremely short period of time. The modern laser scanner provided by the manufacturer shows a significant higher speed of data acquisition, compared to the conventional surveying instruments like total station [23]. Moreover, the TLS can also record the intensity of the reflected pulsed laser and RGB color data of the target object.

Let the origin of the Cartesian (rectangular) coordinate system coincide with the center of the laser scanner, where X and Y are two axes perpendicular to each other lying in the horizontal plane, and the

Z axis is oriented upwards and perpendicular to the horizontal plane (Figure 3). Then, the coordinates of the laser point in the there-dimensional (3-D) space can be calculated from Equations (1) and (2) [24].

$$S = c \times \left(\frac{TOF}{2} \right) \tag{1}$$

$$\begin{cases} x = S \cdot \cos\theta \cdot \cos\alpha \\ y = S \cdot \cos\theta \cdot \sin\alpha \\ z = S \cdot \sin\theta \end{cases} \tag{2}$$

where c is the speed of light; *TOF* is the time of flight of the laser pulse; S is the distance from the scanner to the reflecting surface; θ is the angle between the line *OP* and the *XY* plane; and α is the angle between *X* axis and the orthogonal projection of the *OP* onto the plane *XY*.

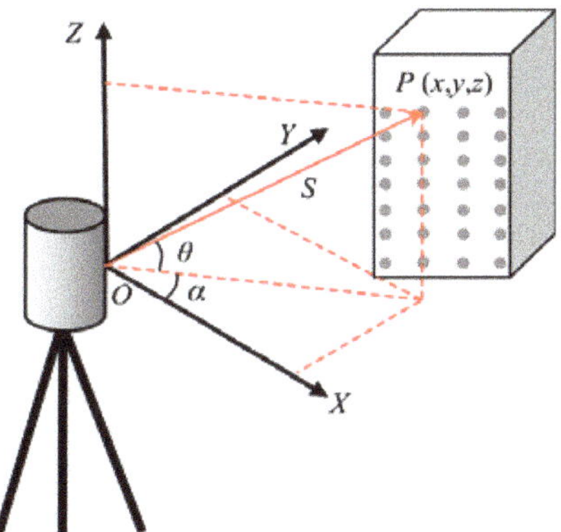

Figure 3. Measuring principle of the terrestrial laser scanning (TLS).

In this paper, a pulse-based scanner, Riegl VZ-400 (provided by RIEGL Laser Measurement Systems, Horn, Austria), is applied to survey the study area. The scanner is capable of measuring the distance ranging from 1.5 m to 600 m. Due to the echo digitization and online waveform processing technique, the highest angular resolution that this instrument can achieve is 0.0005°, and its efficient measurement rate is up to 120,000 points per second. Besides, it offers a wide field of view up to 100° vertical and 360° horizontal. As such, this equipment is suitable for the data acquisition in the vested condition.

3.2. Data Acquisitions and Preprocessing

Measurement campaigns were performed after the filling of the dam. Due to the great size of the dam, it is impossible to acquire the point cloud data of the entire dam surface with only one scan. In addition, a relatively larger number of scans can increase the density of target point clouds, making sure that the overlap between different scans is enough for the alignment of different data sets.

Here, both the particular geologic of this case and the requirement of data processing, nine scans were performed at nine different stations distributed in the surrounding of the dam for the first time in October 2016, when the project began to impound. For each scan, the scanner was placed on the observation station. Then the coordinates of the scanner were measured based on the geodetic network. In this way, it was possible to get more precise coordinates of the scanner and reduce alignment error. It took approximately nine minutes to complete each 360° scan. The whole data set acquired in the first measurement campaign comprises one hundred and thirty-six million points in total, which is treated as the reference point cloud. The second measurement campaign was performed in April 2017 in a similar way as that in the first campaign, and both measurements were referred to the geodetic coordinate system.

These two sets of TLS data were preprocessed using the RISCAN PRO software. First, the geodetic coordinates of each scanner were extracted from the total station based on the geodetic network. By simply adding the instrument height, the geodetic coordinates of the scanner center were acquired. Then the geodetic coordinate data was input into the RISCAN PRO software for the raw scans by the backsighting orientation. In this way, the location of each scan center was determined. After the input of the coordinates, the orientation of each scan remained random. Therefore, a manual modification of the orientations was carried out to make the orientations in space correct. The second process was multi station adjustment. This process was performed using the iterative closet point (ICP) algorithm proposed by Besl and McKay [25]. In the first place scans were aligned pair by pair by means of lowering the "search distance" parameter from meters to some centimeters step by step. After a few adjustments, the alignment of the pair of scans led to an optimal rota-translation alignment matrix. Following this procedure a global alignment was employed for the whole scans to obtain a best fit alignment [26]. Both steps were based on the ICP algorithm and the latter step was executed for the purpose of distributing the residual registration error more homogeneously across the scans [25]. The global alignment of the nine scans resulted in the overall standard deviation of 0.0017 m, which meant the average distance between two closest points. The alignment error is mainly owing to the point spacing of the datasets in the overlap area and the point measurement error in practice. By increasing the point density in the overlap area and reducing the distance between the scanner and the overlap area, the standard deviation can be significantly lowered. In consideration of the high point density of the point clouds, this alignment error is primarily caused by point measurement error which can be reduced by fitting technique. Thus, the alignment error is acceptable for the deformation monitoring of the dam. Meanwhile, the four circular reflector targets were used as tie point for the calibration of the alignment. Similarly, the geodetic coordinates of its center points were measured by total station. Moreover, the geodetic coordinates of the center points were extracted from the scanning data. Then there were two sets of geodetic coordinates of the center points. By taking the geodetic coordinates acquired from the total station as datum, the mean square error of the four center points was calculated out. The mean square error of the points was no more than 0.0013 m, which proved that the alignment of the scans was successful. The alignment of the partial scans leaded to a single point cloud data set consisting of all scanned points.

After alignment of all nine scans, there were some unneeded objects such as dust and vegetation in the point cloud. Via automatic and manual operation, the data was run through terrain filter to remove such points. Eventually, a single point cloud colored with RGB information obtained from the calibrated camera on top of the terrestrial laser scanner came into being, whose partial view is shown in Figure 4 below.

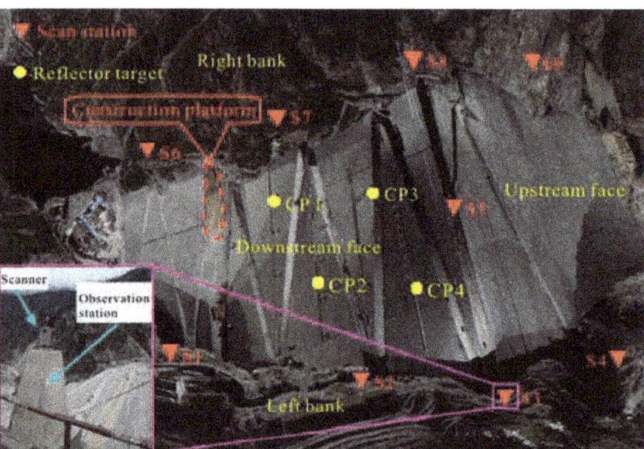

Figure 4. Aligned 3D model of the Changhheba dam (rendered view).

After the alignment of the point cloud, it comes to the stage of the triangulation. The basic input data that can be processed for NURBS surface reconstruction is the triangulated irregular network (TIN) rather than the point cloud. Hence, the TIN should be constructed based on the point cloud first, i.e., triangulation. The triangulation of points is also called "tessellation". By using the Delaunay triangulation algorithm, each set of three closest points in the point cloud are connected to form a triangle, resulting in a non-overlapping triangulation as a whole. However, the initial triangulation generated automatically by the Delaunay triangulation algorithm usually has deficiencies, like acute angled triangles and holes. With automatic analysis, inconsistencies are detected and then repaired through automatic or manual operation. The final operation carried out to refine the TIN is the smoothing. The purpose of smoothing operation is to reduce the noise and therefore to ensure both the quality and accuracy of NURBS surface that will be constructed later. The downstream face of the Changheba dam is made of masonry; the dam surface is flat macroscopically. Thus, the smoothing operation is of central significance to the surface reconstruction.

The point clouds for the Changheba dam are converted into a polygon object with 3.48 billion triangles in total (Figure 5). Only points representing the dam surface are employed and converted into TIN, and other points beyond the survey area are not processed in this section.

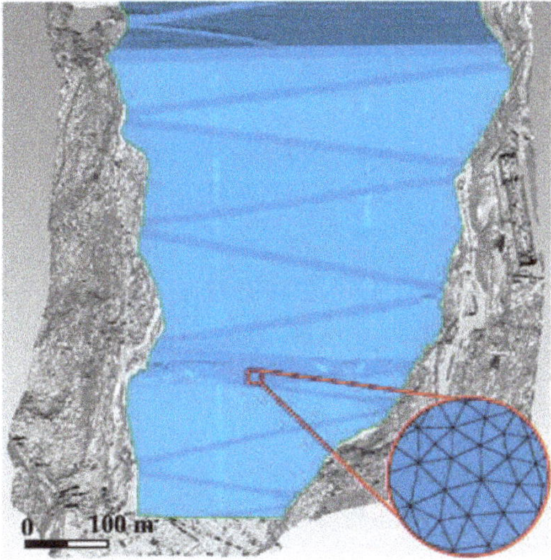

Figure 5. Triangulated irregular network (TIN) model of the Changheba dam.

The previously-constructed TIN is exported and then processed into a NURBS fitting surface model of the Changheba dam. With high accuracy and high spatial resolution, this NURBS surface model can be applied to detect the deformation of the dam surface for further study.

3.3. NURBS Modeling

NURBS is the abbreviation of non-uniform rational B-splines, in which, the non-uniform means that the spacing of the knots is uneven, the Rational implies that the control point can be weighted, and the B-spline represents that B-spline is used as basis function. NURBS is a piecewise rational vector polynomial function advocated by Versprille in 1975 [27]. It can generate and represent arbitrary curves and surfaces better than other functions like radial basis function [28], which can be in either standard shapes or free-form shapes. The NURBS is widely used in computer graphics and the CAD/CAM industry, due to its great flexibility and precision. Besides, fitting surface is a useful means of reducing noise. Thus, NURBS is a useful tool to build surface model of natural land relief [29].

A NURBS surface $P(u, v) = \{x(u, v), y(u, v), z(u, v)\}$ is a piecewise rational surface defined by Equation (3):

$$P(u,v) = \frac{\sum_{i=0}^{m} \sum_{j=0}^{n} w_{i,j} N_{i,k}(u) N_{j,l}(v) d_{i,j}}{\sum_{i=0}^{m} \sum_{j=0}^{n} w_{i,j} N_{i,k}(u) N_{j,l}(v)} \tag{3}$$

where, the $d_{i,j}$ ($i = 0, 1, \ldots, m; j = 0, 1, \ldots, n$) are the control points representing a topological mesh, $w_{i,j}$ is the so-called weights, and the $N_{i,k}(u)$, $N_{j,l}(v)$ are the normalized B-spline basis functions defined on the non-periodic knot. For instance, the mathematical expression of the $N_{i,k}(u)$ can be defined recursively as follow,

$$N_{i,0}(u) = \begin{cases} 1, & u_i \leq u \leq u_{i+1} \\ 0, & otherwise \end{cases} \tag{4}$$

$$N_{i,k}(u) = \frac{u - u_i}{u_{i+k} - u_i} N_{i,k-1}(u) + \frac{u_{i+k+1} - u}{u_{i+k+1} - u_{i+1}} N_{i+1,k-1}(u) \tag{5}$$

To generate a NURBS surface, three groups of parameters, control points $d_{i,j}$, weight factors $w_{i,j}$ and the knot vectors U and V must be determined. On the grounds of the NURBS theory, the dam surface is constructed from the TIN following the four procedures: panel demarcation, surface patch insertion, grid generation, and NURBS surface construction. The four procedures are executed in sequence to construct the NURBS surface, as explained in the following.

Data segmentation must be carried out to precisely construct the NURBS model of dam surface primarily. In this step, the characteristic lines are extracted from the TIN model. The surface curvature detection technique is adhibited for data segmentation. After setting the proper curvature level, the curvature detecting is performed and the contour lines are placed in the areas of curvature. In the meantime, Markers are highlighted on the contour lines. There are two different kinds of markers. A red marker indicates a corner, while a yellow marker represents a non-corner point of inflection. The contour lines and the boundary lines are used as panel demarcation lines dividing the TIN model into forty-three panels (Figure 6a). If a high curvature level was specified, fewer panels would be acquired. Then the precision the NURBS fitting surface would be affected. But more panels require more processing power and lower the operability. In this study, both the precision requirement and operability, the forty-three panels are proper for the NURBS fitting surface. The demarcated panels will be the containers for the surface patches in the next step.

On the basis of the result of the panel demarcation, the patch boundary structure is generated, which means that four-sided patches are inserted into each panel (Figure 6b). A surface patch is a four-sided subdivision of a panel that is approximately equilateral. Based on the auto estimate technique, the target patch count is automatically calculated depending on the size and the smoothness of the panels. As the panels are very irregular and the patches are relatively regular, the adaptive insertion technique is applied in order to acquire fairly uniform patches, which contributes to the precision of the NURBSB model. Using the adaptive surface patch insertion technique, the shape of the surface patches is determined. The model consists of eight hundred and thirty-nine surface patches. From Figure 6b, it can be seen that the patches that comprise a panel differ in size, and the shapes of most patches are close to the rectangle. The inserted patches will be the containers for the grids in the following procedure.

After the surface patches are inserted, a further process is grid generation. A grid is a quadrangular mesh constructed in every patch. The mesh is made up of a forty-by-forty set of rectangles, which means that each panel consists of one thousand and six hundred grids. The quadrangular mesh density can be specified by the user. Needless to say, a finer grid produces greater precision in the eventual NURBS surface. The grid generation process creates an ordered u-v grid in every patch on the model (Figure 6c).

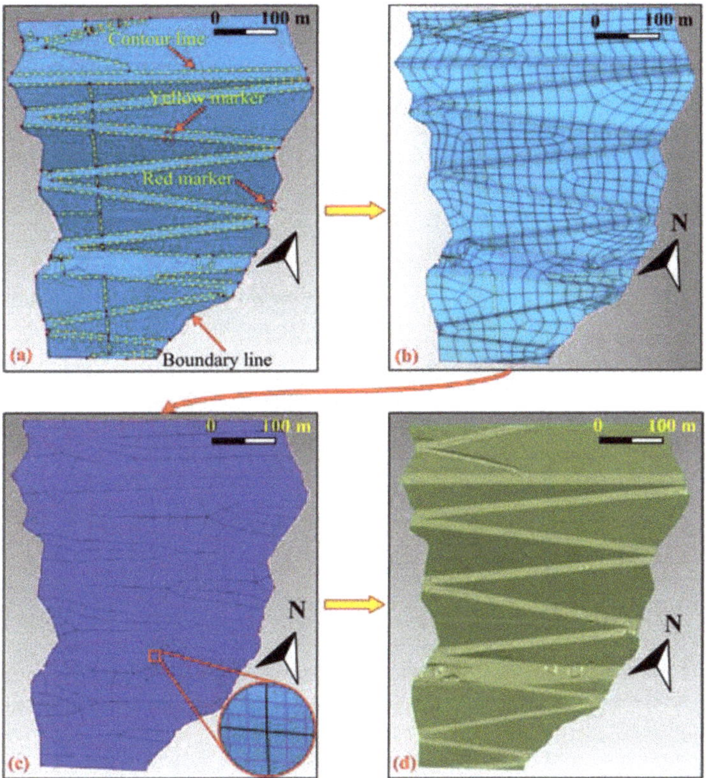

Figure 6. Non-uniform rational B-splines (NURBS) surface construction procedures: (**a**) panel demarcation; (**b**) surface patch insertion; (**c**) grid generation; (**d**) NURBS surface construction.

For the free-form surface of the dam, the exact NURBS surface can be ultimately generated on the TIN using the surface fitting technique. The NURBS surface of the dam is shown in Figure 6d. The consequent high accuracy and high spatial resolution surface model captures very well the morphology and geometry features of downstream face of the Changheba dam, which can be then inquired into the deformation characteristics of the dam [30,31].

3.4. Deformation Measurement by Shortest Distance (SD) Comparison

Deformation can be detected by making geometrical comparison between multi-temporal surface models. In this study, the shortest distance algorithm is applied to the deformation measurement. It is noted that the normal directions of the two surfaces should be generally consistent before the comparison. The algorithm can still work even when the surface normal vector is biased. For each point i $(x_{i.tes}, y_{i.tes}, z_{i.tes})^T$ in the test model, the algorithm searches for its nearest corresponding point j $(x_{j.ref}, y_{j.ref}, z_{j.ref})^T$ in the reference model and computes the SD vector, V_i, that starts in point i and ends in point j (Equation (6)) [32].

$$V_i = (\Delta x_i, \Delta y_i, \Delta z_i)^T = (x_{i.tes}, y_{i.tes}, z_{i.tes})^T - \left(x_{j.ref}, y_{j.ref}, z_{j.ref} \right)^T \tag{6}$$

For the NURBS fitting surface, the central points of grid are used to calculate the shortest distance. The calculated SD vectors do not necessarily represent the real displacements of the object. They are the shortest distances from the point in the test model to the nearest neighbor point in the reference model mathematically (Figure 7). Yet, the shortest distances (SDs) are useful for deformation measurement, since they allow for the detection of the vertical, horizontal, and oblique distances. In this paper, the sign conventions are defined as follow. Positive SDs indicated that the points in the test model are in front of or above those in the reference model. They may not appear to be continuous and can be interpreted as material stack owing to human activity. Negative SDs signified that the detecting part is behind or below the reference dataset, which are related to the vertical settlement and subsidence.

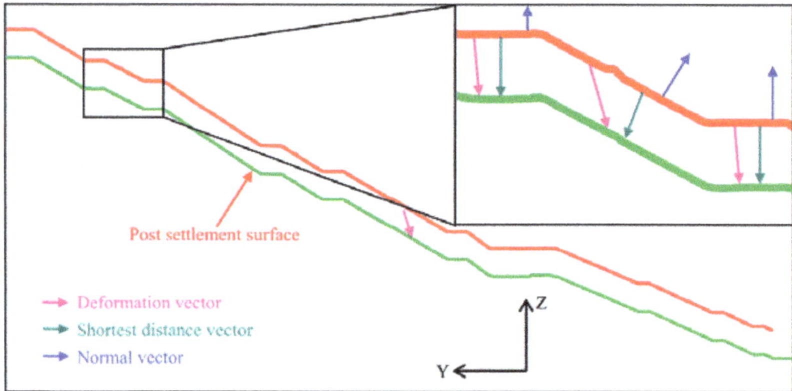

Figure 7. Conceptual comparison of the profiles.

To analyze the deformation of the structure, it is necessary to choose the proper coordinate system [33]. Herein, the main deformation of the dam surface taking place during the operation period is the settlement along the vertical direction. Hence, the Z direction is set as the vertical direction. And the Y direction is set in the direction parallel to the river. The X direction is set in the direction perpendicular to the river. Then, the shortest distance calculated from multi-temporal models can represent the deformation. The calculated data is useful for the dam deformation monitoring [32], with which the deformation of the downstream face can be analyzed in detail.

4. Results

4.1. Deformation Analysis

It is well known that the earth-rock dams are usually subjected to deformation over a relatively long time after the filling. Obviously, the deformation distribution is an essential indicator of the potential instability of the dam. Thus, periodic monitoring of the dam deformation during operation period is of great importance. The deformation of the surface is always larger than the interior due to the effect of displacement cumulation. Therefore, monitoring the surface deformation is an effective way of determining the serviceability of the dam. By introducing the TLS and NURBS technology, making comparison between different multi-temporal scans provides the plentiful and vital information for analysis of the deformation distribution of the dam.

For the Changheba dam studied herein, the NURBS model in October 2016 was treated as reference model, while the NURBS model in April 2017 was regarded as the test model. The reference model was subtracted from the test model. In this way, more than eight hundred thousand central points of grid in the test model were calculated and color-coded mappings of the differences were generated, showing the deformation distribution over the monitored time interval during the operation period.

The deformation distribution of the Changheba dam from October 2016 to April 2017 is shown in Figure 8. On account of the high water level in April 2017, the point cloud data of the upstream face was not obtained, at which the deformation cannot be calculated with only point cloud data in October 2016. As a result, only the deformation of the downstream face is presented. Negative changes representing the dam settlement are shown in cold colors. As shown, the dam has experienced settlements continuously after the filling, which can be interpreted as consolidation. The differential deformation is relatively significant. On top of the dam, the maximum deformation value is -0.0976 m, while the deformation values near the dam toe are close to zero. Along the stream direction, the deformation values of the dam surface get gradually smaller from the top to the toe. In the direction perpendicular to the river, the envelope of deformation exhibits a counter-arch shape (Figure 8). That is to say, the deformation in the middle of the dam is larger than that at the sides of the dam at the same elevation. The deformation of the zigzag road is much larger than its near regions due to the intense human activities.

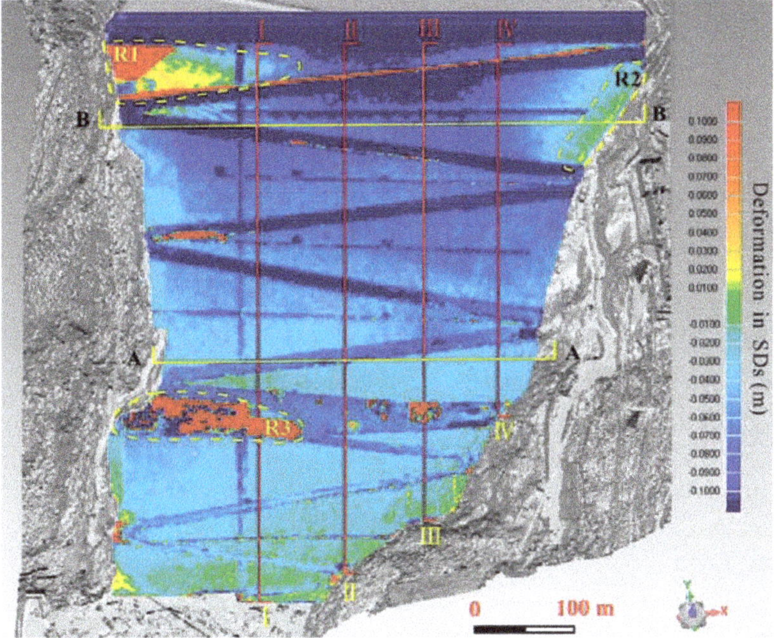

Figure 8. Deformation distribution of the Changheba dam. SDs: shortest distances.

There are three main regions with positive changes in Figure 8, marked as R1, R2, and R3. R3 was in the construct platform in elevation 1551 m in virtue of the construction activities. The other two parts R1 and R2 are located near two abutments. The point cloud data here is missing in the second measurement campaign in April 2017. The hole was filled manually in the process of NURBS model construction. Thus, the calculated result does not indicate the real changes happening in the site.

4.2. Deformation Mechanism

Figure 9 shows some deformation distribution along the cross sections A-A, B-B (identified in Figure 8). As shown in Figures 8 and 9, this dam continued to undergo deformation after the completion of filling of the dam, with the maximum deformation value up to virtually 90 mm in the two cross sections. In the meanwhile, the deformations in the middle at both cross-sections are normally larger than those near the banks, which caused by the effects of constraint from the banks on both sides. The envelopes of deformations of both cross sections are asymmetric, reflecting the fact that the earth-rock dam has been constructed in an asymmetric canyon.

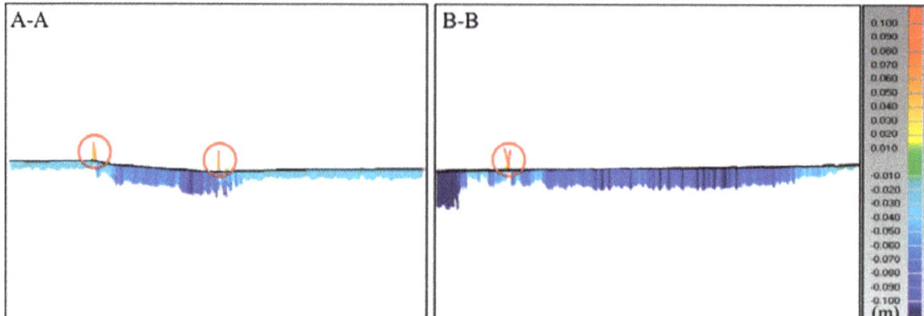

Figure 9. The deformation distribution along the two cross-sections.

It is worth mentioning that the deformations in cross section A-A are much smaller compared to those in cross section B-B. One of the reasons is that the lower part of the dam has been subjected to settlement over a relatively longer time than the upper part of the dam. The lower part of dam started to deform right after its filling, at this time the upper part has not been filled. Another reason is that, after filling of the upper part, additional settlement of the lower part will be induced due to the addition of gravity load on the top. Thus, the relative smaller deformation of lower part is the consequence of combination of a longer period of consolidation and higher stress conditions.

Figure 10 show some deformation distribution along the longitudinal sections I-I, II-II, III-III, and IV-IV, as marked in Figure 8. The magnitudes of deformations increase nonlinearly with the increasing elevation. The relatively larger deformations of the upper part demonstrate that the inner part of the dam has experienced a rapid consolidation during the investigation period. It can also been seen that the deformations of longitudinal sections II-II, III-III are always larger than those at the other two sections I-I, IV-IV, at the same elevation, which means the deformations in the middle are generally bigger than those on the sides at the same elevation. This corresponds with the conclusion drawn from Figure 9. In the meanwhile, there exist several abnormal regions circled in red in Figure 10, which represent the positive changes taking place on the dam surface. This phenomenon results from the human activities and the data missing.

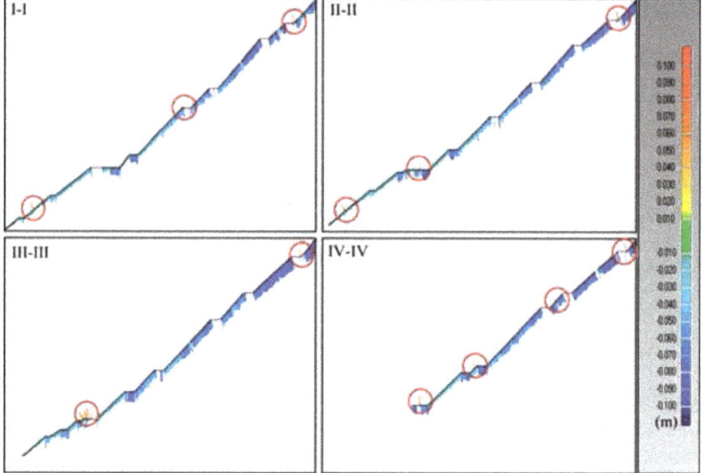

Figure 10. The deformation distribution along longitudinal sections.

5. Discussions

In this paper, a new deformation monitoring method for large structure is formulated. The field experiments are carried out to test its feasibility. In practical works, reducing error is of vital importance for deformation monitoring. According to the different stages of the methodology, the possible errors can be divided into three parts: instrument error, alignment error and modeling error. Many researches have been done on the instrument error and alignment error for the past decades [34,35]. Here, the modeling error of the NURBS technology is analyzed. The NURBS surface model constructed is a kind of fitting surface. There exist some errors in the constructed surfaces, and the error distribution is shown in Figure 11 by making comparison between the original point cloud and the NURBS fitting surface. The overall error of the fitting surface is relatively small, and the distribution of the errors is quite uniform. Errors in most parts are within ±0.002 m. The relatively larger errors are concentrated mainly in the places with large curvatures. One example is in the area of the construction platform including D1 and D2. Since the construction platform is not the focus of this current study, where the NURBS fitting surface is constructed roughly in order to reduce the size of the NURBS surface file. Nevertheless, the fitting surface may not be subtle enough to represent surface of objects with the large curvatures where large errors are likely to be induced. Another area with relatively large errors is at the two sides of the zigzag road form dam crest to dam toe. The NURBS surface is about 0.006 m higher than the original points inside the zigzag road. On the contrary, the NURBS surface is approximately 0.006 m lower than the original points outside the zigzag road. This is also due to the large curvatures of the steps and the drainage ditch. As the larger errors don't exist in the main study reaches, the constructed NURBS surface can be applied to the analysis of the deformation distribution. Error analysis above proves that the NURBS fitting surface is useful way of surface reconstruction.

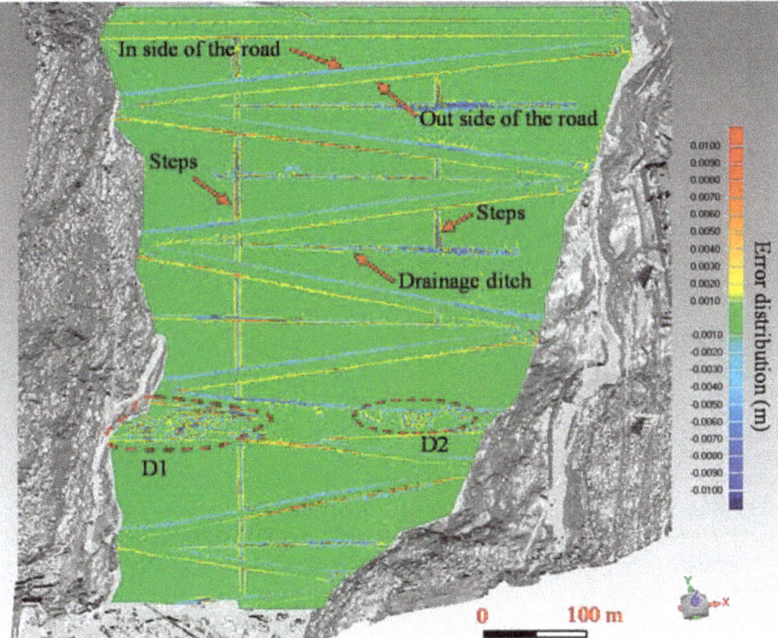

Figure 11. Error distribution in the NURBS modeling process.

6. Conclusions

The deformation distribution has a significant influence on the stability and safety of large artificial and natural structures. By taking full advantage of the TLS and NURBS technology, a new methodology of deformation monitoring and analysis is presented. Using the earth-rock dam as an example, the point cloud gained by TLS allows the detection of small deformation for its high accuracy and capability of target acquisition. The NURBS model based on the point cloud is characterized by high precision and high spatial resolution. Eventually, the holistic deformation distribution of the downstream face is shown in the cloud chart. The deformation monitoring achieved great success with the proposed methodology. The TLS has proved to be a valid solution for the deformation acquisition in three-dimensional space. The NURBS modeling technology is capable of dealing with a huge number of points and making use of them. In comparison with the traditional monitoring, the methodology that integrates TLS and NURBS technologies permits a better grasp of the deformation distribution of the large structures.

The millimeter level measurement requires that the data acquisition is performed with great patience and plenty of time. In the future works, data acquisition in the field should be optimized to reduce the field working time. Besides, the data processing is also time-consuming and complicated. Several procedures require personal experience and expertise. These are several potential areas in terms of improving the performance of the presented methodology, in which further research will be conducted.

Author Contributions: H.X. and H.L. conceived the original ideas, designed the experiments, processed the data and drafted the manuscript. S.Q. and J.Z. helped to write and edit the manuscript. J.Z. and X.Y. supervised the project.

Funding: This research was funded by the Key Project of the Power Construction Corporation of China (ZDZX-5) and the Youth Science and Technology Fund of Sichuan Province (2016JQ0011).

Conflicts of Interest: The authors declare no conflict of interest.

References

1. Xu, D.S.; Zhao, Y.M.; Liu, H.B.; Zhu, H.H. Deformation monitoring of metro tunnel with a new ultrasonic-based system. *Sensors* **2017**, *17*, 1758. [CrossRef] [PubMed]
2. Gan, W.; Hu, W.; Liu, F.; Tang, J.; Li, S.; Yang, Y. Bridge continuous deformation measurement technology based on fiber optic gyro. *Photonic Sens.* **2016**, *6*, 71–77. [CrossRef]
3. Angeli, M.G.; Pasuto, A.; Silvano, S. A critical review of landslide monitoring experiences. *Eng. Geol.* **2010**, *55*, 133–147. [CrossRef]
4. Zhao, Q.; Lin, H.; Jiang, L.; Chen, F.; Cheng, S. A study of ground deformation in the Guangzhou urban area with persistent scatterer interferometry. *Sensors* **2009**, *9*, 503–518. [CrossRef] [PubMed]
5. He, X.; Yang, G.; Ding, X.; Chen, Y. Application and evaluation of a GPS multi-antenna system for dam deformation monitoring. *Earth Planets Space* **2004**, *56*, 1035–1039. [CrossRef]
6. Lackner, S.; Lienhart, W.; Supp, G.; Marte, R. Geodetic and fibre optic measurements of a full-scale bi-axial compressional test. *Empire Surv. Rev.* **2016**, *48*, 86–93. [CrossRef]
7. Zhou, W.; Li, S.; Zhou, Z.; Chang, S. Remote sensing of deformation of a high concrete-faced rockfill dam using InSAR: A study of the Shuibuya dam, China. *Remote Sens.* **2016**, *8*, 255. [CrossRef]
8. Huang, Q.; Luzi, G.; Monserrat, O.; Crosetto, M. Ground-based synthetic aperture radar interferometry for deformation monitoring: A case study at geheyan dam, China. *J. Appl. Remote Sens.* **2017**, *11*, 036030. [CrossRef]
9. Milillo, P.; Perissin, D.; Salzer, J.T.; Lundgren, P.; Lacava, G.; Milillo, G.; Serio, C. Monitoring dam structural health from space: Insights from novel insar techniques and multi-parametric modeling applied to the pertusillo dam basilicata, Italy. *Int. J. Appl. Earth Obs. Geoinf.* **2016**, *52*, 221–229. [CrossRef]
10. Wang, T.; Perissin, D.; Rocca, F.; Liao, M.S. Three gorges dam stability monitoring with time-series insar image analysis. *Sci. China-Earth Sci.* **2011**, *54*, 720–732. [CrossRef]

11. Lichti, D.D.; Gordon, S.J. Error propagation in directly georeferenced terrestrial laser scanner point clouds for cultural heritage recording. In Proceedings of the FIG Working Week, Athens, Greece, 22–27 May 2004.

12. Vezočnik, R.; Ambrožič, T.; Sterle, O.; Bilban, G.; Pfeifer, N.; Stopar, B. Use of terrestrial laser scanning technology for long term high precision deformation monitoring. *Sensors* **2009**, *9*, 9873–9895. [CrossRef] [PubMed]

13. Rosser, N.J.; Petley, D.N.; Lim, M.; Dunning, S.A.; Allison, R.J. Terrestrial laser scanning for monitoring the process of hard rock coastal cliff erosion. *Q. J. Eng. Geol. Hydrogeol.* **2005**, *38*, 363–375. [CrossRef]

14. Sturzenegger, M.; Stead, D. Quantifying discontinuity orientation and persistence on high mountain rock slopes and large landslides using terrestrial remote sensing techniques. *Nat. Hazards Earth Syst. Sci.* **2009**, *9*, 267–287. [CrossRef]

15. Lato, M.; Diederichs, M.S.; Hutchinson, D.J.; Harrap, R. Optimization of LiDAR scanning and processing for automated structural evaluation of discontinuities in rock masses. *Int. J. Rock Mech. Min. Sci.* **2009**, *46*, 194–199. [CrossRef]

16. Cheng, Y.J.; Qiu, W.; Lei, J. Automatic extraction of tunnel lining cross-sections from terrestrial laser scanning point clouds. *Sensors* **2016**, *16*, 1648. [CrossRef] [PubMed]

17. Little, M.J. Slope monitoring strategy at PPRust open pit operations. In Proceedings of the International Symposium on Stability of Rock Slopes in Open Pit Mining & Civil Engineering, Johannesburg, South Africa, 3–6 April 2006.

18. Girardeau-Montaut, D.; Roux, M.; Marc, R.; Thibault, G. Change detection on points cloud data acquired with a ground laser scanner. In Proceedings of the ISPRS Workshop Laser Scanning, Enschede, The Netherlands, 12–14 September 2005.

19. Facundo, M.; Sapiro, G. Comparing point clouds. In Proceedings of the 2004 Eurographics/ACM SIGGRAPH Symposium on Geometry Processing, Nice, France, 8–10 July 2004; pp. 32–40.

20. Sun, W.; Bradley, C.; Zhang, Y.F.; Loh, H.T. Cloud data modelling employing a unified, non-redundant triangular mesh. *Comput. Aided Des.* **2001**, *33*, 183–193. [CrossRef]

21. Chang, M.C.; Leymarie, F.F.; Kimia, B.B. Surface Reconstruction from Point Clouds by Transforming the Medial Scaffold. *Comput. Vis. Image Understand.* **2009**, *113*, 1130–1146. [CrossRef]

22. Kuo, C.W.; Brierley, G.; Chang, Y.H. Monitoring channel responses to flood events of low to moderate magnitudes in a bedrock-dominated river using morphological budgeting by terrestrial laser scanning. *Geomorphology* **2015**, *235*, 1–14. [CrossRef]

23. Abellán, A.; Vilaplana, J.M.; Calvet, J.; Garcíasellés, D.; Asensio, E. Rockfall monitoring by Terrestrial Laser Scanning—Case study of the basaltic rock face at Castellfollit de la Roca (Catalonia, Spain). *Nat. Hazards Earth Syst. Sci.* **2011**, *11*, 829–841. [CrossRef]

24. Yue, D.; Wang, J.; Zhou, J.; Chen, X.; Ren, H. Monitoring slope deformation using a 3-D laser image scanning system: A case study. *Min. Sci. Tech.* **2010**, *20*, 898–903. [CrossRef]

25. Besl, P.J.; McKay, N.D. A method for registration of 3D shapes. *IEEE Trans. Pattern Anal. Mach. Intell.* **1992**, *14*, 239–256. [CrossRef]

26. Zhang, Z. Iterative point matching for registration of freeform curves and surfaces. *Int. J. Comput. Vis.* **1994**, *13*, 119–152. [CrossRef]

27. Versprille, K.J. Computer-Aided Design Applications of the Rational B-Spline Approximation Form. Ph.D. Thesis, Syracuse University, Syracuse, NY, USA, 1975.

28. Gonzálezaguilera, D.; Gómezlahoz, J.; Sánchez, J. A new approach for structural monitoring of large dams with a three-dimensional laser scanner. *Sensors* **2008**, *8*, 5866–5883. [CrossRef] [PubMed]

29. Meng, Y.; Cai, Z.; Xu, W.; Tian, B.; Zhou, J. A method for three-dimensional nephogram real-time dynamic visualization of safety monitoring data field in slope engineering. *Chin. J. Rock Mech. Eng.* **2012**, *31*, 3482–3490. (In Chinese)

30. Gruen, A.; Akca, D. Least squares 3D surface and curve matching. *ISPRS-J. Photogramm. Remote Sens.* **2005**, *59*, 151–174. [CrossRef]

31. Lemmens, M. Product survey 3D laser mapping. *GIM Int.* **2004**, *18*, 44–47.

32. Oppikofer, T.; Jaboyedoff, M.; Blikra, L.; Derron, M.H.; Metzger, R. Characterization and monitoring of the Åknes rockslide using terrestrial laser scanning. *Nat. Hazards Earth Syst. Sci.* **2009**, *9*, 1003–1019. [CrossRef]

33. Dąbek, P.B.; Patrzalek, C.; Ćmielewski, B.; Żmuda, R. The use of terrestrial laser scanning in monitoring and analyses of erosion phenomena in natural and anthropogenically transformed areas. *Cogent Geosci.* **2018**, *4*, 1437684. [CrossRef]

34. Lichti, D.D.; Jamtsho, S. Angular resolution of terrestrial laser scanners. *Photogramm. Rec.* **2006**, *21*, 141–160. [CrossRef]

35. Olsen, M.J.; Johnstone, E.; Driscoll, N.; Ashford, S.A.; Kuester, F. Terrestrial laser scanning of extended cliff sections in dynamic environments: Parameter analysis. *J. Surv. Eng.* **2009**, *135*, 161–169. [CrossRef]

Article

Adaptive Multiclass Mahalanobis Taguchi System for Bearing Fault Diagnosis under Variable Conditions

Ning Wang [1,2,3], Zhipeng Wang [1,2,3,*], Limin Jia [1,2,3,*], Yong Qin [1,2,3], Xinan Chen [1,2,3] and Yakun Zuo [1,2,3]

1   State Key Lab of Rail Traffic Control and Safety, Beijing Jiaotong University, Beijing 100044, China; 17114235@bjtu.edu.cn (N.W.); yqin@bjtu.edu.cn (Y.Q.); 15114217@bjtu.edu.cn (X.C.); 18120772@bjtu.edu.cn (Y.Z.)
2   National Engineering Laboratory for System Safety and Operation Assurance of Urban Rail Transit, Guangzhou 510000, China
3   Beijing Research Center of Urban Traffic Information Sensing and Service Technologies, Beijing Jiaotong University, Beijing 100044, China
*   Correspondence: zpwang@bjtu.edu.cn (Z.W.); lmjia@bjtu.edu.cn (L.J.); Tel.: +86-010-5168-4281(Z.W.); +86-010-5168-4639 (L.J.)

Received: 16 November 2018; Accepted: 19 December 2018; Published: 21 December 2018

Abstract: Bearings are vital components in industrial machines. Diagnosing the fault of rolling element bearings and ensuring normal operation is essential. However, the faults of rolling element bearings under variable conditions and the adaptive feature selection has rarely been discussed until now. Thus, it is essential to develop a practicable method to put forward the disposal of the fault under variable conditions. Considering these issues, this paper uses the method based on the Mahalanobis Taguchi System (MTS), and overcomes two shortcomings of MTS: (1) MTS is an effective tool to classify faults and has strong robustness to operating conditions, but it can only handle binary classification problems, and this paper constructs the multiclass measurement scale to deal with multi-classification problems. (2) MTS can determine important features, but uses the hard threshold to select the features, and this paper selects the proper feature sequence instead of the threshold to overcome the lesser adaptivity of the threshold configuration for signal-to-noise gain. Hence, this method proposes a novel method named adaptive Multiclass Mahalanobis Taguchi system (aMMTS), in conjunction with variational mode decomposition (VMD) and singular value decomposition (SVD), and is employed to diagnose the faults under the variable conditions. Finally, this method is verified by using the signal data collected from Case Western Reserve University Bearing Data Center. The result shows that it is accurate for bearings fault diagnosis under variable conditions.

Keywords: fault diagnosis; bearing; SVD; VMD; adaptive Multiclass Mahalanobis Taguchi System

1. Introduction

Rolling element bearings have wide applications in industrial machines and are one of the most critical components. If faults occur in bearings, equipment could be damaged and disasters might happen consequently. Therefore, it is essential to monitor the health conditions of bearings. The analysis of vibration signals has been a hot research pot and used to detect faults of bearings. It is crucial to recognize faults occurring in bearings and avoid fatal breakdowns as early as possible. For decades, many researchers have conducted extensive research on fault diagnosis. At present, the fault diagnosis methods are divided into model-based methods and data-driven methods. The model-based methods generally build on the physics of the process, generating the residuals between the measure process variables and estimates [1], such as Hidden Markov Modeling (HMM) which is successfully applied to bearing fault detection and diagnosis [2]. Autoregressive modelling [3] also has had excellent performance in bearing fault diagnosis. The accelerated degradation testing (ADT) [4] method is useful in fault diagnosis and lifespan prediction, and reference [5] presents a new approach using observer-based residual generation with no complicated design constraints to establish the relationship between the state estimation error and the fault signal. For the data-driven method, which is based on the historical data and does not need accurate mathematic and priori knowledge, it has wide applications in fault diagnosis. For example, Bayesian network is an excellent data-driven diagnosis method [6–8]. Machine learning algorithms, such as K-nearest Neighbor (KNN), Deep Convolutional Neural Networks (DCNN), and auto-encoders are effective in the fault diagnosis of various bearings [9–12]. There is also decision tree [13], Support Vector Machine (SVM) [14], and wavelet transform [15]. Artificial neural network (ANN) and Trace Ratio Linear Discriminant Analysis are other methods of diagnosing the bearings fault [16,17]. In addition, the vibration signal under the various operating conditions (especially in low rotational speed) is non-stationary and non-linear, the characteristic defect frequencies move continuously with the change of rotating speed, and is the same as the bearing that is going to break down. If the bearing progresses toward failure, the nonlinear features also start to be dominated by stochastic signal. With the vibration signal measured for diagnosing the bearing fault under variable condition, it will be difficult to diagnose the fault of bearing by using the traditional methods. Reference [18] proposed a signal selection scheme based upon two order tracking techniques from complicated non-stationary operational measured vibrations. Reference [19] reviewed features extraction methods and its application on bearing vibration signal and presents an empirical study of feature extraction methods in low rotational speed. Reference [20] proposes the estimation of instantaneous speed relative fluctuation (ISRF) in a vibration speed. Reference [21] proposes Stacked Convolutional Autoencoders (SCAE) together with DCNN in stationary and non-stationary speed operation. Graph-based rebalance semi-supervised learning (GRSSL) [22], weighted self-adaptive evolutionary extreme learning machine (WSaE-ELM) [23] and Singular Spectrum Analysis [24] are effective in diagnosing the fault under variable conditions.

However, although the aforementioned methods are effective for bearing faults diagnosis, the feature selection part is often non-adaptive or unexplainable. The conventional methods mainly extract and select features manually, which relies heavily on the experts' knowledge and experience. Since the signals acquired in the real world might be various in many different aspects, the features selected manually might be sensitive in the variations of operation conditions and import inevitable errors for fault diagnosis. The deep learning algorithms can extract features automatically and overcome this drawback. However, deep learnings are data-hungry and require plenty of training data which are hardly acquired in practice, especially the faulty data under different conditions. Besides, the features acquired by deep learnings are unexplainable. Therefore, the aforementioned methods can hardly diagnose the faults under various operation conditions in practice. In contrast, the Mahalanobis Taguchi System (MTS) is more robust than the other methods in various operating conditions [25].

MTS offers a tool to determine important features and optimize the system. It is a different form compared with the other classification methods, because this classification model of measurement scale is constructed by using the class samples. It is useful to diagnose bearing faults under various conditions because the different pattern could be identified by using the Mahalanobis distance (MD) and Taguchi method. In this paper, MD is used to calculate the distance of the correlations between the benchmark and others, and the distance could be measured without the volatility of data. The advantage of MD is that it takes into consideration the correlations between the features and this consideration is very important in pattern analysis, which is why MTS is suitable for bearing fault diagnosis under various conditions [26]. On the other hand, the Taguchi method is used to select features without manual intervention, which could improve the robustness of the algorithm. MTS also offers an effective tool for multivariate analysis [27], considering that the bearing faults can be classified according to locations, such as inner race, outer race and rolling element [28]. However, when the conventional MTS is used for bearing fault diagnosis, misclassifications might occur due to less adaptivity of the threshold configuration for signal-to-noise gain. During the feature selection, the threshold is normally set as a constant, which might result in the overfitting problem. If the threshold value is too large, some critical features might be eliminated. On the contrary, if the threshold value is too low, some useless or harmful feature might be selected. Therefore, if the threshold value does not march the training data sufficiently, misclassifications emerge.

To overcome this drawback, this paper presents a novel method named adaptive Multiclass Mahalanobis Taguchi system (aMMTS) for bearing fault diagnosis. This method employs the MTS for multi-classification by considering different conditions as different benchmarks respectively. The results are based on the minimum MDs between the data and each benchmark data, and the label of the data is determined to be consistent with the label of the benchmark data whose MD is minimum. The method can be described briefly as follows: Firstly, after the Mahalanobis space (MS) is constructed by the two-level orthogonal array of the Taguchi method, aMMTS calculates the MDs from the data to the benchmark data and obtains features' signal-to-noise ratios (SNRs) and gain values by using two-level orthogonal array. Secondly, the features are selected adaptively by recalculating the MDs via rearranging the order of features' gain values by ascending and descending. Therefore, the proposed method is able to select the best classification result according to the adaptive chosen sequence of features' SNRs instead of a hard threshold. Here, the sequence of SNR is determined by a function, which selects several maximum or minimum features to calculated MDs. Finally, a set of features with the best results is selected as the final feature vector. In this method, two different sets of training samples are employed to calculate the SNRs respectively and obtain the final feature vectors respectively. By the aforementioned improvement, the proposed aMMTS is capable to overcome the drawback of the conventional MTS and prevent the over-fitting problem. Therefore, the aMMTS is insensitive to the operation conditions and can be employed for bearing fault diagnosis.

Moreover, this method is combined with variational mode decomposition (VMD) [29] and singular value decomposition (SVD) to diagnose the faults. VMD is an entirely non-recursive algorithm, and is used to decompose the signal. It has been proven that due to the characteristics of nonlinear vibration in the bearings, VMD is more efficient than empirical mode decomposition (EMD) and Fourier transform (FT) under variable condition. SVD is used to extract the features. Therefore, VMD and SVD are employed in this paper.

The rest of this article is organized as follows: Section 2 introduces the algorithms involved in this paper. Section 3 illustrates the experiments to validate the proposed method. Section 4 is the conclusions.

2. Methodology

In this paper, the main steps of fault diagnosis are signal decomposition, feature extraction and fault detection. The detailed process and method are as follows:

Step1: The VMD in conjunction with wavelet denoising is employed to eliminate the noises and decompose the raw signals;

Step2: Extracting features from the decomposed signals by SVD;

Step3: The proposed aMMTS is employed for the fault diagnosis.

The steps of this method are shown in Figure 1.

2.1. VMD

After the wavelet denoising is used to estimate the noise of the raw signal, this paper employs VMD to decompose non-stationary signals. VMD can decompose a signal into different simple intrinsic mode functions, whose frequency center and bandwidth are band-limited and determined by iterative searching for the optimal solution of the variational model. The constrained formula is given as:

$$\min_{\mu_k,\omega_k}\left\{\sum_k||\partial_t[(\delta(t)+\frac{j}{\pi t})*\mu_k()t]|e^{-j\omega_k t}||_2^2\right\} \qquad s.t.\sum\mu_k = f \tag{1}$$

where μ_k is the sub-signals, ω_k represents the center frequency of sub-modes. The optimal solution can be solved as the minimization problem, which could be addressed by introducing a quadratic penalty and Lagrangian multipliers [30]:

$$L(\{\mu_k\}\{\omega_k\},\lambda) = \alpha\sum_k ||\partial_t[(\delta(t) + \tfrac{j}{\pi t}) * \mu_k(t)]|e^{-j\omega_k t}||_2^2 + ||f(t) - \sum_k \mu_k(t)||_2^2 + \langle\lambda(t), f(t) - \sum_k \mu_k(t)\rangle \tag{2}$$

α denotes the balancing parameter of the constraint.

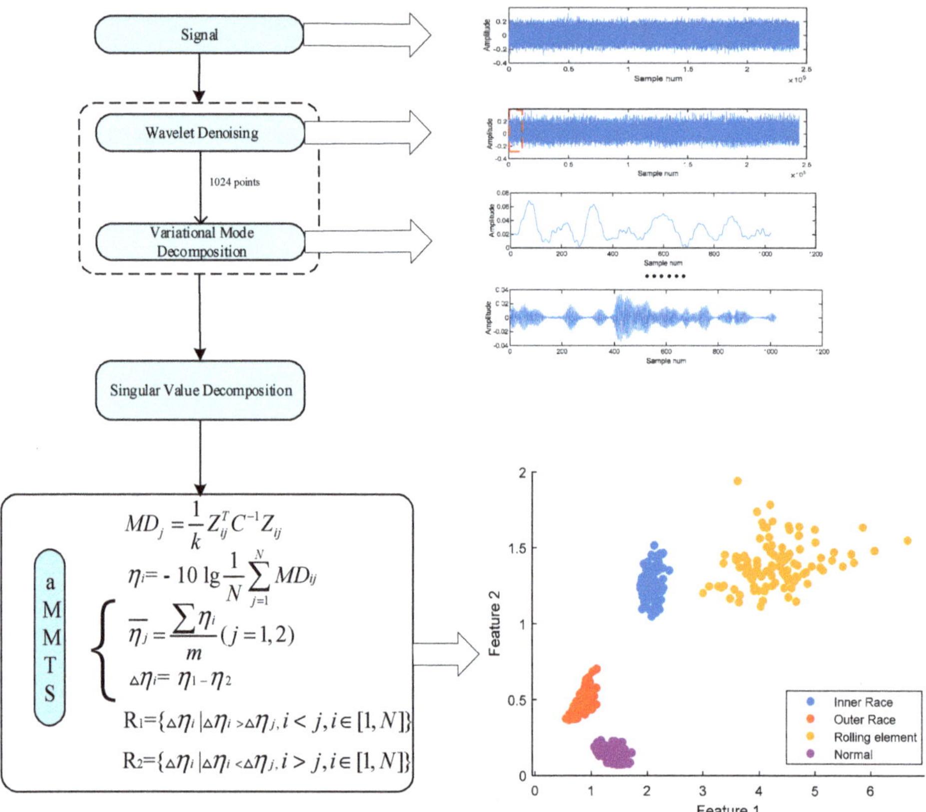

Figure 1. The scheme of the proposed fault diagnosis method.

All sub-signals μ_k are updated for all $\omega \geq 0$ as follow:

$$\hat{\mu}_k^{n+1} \leftarrow \frac{\hat{f} - \sum_{i<k}\hat{\mu}_i^{n+1} - \sum_{i>k}\hat{\mu}_i^n + \frac{\hat{\lambda}^n}{2}}{1 + 2\alpha(\omega - \omega_k^n)^2} \tag{3}$$

$\hat{\mu}_k^1$, $\hat{\omega}_k^1$ and $\hat{\lambda}^1$ are initialized to all zeroes.

All center frequency of sub-modes ω_k are updated as follow:

$$\omega_k \leftarrow \frac{\int_0^\infty \omega\left|\hat{\mu}_k^{n+1}(\omega)\right|^2 d\omega}{\int_0^\infty \left|\hat{\mu}_k^{n+1}(\omega)\right|^2 d\omega} \tag{4}$$

End for

$$\hat{\lambda}^{n+1} \leftarrow \hat{\lambda}^n + \tau(\hat{f} - \sum_k \hat{\mu}_k^{n+1}) \tag{5}$$

Until:

$$\frac{\sum_k \left\| \hat{\mu}_k^{n+1} - \hat{\mu}_k^n \right\|_2^2}{\|\hat{\mu}_k^n\|_2^2} < \varepsilon \tag{6}$$

2.2. SVD

After the signals are decomposed into several modes by VMD, the features are extracted from the modes by SVD, and can be constructed as a feature matrix. SVD is a powerful tool for feature extraction in linear algebra. According to SVD, the matrix could be decomposed as follow:

$$X = U\omega V^T \tag{7}$$

where X represents a $m \times n$ matrix. There are two orthogonal matrices: matrix $U(m \times m)$ and $V(n \times n)$, and a singular diagonal matrix $\omega(\omega_{ij} \neq 0, i = j$ and $\omega_{11} \geq \omega_{22} \geq \cdots \geq 0)$, the diagonal element $\omega_{11}, \omega_{22}, \cdots, \omega_{mm}$ is the singular value of X. U is called left singular vector, and the columns of U. V is called right singular vector. The columns of U and V especially are orthogonal to each other, and are base vector. To obtain more intrinsic information in the matrix, the singular vectors are selected. As a consequence, SVD is employed to decompose the eigenmatrix, and obtained the singular value vectors $(\omega_{11}, \omega_{22}, \cdots, \omega_{mm})$.

2.3. Mahalanobis–Taguchi System

Mahalanobis–Taguchi System is a pattern recognition method integrated by the MD, orthogonal table and other tools such as SNR that are proposed by Taguchi [31], who introduces the experimental design of the field SNR to pattern recognition, which can reduce the dimensions of data, and use the orthogonal table to construct the MS. The MSs are used to calculate the MDs of the experimental data, and the valid features are distinguished by the SNR. Then, the MSs are recalculated by using the valid features. Finally, the results are obtained. The calculation of the MD is described as follows:

2.3.1. Mahalanobis Distance

The MD is a method of using normal data to normalize the fault data to compute the average distance between points and groups using normal data, the calculation formula of MD is as follows:

$$MD_j = \frac{1}{k} Z_{ij}^T C^{-1} Z_{ij} \tag{8}$$

$$Z_{ij} = \left\{ z_{1j}, z_{2j}, z_{3j}, \ldots, z_{kj} \right\}, z_{ij} = \frac{x_{ij} - \overline{x}_i}{s_i} \tag{9}$$

MD_j represents the MD of the jth sample, k represents the number of the feature, x_{ij} represents the ith feature's value of the of the jth sample $\overline{x}_i$ represents the mean of ith feature, s_i represents the Standard deviation of ith feature, C^{-1} represents the inverse matrix of the correlation coefficient matrix.

2.3.2. Taguchi Method

In the Mahalanobis–Taguchi System, the MD measures the deviation of the test value from the normal value. The Taguchi method is able to select the features which have a larger contribution to identifying bearing faults, and then use the selected valid features to calculate the MD; the method is as below:

Orthogonal Array

Selecting the appropriate two-level orthogonal array, and then the k-original features obtained by VMD and SVD, are arranged into each column of the orthogonal array. In the orthogonal array, "1" indicates that the feature is selected, "2" indicates that the feature is not selected, and a MS is generated according to each row of the orthogonal array.

SNR and Its Gain

The main function of signal noise ratio is to select a valid feature, the calculation formula of generating SNR η_i in the i line based on the orthogonal array.

$$\eta_i = -10 \lg \frac{1}{N} \sum_{j=1}^{N} MD_{ij} \tag{10}$$

$j \in [1, N]$ represents the number of training samples.

η_i represents the recognition effects of the characteristic feature, the valid feature is selected by comparing the mean of SNR of each characteristic feature at two levels. The formula is as follow:

$$\overline{\eta_j} = \frac{\sum \eta_i}{m} j = 1,2 \tag{11}$$

$$\overline{\Delta \eta_j} = \sum \left(\frac{\overline{\eta_j}}{m} \right) \tag{12}$$

$j = 1, 2$ represents two levels, i represents the number of rows in the MS, $\overline{\eta_1}$ represents that in the level of '1', the recognition effects to identify abnormal conditions of using this feature. $\overline{\eta_2}$ represents that in the level of '2', the recognition effects to identify abnormal conditions of not using this feature. If $\overline{\Delta \eta_j}$ represents the SNR gain, $R = \left\{ \overline{\Delta \eta_j} \middle| \overline{\Delta \eta_i} > 0 \right\}$ that indicates that this feature is a valid feature, if $\overline{\Delta \eta_j} < 0$ that indicates that this feature is not a valid feature.

2.3.3. Adaptive Multiclass MTS

After SNR gain is calculated, to overcome the drawback of the conventional MTS during the threshold selection, this paper presents the adaptive multiclass MTS. There are several following improvements:

(1) Solving the multiple-classification problem. Selecting samples from each kind as the benchmark data. Then, the distances between other data and each benchmark data are calculated by MTS. Therefore, the label of the benchmark data with the minimum distance is selected as the label of the training data.

(2) Selecting the features adaptively. The feature sequence is selected several times, and the best fault recognition is the one which is minus MDs between it and benchmark. It solves the error problem caused by hard threshold selection.

(3) Avoiding the overfitting problem. Since the SNR gains are calculated by training samples, different training samples are employed to calculate the MDs and identify bearing faults, and the difference validation samples are set to validate the identified result and recalculate MDs.

This method is shown in Figure 2.

This adaptive multiclass MTS can be described as follow:

First, the data are labeled, the multi-class MSs are constructed, MDs between MSs are calculated and SNR gains are obtained. This step is shown as Figure 3; m is the number of samples, n is the number of features, and j is the number of the label. The training data are divided into three parts, and one of them is named as Benchmark. The data A_j and C_j represents one kind of data respectively (such as normal, fault of inner race, outer race, rolling element), and data B includes all kinds of data.

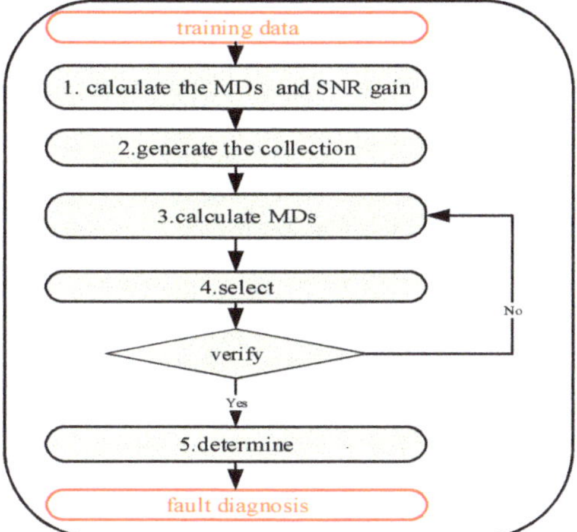

Figure 2. The step of Multiclass–Mahalanobis–Taguchi system (aMMTS).

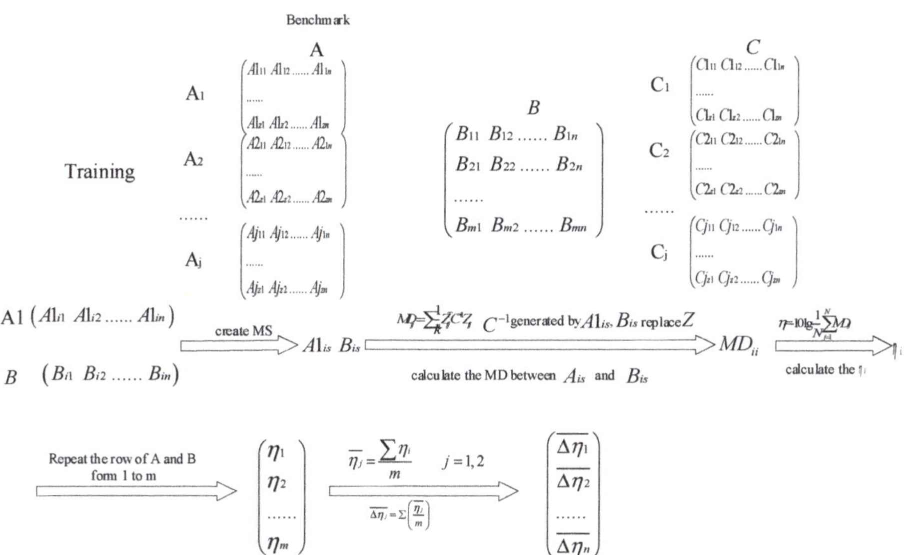

Figure 3. The first step of aMMTS.

Second, new sequences of feature parameters are generated by the sequence of features' SNR gains in ascending and descending order, then two collections are obtained from the above sequences based on the ascending or descending order, with the ascending and descending collection as follows:

$$r_k = \left(\overline{\triangle \eta_{ki}}, \ldots, \overline{\triangle \eta_{kj}} \middle| \overline{\triangle \eta_{ki}} > \overline{\triangle \eta_{kj}}, i < j, i \in [1, N] \right) \tag{13}$$

$$q_k = \left(\overline{\triangle \eta_{ki}}, \ldots, \overline{\triangle \eta_{kj}} \middle| \overline{\triangle \eta_{ki}} < \overline{\triangle \eta_{kj}}, i > j, i \in [1, N] \right) \tag{14}$$

$$R = \{ r_k | k \in [1, N] \} \tag{15}$$

$$Q = \{ q_k | k \in [1, N] \} \tag{16}$$

This step is shown in Figure 4:

$$R = \left\{ \begin{array}{l} \left(\overline{\Delta\eta_3} \right) \\ \left(\overline{\Delta\eta_3} \; \overline{\Delta\eta_7} \right) \\ \left(\overline{\Delta\eta_3} \; \overline{\Delta\eta_7} \; \overline{\Delta\eta_m} \right) \\ \cdots\cdots \\ \left(\overline{\Delta\eta_3} \; \overline{\Delta\eta_7} \; \overline{\Delta\eta_m} \cdots\cdots \overline{\Delta\eta_1} \right) \end{array} \right.$$

$$\begin{pmatrix} \overline{\Delta\eta_1} \\ \overline{\Delta\eta_2} \\ \cdots\cdots \\ \overline{\Delta\eta_n} \end{pmatrix} \xrightarrow{r_k=(\Delta\eta_{ki},\dots,\Delta\eta_{kj} \,|\, \Delta\eta_{ki} > \Delta\eta_{kj}, i < j, i \in [1,N])}$$

$$eg: \overline{\Delta\eta_3} > \overline{\Delta\eta_7} > \overline{\Delta\eta_n} > \cdots\cdots > \overline{\Delta\eta_i} > \overline{\Delta\eta_6} > \overline{\Delta\eta_1}$$

$$\xrightarrow{q_k=(\Delta\eta_{ki},\dots,\Delta\eta_{kj} \,|\, \Delta\eta_{ki} < \Delta\eta_{kj}, i > j, i \in [1,N])}$$

$$Q = \left\{ \begin{array}{l} \left(\overline{\Delta\eta_1} \right) \\ \left(\overline{\Delta\eta_1} \; \overline{\Delta\eta_6} \right) \\ \left(\overline{\Delta\eta_1} \; \overline{\Delta\eta_6} \; \overline{\Delta\eta_i} \right) \\ \cdots\cdots \\ \left(\overline{\Delta\eta_1} \; \overline{\Delta\eta_6} \; \overline{\Delta\eta_i} \cdots\cdots \overline{\Delta\eta_3} \right) \end{array} \right.$$

Figure 4. The second step of aMMTS.

Third, and the positions of feature's SNR gain are the same as the positions of corresponding features in the sequences, the features that are used to recalculate the MDs are selected by the corresponding feature's SNR gain. This step is shown in Figure 5. The MDs that are between each kind of A and Ci are calculated.

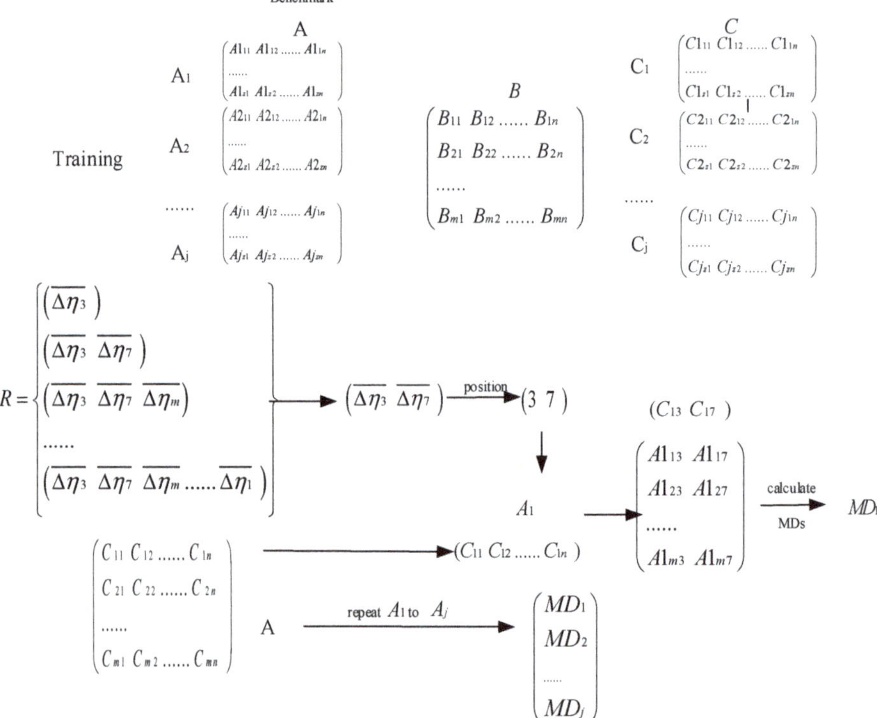

Figure 5. The third step of aMMTS.

Forth, the proper sequence of SNR gain is chosen by a function, which is according to the minimum MD. If two labels of data corresponding to minimum MD are the same, the recognition result is right, and accumulate the number of right recognition results. S is the recognition result. This step is shown in Figure 6.

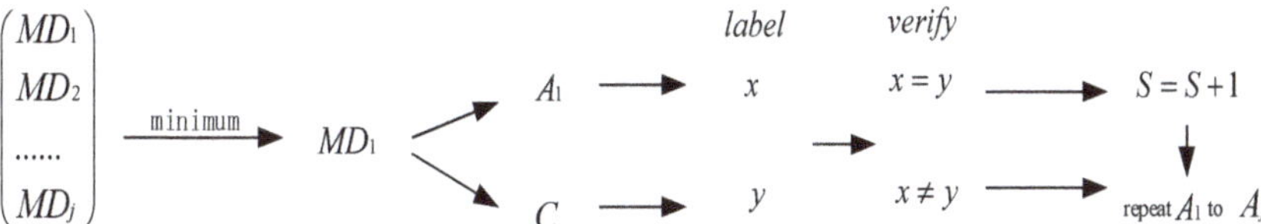

Figure 6. The forth step of aMMTS.

Finally, the recognition result is verified if there is a unique optimal recognition result, and the SNR gain's position in the sequence is the feature's order. If there are several optimal recognition results, repeat step 3 to recalculate the result. Afterwards, a set of sequences with the best recognition effect is determined as the feature sequence, which solves the self-adaptation problem of thresholds.

3. Results

In this paper, the experimental data are from Case Western Reserve University Bearing Data Center. This experiment involved three different faults that occurred on three components: inner race, outer race and rolling element. The vibration signals were acquired under four different speeds: 1797 r/min, 1772 r/min, 1750 r/min, and 1730 r/min, and the sampling frequency was set to 12 kHz. To demonstrate the aMMTS, this study randomly selected the data in the dataset under the defect of 0.07 inches. The number of samples are shown in Table 1.

Table 1. The number of samples.

	Label	Motor Load (hp)	Speed (r/min)	Training Samples BenchMark	Group A	Group B	Validation Samples	Test Samples
Inner Race	1	0 (0W)	1797	27	27	29	27	30
	1	1 (735W)	1772	27	27	29	27	30
	1	2 (1470W)	1750	27	27	29	27	30
	1	3 (2205W)	1730	27	27	29	27	30
Outer Race	2	0 (0W)	1797	27	27	29	27	30
	2	1 (735W)	1772	27	27	29	27	30
	2	2 (1470W)	1750	27	27	29	27	30
	2	3 (2205W)	1730	27	27	29	27	30
Rolling Element	3	0 (0W)	1797	27	27	29	27	30
	3	1 (735W)	1772	27	27	29	27	30
	3	2 (1470W)	1750	27	27	29	27	30
	3	3 (2205W)	1730	27	27	29	27	30
Normal	0	0 (0W)	1797	27	27	29	27	30
	0	1 (735W)	1772	27	27	29	27	30
	0	2 (1470W)	1750	27	27	29	27	30
	0	3 (2205W)	1730	27	27	29	27	30

There were 2192 samples; 548 for inner race, 548 for outer race, 548 for rolling element and 548 for normal. The data are divided into three parts: training data, validation data and test data. In order to avoid the overfitting caused by the training data, training samples were used to construct MS, generate the SNR gain and calculate MDs by using the sequences of SNR gains, and were divided into three parts, with one of the parts set as benchmark group. To avoid the over-fitting problem, group A was used to construct MS and generate the SNR gain, and group B were used to calculate the MDs with the sequences of SNR gain and identify faults.

Validation samples were used to verify the recognition result if there exists the same minimum MDs, and the sequence was selected according to the best result.

Test samples were used to validate the proposed method.

3.1. Signal Decomposition by Using VMD and Wavelet Denosing

Above all, this study employed wavelet denoising to remove the noise from the raw signals. First, the Daubechies 5 (db5) was used to decompose the signal, and obtained the wavelet decomposition vector and the bookkeeping vector. Second, thresholds wavelet coefficient was calculated by setting the detail vector which would be compressed as [1–3] and the vector which is the corresponding percentages of lower coefficients as [100,90,80], and using the wavelet decomposition vector and the bookkeeping vector. Lastly, the thresholds, Daubechies 5 (db5) and decomposed signals were used to reconstruct the denoising signal. Then the VMD was used to decompose the signal, and was needed to give the preset IMF component number K and penalty parameter α which constrained the moderate bandwidth. The value of α toke the default value 1024, the value of K was 8. An example is shown in Figure 7.

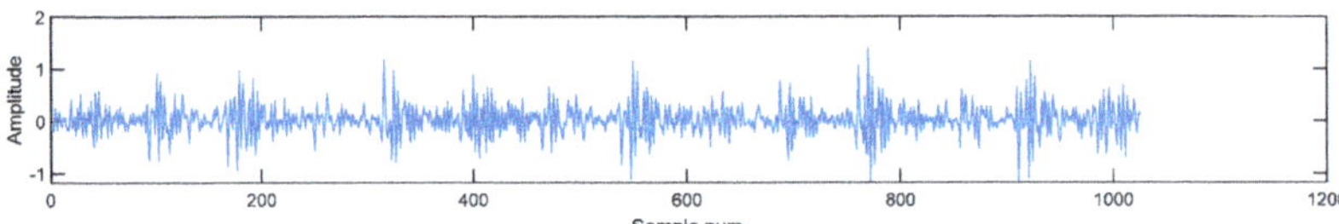

Figure 7. The intercepted signal.

IMFs are as shown in Figure 8.

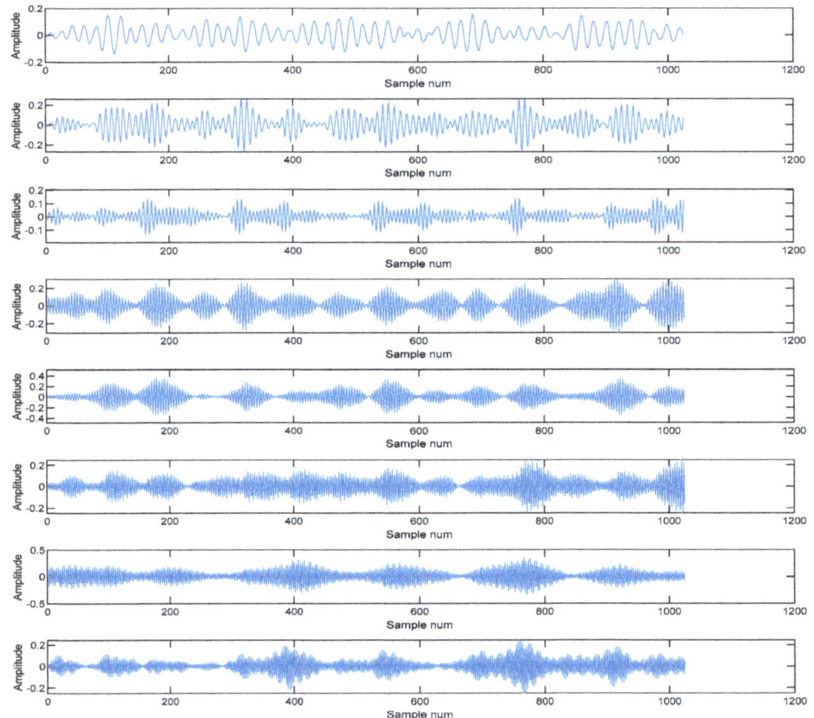

Figure 8. The IMFs.

3.2. Feature Extraction by Using SVD

SVD was used to analyze the IMFs. After the signal decomposition, the IMF matrix was decomposed by SVD, and obtained singular value vectors. The singular value vectors were considered as features and formed the feature matrix. Then, the feature matrix was used to diagnose the fault by aMMTS. To avoid the over-fitting problem, the features were divided into training samples, validation samples and test samples. The features of the above IMFs of those were shown in Table 2.

Table 2. The feature of IMFs.

	IMF1	IMF2	IMF3	IMF4	IMF5	IMF6	IMF7	IMF8
Features	2.019	1.968	1.836	1.582	1.491	1.459	1.264	1.168

The features obtained by SVD are shown in Table 3.

Table 3. The features of decomposed signals.

	Features							
	1	2	3	4	5	6	7	8
Normal	1.554	1.116	0.848	0.479	0.412	0.324	0.214	0.095
	1.472	1.230	0.473	0.425	0.292	0.258	0.197	0.083
	1.094	1.034	0.897	0.428	0.321	0.270	0.185	0.082
	1.173	0.939	0.931	0.683	0.387	0.287	0.229	0.080
Inner Race	2.020	1.968	1.836	1.582	1.491	1.459	1.264	1.168
	2.034	1.993	1.916	1.603	1.583	1.306	1.195	0.913
	2.115	2.023	1.918	1.763	1.570	1.410	1.376	1.143
	1.941	1.841	1.798	1.592	1.482	1.430	1.393	1.151
Rolling Element	0.811	0.793	0.667	0.577	0.562	0.542	0.451	0.223
	0.932	0.702	0.585	0.583	0.520	0.505	0.464	0.408
	0.740	0.657	0.556	0.515	0.501	0.487	0.435	0.356
	0.968	0.773	0.686	0.623	0.591	0.528	0.476	0.467
Outer Race	5.349	4.209	3.833	3.772	3.146	1.943	1.469	1.163
	5.312	4.468	4.003	3.505	2.988	2.751	1.416	1.145
	4.141	3.274	3.024	2.945	2.440	1.743	1.631	1.019
	3.334	2.901	2.560	2.382	2.036	1.697	1.236	0.941

3.3. Fault Diagnosis Using aMMTS

After the feature extraction, the aMMTS was used to identify and diagnose fault modes. The steps of aMMTS are as follow:

Firstly, the MS of training and benchmark were constructed, the eight-factor and two-level orthogonal array is shown in Table 4, and the MS based on Table 2 is shown in Table 5;

Table 4. The eight-factor and two-level orthogonal array.

	A	B	C	D	E	F	G	H
1	1	1	1	1	1	1	1	1
2	1	1	1	1	1	2	2	2
3	1	1	2	2	2	1	1	1
4	1	2	1	2	2	1	2	2
5	1	2	2	1	2	2	1	2
6	1	2	2	2	1	2	2	1
7	2	1	2	2	1	1	2	2
8	2	1	2	1	2	2	2	1
9	2	1	1	2	2	2	1	2

Table 5. The Mahalanobis space (MS) based on the inner race.

	A	B	C	D	E	F	G	H
1	2.020	1.968	1.836	1.582	1.491	1.459	1.264	1.168
2	2.020	1.968	1.836	1.582	1.491			
3	2.020	1.968				1.459	1.264	1.168
4	2.020		1.836			1.459		
5	2.020			1.582			1.264	
6	2.020				1.491			1.168
7		1.968			1.491	1.459		
8		1.968		1.582				1.168
9		1.968	1.836				1.264	

Secondly, the MD was calculated, and SNR gain was also obtained by benchmark samples and training samples. The SNR gain is shown in Table 6;

Table 6. The signal-to-noise ratio (SNR) gain of features.

	Features							
	1	2	3	4	5	6	7	8
Normal	2.512	−0.471	0.390	0.726	0.789	−0.049	1.519	1.469
Inner Race	2.512	−0.471	0.390	0.726	0.789	−0.049	1.519	1.469
Rolling element	2.397	0.137	1.007	0.432	1.078	0.482	1.283	0.770
Outer Race	2.439	0.970	3.801	−0.104	0.676	2.216	1.501	−1.809

Thirdly, the MDs between the benchmark samples and validation samples were calculated by using the ascending and descending order of SNR;

Fourthly, the validation samples were used to verify the correctness of feature selection which existed more than one smallest MD;

Fifthly, the best sequence was chosen and set as the sequence of features.

Lastly, the best sequence was used to identify the test samples. Took the benchmark is outer race as the example shown in Figure 9.

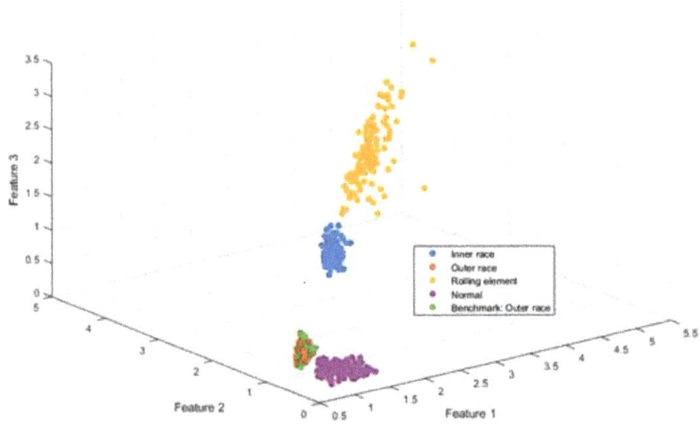

Figure 9. The classification result of outer race.

Finally, the test sample was used to test the result of the method, and the benchmarks were inner race, rolling element, outer race and normal. The results are shown in Table 7 and the MDs between benchmark and test sample are shown in Figure 10.

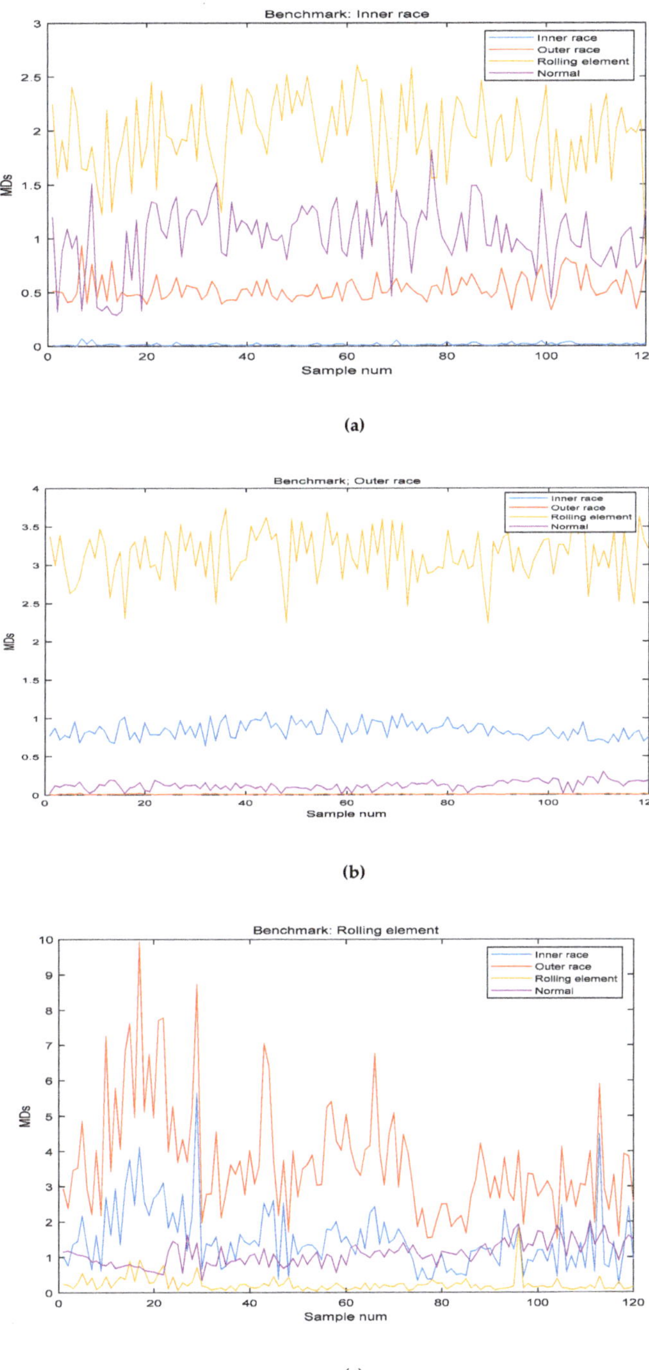

(a)

(b)

(c)

Figure 10. *Cont.*

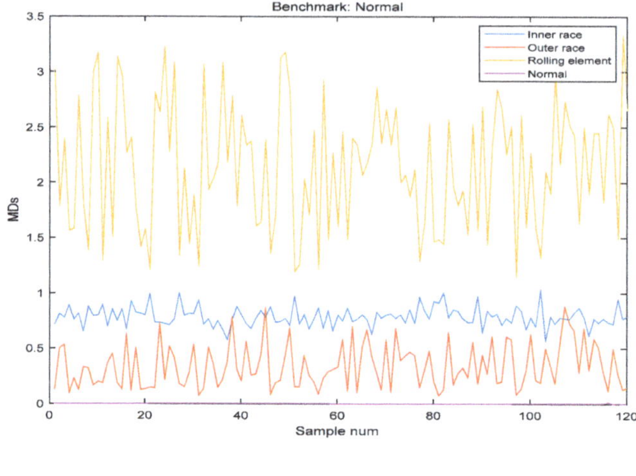

(d)

Figure 10. The Mahalanobis distances (MDs) between benchmark and testing data: (**a**) The MDs between benchmark (Inner race) and testing data; (**b**) The MDs between benchmark (Outer race) and testing data; (**c**) The MDs between benchmark (Rolling element) and testing data; (**d**) The MDs between benchmark (Normal) and testing data.

Table 7. The recognition result.

	Inner Race	**Outer Race**	**Rolling Element**	**Normal**	**Total**
Result	100%	99.16%	95%	100%	98.54%

As shown in Table 7 and Figure 10, this method accurately classified and diagnosed the fault of the bearing by using the different benchmarks. The recognition results of normal and outer race reached 100%. However, it is not accurate enough to diagnose the fault of inner race and rolling element. However, in the normal and the fault of inner race, it is effective in industrial application.

4. Discussion

Rolling element bearings are one of the most frequently used components in rotating machineries. This paper presents the method based on the wavelet denoising VMD-SVD-aMMTS to diagnose the fault of bearings under the variable conditions. Firstly, VMD is used to decompose the signal. Secondly, SVD is used to extract the feature. The adaptive aMMTS uses the feature sequences and multi-benchmarks to overcome the drawback of MTS for adaptive feature selection, multi-classification and over-fitting. The experimental result shows that the method could accurately diagnose faults effectively.

However, in the actual situation, there is an imbalance between fault data and normal data. In this method, aMMTS lacks research on the imbalanced study. The absence of faulty data may create a new problem, such over-fitting. Therefore, additional experiments under imbalanced data should be done to improve the method.

Author Contributions: N.W. collected and analyzed the data, made charts and diagrams, conceived and performed the experiments and wrote the paper; Z.W. and L.J. conceived the structure, provided guidance and modified the manuscript; X.C. analyzed the data and contributed analysis tools. Y.Q. provided guidance. Y.Z. revised the reviews.

Funding: This research was funded by National Key R&D Program of China grant number (No.2016YFB1200402).

Conflicts of Interest: The authors declare no conflict of interest.

References

1. Tidriri, K.; Chatti, N.; Verron, S.; Tiplica, T. Bridging data-driven and model-based approaches for process fault diagnosis and health monitoring: A review of researches and future challenges. *Annu. Rev. Control* **2016**, *42*, 63–81. [CrossRef]
2. Xin, G.; Hamzaoui, N.; Antoni, J. Semi-automated diagnosis of bearing faults based on a hidden Markov model of the vibration signals. *Measurement* **2018**, *127*, 141–166. [CrossRef]
3. Al-Bugharbee, H.; Trendafilova, I. A fault diagnosis methodology for rolling element bearings based on advanced signal pretreatment and autoregressive modelling. *J. Sound Vib.* **2016**, *369*, 246–265. [CrossRef]
4. Park, B.; Jeong, H.; Huh, H.; Kim, M.; Lee, S. Experimental study on the life prediction of servo motors through model-based system degradation assessment and accelerated degradation testing. *J. Mech. Sci. Technol.* **2018**, *32*, 5105–5110. [CrossRef]
5. Jeong, H.; Park, B.; Park, S.; Min, H.; Lee, S. Fault detection and identification method using observer-based residuals. *Reliab. Eng. Syst. Saf.* **2018**. [CrossRef]
6. Cai, B.; Huang, L.; Xie, M. A data-driven fault diagnosis methodology in three-phase inverters for PMSM drive systems. *IEEE Trans. Power Electron.* **2017**, *32*, 5590–5600. [CrossRef]
7. Cai, B.; Huang, L.; Xie, M. Bayesian Networks in Fault Diagnosis. *IEEE Trans. Ind. Inform.* **2017**, *13*, 2227–2240. [CrossRef]
8. Cai, B.; Kong, X.; Liu, Y.; Lin, J.; Yuan, X.; Xu, H.; Ji, R. Application of Bayesian networks in reliability evaluation. *IEEE Trans. Ind. Inform.* **2018**. [CrossRef]
9. Chen, Z.; Deng, S.; Chen, X.; Li, C.; Sanchez, R.-V.; Qin, H. Deep neural networks-based rolling bearing fault diagnosis. *Microelectron. Reliab.* **2017**, *75*, 327–333. [CrossRef]
10. Pandya, D.H.; Upadhyay, S.H.; Harsha, S.P. Fault diagnosis of rolling element bearing with intrinsic mode function of acoustic emission data using APF-KNN. *Expert Syst. Appl.* **2013**, *40*, 4137–4145. [CrossRef]
11. Shao, H.; Jiang, H.; Lin, Y.; Li, X. A novel method for intelligent fault diagnosis of rolling bearings using ensemble deep auto-encoders. *Mech. Syst. Signal Process.* **2018**, *102*, 278–297. [CrossRef]
12. Jia, F.; Lei, Y.; Guo, L.; Lin, J.; Xing, S. A neural network constructed by deep learning technique and its application to intelligent fault diagnosis of machines. *Neurocomputing* **2018**, *272*, 619–628. [CrossRef]
13. Jagath Sri Lal Senanayaka, H.V.K.; Kjell, G. Robbersmyr Towards online bearing fault detection using envelope analysis of vibration signal and decision tree classification algorithm. In Proceedings of the 2017 20th International Conference on Electrical Machines and Systems (ICEMS), Sydney, Australia, 11–14 August 2017. [CrossRef]
14. Rama Krishna, K.; Ramachandran, K.I. Machinery Bearing Fault Diagnosis Using Variational Mode Decomposition and Support Vector Machine as a Classifier. *IOP Conf. Ser. Mater. Sci. Eng.* **2018**, *310*, 012076. [CrossRef]
15. Yan, R.; Gao, R.X.; Chen, X. Wavelets for fault diagnosis of rotary machines: A review with applications. *Signal Process.* **2014**, *96*, 1–15. [CrossRef]
16. Ben Ali, J.; Fnaiech, N.; Saidi, L.; Chebel-Morello, B.; Fnaiech, F. Application of empirical mode decomposition and artificial neural network for automatic bearing fault diagnosis based on vibration signals. *Appl. Acoust.* **2015**, *89*, 16–27. [CrossRef]
17. Muruganatham, B.; Sanjith, M.A.; Krishnakumar, B.; Satya Murty, S.A.V. Roller element bearing fault diagnosis using singular spectrum analysis. *Mech. Syst. Signal Process.* **2013**, *35*, 150–166. [CrossRef]
18. Ke, F.; KeSheng, W.; Mian, Z.; Qing, N.; Ming, J.Z. A diagnostic signal selection scheme for planetary gearbox vibration monitoring under non-stationary operational conditions. *Meas. Sci. Technol.* **2017**, *28*, 035003.
19. Caesarendra, W.; Tjahjowidodo, T. A Review of Feature Extraction Methods in Vibration-Based Condition Monitoring and Its Application for Degradation Trend Estimation of Low-Speed Slew Bearing. *Machines* **2017**, *5*, 21. [CrossRef]
20. Francois Combet, R.Z. A new method for the estimation of the instantaneous speed relative fluctuation in a vibration signal based on the short time scale transform. *Mech. Syst. Signal Process.* **2009**, *23*, 16. [CrossRef]
21. Cabrera, D.; Sancho, F.; Li, C.; Cerrada, M.; Sánchez, R.V.; Pacheco, F.; de Oliveira, J.V. Automatic feature extraction of time-series applied to fault severity assessment of helical gearbox in stationary and non-stationary speed operation. *Appl. Soft Comput.* **2017**, *58*, 53–64. [CrossRef]

22. Chen, X.; Wang, Z.; Zhang, Z.; Jia, L.; Qin, Y. A Semi-Supervised Approach to Bearing Fault Diagnosis under Variable Conditions towards Imbalanced Unlabeled Data. *Sensors* **2018**, *18*, 2097. [CrossRef]

23. Wang, Z.; Jia, L.; Kou, L.; Qin, Y. Spectral Kurtosis Entropy and Weighted SaE-ELM for Bogie Fault Diagnosis under Variable Conditions. *Sensors* **2018**, *18*, 1705. [CrossRef]

24. Ciabattoni, L.; Ferracuti, F.; Freddi, A.; Monteriù, A. Statistical Spectral Analysis for Fault Diagnosis of Rotating Machines. *IEEE Trans. Ind. Electron.* **2018**, *65*, 4301–4310. [CrossRef]

25. Chao-Ton Su, Y.-H.H. An Evaluation of the Robustness of MTS for Imbalanced Data. *IEEE Trans. Knowl. Data Eng.* **2007**, *19*, 12. [CrossRef]

26. Cudney, E.A.; Hong, J.; Jugulum, R.; Paryani, K.; Ragsdell, K.M.; Taguchi, G. An Evaluation of Mahalanobis-Taguchi System and Neural Network for Multivariate Pattern Recognition. *J. Ind. Syst. Eng.* **2007**, *1*, 12.

27. Soylemezoglu, A.; Jagannathan, S.; Saygin, C. Mahalanobis Taguchi System (MTS) as a Prognostics Tool for Rolling Element Bearing Failures. *J. Manuf. Sci. Eng.* **2010**, *132*. [CrossRef]

28. Idriss El-Thalji, E.J. A summary of fault modelling and predictive health monitoring of rolling element bearings. *Mech. Syst. Signal Process.* **2015**, *60*, 252–272. [CrossRef]

29. Dragomiretskiy, K.; Zosso, D. Variational Mode Decomposition. *IEEE Trans. Signal Process.* **2014**, *62*, 531–544. [CrossRef]

30. Wang, Y.; Markert, R.; Xiang, J.; Zheng, W. Research on variational mode decomposition and its application in detecting rub-impact fault of the rotor system. *Mech. Syst. Signal Process.* **2015**, *60–61*, 243–251. [CrossRef]

31. Taguchi, G.; Konishi, S.; Konishi, S. *Taguchi Methods: Orthogonal Arrays and Linear Graphs. Tools for Quality Engineering*; American Supplier Institute: Dearborn, MI, USA, 1987.

Article

Variational Bayesian Based Adaptive Shifted Rayleigh Filter for Bearings-Only Tracking in Clutters

Jing Hou [ID], Yan Yang * and Tian Gao

School of Electronic and Information, Northwestern Polytechnical University, Xi'an 710072, China; jhou0825@nwpu.edu.cn (J.H.); tiangao@nwpu.edu.cn (T.G.)
* Correspondence: yangyan7003@nwpu.edu.cn

Received: 11 February 2019; Accepted: 26 March 2019; Published: 28 March 2019

Abstract: This paper considers bearings-only target tracking in clutters with uncertain clutter probability. The traditional shifted Rayleigh filter (SRF), which assumes known clutter probability, may have degraded performance in challenging scenarios. To improve the tracking performance, a variational Bayesian-based adaptive shifted Rayleigh filter (VB-SRF) is proposed in this paper. The target state and the clutter probability are jointly estimated to account for the uncertainty in clutter probability. Performance of the proposed filter is evaluated by comparing with SRF and the probability data association (PDA)-based filters in two scenarios. Simulation results show that the proposed VB-SRF algorithm outperforms the traditional SRF and PDA-based filters especially in complex adverse scenarios in terms of track continuity, track accuracy and robustness with a little higher computation complexity.

Keywords: bearings-only tracking; clutter; variational Bayesian; Shifted Rayleigh Filter

---

1. Introduction

Bearings-only target tracking is to estimate the current position and velocity of a target using only the noise-corrupted bearing measurements from one or multiple observer platforms. It is an important tracking problem that arises in both military and civilian applications, such as underwater sonar tracking, bistatic radar, air traffic control and computer vision [1].

Because of the intrinsic nonlinearities in the measurement models, it is difficult to acquire an optimal solution of this problem. Several suboptimal algorithms have been developed for bearings-only tracking in the literature. The extended Kalman filter (EKF) [2] in the Cartesian coordinate system is an early attempt. However, it is easy to diverge. To improve the stability of the EKF, the bearings-only tracking problem was formulated in modified polar coordinates, resulting in the modified polar coordinate EKF (MPEKF) [3]. However, it requires good initialization to guarantee convergence. The well-known pseudo-linear estimator (PLE) [4] was also developed to solve the bearings-only tracking problem. However, it gives a biased estimate at long ranges. In recent years, some bias compensation techniques were developed to improve the performance of PLE [5–7]. In addition, more sophisticated nonlinear Kalman filtering algorithms, such as unscented Kalman filter (UKF) [8,9], cubature Kalman filter (CKF) [1] and particle filter (PF) [10,11] were applied for bearings-only target tracking. PF can provide good performance but at the price of heavy computation load. Noteworthily, Clark et al. [12,13] proposed a novel shifted Rayleigh filter (SRF) for bearings-only tracking, which is still based on the approximation of conditional expectations but with novel feature which is performing a calculation to exploit the essential structure of the nonlinearities in a new way. It is shown to exhibit similar performance to the PF in certain challenging scenarios with much lower computational complexity [14].

However, these algorithms do not consider the impact of clutter which makes the bearings-only tracking problem more intractable. Apparently, the classical treatment of clutter in target tracking problem can be extended to the bearings-only tracking problem. For example, Reference [15] integrated the maximum entropy fuzzy probabilistic data association (MEFPDA) with the square-root cubature Kalman filter (SCKF) to deal with the clutter in bearings-only tracking. Clark et al. also included the effect of clutter measurements into the SRF algorithm in [12]. However, in the classical algorithms, the clutter probability is usually assumed known and constant, which maybe time-varying or hard to determine in advance especially in adverse scenarios. Use of incorrect value of the clutter probability may lead to track accuracy degradation even track loss. A straight-forward idea is to account for the unknown clutter probability in the process of estimation of the state. That is, we need to solve the problem of bearings-only target tracking with uncertain parameter of clutter probability.

As we know, the Bayesian approach is the most general approach of solving the problem with uncertain parameters. However, it is not trivial to get the analytical solution for most Bayesian approaches due to complex nonlinear probability density function or high dimension of integration. Recently, the variational Bayesian (VB)-based adaptive filters [16–18] have drawn extensive attention, which utilize a new simpler, analytically tractable distribution to approximate the true posterior distribution so that avoiding direct calculations of complex integrals. Its adaptive strategy has a strong ability of tracking time-varying parameters. Therefore, we adopt the VB method to jointly estimate the target state and the clutter probability within the framework of SRF in this paper for bearings-only tracking in clutters. By establishing a conjugate exponential model for clutter probability and data association indicator, the proposed filter is derived using the iterative filtering framework. The tracking performance of the proposed VB-SRF is evaluated by comparing with SRF, PDA-SCKF [15] and MEFPDA-SCKF [15] via two simulation examples. It shows that the proposed filter outperforms the traditional SRF and two PDA-SCKF-based filters in complex mismatched scenarios in terms of track continuity and track accuracy but at the cost of higher computation complexity.

The remainder of the paper is organized as follows. Section 2 gives the problem formulation. In Section 3, the variational Bayesian filtering is described. Section 4 derives the VB-based adaptive SRF. Section 5 provides simulation results and performance evaluation of the proposed approach, followed by the conclusions in Section 6.

2. The Shifted Rayleigh Filter Algorithm

2.1. The Bearing Model

Considering the shifted Rayleigh filter (SRF) for bearings-only tracking in R^2, the state equation and the measurement equation are described as [13]:

$$\mathbf{x}_k = \mathbf{F}_{k-1}\mathbf{x}_{k-1} + \mathbf{u}^s_{k-1} + \mathbf{v}_{k-1} \tag{1}$$

$$\mathbf{y}_k = \mathbf{H}_k\mathbf{x}_k + \mathbf{u}^m_k + \mathbf{w}_k$$
$$\mathbf{b}_k = \Pi(\mathbf{y}_k) \tag{2}$$

where, $\mathbf{x}_k$ is the state vector which describes the position and velocity of the target; $\mathbf{b}_k$ is the noisy bearing measurement. Π denotes the projection of the plane onto the unit circle. That is taking a 2-vector $\mathbf{y}_k$ into its normalized form $\mathbf{y}_k/\|\mathbf{y}_k\|$. Then $\mathbf{b}_k = (sin\theta_k, cos\theta_k)^T$, where θ_k is the bearing of the target position relative to the sensor platform. $\mathbf{F}_{k-1}$ and $\mathbf{H}_k$ are the state transition matrix and measurement matrix, respectively. $\mathbf{u}^s_{k-1}$ and $\mathbf{u}^m_k$ are the inputs to the system to increase the versatility of the model, for example, to reflect known perturbations to the dynamics and changes in sensor location; $\mathbf{v}_{k-1}$ is the Gaussian process noise with zero mean and covariance $\mathbf{Q}_v$, and $\mathbf{w}_k$ is the Gaussian measurement noise with zero mean and covariance $\mathbf{Q}_w$ and independent of $\mathbf{v}_k$.

The unusual point in the measurement model is that the noise $\mathbf{w}_k$ is modelled as additive noise present in an "augmented" measurement, $\mathbf{y}_k$, of the Cartesian coordinates of target relative to the sensor platform, which is projected onto the plane to generate the actual bearing measurement $\mathbf{b}_k$. It is different with the traditional "angle plus white noise" model, expressed as

$$\theta_k = arctan(d_1/d_2) + \epsilon_k \tag{3}$$

where, d_1, d_2 are the components of the displacement vector $\mathbf{d}_k = \mathbf{H}_k\mathbf{x}_k + \mathbf{u}_k^m$. ϵ_k is the sensor noise with Gaussian distribution $\mathcal{N}(0, \sigma^2)$ and independent of the displacement vector $\mathbf{d}_k$.

Actually, as explained in [13], the shifted Rayleigh bearing model (2) can be related with the traditional model (3) by making a variant on the shifted Rayleigh noise model

$$\begin{aligned} \mathbf{y}'_k &= \mathbf{d}_k + ||\mathbf{d}_k||\mathbf{e}' \\ \mathbf{b}'_k &= \Pi(\mathbf{y}'_k) \end{aligned} \tag{4}$$

where, $\mathbf{e}'$ is an $\mathcal{N}(\mathbf{0}, \sigma^2\mathbf{I}_{2\times2})$ distributed random variable. The only difference with model (2) is the noise term $\mathbf{w}_k$ used to construct the augmented measurement $\mathbf{y}_k$ is replaced by $||\mathbf{d}_k||\mathbf{e}'$. $||\mathbf{d}_k||\mathbf{e}'$ differs from $\mathbf{w}_k$, but has identical first and second moments, and is uncorrelated with $\mathbf{d}_k$.

The angular θ'_k of the modified vector bearing $\mathbf{b}'_k$ can be represented as

$$\theta'_k = arctan(d_1/d_2) + \epsilon'_k \tag{5}$$

where ϵ'_k is a zero mean random variable, restricted to $[-\pi, \pi]$, independent of $\mathbf{d}_k$, with density $\alpha_\sigma(\cdot)$:

$$\alpha_\sigma(\theta) = \frac{e^{-1/2\sigma^2}}{2\pi}\left(1 + \sqrt{2\pi}\frac{cos\theta}{\sigma}F_{normal}\left(\frac{cos\theta}{\sigma}\right)e^{1/2(cos\theta/\sigma)^2}\right) \tag{6}$$

where, $F_{normal}(\cdot)$ is the cumulative distribution function of a standard $\mathcal{N}(0, 1)$ variable.

Note that θ' given by (5), is very close to the bearing θ in the standard model (3). The only difference is that the densities of the noise terms used in their construction are $\alpha_\sigma(\theta)$ and the normal $\mathcal{N}(0, \sigma^2)$ density, respectively. Reference [13] plots the two densities for $\sigma^2 \leq 1$. It shows the two density functions are virtually indistinguishable.

Given the bearing model, the SRF is to calculate the estimates of the conditional mean and covariance of the target state $\mathbf{x}_k$, given measurements up to time k, $\mathbf{b}_{1:k}$. That is,

$$\hat{\mathbf{x}}_{k|k} = E[\mathbf{x}_k|\mathbf{b}_{1:k}], \quad \mathbf{P}_{k|k} = cov[\mathbf{x}_k|\mathbf{b}_{1:k}] \tag{7}$$

The formulas for the SRF algorithm can be seen in [13].

2.2. The Treatment of Clutter

Accounting for the effects of clutter on the bearing measurements, we represent a cluttered bearing measurement as

$$z_k = (1 - r_k)\theta_k + r_k U_k \tag{8}$$

where U_k is the bearing measurement of the clutter, which is assumed to be an uniform random variable on $[-\pi, \pi]$; θ_k is the bearing measurement of the actual target from the sensor, r_k is defined as data association indicator at time k.

$$r_k = \begin{cases} 1 & \text{if the measurement is from clutter} \\ 0 & \text{if the measurement is from target} \end{cases} \tag{9}$$

The prior of r_k is assumed independent of the previous data associations and can be described as

$$p(r_k) = \begin{cases} \varsigma & \text{if } r_k = 1 \\ 1 - \varsigma & \text{if } r_k = 0 \end{cases} \tag{10}$$

Then the likelihood of the measurement is

$$p(z_k|\mathbf{x}_k, r_k) = \begin{cases} 1/2\pi & if \quad r_k = 1 \\ f(\theta_k|\mathbf{x}_k) & if \quad r_k = 0 \end{cases} \tag{11}$$

It can be represented in two forms:

$$p(z_k|\mathbf{x}_k, r_k) = \frac{1}{2\pi}r_k + f(\theta_k|\mathbf{x}_k)(1 - r_k) \tag{12}$$

$$= (\frac{1}{2\pi})^{r_k} f(\theta_k|\mathbf{x}_k)^{1-r_k} \tag{13}$$

Note that the two forms are used for different purpose in the later section. $f(\theta_k|\mathbf{x}_k)$ is the likelihood of the target bearing measurement, whose expression is given in Appendix A.

Here, we need to explain the reason for using this representation (8) of the cluttered measurement. The information carried in (8) is identical with the projection measurement $\mathbf{z}_k^b = [sinz_k, cosz_k]$. So the mean and covariance of the target state $\mathbf{x}_k$ conditioned on cluttered measurement $\mathbf{z}_{k'}^b$ which are the aims needs to be achieved in the SRF, are equivalent with these conditioned on z_k. Furthermore, this representation (8) is more convenient for calculation. Thus, z_k instead of $\mathbf{z}_k^b$ is adopted to represent the clutter measurement.

Based on the cluttered measurement model, given clutter probability ξ, the state estimate and covariance can be calculated as follows.

Suppose the density of $\mathbf{x}_{k-1}$ given measurements up to $k-1$, $p_{k-1|k-1}(\mathbf{x}_{k-1})$, is normal with mean $\bar{\mathbf{x}}_{k-1|k-1}$ and covariance $\bar{\mathbf{P}}_{k-1|k-1}$. The posterior density of state $\mathbf{x}_k$ conditioned on cluttered bearing measurements $\mathbf{z}_{1:k}$ is given as

$$p(\mathbf{x}_k|\mathbf{z}_{1:k}) = p(\mathbf{x}_k|z_k, \mathbf{x}_{k-1} \sim p_{k-1|k-1}(\mathbf{x}_{k-1}))$$
$$= q_k(0)p_{k|k}(\mathbf{x}_k) + q_k(1)p_{k|k-1}(\mathbf{x}_k) \tag{14}$$

where $q_k(i) = p(r_k = i|z_k, \mathbf{z}_{1:k-1})$, $p_{k|k}(\mathbf{x}_k)$ is the non-normal density of $\mathbf{x}_k$ conditioned on $r_k = 0$ and z_k, or, equivalently, on θ_k, and $p_{k|k-1}(\mathbf{x}_k)$ is the density of $\mathbf{x}_k$ when there is no target measurement at time k. It is normal with mean $\hat{\mathbf{x}}_{k|k-1}$ and covariance $\mathbf{P}_{k|k-1}$. Thus, the state estimate and covariance at time k can be obtained as

$$\bar{\mathbf{x}}_{k|k} = E[\mathbf{x}_k|z_k, \mathbf{x}_{k-1} \sim p_{k-1|k-1}(\mathbf{x}_{k-1})]$$
$$= q_k(0)\hat{\mathbf{x}}_{k|k} + q_k(1)\hat{\mathbf{x}}_{k|k-1} \tag{15}$$

$$\bar{\mathbf{P}}_{k|k} = cov[\mathbf{x}_k|z_k, \mathbf{x}_{k-1} \sim p_{k-1|k-1}(\mathbf{x}_{k-1})]$$
$$= q_k(0)(\mathbf{P}_{k|k} + (\hat{\mathbf{x}}_{k|k} - \bar{\mathbf{x}}_{k|k})(\hat{\mathbf{x}}_{k|k} - \bar{\mathbf{x}}_{k|k})^T)$$
$$+ q_k(1)(\mathbf{P}_{k|k-1} + (\hat{\mathbf{x}}_{k|k-1} - \bar{\mathbf{x}}_{k|k})(\hat{\mathbf{x}}_{k|k-1} - \bar{\mathbf{x}}_{k|k})^T) \tag{16}$$

where $\hat{\mathbf{x}}_{k|k}$ and $\mathbf{P}_{k|k}$ are the state estimate and its covariance based on the actual target measurement. $\hat{\mathbf{x}}_{k|k-1}$ and $\mathbf{P}_{k|k-1}$ are the predicted target state estimate and covariance. All of these can be obtained using the basic formulas of SRF.

The conditional densities $q_k(0)$ and $q_k(1)$ are given by the equations

$$q_k(0) = c_k(1 - \xi)f_k(\theta_k|\mathbf{z}_{1:k-1}) \tag{17}$$

$$q_k(1) = \frac{c_k\xi}{2\pi} \tag{18}$$

where c_k is the normalizing constant and $f_k(\theta_k|\mathbf{z}_{1:k-1})$ is the density of the actual target bearing θ_k conditioned on measurements $\mathbf{z}_{1:k-1}$. The expression is given in Appendix B.

However, in a complex environment, the probability of clutter is time-varying or hard to determine in advance. In this case, the probability ξ is unknown, so the above formulas are not applicable. In this paper, we resort the VB method to find the joint posterior density of $\mathbf{x}_k$ and ξ so as to account for the uncertainty in clutter probability.

3. Variational Bayesian Filtering

In this section, we first review the conjugate exponential (CE) model, and then gives the variational Bayesian solution of the CE model.

3.1. Conjugate Exponential Model

Given measurements $\mathbf{z}_{1:k-1}$, the posterior of the system state $p(\mathbf{x}_{k-1}|\mathbf{z}_{1:k-1})$ and the posterior of the parameter $p(\mathbf{r}_{k-1}|\mathbf{z}_{1:k-1})$, we assume the complete-data likelihood in the exponential family:

$$p(\mathbf{x}_k, \mathbf{z}_k|\mathbf{r}_k, \mathbf{z}_{1:k-1}) = g(\mathbf{r}_k)f(\mathbf{x}_k, \mathbf{z}_k)e^{\phi(\mathbf{r}_k)^T u(\mathbf{x}_k, \mathbf{z}_k)} \tag{19}$$

where, $\phi(\mathbf{r}_k)$ is the vector of natural parameters $\mathbf{r}_k$, u and f are known functions, and g is a normalization constant:

$$g(\mathbf{r}_k)^{-1} = \int f(\mathbf{x}_k, \mathbf{z}_k)e^{\phi(\mathbf{r}_k)^T u(\mathbf{x}_k, \mathbf{z}_k)} d\mathbf{x}_k d\mathbf{z}_k$$

The parameter prior is conjugate to the complete-data likelihood:

$$p(\mathbf{r}_k|\alpha_k^-, \beta_k^-) = h(\alpha_k^-, \beta_k^-)g(\mathbf{r}_k)^{\beta_k^-} e^{\phi(\mathbf{r}_k)^T \alpha_k^-} \tag{20}$$

where α_k^- and β_k^- are hyperparameters of the prior, and h is a normalization constant. Note the prior $p(\mathbf{r}_k|\alpha_k^-, \beta_k^-)$ is said to be conjugate to the likelihood $p(\mathbf{x}_k, \mathbf{z}_k|\mathbf{r}_k)$ if and only if the posterior

$$p(\mathbf{r}_k|\alpha_k, \beta_k) \propto p(\mathbf{r}_k|\alpha_k^-, \beta_k^-)p(\mathbf{x}_k, \mathbf{z}_k|\mathbf{r}_k)$$

is of the same parametric form as the prior. Then we call models that satisfy Equations (19) and (20) conjugate-exponential.

3.2. VB Approximation Method

Applying Bayes' rule, we have the joint posterior of $\mathbf{x}_k$ and r_k as

$$p(\mathbf{x}_k, \mathbf{r}_k|\mathbf{z}_{1:k}) \propto p(\mathbf{x}_k, \mathbf{z}_k|\mathbf{r}_k, \mathbf{z}_{1:k})p(\mathbf{r}_k|\alpha_k^-, \beta_k^-) \tag{21}$$

The analytic solution to (21) would be difficult to calculate. Here, we use the VB method to approximate the true posterior distribution with a product of tractable marginal posteriors [17].

$$p(\mathbf{x}_k, \mathbf{r}_k|\mathbf{z}_{1:k}) \approx Q(\mathbf{x}_k, \mathbf{r}_k) = Q_x(\mathbf{x}_k)Q_r(\mathbf{r}_k) \tag{22}$$

where $Q_x(\mathbf{x}_k)$ and $Q_r(\mathbf{r}_k)$ are unknown approximating marginal densities of $\mathbf{x}_k$ and $\mathbf{r}_k$.

The basic idea of VB approximation is to minimize the Kullback- Leibler (KL) divergence between the approximating posterior and the true posterior:

$$KL\left[Q_x(\mathbf{x}_k)Q_r(\mathbf{r}_k)||p(\mathbf{x}_k, \mathbf{r}_k|\mathbf{z}_{1:k})\right] = \int Q_x(\mathbf{x}_k)Q_r(\mathbf{r}_k) \times \log\left(\frac{Q_x(\mathbf{x}_k)Q_r(\mathbf{r}_k)}{p(\mathbf{x}_k, \mathbf{r}_k|\mathbf{z}_{1:k})}\right) d\mathbf{x}_k d\mathbf{r}_k \tag{23}$$

Given the measurements $\mathbf{z}_{1:k}$, we can minimize the KL divergence with respect to the probability densities $Q_x(\mathbf{x}_k)$ and $Q_r(\mathbf{r}_k)$ in turn, while keeping the other fixed. Then, the following equations can be given as:

$$Q_x(\mathbf{x}_k) \propto exp(\langle \ln p(\mathbf{x}_k, \mathbf{r}_k, \mathbf{z}_k|\mathbf{z}_{1:k-1})\rangle_{\mathbf{r}_k}) \tag{24}$$

$$Q_r(\mathbf{r}_k) \propto exp(\langle \ln p(\mathbf{x}_k, \mathbf{r}_k, \mathbf{z}_k | \mathbf{z}_{1:k-1}) \rangle_{\mathbf{x}_k}) \tag{25}$$

where $\langle \cdot \rangle_{\mathbf{x}_k}$ and $\langle \cdot \rangle_{\mathbf{r}_k}$ denote the expectations with respect to $Q_x(\mathbf{x}_k)$ and $Q_r(\mathbf{r}_k)$, respectively. Obviously, it is not an explicit solution since the distribution of each parameter is dependent on the other and neither distributions is known. The mechanism of VB method is to firstly give the initial values of the parameters and then use expectation-maximum (EM) algorithm to iteratively calculate $Q_x(\mathbf{x}_k)$ and $Q_r(\mathbf{r}_k)$ until convergence. For the above CE models, $Q_x(\mathbf{x}_k)$ and $Q_r(\mathbf{r}_k)$ can be obtained from the following procedure [19]:

(1)   The VB expectation step yields:

$$Q_x(\mathbf{x}_k) \propto f(\mathbf{x}_k, \mathbf{z}_k) e^{\langle \phi(\mathbf{r}_k) \rangle_{\mathbf{r}_k}^T u(\mathbf{x}_k, \mathbf{z}_k)} = p(\mathbf{x}_k | \mathbf{z}_k, \langle \phi(\mathbf{r}_k) \rangle_{\mathbf{r}_k}) \tag{26}$$

(2)   The VB maximization step yields that $Q_r(\mathbf{r}_k)$ is conjugate and of the form

$$Q_r(\mathbf{r}_k) = h(\alpha_k, \beta_k^-) g(\mathbf{r}_k)^{\beta_k} e^{\phi(\mathbf{r}_k)^T \alpha_k} \tag{27}$$

where, α_k and β_k are the hyper-parameters, and

$$\alpha_k = \alpha_k^- + \langle u(\mathbf{x}_k, \mathbf{z}_k) \rangle_{\mathbf{x}_k} \tag{28}$$

$$\beta_k = \beta_k^- + n \tag{29}$$

where n is the dimension of the measurement.

4. VB Based Adaptive Shifted Rayleigh Filter with Unknown Clutter Probability

Considering the system model (1) and the measurement model (8) described in Section 2, we adopt the VB method within the SRF framework to get the joint estimation of the target state and the clutter probability.

The core is to determine the posterior approximation $Q(\mathbf{x}_k, r_k, \xi)$. Assume factorization $Q(\mathbf{x}_k, r_k, \xi) \approx Q_x(\mathbf{x}_k) Q(r_k, \xi)$, we can obtain $Q_x(\mathbf{x}_k)$, $Q_r(r_k)$ and $Q_\xi(\xi_k)$ at each time k through the following procedure.

(1)   Optimization of $Q_x(\mathbf{x}_k)$ for fixed $Q(r_k, \xi)$.

First, by using the first form of the measurement likelihood (12), the complete-data likelihood is presented as

$$p(\mathbf{x}_k, z_k | r_k, \mathbf{z}_{1:k-1}) = p(z_k | \mathbf{x}_k, r_k) p(\mathbf{x}_k | r_k, \mathbf{z}_{1:k-1})$$
$$= [\frac{1}{2\pi} r_k + f(\theta_k | \mathbf{x}_k)(1 - r_k)] \mathcal{N}(\mathbf{x}_k; \hat{\mathbf{x}}_{k|k-1}, \mathbf{P}_{k|k-1}) \tag{30}$$

Then according to (24), we can get

$$Q_x(\mathbf{x}_k) \propto exp\{\langle \ln p(z_k, \mathbf{x}_k | r_k, \mathbf{z}_{1:k-1}) \rangle_{r_k, \xi}\}$$
$$= [\frac{1}{2\pi} \langle r_k \rangle_{r_k} + f(\theta_k | \mathbf{x}_k)(1 - \langle r_k \rangle_{r_k})] \mathcal{N}(\mathbf{x}_k; \hat{\mathbf{x}}_{k|k-1}, \mathbf{P}_{k|k-1})$$
$$= \frac{1}{2\pi} \mathcal{N}(\mathbf{x}_k; \hat{\mathbf{x}}_{k|k-1}, \mathbf{P}_{k|k-1}) \langle r_k \rangle_{r_k} + f(\theta_k | \mathbf{x}_k) \mathcal{N}(\mathbf{x}_k; \hat{\mathbf{x}}_{k|k-1}, \mathbf{P}_{k|k-1})(1 - \langle r_k \rangle_{r_k}) \tag{31}$$
$$\approx \frac{1}{2\pi} \mathcal{N}(\mathbf{x}_k; \hat{\mathbf{x}}_{k|k-1}, \mathbf{P}_{k|k-1}) \langle r_k \rangle_{r_k} + \mathcal{N}(\mathbf{x}_k; \hat{\mathbf{x}}_{k|k}, \mathbf{P}_{k|k}) f(\theta_k | \mathbf{z}_{1:k-1})(1 - \langle r_k \rangle_{r_k}) \tag{32}$$

The approximation sign in (32) is because the following:

$$f(\theta_k | \mathbf{x}_k) \mathcal{N}(\mathbf{x}_k; \hat{\mathbf{x}}_{k|k-1}, \mathbf{P}_{k|k-1}) = p(\theta_k | \mathbf{x}_k, \mathbf{z}_{1:k-1}) p(\mathbf{x}_k | \mathbf{z}_{1:k-1})$$
$$= p(\mathbf{x}_k | \theta_k, \mathbf{z}_{1:k-1}) p(\theta_k | \mathbf{z}_{1:k-1})$$
$$\approx \mathcal{N}(\mathbf{x}_k; \hat{\mathbf{x}}_{k|k}, \mathbf{P}_{k|k}) f(\theta_k | \mathbf{z}_{1:k-1}) \tag{33}$$

where $f(\theta_k|\mathbf{z}_{1:k-1})$ is derived in Appendix B.

Comparing (32) with the posterior density (14) of SRF, the difference lies in the weights. Except for a normalization constant, the clutter probability ξ used before in (14) has been replaced by $\langle r_k\rangle_{r_k}$ in (32), which is updated online.

(2)   Optimization of $Q(r_k, \xi)$ for fixed $Q_x(\mathbf{x}_k)$.

We use the VB method again by factorizing $Q(r_k, \xi) \approx Q_r(r_k)Q_\xi(\xi)$. Assume the conjugate prior of r_k is binomial distributed with parameter ξ. That is,

$$p(r_k|\xi) = \xi^{r_k}(1 - \xi)^{1-r_k} \tag{34}$$

and $p(\xi)$ follows beta distribution with parameters α_1 and α_2.

$$p(\xi; \alpha_1, \alpha_2) = \frac{1}{B(\alpha_1, \alpha_2)}\xi^{\alpha_1-1}(1 - \xi)^{\alpha_2-1} \tag{35}$$

where $B(\alpha_1, \alpha_2) = \Gamma(\alpha_1)\Gamma(\alpha_2)/\Gamma(\alpha_1 + \alpha_2)$.

To have the form of (19), the complete-data likelihood $p(\mathbf{x}_k, z_k|r_k, \mathbf{z}_{1:k-1})$ is re-derived using the second form (13) of $p(z_k|\mathbf{x}_k, r_k)$ as

$$
\begin{aligned}
p(\mathbf{x}_k, z_k|r_k, \mathbf{z}_{1:k-1}) &= (1/2\pi)^{r_k}[f(\theta_k|\mathbf{x}_k)]^{1-r_k}\mathcal{N}(\mathbf{x}_k; \hat{\mathbf{x}}_{k|k-1}, \mathbf{P}_{k|k-1}) \\
&= f(\theta_k|\mathbf{x}_k)\mathcal{N}(\mathbf{x}_k; \hat{\mathbf{x}}_{k|k-1}, \mathbf{P}_{k|k-1})exp\{r_k[ln(1/2\pi) - ln[f(\theta_k|\mathbf{x}_k)]]\}
\end{aligned}
\tag{36}
$$

Rewriting the conjugate prior of r_k in the form of (20), we can get

$$
\begin{aligned}
p(r_k|\xi) &= \xi^{r_k}(1 - \xi)^{1-r_k} \\
&= (1 - \xi)exp[r_k ln(\frac{\xi}{1-\xi})]
\end{aligned}
\tag{37}
$$

Then applying (27), $Q_r(r_k)$ can be obtained as

$$Q_r(r_k) = (1 - \eta_k)exp[r_k ln(\frac{\eta_k}{1-\eta_k})] \tag{38}$$

where, η_k is the hyper-parameter and updated as

$$ln(\frac{\eta_k}{1-\eta_k}) = \langle ln(\frac{\xi}{1-\xi})\rangle_\xi + \langle [ln(1/2\pi) - ln[f(\theta_k|\mathbf{x}_k)]]\rangle_{\mathbf{x}_k} \tag{39}$$

Likewise, rewriting the conjugate prior of ξ in the form of (20), we can get

$$
\begin{aligned}
p(\xi) &= \frac{1}{B(\alpha_1, \alpha_2)}\xi^{\alpha_1-1}(1 - \xi)^{\alpha_2-1} \\
&= \frac{1}{B(\alpha_1, \alpha_2)}(1 - \xi)^{\alpha_1+\alpha_2-2}(1 - \xi)^{-(\alpha_1-1)}\xi^{\alpha_1-1} \\
&= \frac{1}{B(\alpha_1, \alpha_2)}(1 - \xi)^{\alpha_1+\alpha_2-2}exp[(\alpha_1 - 1)ln(\frac{\xi}{1-\xi})]
\end{aligned}
\tag{40}
$$

Then applying (27), $Q_\xi(\xi)$ can be obtained as

$$Q_\xi(\xi) = \frac{1}{B(\alpha_1', \alpha_2')}(1 - \xi)^{\alpha_1'+\alpha_2'-2}exp[(\alpha_1' - 1)ln(\frac{\xi}{1-\xi})] \tag{41}$$

with hyper-parameters

$$\alpha_1' = \alpha_1 + \langle r_k \rangle_{r_k} \tag{42}$$

$$\alpha_2' = \alpha_2 + n - \langle r_k \rangle_{r_k} \tag{43}$$

where n is the dimension of the measurement.

According to the approximated posteriors of $Q_r(r_k)$ and $Q_\zeta(\zeta)$, $\langle r_k \rangle_{r_k}$ and $\langle ln(\frac{\zeta}{1-\zeta}) \rangle_\zeta$ can be obtained as:

$$\langle r_k \rangle_{r_k} = \eta_k \tag{44}$$

$$\langle ln(\frac{\zeta}{1-\zeta}) \rangle_\zeta = \psi(\alpha_1') - \psi(\alpha_2') \tag{45}$$

where $\psi(\cdot)$ is the digamma function.

Taking expectation and covariance on the posterior $Q_x(\mathbf{x}_k)$, the conditional mean and covariance of the target state can then be obtained. We summarize the entire filtering procedure of the VB-based SRF (VB-SRF) in Algorithm 1.

Algorithm 1 : VB-SRF.

---

(1) Initialization: $\bar{\mathbf{x}}_{0|0}$, $\bar{\mathbf{P}}_{0|0}$, $\mathbf{Q}_v$, $\mathbf{Q}_w$, η_0, $\alpha_{1,0}$, $\alpha_{2,0}$

(2) Prediction:

$\hat{\mathbf{x}}_{k|k-1} = \mathbf{F}_{k-1}\bar{\mathbf{x}}_{k-1|k-1} + \mathbf{u}_{k-1}^s$

$\mathbf{P}_{k|k-1} = \mathbf{F}_{k-1}\bar{\mathbf{P}}_{k-1|k-1}\mathbf{F}_{k-1}^T + \mathbf{Q}_v$

$\mathbf{S}_k = \mathbf{H}_k\mathbf{P}_{k|k-1}\mathbf{H}_k^T + \mathbf{Q}_k^m$

$\eta_{k|k-1} = \rho\eta_{k-1}, \quad \alpha_{1,k|k-1} = \rho\alpha_{1,k-1}, \quad \alpha_{2,k|k-1} = \rho\alpha_{2,k-1}$

where ρ is the scale factor and $0 < \rho \leq 1$.

(3) Update: the update of VB-SRF utilizes iterate filtering framework.

(3.a) First set: $\bar{\mathbf{x}}_{k|k}^{(0)} = \hat{\mathbf{x}}_{k|k-1}$, $\bar{\mathbf{P}}_{k|k}^{(0)} = \mathbf{P}_{k|k-1}$, $\eta_k^{(0)} = \eta_{k|k-1}$, $\alpha_{1,k}^{(0)} = \alpha_{1,k|k-1}$, $\alpha_{2,k}^{(0)} = \alpha_{2,k|k-1}$

(3.b) Calculate state estimation and its covariance using SRF when the measurement is from the target:

$\mathbf{K}_k = \mathbf{P}_{k|k-1}\mathbf{H}_k^T\mathbf{S}_k^{-1}$

$\varepsilon_k = (\mathbf{b}_k^T\mathbf{S}_k^{-1}\mathbf{b}_k)^{-1/2}\mathbf{b}_k^T\mathbf{S}_k^{-1}(\mathbf{H}_k\hat{\mathbf{X}}_{k|k-1} + \mathbf{u}_k^m)$

$\gamma_k = (\mathbf{b}_k^T\mathbf{S}_k^{-1}\mathbf{b}_k)^{-1/2}\rho_n(\varepsilon_k)$

$\delta_k = (\mathbf{b}_k^T\mathbf{S}_k^{-1}\mathbf{b}_k)^{-1/2}[2 + \varepsilon_k\rho_2(\varepsilon_k) - \rho_2^2\varepsilon_k]$

$\rho_2(\varepsilon_k) = \frac{\varepsilon_k e^{-\varepsilon_k^2/2} + \sqrt{2\pi}(\varepsilon_k^2+1)F_{normal}(\varepsilon_k)}{e^{-\varepsilon_k^2/2} + \sqrt{2\pi}(\varepsilon_k)F_{normal}(\varepsilon_k)}$

$\hat{\mathbf{x}}_{k|k} = (\mathbf{I} - \mathbf{K}_k\mathbf{H}_k)\hat{\mathbf{x}}_{k|k-1} - \mathbf{K}_k\mathbf{u}_k^m + \gamma_k\mathbf{K}_k\mathbf{b}_k$

$\mathbf{P}_{k|k} = (\mathbf{I} - \mathbf{K}_k\mathbf{H}_k)\mathbf{P}_{k|k-1} + \delta_k\mathbf{K}_k\mathbf{b}_k\mathbf{b}_k^T\mathbf{K}_k^T$

(3.c) For $j = 1 : N$, iterate the following N (N denotes iterated times) steps:

- **Calculate the fused state estimation and its covariance:**

$\bar{\mathbf{x}}_{k|k}^{(j)} = \frac{1}{2\pi c}\eta_k^{(j-1)}\hat{\mathbf{x}}_{k|k-1} + \frac{1}{c}(1 - \eta_k^{(j-1)})f(\theta_k|\mathbf{z}_{1:k-1})\hat{\mathbf{x}}_{k|k}$

$\bar{\mathbf{P}}_{k|k}^{(j)} = \frac{1}{2\pi c}\eta_k^{(j-1)}(\mathbf{P}_{k|k} + (\hat{\mathbf{x}}_{k|k} - \bar{\mathbf{x}}_{k|k})(\hat{\mathbf{x}}_{k|k} - \bar{\mathbf{x}}_{k|k})^T)$

$\qquad + \frac{1}{c}(1 - \eta_k^{(j-1)})f(\theta_k|\mathbf{z}_{1:k-1})(\mathbf{P}_{k|k-1} + (\hat{\mathbf{x}}_{k|k-1} - \bar{\mathbf{x}}_{k|k}^{(j)})(\hat{\mathbf{x}}_{k|k-1} - \bar{\mathbf{x}}_{k|k}^{(j)})^T)$

where $c = \frac{1}{2\pi}\eta_k^{(j-1)} + f(\theta_k)(1 - \eta_k^{(j-1)})$ is a normalization term, and $f(\theta_k|\mathbf{z}_{1:k-1})$ can be obtained using (A6).

- **Update parameters:**

$ln(\frac{\eta_k^{(j)}}{1-\eta_k^{(j)}}) = \psi(\alpha_{1,k}^{(j-1)}) - \psi(\alpha_{2,k}^{(j-1)}) + ln(1/2\pi) - lnf(\theta_k|\bar{\mathbf{x}}_{k|k}^{(j)})$

$\alpha_{1,k}^{(j)} = \alpha_{1,k}^{(j-1)} + \eta_k^{(j)}$

$\alpha_{2,k}^{(j)} = \alpha_{2,k}^{(j-1)} - \eta_k^{(j)} + 1$

- **End for and set** $\bar{\mathbf{x}}_{k|k} = \bar{\mathbf{x}}_{k|k}^{(N)}$, $\bar{\mathbf{P}}_{k|k} = \bar{\mathbf{P}}_{k|k}^{(N)}$, $\eta_k = \eta_k^{(N)}$, $\alpha_{1,k} = \alpha_{1,k}^{(N)}$, $\alpha_{2,k} = \alpha_{2,k}^{(N)}$.

---

5. Simulation Results

To evaluate the performance of the VB-SRF algorithm, two scenarios which are almost the same with these in [12,13] are utilized. The differences lie in the clutter probability in scenario 1 and the sensor tracks in scenario 2 which were not detailed in [12]. The two scenarios are very representative. In scenario 1, a maneuvering sensor is used in order to satisfy the condition of observability in bearings-only tracking. In scenario 2, multiple distributed sensors with large noise variance are utilized, which make the problem more challenging. The tracking performance of the VB-SRF algorithm was compared with the SRF algorithm, the MEFPDA-SCKF algorithm and PDA-SCKF algorithm in terms of track loss, track accuracy and computation complexity. The track loss is declared when the track error is large enough that making the filter diverge. Root mean square (RMS) error is used to show the track accuracy. In addition, the computation complexity is reflected by the computation time of each filter. The simulation codes can be downloaded through Github [20].

5.1. Scenario 1

In scenario 1, a target moves along a horizontal track, with zero vertical displacement, according to a white noise acceleration model. The state of the target is represented as $\mathbf{x}_k = [x_{1,k}, \dot{x}_{1,k}]$, where $x_{1,k}$ and $\dot{x}_{1,k}$ are the horizontal distance and velocity at time k, respectively. The observer platform follows an approximately parallel track at a constant average speed. The horizontal and vertical displacements of the platform $x_k^p = [x_{1,k}^p, x_{2,k}^p]^T$ are governed by the equations:

$$x_{1,k}^p = 4k + \tilde{x}_{1,k}^p \tag{46}$$

$$x_{2,k}^p = 20 + \tilde{x}_{2,k}^p \tag{47}$$

in which $\tilde{x}_{1,k}^p$ and $\tilde{x}_{2,k}^p$ are zero mean Gaussian white noise processes, both with variance $q = 1$.

The measurement is the angle (in radians) of the line-of-sight of this target from the platform. The sensor noise is Gaussian white noise with variance $\sigma^2 = (0.05)^2 \text{rad}^2 = 2.86^2 \text{deg}^2$. The true clutter probability is 0.8. Other parameters are detailed in [13]. The configuration of the observer platform and target is illustrated in Figure 1. The bearing measurement of the target is presented in Figure 2. It can be seen that there is no abrupt change.

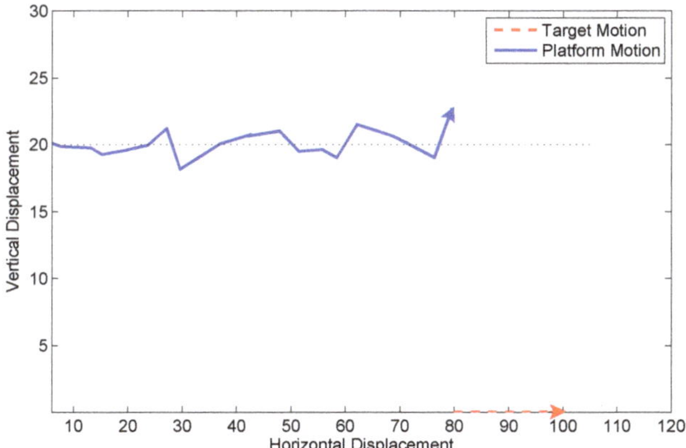

Figure 1. Target-observer geometry in Scenario 1.

The clutter probability used in SRF is set as $p_c = 0.8$, which is the same with the true clutter probability. The initial values of VB-SRF parameters are $\eta_0 = 0.8$, $\alpha_{1,0} = 2$, $\alpha_{2,0} = 10$. The clutter density λ is calculated using $-kln(1 - p_c)/2\pi$ in MEFPDA-SCKF and PDA-SCKF. Figure 3 presents the RMS target position errors of the four filters using 1000 Monte Carlo runs. It can be seen that SRF and VB-SRF have comparable performance under the correct clutter probability. MEFPDA-SCKF and PDA-SCKF have a little better tracking accuracy than SRF and VB-SRF.

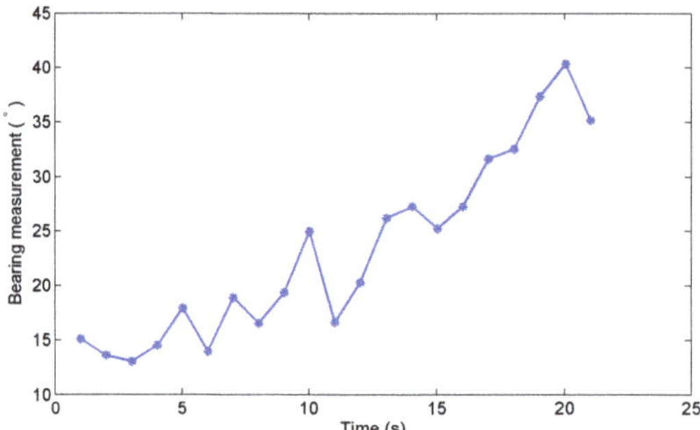

Figure 2. The measurement of the target in Scenario 1.

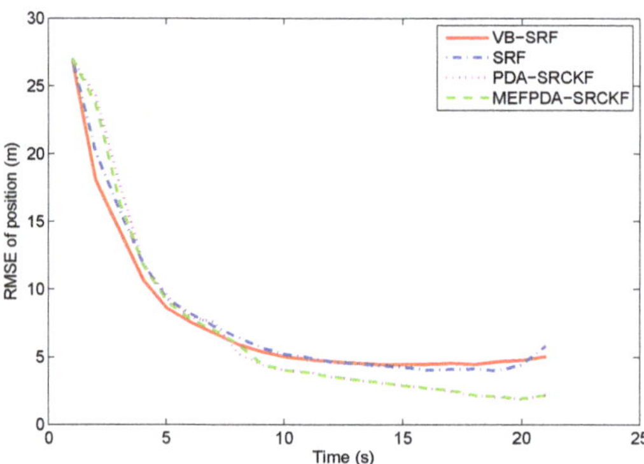

Figure 3. RMS target position errors with correct clutter probability in Scenario 1.

In challenging scenarios, the clutter probability maybe unknown or time varying. The pre-set parameters are probably inaccurate. So here we set mistuned clutter probabilities for the four filters: (1) $p_c = \eta_0 = 0.7$; (2) $p_c = \eta_0 = 0.5$; (3) $p_c = \eta_0 = 0.3$. The percentages of track losses in 1000 Monte Carlo runs are given in Table 1. Clearly, the VB-SRF algorithm outperforms the SRF algorithm due to fewer track losses. Moreover the proportion of track losses increases as the mistuning aggravates. Especially, there are 2.9% tracks are lost for SRF while only 0.1% tracks are lost for VB-SRF in the worst case. In addition, we can see that MEFPDA-SCKF and PDA-SCKF have no track loss in all the three mistuned cases.

The RMS position errors of the four filters with different mistuned clutter probabilities are shown in Figure 4. We only consider the runs without track loss. From the figure, we can see that the tracking accuracy of VB-SRF is slightly better than SRF when $\eta_0 \geq 0.5$. However, the performance difference is not obvious. When $p_c = \eta_0 = 0.3$, VB-SRF exhibits distinct superiority over SRF. It implies that VB-SRF is more robust than SRF. Meanwhile, MEFPDA-SCKF and PDA-SCKF show better tracking accuracy than SRF and VB-SRF under all the three mistuned cases. They are almost not affected by the mistuning. It shows MEFPDA-SCKF and PDA-SCKF are more accurate and robust than SRF and VB-SRF under this simple scenario.

Table 2 shows the computation time of the four filters with 100 Monte Carlo runs. It is clear that VB-SRF has the maximum time of computation, which is about 2 times of that of SRF. The computation time of MEFPDA-SCKF and PDA-SCKF are comparable and both smaller than SRF and VB-SRF.

On the whole, the VB-SRF algorithm outperforms the SRF in terms of track continuity, track accuracy and robustness especially in severely mismatched scenarios but with higher computation complexity. MEFPDA-SCKF and PDA-SCKF perform better than SRF and VB-SRF in all aspects. It illustrates that the PDA-SCKF-based strategy has superiority over the SRF-based strategy in handing the clutters in simple scenarios.

Table 1. The percentages of track losses of four filters in two scenarios.

	Scenario 1			Scenario 2		
	$p_c = 0.7$	$p_c = 0.5$	$p_c = 0.3$	$p_c = 0.667$	$p_c = 0.5$	$p_c = 0.3$
VB-SRF	0	0	0.1%	0	0	0
SRF	0.9%	1.6%	2.9%	0	0	2.7%
MEFPDA-SCKF	0	0	0	13.5%	13.3%	14.2%
PDA-SCKF	0	0	0	20.1%	20.8%	22.4%

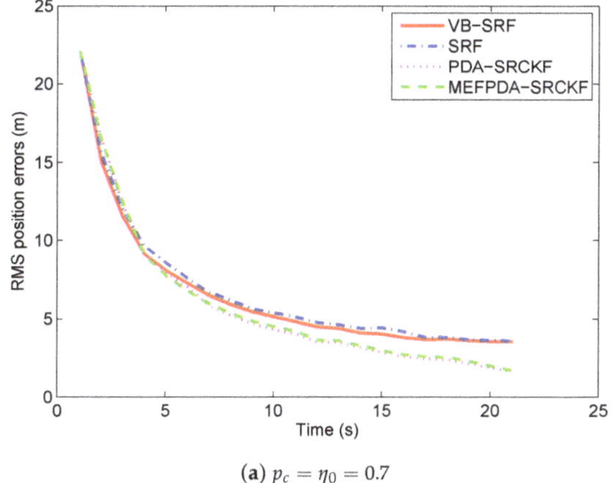

(**a**) $p_c = \eta_0 = 0.7$

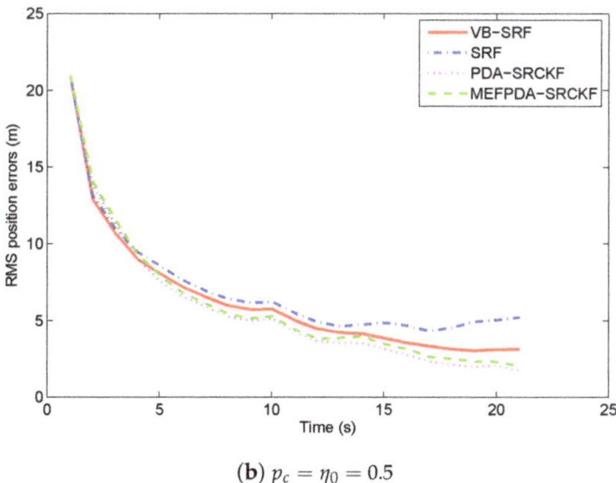

(**b**) $p_c = \eta_0 = 0.5$

Figure 4. *Cont.*

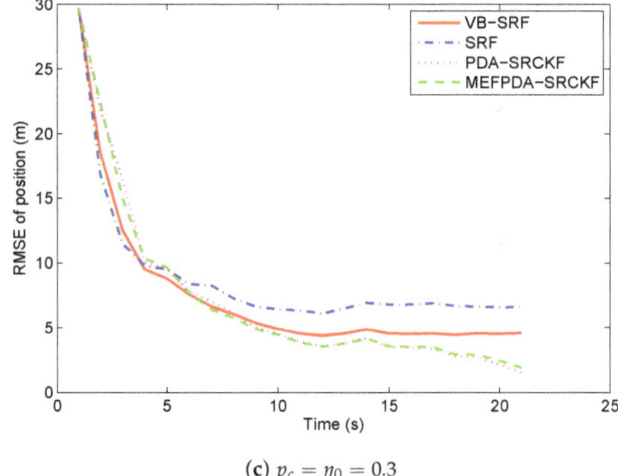

(c) $p_c = \eta_0 = 0.3$

Figure 4. RMS target position errors with different mistuned clutter probabilities in Scenario 1.

Table 2. Computation time of the four filters with 100 Monte Carlo runs for two scenarios.

	Scenario 1	Scenario 2
VB-SRF	0.7406 s	1.0236 s
SRF	0.3690 s	0.5779 s
MEFPDA-SCKF	0.2066 s	0.3314 s
MEFPDA-SCKF	0.2092 s	0.3128 s

5.2. Scenario 2

For scenario 2, the aim is to track a single target from several drifting sonobuoys. A monitoring aircraft estimates the positions of the drifting sonobuoys by observing the direction of arrival of sensor transmissions. The sonobuoys track the position of the target by means of noisy bearings measurements. The state is 12-dimensional:

$$\mathbf{x}_k = [x_{0,k}, \dot{x}_{0,k}, y_{0,k}, \dot{y}_{0,k}, x_{1,k}, y_{1,k}, x_{2,k}, y_{2,k}, x_{3,k}, y_{3,k}, u_{1,k}, u_{1,k}]^T \tag{48}$$

the first four components represent the (x, y) coordinates of the position and velocity of the target, the next six, the coordinates of the positions of the three sonobuoys, and the last two, those of the drift current effecting all three sonobuoys.

Six simultaneous measurements are made at each time step. Three of these are measurements of the bearing angles of the sonobuoys from the monitoring platform and they are uncluttered. Three are the bearing angles of the target from the sonobuoys, which are subject to clutter. The standard deviation of monitoring sensor bearing noise and sonobuoy sensor bearing noise are $0.8°$ and $16°$, respectively. The true probability of clutter is set as 0.667. In addition, the bearing of the clutter is uniformly distributed over $[-\pi, \pi]$. Other simulation parameters can be referred to [12]. 200 Monte Carlo runs are performed to evaluate the performance of the proposed filter. Figure 5 shows the behaviour of the estimates of target and sonobuoy positions provided by both the SRF and VB-SRF, for a typical simulation.

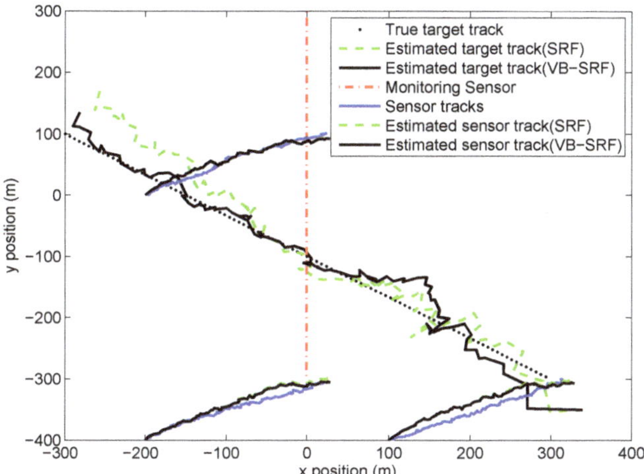

Figure 5. Typical tracks of target, drifting sonobuoy sensors, together with the estimated tracks.

The RMS target position errors of the four filters with correct clutter probability assumption are given in Figure 6. Compared with scenario 1, the differences between the four filters are more dramatic in scenario 2. This is probably because scenario 2 is more complex in which multiple sensors are used to observe the target and give abruptly changing and severely noise-corrupted bearing measurements of the target, shown in Figure 7. Thus, even a minor change in filtering strategy could result in large variations in performance. Meanwhile, seen from Figure 7, an abrupt change (almost from $+180°$ to $-180°$) occurs in the target bearing measurement from sonobuoy sensor 3 at $k = 62$ s, which leads to several track losses shown in Table 1 and much larger position errors of MEFPDA-SCKF and PDA-SCKF. Whereas, SRF and VB-SRF are less affected by the abrupt bearing variation since the value of the projected measurement $b_k = (sin\theta_k, cos\theta_k)^T$ is invariant when there is a $360°$ change in bearing θ_k. They have comparable tracking accuracy. It is hard to decide which one is better.

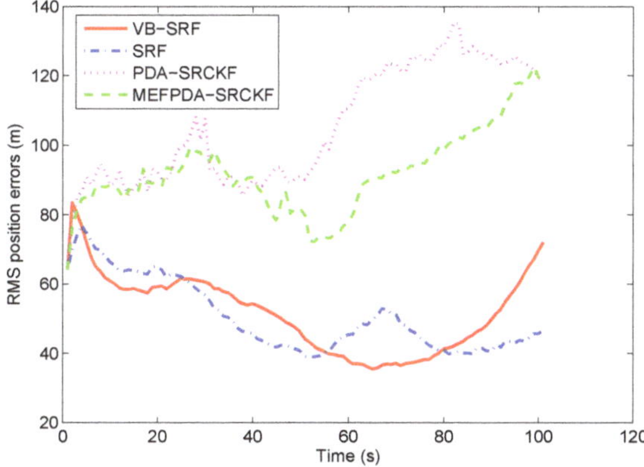

Figure 6. RMS target position errors with correct clutter probability in Scenario 2.

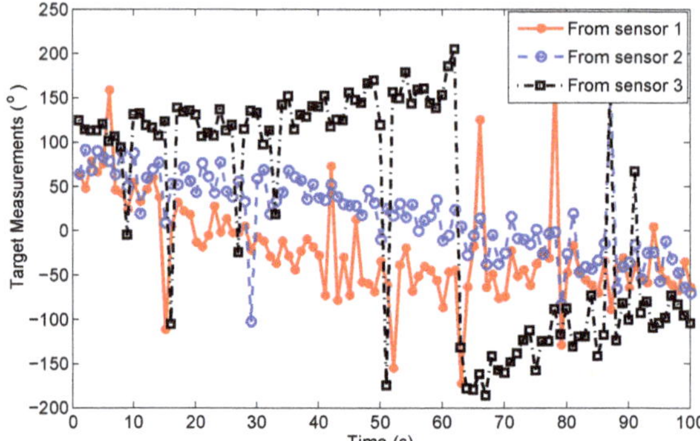

Figure 7. The target measurements from three sonobuoy sensors in Scenario 2.

To compare the performance under adverse scenarios, we set mistuned clutter probabilities as: (1) $p_c = \eta_0 = 0.5$; (2) $p_c = \eta_0 = 0.3$. In case (1), there is no track loss in SRF and VB-SRF and the RMS position errors of the two filters are shown in Figure 8a. We can see that the RMS position errors of SRF are slightly increased compared with the case with no mistuning, while the RMS position errors of VB-SRF remain almost unchanged. In case (2), as shown in Table 1, 2.7% tracks are lost for SRF while no track is lost for VB-SRF. Meanwhile, as can be seen in Figure 8b, VB-SRF has much smaller RMS errors than the SRF. For MEFPDA-SCKF and PDA-SCKF, unlike with scenario 1, they have higher percentage of track losses and larger RMS position errors than VB-SRF in both mistuned cases. It shows that VB-SRF is superior to PDA-SCKF-based algorithms in more challenging scenarios. In addition, from Table 2, we can see that the computation time of VB-SRF is twice of the SRF and three times of MEFPDA-SCKF and PDA-SCKF. Overall, all these reveal that the proposed VB-SRF algorithm has significant performance superiority in severely mismatched and complex cases at the cost of a little higher computation complexity.

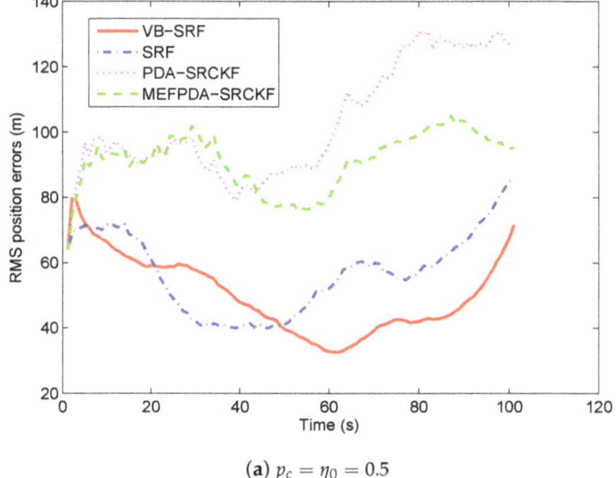

(**a**) $p_c = \eta_0 = 0.5$

Figure 8. *Cont.*

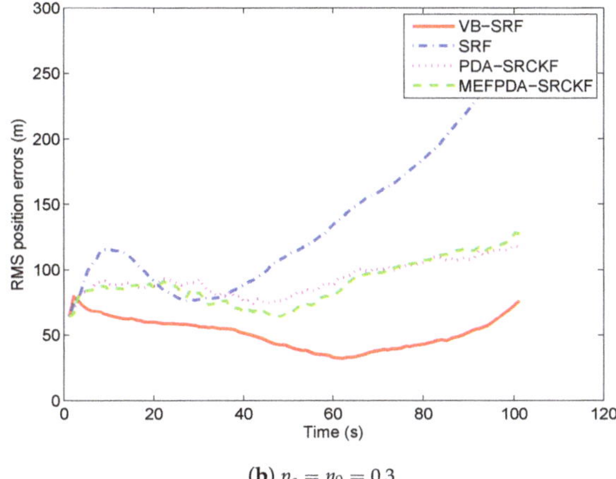

(**b**) $p_c = \eta_0 = 0.3$

Figure 8. RMS target position errors with mistuned clutter probability in Scenario 2.

6. Conclusions

Bearings-only target tracking in the presence of clutter is a difficult problem because of the nonlinearity of the measurement model, the measurement origin uncertainty and the observability of the target. The Shifted Rayleigh filter (SRF) is shown to exhibit good performance for bearings-only target tracking in certain challenging scenarios through exploiting the essential structure of the nonlinearities in a new way. However, the clutter probability is assumed known and constant in SRF, which may not match with the truth especially in adverse scenarios. Therefore, to handle the bearings-only target tracking in clutters with uncertain clutter probability, a variational Bayesian-based adaptive shifted Rayleigh filter (VB-SRF) is proposed in this paper. By establishing a conjugate exponential model of the clutter probability and the data association indicator, the approximated posterior probability densities of the target state and the clutter parameters are iteratively calculated using the VB expectation and maximization steps. Finally, joint estimation of the target state and the clutter probability are achieved in the framework of SRF. The tracking performance of the proposed filter is compared with SRF, PDA-SCKF and MEFPDA-SCKF via two simulation examples. It shows that the proposed filter outperforms the other three filters in terms of track continuity and track accuracy with a little higher computation complexity in complex adverse scenarios. In addition, it also reveals that the proposed VB-SRF exhibits better robustness than the traditional SRF.

Author Contributions: J.H. conceived this paper, derived the method, and wrote the original draft; Y.Y. made the investigation; and T.G. revised the paper and provided some valuable suggestions.

Funding: This research was funded by Aeronautical Science Foundation of China (No. 2016ZC53033), and Shaanxi Provincial Key Research and Development Programs (No. 2017ZDXM-GY-06, No. 2017GY-057), and Xi'an Science and Technology Planning Project–Scientific and Technological Innovation Guidance Project (No. 201805042YD20CG26 (8)).

Conflicts of Interest: The authors declare no conflict of interest.

Abbreviations

The following abbreviations are used in this manuscript:

VB	Variational Bayesian
SRF	Shifted Rayleigh Filter
PDA	Probability Data Association
EKF	Extended Kalman Filter

MPEKF	Polar Coordinate EKF
PLE	Pseudo-Linear Estimator
UKF	Unscented Kalman Filter
CKF	Cubature Kalman Filter
PF	Particle Filter
MEFPDA	Maximum Entropy Fuzzy Probabilistic Data Association
SCKF	Square-root Cubature Kalman Filter
CE	Conjugate Exponential
KL	Kullback- Leibler
EM	Expectation-Maximum
RMS	Root Mean Square

Appendix A. Derivation of $f(\theta_k|\mathbf{x}_k)$

Given the true state $\mathbf{x}_k$, the "augmented" measurement $\mathbf{y}_k$ is assumed to be $\mathcal{N}(\mathbf{m}_k, \mathbf{Q}_w)$ variable, where $\mathbf{m}_k = \mathbf{H}_k\mathbf{x}_k + \mathbf{u}_k^m$. θ_k is the bearing of $\mathbf{y}_k$.

Let $\mathbf{y}'_k = \mathbf{Q}_w^{-1/2}\mathbf{y}_k$, then $\mathbf{y}'_k \sim \mathcal{N}(\mathbf{Q}_w^{-1/2}\mathbf{m}_k, \mathbf{I})$. Its bearing has a density about its mean of the form $\alpha_\sigma(\theta)$. More precisely, let

$$g_k(\theta) = [(sin\theta, cos\theta)\mathbf{Q}_w^{-1}(sin\theta, cos\theta)^T]^{-1/2} \tag{A1}$$

and define

$$h_k(\theta) = arctan\left(\frac{a_{11}sin\theta + a_{12}cos\theta}{a_{21}sin\theta + a_{22}cos\theta}\right) \tag{A2}$$

where $[a_{ij}] = \mathbf{Q}_w^{-1/2}$.

Actually, $g_k(\theta)$ and $h_k(\theta)$ are the reciprocal length and bearing of the transformed unit vector $\frac{\mathbf{Q}_w^{-1/2}\mathbf{y}_k}{||\mathbf{y}_k||}$. The bearing $h_k(\theta_k)$ then has $\alpha_{g_k(\theta_k^m)}(h_k(\theta_k) - h_k(\theta_k^m))$ as its density, where θ_k^m is the bearing of $\mathbf{m}_k$. Inserting a Jacobian term, we can obtain the likelihood of measurement θ_k:

$$f(\theta_k|\mathbf{x}_k) = \frac{g_k^2(\theta_k)}{(det\mathbf{Q}_w)^{1/2}}\alpha_{g_k(\theta_k^m)}(h_k(\theta_k) - h_k(\theta_k^m)) \tag{A3}$$

Appendix B. Derivation of $f(\theta_k|\mathbf{z}_{1:k-1})$

Given the previous measurements $\mathbf{y}_{1:k-1}$, the "augmented" measurement $\mathbf{y}_k$ is assumed to be $\mathcal{N}(\hat{\mathbf{y}}_k, \mathbf{S}_k)$ variable, where $\hat{\mathbf{y}}_k = \mathbf{H}_k\hat{\mathbf{x}}_{k|k-1} + \mathbf{u}_k^m$, $\mathbf{S}_k = \mathbf{H}_k\mathbf{P}_{k|k-1}\mathbf{H}_k^T + \mathbf{Q}_w$.

Let $\mathbf{y}'_k = \mathbf{S}_k^{-1/2}\mathbf{y}_k$, then $\mathbf{y}'_k \sim \mathcal{N}(\mathbf{S}_k^{-1/2}\hat{\mathbf{y}}_k, \mathbf{I})$. The bearing of $\mathbf{y}'_k$ has a density about its mean of the form $\alpha_\sigma(\theta)$.

Let

$$g'_k(\theta) = [(sin\theta, cos\theta)\mathbf{S}_k^{-1}(sin\theta, cos\theta)^T]^{-1/2} \tag{A4}$$

$$h'_k(\theta) = arctan\left(\frac{s_{11}sin\theta + s_{12}cos\theta}{s_{21}sin\theta + s_{22}cos\theta}\right) \tag{A5}$$

where $[s_{ij}] = \mathbf{S}_k^{-1/2}$.

$g'_k(\theta)$ and $h'_k(\theta)$ are the reciprocal length and bearing of the transformed unit vector $\frac{\mathbf{S}_k^{-1/2}\mathbf{y}_k}{||\mathbf{y}_k||}$. The bearing $h'_k(\theta_k)$ then has $\alpha_{g'_k(\hat{\theta}_k)}(h'_k(\theta_k) - h'_k(\hat{\theta}_k))$ as its density, where $\hat{\theta}_k$ is the bearing of $\hat{\mathbf{y}}_k$. Inserting a Jacobian term, we can obtain the probability density function of measurement θ_k given previous measurements $\mathbf{z}_{1:k-1}$:

$$f(\theta_k|\mathbf{z}_{1:k-1}) = \frac{g_k'^2(\theta_k)}{(det\mathbf{S}_k)^{1/2}}\alpha_{g'_k(\hat{\theta}_k)}(h'_k(\theta_k) - h_k(\hat{\theta}_k)) \tag{A6}$$

References

1. Leong, P.H.; Arulampalam, S.; Lamahewa, T.A.; Abhayapala, T.D. A Gaussian-sum based cubature Kalman filter for bearings-only tracking. *IEEE Trans. Aerosp. Electron. Syst.* **2013**, *49*, 1161–1176. [CrossRef]
2. Aidala, V.J. Kalman filter behavior in bearings-only tracking applications. *IEEE Trans. Aerosp. Electron. Syst.* **1979**, *15*, 29–39. [CrossRef]
3. Aidala, V.J.; Hammel, S. Utilization of modified polar coordinates for bearings-only tracking. *IEEE Trans. Autom. Control* **1983**, *28*, 283–294. [CrossRef]
4. Aidala, V.J.; Nardone, S.C. Biased estimation properties of the pseudo linear tracking filter. *IEEE Trans. Aerosp. Electron. Syst.* **1982**, *18*, 432–441. [CrossRef]
5. Doğançay, K. On the efficiency of a bearings-only instrumental variable estimator for target motion analysis. *Signal Process.* **2005**, *85*, 481–490. [CrossRef]
6. Doğançay, K. Bias compensation for the bearings-only pseudolinear target track estimator. *IEEE Trans. Signal Process.* **2006**, *54*, 59–68. [CrossRef]
7. Nguyen, N.H.; Doğançay, K. Improved pseudolinear Kalman filter algorithms for bearings-only target tracking. *IEEE Trans. Signal Process.* **2017**, *65*, 6119–6134. [CrossRef]
8. Wang, W.P.; Liao, S.; Xing, T.W. The unscented Kalman filter for state estimation of 3-dimension bearing-only tracking. In Proceedings of the 2009 International Conference on Information Engineering and Computer Science, Wuhan, China, 19–20 December 2009; pp. 1–5.
9. Yang, R.; Ng, G.W.; Bar-Shalom, Y. Bearings-only tracking with fusion from heterogenous passive sensors: ESM/EO and acoustic. In Proceedings of the 2015 18th International Conference on Information Fusion (Fusion), Washington, DC, USA, 6–9 July 2015; pp. 1810–1816.
10. Hong, S.H.; Shi, Z.G.; Chen, K.S. Novel roughening algorithm and hardware architecture for bearings-only tracking using particle filter. *J. Electromagn. Waves Appl.* **2008**, *22*, 411–422. [CrossRef]
11. Chang, D.C.; Fang, M.W. Bearing-only maneuvering mobile tracking with nonlinear filtering algorithms in wireless sensor networks. *IEEE Syst. J.* **2014**, *8*, 160–170. [CrossRef]
12. Clark, J.M.C.; Vinter, R.B.; Yaqoob, M.M. The shifted Rayleigh filter for bearings only tracking. In Proceedings of the 2005 7th International Conference on Information Fusion, Philadelphia, PA, USA, 25–28 July 2005; pp. 93–100.
13. Clark, J.M.C.; Vinter, R.B.; Yaqoob, M.M. Shifted Rayleigh filter: A new algorithm for bearings-only tracking. *IEEE Trans. Aerosp. Electron. Syst.* **2007**, *43*, 1373–1384. [CrossRef]
14. Arulampalam, S.; Clark, M.; Vinter, R. Performance of the shifted Rayleigh filter in single-sensor bearings-only tracking. In Proceedings of the 2007 10th International Conference on Information Fusion, Quebec, QC, Canada, 9–12 July 2007; pp. 1–6.
15. Mei, D.; Liu, K.; Wang, Y. MEFPDA-SCKF for underwater single observer bearings-only target tracking in clutter. In Proceedings of the 2013 OCEANS—San Diego, San Diego, CA, USA, 23–27 September 2013; pp. 1–6.
16. Sarkka, S.; Nummenmaa, A. Recursive noise adaptive Kalman filtering by variational Bayesian approximations. *IEEE Trans. Autom. Control* **2009**, *54*, 596–600. [CrossRef]
17. Li, K.; Chang, L.; Hu, B. A variational Bayesian-based unscented Kalman filter with both adaptivity and robustness. *IEEE Sens. J.* **2016**, *16*, 6966–6976. [CrossRef]
18. Sun, J.; Zhou, J.; Li, X.R. State estimation for systems with unknown inputs based on variational Bayes method. In Proceedings of the 2012 15th International Conference on Information Fusion, Singapore, 9–12 July 2012; pp. 983–990.
19. Beal, M.J. Variational Algorithms for Approximate Bayesian Inference. Ph.D. Dissertation, Gatsby Computational Neuroscience Unit, University College London, London, UK, 2003.
20. Hou, J. jaylin254/VB-SRF1. Available online: https://github.com/jaylin254/VB-SRF1 (accessed on 23 March 2019).

Article

DOA Estimation and Self-calibration under Unknown Mutual Coupling

Dong Qi *, **Min Tang, Shiwen Chen, Zhixin Liu** and **Yongjun Zhao**

National Digital Switching System Engineering and Technological Research Center (NDSC),
Zhengzhou 86-450001, China; tangminmvp@126.com (M.T.); ndsccsw@126.com (S.C.);
liuzhixin54@sina.com (Z.L.); zhaoyjzz@163.com (Y.Z.)
* Correspondence: qidong1027@126.com; Tel.: +86-17513261827

Received: 22 January 2019; Accepted: 21 February 2019; Published: 25 February 2019

Abstract: In practical applications, the assumption of omnidirectional elements is not effective in general, which leads to the direction-dependent mutual coupling (MC). Under this condition, the performance of traditional calibration algorithms suffers. This paper proposes a new self-calibration method based on the time-frequency distributions (TFDs) in the presence of direction-dependent MC. Firstly, the time-frequency (TF) transformation is used to calculate the space-time-frequency distributions (STFDs) matrix of received signals. After that, the estimated steering vector and corresponding noise subspace are estimated by the steps of noise removing, single-source TF points extracting and clustering. Then according to the transformation relationship between the MC coefficients, steering vector and MC matrix, we deduce a set of linear equations. Finally, with two-step alternating iteration, the equations are solved by least square method in order to estimate DOA and MC coefficients. Simulations results show that the proposed algorithm can achieve direction-dependent MC self-calibration and outperforms the existing algorithms.

Keywords: DOA estimation; direction-dependent mutual coupling; time-frequency distribution; self-calibration

---

1. Introduction

DOA estimation, as an important branch of array signal processing, is widely used in radar, sonar, radio astronomy and other fields [1]. In the past decades, many classical algorithms have been proposed which perform well in ideal situations, such as MUSIC, ESPRIT and other subspace-based algorithms. However, in practical applications, the performance of above algorithms suffers due to the effect of gain/phase uncertainties [2], sensor position perturbation [3] and mutual coupling (MC) [4], among which the mutual coupling caused by mutual excitation of array elements are common in engineering and makes the estimation accuracy deteriorate seriously.

A number of methods have been proposed for DOA estimation in the presence of mutual coupling, which can be classified into two types: active-calibration [5] and self-calibration [6]. The active-calibration method which makes use of the calibration sources whose DOAs are exactly known, can achieve high DOA accuracy with low computation complexity. However, the existence of the calibration sources increases the additional cost of the system, and the performance of the algorithm deteriorates rapidly in the presence of DOA errors of the calibration sources, which is inevitable in practice.

On the contrary, self-calibration method is preferable since it does not require any prior knowledge of source locations and accomplishes the DOA estimation and error calibration online. In [7], an iterative algorithm is proposed for the estimation of DOA and MC coefficients, however, the result will converge to the local optimum if the initial values deviate far from the real ones. For uniform circular arrays, a self-calibration method is proposed based on rank-reduction estimator by using the complex symmetric Toeplitz property in [8]. The algorithm only needs one-dimensional search which lowers the computational complexity, but its parameter estimation is prone to be ambiguous. Moreover, reference [9,10] utilize alternating iteration and recursive estimation to solve this problem, respectively. Recently the methods which make use of instrumental sensors for array calibration has also been developed [11,12]. They exploit the fact that only part of the new array has mutual coupling or other errors after adding instrumental sensors into the original array. But in practice, it is impossible to obtain the ideal instrumental sensors.

The above algorithms are only adopted to direction-independent mutual coupling which is modelled with a single matrix. As it is established under the assumption of the omnidirectional antenna array, the model becomes ineffective when the array elements are not omnidirectional antennas. However, in practical engineering, due to the limitations of the manufacture and the working environment of the antenna, the array elements have directional beam pattern in general [13]. As a result, the mutual coupling is direction-dependent, leading to performance degradation of the existing algorithms. Thus, it is of great practical significance to study the DOA estimation in the presence of direction-dependent mutual coupling. Few papers are proposed to solve this problem. Based on rank-rare theory, a method of 2D-DOA estimation for direction-dependent MC is proposed in [14]. However, in order to estimate the direction-dependent MC coefficients, the algorithm adopts the idea of receiving mutual-impedance method proposed in [15], which is an off-line measurement algorithm. Therefore, the algorithm fails in the time-varying systems. In [16] Ahmet proposed a method to calibrate the direction-dependent mutual coupling and estimate the DOAs. The algorithm divides the angle search range into several sectors by means of angle sector. By comparing the spectral peaks in each sector, the angle interval of initial angle and the corresponding MC coefficients are estimated, and then self-calibration is completed by iterations. However, this method is easy to fail when the angular spacing between the DOAs of the incident signals is small or the initial value deviation is large.

Motivated by these facts, in this paper a new algorithm for estimation of DOAs and MC coefficients is proposed based on the idea of time-frequency distributions (TFDs) which has been widely applied in blind source separation [17,18]. Firstly, the space-time-frequency distributions (STFDs) matrix of the received signal is solved. Then, the single-source time-frequency (TF) points are extracted by denoising and removing the cross-terms at each TF point, and the optimal STFDs matrix of each signal is estimated by clustering the single-source TF points. Finally, the STFDs matrix is decomposed into eigenvectors and noise subspaces of each signal, and an alternating iteration method based on least squares is proposed for estimation of DOA and MC coefficients. The simulation results show that the algorithm can provide satisfactory performance in case of direction-dependent MC. The main contributions are as follows:

(1) Time-frequency analysis is utilized to solve the problem of direction-dependent mutual coupling in proposed approach.
(2) Compared with the existing algorithms, the proposed method is improved in estimation accuracy and robustness against mutual coupling.
(3) The proposed method can achieve DOA estimation under multipath or underdetermined conditions.

The rest of the paper is organized as follows: Section 2.1 is devoted to the problem formulation. Then an approach based on the TFDs is proposed to obtain the steering vector and noise subspace of each signal by separating the mixed signals, and DOAs and MC coefficients are estimated by a two-step alternate iteration in Section 2.2. Next the algorithm analysis is provided in Section 2.3. Simulations are conducted to illustrate the effectiveness of the proposed methods in Section 3 and conclusions are finally drawn in Section 4.

2. Models and Methods

2.1. Array Signal Model

Considering K far-field narrow-band signals $s_k(t)(k = 1, 2, \cdots, K)$ impinging on the array which is composed of M elements, and the directions of arrival are $\{\theta_1, \theta_2, \cdots, \theta_K\}$, respectively. Then the received signals at the t-th sample can be expressed as:

$$X(t) = A(\theta)s(t) + N(t)t = 1, 2, \cdots, T \tag{1}$$

where $X(t) = [x_1(t), x_2(t), \ldots, x_M(t)]^T$ are the array outputs. $N(t) = [n_1(t), n_2(t), \ldots, n_M(t)]^T$ denotes zero-mean additive white Gaussian noise. $A(\theta) = [a(\theta_1), a(\theta_2), \cdots, a(\theta_K)]$ is the ideal manifold matrix, $a(\theta_i)$ is the steering vector of the i-th signal.

In the presence of direction-dependent MC, the real steering vector is written as:

$$b(\theta) = C(\theta)a(\theta) \tag{2}$$

where $C(\theta) \in \mathbb{C}^{M \times M}$ is the MC matrix.

For a uniform linear array or a uniform circular array model, the MC matrix can be expressed as a band complex symmetric Toeplitz matrix or a three-band complex cyclic matrix, respectively. The following transformation form is used to represent the MC matrix uniformly:

$$C(\theta_k) = CMC(c(\theta_k)) \tag{3}$$

where $CMC(\cdot)$ is the operation of constructing the MC matrix using MC coefficients. $c(\theta_k) = [c_{1k}, c_{2k}, \cdots, c_{pk}, \mathbf{0}_{M-p}]^T \in \mathbb{C}^{M \times 1}$ are the MC coefficients, $\mathbf{0}_{M-p}$ is a $1 \times (M{-}p)$ zero vector and p is the degree of freedom of MC, which equals the number of non-zero elements of $c(\theta_k)$, so in the calculation process, only the first p elements are solved. Thus, the simplified vector $[c_{1k}, c_{2k}, \cdots, c_{pk}]^T$ is used in the formula derivation and simulation conditions. Taking the uniform linear array as an example, the MC matrix of θ_k can be expressed as:

$$C(\theta_k) = \begin{bmatrix} c_{1k} & \cdots & c_{pk} & & \\ \vdots & \ddots & & \ddots & \\ c_{pk} & & c_{1k} & & c_{pk} \\ & \ddots & & \ddots & \vdots \\ & & c_{pk} & \cdots & c_{1k} \end{bmatrix}_{M \times M} \tag{4}$$

Then the array receiving model is expressed as

$$X(t) = B(\theta)s(t) + N(t) \tag{5}$$

where $B(\theta) = [b(\theta_1), b(\theta_2), \cdots, b(\theta_K)] = [C(\theta_1)a(\theta_1), C(\theta_2)a(\theta_2), \cdots, C(\theta_K)a(\theta_K)]$ is the real array manifold matrix containing direction-dependent MC.

2.2. Method of DOA and MC Coefficients Estimation

In this section, the TFDs are introduced to obtain the true steering vector and noise-subspace of each signal, with which DOA and MC coefficients are estimated. The algorithm flow is shown in Figure 1.

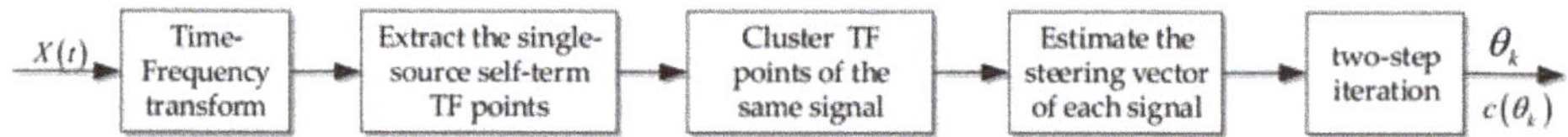

Figure 1. The flow diagram of the proposed algorithm.

The first step is obtaining the STFD matrix by quadric time-frequency transform of $X(t)$. Secondly the single-source self-term TF points are extracted by denoising and removing the cross-term and common-term generated by the signals. After Step 2, the optional STFD matrix of each signal is obtained by clustering TF points of the same signal. Then the steering vector and noise-subspace of each signal are estimated by the eigen-decomposition of STFD matrix. Finally, the two-step iteration is performed to estimate the DOAs and corresponding MC coefficients. The following is the method of DOA and MC coefficient estimation. Firstly, the basic concept of the TFDs and the steps of steering vector estimation are introduced.

2.2.1. Steering Vector Estimation Based on TFDs

For a single signal $x(t)$, the TFDs in discrete time form can be expressed as:

$$\rho_{xx}(t,f) = \sum_{k=-\infty}^{+\infty} \sum_{l=-\infty}^{+\infty} x(t+k+l)x^*(t+k-l)\varphi(k,l)e^{-j4\pi fl} \tag{6}$$

where $\varphi(k,l)$ is the kernel function, $(\cdot)^*$ denotes the conjugate operator, $\rho_{xx}(t,f)$ is the self-term TF point of $x(t)$.

For two signals $x_1(t)$ and $x_2(t)$, the cross-time-frequency distributions in discrete time form can be expressed as:

$$\rho_{x_1x_2}(t,f) = \sum_{k=-\infty}^{+\infty} \sum_{l=-\infty}^{+\infty} x_1(t+k+l)x_2^*(t+k-l)\varphi(k,l)e^{-j4\pi fl} \tag{7}$$

$\rho_{x_1x_2}(t,f)$ is the cross-term TF point of $x_1(t)$ and $x_2(t)$.

Therefore, the STFDs matrix of $X(t)$ is defined as

$$D_{XX}(t,f) = \sum_{k=-\infty}^{+\infty} \sum_{l=-\infty}^{+\infty} X(t+k+l)X^H(t+k-l)\varphi(k,l)e^{-j4\pi fl} \tag{8}$$

According to the definition in paper [17], TF points in time-frequency domains can be divided into three classes including self-term TF points, single-source self-term TF points and cross-term TF points which are expressed as $(t_a, f_a), (t_{as}, f_{as})$ and (t_c, f_c), respectively.

The energy of self-term TF points is generated by one or more sources whose cross-term energy is approximately zero. Single-source self-term TF points are generated by only the self-term of single signal, and the cross-term TF points are generated by the cross-term of the signals whose self-term energy is nearly zero.

Under the noiseless condition, substituting Equation (5) into Equation (8), we obtain:

$$\begin{aligned} D_{XX}(t,f) &= \sum_{k=-\infty}^{+\infty} \sum_{l=-\infty}^{+\infty} Bs(t+k+l)s^H(t+k-l)f(k,l)e^{-j4\pi fl}B^H \\ &= B\left(\sum_{k=-\infty}^{+\infty} \sum_{l=-\infty}^{+\infty} S(t+k+l)S^H(t+k-l)f(k,l)e^{-j4\pi fl} \right)B^H \\ &= BD_{ss}(t,f)B^H \end{aligned} \tag{9}$$

where D_{XX} is the STFDs matrix of array outputs. D_{ss} is the STFDs matrix of the incident signals whose principal diagonal elements are generated by the self-term of incident signals and the non-principal diagonal elements are corresponding to the cross-terms between the signals.

For the self-term TF points, since the signal energy is mainly generated by the self-term of sources, the non-diagonal element of D_{ss} is approximately zero, so it can be expressed as a diagonal matrix, so the STFDs matrix can be expressed as:

$$D_{XX}(t_a, f_a) = B \begin{bmatrix} \rho_{s_1 s_1}(t_a, f_a) & 0 & \cdots & 0 \\ 0 & \rho_{s_2 s_2}(t_a, f_a) & & \vdots \\ \vdots & & \ddots & 0 \\ 0 & & \cdots & 0 & \rho_{s_K s_K}(t_a, f_a) \end{bmatrix} B^H \tag{10}$$

On this basis, when the energy at the TF points is generated only by the single source s_i ($i = 1, 2 \cdots K$), $D_{XX}(t_{as}, f_{as})$ can be expressed as:

$$D_{XX}(t_{as}, f_{as}) = \rho_{s_i s_i}(t_{as}, f_{as}) a_i a_i^H \tag{11}$$

It is deduced from Equation (11) that the steering vector of each signal can be obtained by eigen-decomposition of the STFDs matrix at the TF points of single-source self-term. The procedure of extracting TF points from single source term is as follows.

Step 1: Remove the noise: In order to reduce the computation burden and improve the estimation accuracy, it is necessary to set an appropriate threshold to remove the TF points with low energy which may be generated by noise. For each time slice (t_p, f), apply Equation (12) for all frequency f_q points in this slice, and then the TF points with large energy remain:

$$\frac{\|D_{XX}(t_p, f_q)\|}{\max\limits_{f}\{\|D_{XX}(t_p, f)\|\}} > \varepsilon_1 \tag{12}$$

where ε_1 is a small positive real number and $\|\bullet\|$ represents the operator of 2-norm.

Step 2: Extract the TF points of self-term: After removing the noise points, the retaining TF points mainly include self-term TF points and cross-term TF points. At the self-term TF points, the STFDs matrix is approximately diagonal, and the values of the principal diagonal elements are much larger than those of the other elements, so the STFDs matrix at self-term TF points yields:

$$\frac{trace\{D_{XX}(t, f)\}}{\|D_{XX}(t, f)\|} > \varepsilon_2 \tag{13}$$

where ε_2 is a positive real number close to but less than 1.

Step 3: Extract the TF points of single-source self-term: For signals which overlap in time-frequency domain, the self-term TF points may be composed of multiple signals. Therefore, it is necessary to extract the single-source self-term TF points from the self-term TF points, which can be accomplished by Equation (14):

$$\left| \frac{\lambda_{\max}\{D_{XX}(t, f)\}}{trace\{D_{XX}(t, f)\}} - 1 \right| \leq \varepsilon_3 \tag{14}$$

where ε_3 is a small positive real threshold and $\lambda_{\max}\{D_{XX}(t, f)\}$ is the largest eigenvalue of $D_{XX}(t, f)$.

With the three steps above, the single-source self-term TF points are obtained. Then the steering vectors of each signal as well as the noise subspace could be estimated by the eigen-decomposition of their STFDs matrix. However, the above derivation is completed without considering the noise. In the presence of noise, the steering vectors estimated by only a few TF points are biased. Therefore, it is necessary to obtain more accurate information by clustering the multiple single-source self-term TF points of the same signal. A time-frequency clustering method is provided in Step 4.

Step 4: Cluster the TF points of the same signal: The steering vector of each signal can be estimated as the principal eigenvector of STFDs matrix at each TF point. Regarding the steering vector which contains the DOA information as a feature, all self-term TF points can be classified into $Q(Q \geq K)$ categories by the classification algorithm. That is to say, if the following conditions are satisfied, TF points of $(t_1, f_1), (t_2, f_2)$ belong to the same category:

$$d(a(t_1, f_1), a(t_2, f_2)) < \varepsilon_4 \tag{15}$$

where $d(x,y)$ is the Euclidean distance between x and y, and ε_4 is a small positive threshold.

After clustering, we extract the first K categories which contain the most TF points, and obtain the TFDs of the K signals, respectively. Then the STFDs matrices at those TF points belonging to the first K categories are summed and averaged. Finally, the eigen-decomposition of the average STFDs matrix is performed to estimate the steering vector of each signal and the corresponding noise subspace, which are denoted as $\tilde{b}(\theta_k)$ and $E_n(\theta_k)$, respectively, $k = 1, 2, \cdots, K$.

The above steps of removing the noise, extracting self-term TF points, and clustering TF points of same signal aimed to essentially select the TF points that satisfy the specific conditions, and then the matrix at the TF point is processed.

2.2.2. DOA and MC Coefficients Estimation

For a uniform linear array or uniform circular array model, the transformation relationship between the MC coefficients vector and MC matrix can be expressed by Equation (16):

$$b(\theta_k) = C(\theta_k)a(\theta_k) = T(\theta_k)c(\theta_k) k = 1, 2, \cdots, K \tag{16}$$

where $T(\theta_k)$ is transformation matrix which contains the direction information.

For a uniform linear array, the transformation matrix can be expressed as:

$$T(\theta_k) = Q_1(\theta_k) + Q_2(\theta_k) \tag{17}$$

where:

$$[Q_1(\theta_k)]_{ij} = \begin{cases} [a(\theta_k)]_{i+j-1} & i+j \leq M+1 \\ 0 & i+j > M+1 \end{cases}$$
$$[Q_2(\theta_k)]_{ij} = \begin{cases} [a(\theta_k)]_{i-j+1} & i \geq j \geq 2 \\ 0 & otherwise \end{cases} \tag{18}$$

Similarly, for a uniform circular array the transformation matrix can be expressed as:

$$T(\theta_k) = Q_1(\theta_k) + Q_2(\theta_k) + Q_3(\theta_k) + Q_4(\theta_k) \tag{19}$$

where $Q_1(\theta_k), Q_2(\theta_k), Q_3(\theta_k), Q_4(\theta_k)$ yield:

$$[Q_1(\theta_k)]_{ij} = \begin{cases} [a(\theta_k)]_{i+j-1} & i+j \leq M+1 \\ 0 & i+j > M+1 \end{cases}$$
$$[Q_2(\theta_k)]_{ij} = \begin{cases} [a(\theta_k)]_{i-j+1} & i \geq j \geq 2 \\ 0 & otherwise \end{cases}$$
$$[Q_3(\theta_k)]_{ij} = \begin{cases} [a(\theta_i)]_{M+1+i-j} & i < j \leq l \\ 0 & otherwise \end{cases}$$
$$[Q_4(\theta_k)]_{ij} = \begin{cases} [a(\theta_k)]_{i+j-M-1} & 2 \leq i \leq l, i+j \geq M+2 \\ 0 & otherwise \end{cases} \tag{20}$$

As there is a multiple relation between the estimated steering vector and the actual steering vector, we deduce that:

$$\boldsymbol{b}(\theta_k) = \rho_k \widetilde{\boldsymbol{b}}(\theta_k) = \boldsymbol{C}(\theta_k)\boldsymbol{a}(\theta_k) = \boldsymbol{T}(\theta_k)\boldsymbol{c}(\theta_k) \tag{21}$$

where ρ_k is the multiplier. And the MC coefficients $c(\theta_k)$ can be calculated by the least squares method:

$$\widetilde{\boldsymbol{c}}(\theta_k) = \rho_k \left(\boldsymbol{T}^H(\theta_k)\boldsymbol{T}(\theta_k) \right)^{-1} \boldsymbol{T}^H(\theta_k)\widetilde{\boldsymbol{b}}(\theta_k) \tag{22}$$

According to Equation (22), we find that there also exists a corresponding coefficient relationship between the estimated MC coefficients $\widetilde{c}(\theta_k)$ and the true MC coefficients $c(\theta_k)$. Because the first element of $\widetilde{c}(\theta_k)$ should be 1, the actual MC coefficients can be obtained by normalizing the first element of $\widetilde{c}(\theta_k)$ with Equation (23):

$$\hat{c}(\theta_k) = \frac{\widetilde{c}(\theta_k)}{[\widetilde{c}(\theta_k)]_1} \tag{23}$$

where $[\widetilde{c}(\theta_k)]_1$ represents the first element of $\widetilde{c}(\theta_k)$.

Then DOA of the k-th signal is estimated according to Equation (24) (root MUSIC algorithm could be applied to solve the functions) with the new MC matrix $\hat{C}(\theta_k) = \mathrm{CMC}(\hat{c}(\theta_k))$:

$$\min_{\theta_k} \|\boldsymbol{E}_n^H(\theta_k)\boldsymbol{C}(\theta_k)\boldsymbol{a}(\theta_k)\|^2 k = 1, 2, \cdots, K \tag{24}$$

However, as the direction θ_k and the MC coefficients are unknown, the function cannot be solved directly. Based on the above steps, a new method for DOA estimation and MC coefficients is proposed based on the two-step alternating iteration.

Firstly, initialize the MC coefficients and compute the DOAs $\hat{\theta}_1, \hat{\theta}_2, \cdots, \hat{\theta}_K$ through the conventional algorithms. Then construct the transformation matrix $\boldsymbol{T}(\hat{\theta}_1), \boldsymbol{T}(\hat{\theta}_2), \cdots, \boldsymbol{T}(\hat{\theta}_K)$ according to $\hat{\theta}_1, \hat{\theta}_2, \cdots, \hat{\theta}_K$. On this basis, the MC coefficients is estimated by Equations (22) and (23) with which the new MC matrix is constructed. Finally the DOA of each signal is obtained using Equation (24), and transformation matrix with the new angle information is constructed. In this way, iterations are carried out alternately until the estimation bias is less than the threshold or the iterations have been repeated for certain times. The proposed algorithm is summarized in Table 1.

Table 1. Algorithm for estimation of DOA and direction-dependent MC based on TFDs.

Step 1. Collect N snapshots and calculate the STFDs of received signal with (9).
Step 2. Extract the single source TF points based on (12)–(14).
Step 3. Cluster the single source TF points into $Q(Q \geq K)$ categories using (15).
Step 4. Estimate $\widetilde{b}(\theta_k)$ and $E_n(\theta_k)$ of each signal with the K largest classes.
Step 5. Construct the transformation matrix with $\widetilde{b}(\theta_k)$.
Step 6. Estimate the MC coefficients based on Equations (22) and (23).
Step 7. Construct the MC matrix using the new MC coefficients and estimate the DOAs using (24).
Step 8. Repeat Step 5 to Step 7 until the estimation errors is less than the threshold or the iterations have been repeated for certain times.

2.3. Algorithmic Analysis

1) For the selection of the TFD kernel function, it is known that different TFD kernel functions correspond to different TFD transforms, which can be divided into linear time-frequency transforms and quadratic time-frequency transforms. The Wigner-Ville distribution (WVD) is one of the quadratic TFDs that has better performance in TF focusing and resolution. These two elements are significant in the extraction and clustering of TF points. However, due to the interaction between different signals, cross-terms are generated, which cause false time-frequency information and degrade the estimation accuracy. Therefore, the smoothed pseudo-Wigner-Ville distribution (SPWVD) is utilized to suppress the cross-terms between signals by windowing

method in this paper. Similarly, the short-time Fourier transform (STFT) can also provide the accurate time-frequency distribution to solve the problem and we should choose the kernel function flexibly for the different conditions.

2) This algorithm aims at the estimation of direction-dependent MC, but it is also applicable in the presence of a single MC. The single MC coefficients can be obtained by averaging the estimated MC coefficients of different directions.

3) The algorithm is based on the condition that the TFDs of the signals do not completely overlap, otherwise the blind separation will not be effective. This condition is easy to be obtained in practice. Even the coherent signals (co-frequency interference or multipath signals with different arrival time due to different propagation paths) has different TFDs. Therefore, the proposed algorithm is also effective for coherent signals with different arrival times.

4) This algorithm is able to estimate DOA and MC coefficients under undetermined conditions. That means, the algorithm is still effective when the number of array elements is less than the number of signals. This is because the steering vectors of each signal can be estimated using TFDs, which is similar to estimating the DOA and MC coefficients of each signal separately.

5) For the selection of empirical parameters, we find that ε_1 is the threshold for noise points removing which depends on the ratio between the power of noise and signals. The lager the ratio is, the larger ε_1 should be. However, the power of noise distributes in the whole TF domain uniformly while the power of signals mainly distributes on a few TF points. Therefore, the noise to signal ratio is small in general. Usually we set $\varepsilon_1 \leq 0.3$ in this paper. ε_2 is the threshold for the self-term TF points extraction which approaches to 1 and $\varepsilon_2 = 0.9$ in this paper. ε_3 is a small positive threshold for the single-source TF points extraction and is fixed at 0.2 in this paper. ε_4 is set to cluster the TF points into Q categories and we set $\varepsilon_4 = 0.2$.

3. Numerical Simulation

This section illustrates the effectiveness of the algorithm through simulations with other methods as comparison. Without loss of generality, a uniform linear array composed of 7 elements is selected in this paper, and $p = 3$. Assuming that there are three far-field narrowband LFM (linear frequency modulation) signals in the space, and the duration of three signals is the same. The sampling frequency of system is 50 MHz. The parameters of signals are listed in Table 2.

Table 2. Parameters of Linear Frequency Modulation Signals.

Parameters	LFM Signal 1	LFM Signal 2	LFM Signal 3
Frequency	500 MHz–515 MHz	515 MHz–500 MHz	525 MHz–510 MHz
DOA	$-20°$	$-5°$	$10°$
Snapshots	512	512	512
SNR	$10\,dB$	$10\,dB$	$10\,dB$
MC coefficients	$\begin{bmatrix} 1 \\ 0.623 + i0.587 \\ 0.365 + i0.241 \end{bmatrix}$	$\begin{bmatrix} 1 \\ 0.579 + i0.502 \\ 0.314 + i0.321 \end{bmatrix}$	$\begin{bmatrix} 1 \\ 0.427 + i0.604 \\ 0.256 + i0.336 \end{bmatrix}$

3.1. Calibration Results of Proposed Algorithm

In this section, we provide the results of each step shown in Section 3.1 to verify the effectiveness of the algorithm. Figure 2a shows a three-dimensional time-frequency diagram of the received signal. From which, we find that cross-terms exist in the time-frequency domain of the three signals. Although the smooth pseudo-Wigner distribution is used to remove most of the cross-terms, there still remain some cross-terms in the overlapping part. Figure 2b is a two-dimensional time-frequency distribution which has been binarily processed. It can be seen that except for the time-frequency points of the signal itself, there are also noise points and the time-frequency points of cross-terms between different signals.

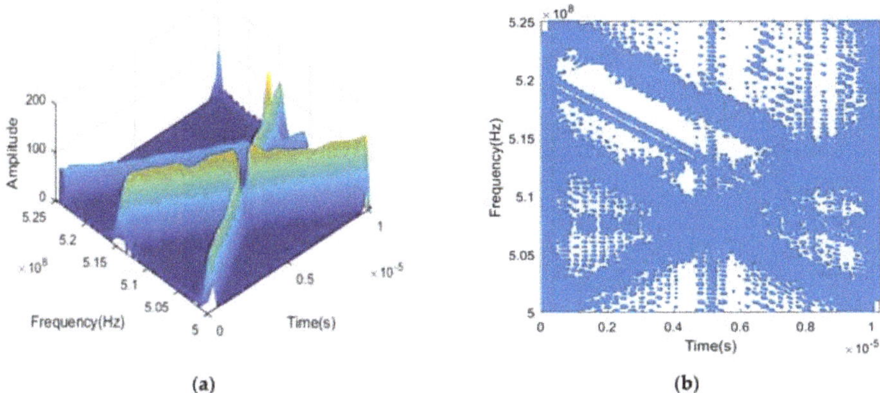

(a)   (b)

Figure 2. Time-Frequency Distribution of received signals. (**a**) Time-Frequency Distribution (3-dimensional); (**b**) Time-Frequency Distribution (2-dimensional).

Figure 3 indicates the processing results introduced in Section 3.1. Figure 3a shows the result of removing the noise points, in which we find that nearly all the noise points can be filtered, while only the TF points with high energy remain. Figure 3b shows the result of the self-term TF points extraction. It is seen that the cross-term TF points are removed but the self-term TF points in the overlapping area are removed as well. Figure 3c shows the result of the single-source self-term TF points extraction. We find that there are only single-source TF points remaining while the overlapping self-term TF points of multiple signals are removed. Figure 3d–f are the TF distribution diagrams of Signals 1–3 after clustering, respectively. It is found that the TF points of each signal can be obtained by clustering with spatial information as feature. And they are matched with the modulation frequency of the real signals.

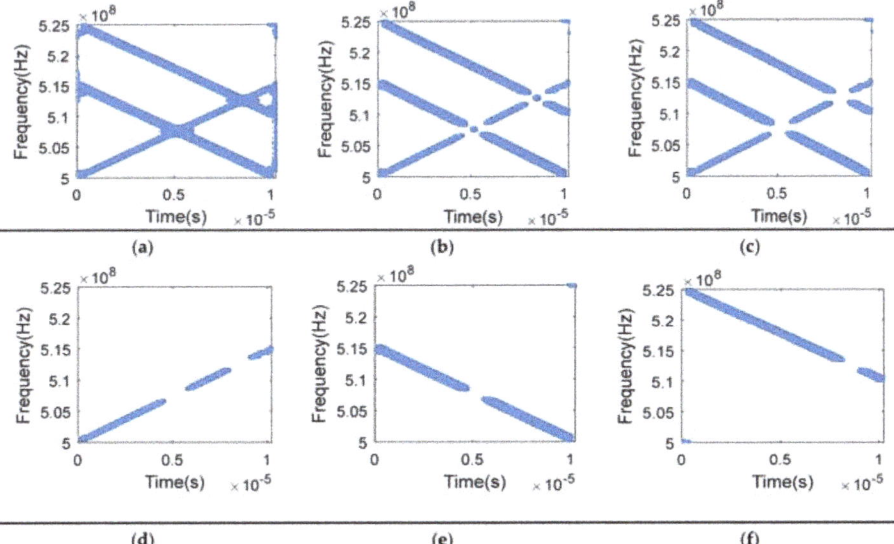

(a)   (b)   (c)

(d)   (e)   (f)

Figure 3. Processing steps of TFDs. (**a**)The TFDs after removing noise points; (**b**) The TFDs of self-term points; (**c**) The TFDs of single-source self-term points; (**d**)The TFDs of signal 1 after clustering; (**e**) The TFDs of signal 2 after clustering; (**f**) The TFDs of signal 3 after clustering.

Figure 4a–c show the variation of the estimated DOAs versus the iteration times of three signals respectively. We find that after a few times of iteration, the estimated DOAs gradually approach the real value and finally stabilize. Figures 5 and 6 indicate variation of the estimated DOAs and MC coefficients versus the iteration times respectively. It can be seen from the figures that the algorithm has high estimation accuracy in DOA and MC coefficients under the condition of $SNR = 10\ dB$ after a few times of iteration.

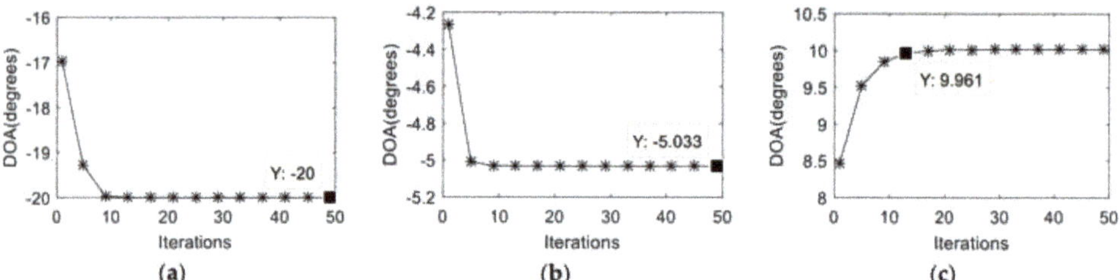

Figure 4. Variation of the estimated DOAs versus the iterations of three signals. (**a**) Variation of the estimated DOAs versus the iterations of signal 1; (**b**) Variation of the estimated DOAs versus the iterations of signal 2;(**c**) Variation of the estimated DOAs versus the iterations of signal 3.

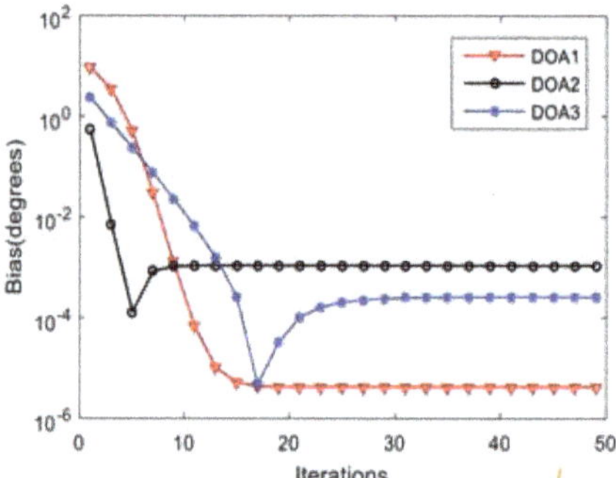

Figure 5. Variation of the estimated DOAs versus the iteration times.

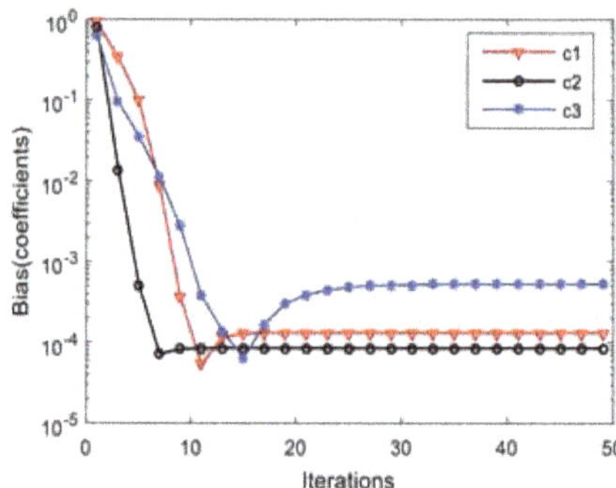

Figure 6. Variation of the estimated MC coefficients versus the iteration times.

3.2. RMSE Comparison versus Input Signal-Noise-Ratio (SNR)

In this section, performance of the proposed algorithm is investigated with different input SNR from -10 dB to 14 dB, which is compared with PEDDMC algorithm [16], conventional MUSIC and the Cramer-Rao lower bound (CRLB) with unknown mutual coupling [7]. The RMSE are obtained through 200 Monte-Carlo simulations. The calculation formula of RMSE is as follows:

$$RMSE = \sqrt{\sum_{n=1}^{N_s} \sum_{i=1}^{K} \left(\theta_i - \hat{\theta}_{i,n}\right)^2 / (KN_s)} \tag{25}$$

As shown in the Figure 7, when the SNR is low, estimation errors of the three algorithms are large. With the increase of SNR, performance of the proposed algorithm and PEDDMC algorithm are improved. However, the proposed algorithm has lower RMSE because the TFDs algorithm has better anti-noise performance, and in addition the DOA estimation and error calibration for each signal are carried out separately. As a result, the proposed method has better performance and its RMSE follows the CRLB closely. As for the MUSIC algorithm, it is a classical super-resolution algorithm with superior estimation performance and robustness under ideal conditions, but it fails in the presence of mutual coupling conditions, since it has no ability to achieve calibration. As a result its performance does not improve as SNR increases.

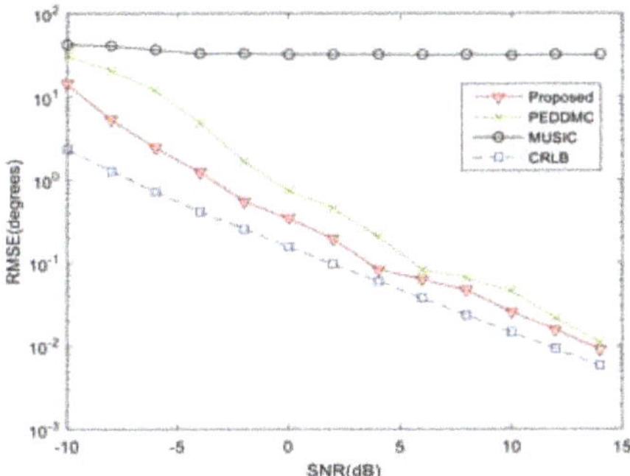

Figure 7. Comparison of DOA estimation performance versus SNR.

3.3. RMSE Comparison versus Input Snapshots

This section compares the performance of the algorithms with different number of snapshots from 100 to 1000. We fix the SNR at 15 dB and calculate the output RMSE through 200 Monte Carlo simulations. From Figure 8, it is known that the estimation performance of proposed algorithm is poor under the condition of small snapshots. This is because small snapshots will lead to the TF resolution deterioration of time-frequency distribution, which affects the extracting and clustering of TF points from single-source self-term. With the increase of snapshots, the RMSE of the proposed algorithm gradually becomes lower than that of the PEDDMC algorithm and becomes very close to the CRLB.

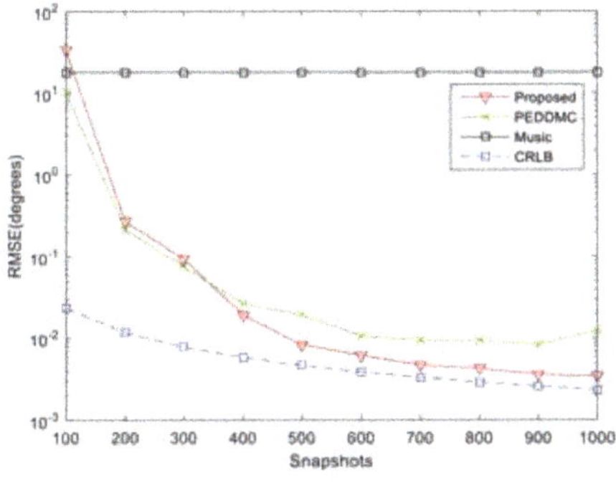

Figure 8. Comparison of DOA estimation performance versus snapshots.

4. Conclusions

Considering the problem that the performance of traditional calibration algorithms degrades in the presence of direction-dependent mutual coupling, this paper introduces the time-frequency analysis method into array signal processing and proposes a new self-calibration algorithm based on alternating iteration. The simulation results show that the proposed algorithm is effective in the presence of direction-dependent mutual coupling and outperforms the existing algorithms. Though the computational complexity of the proposed algorithm is higher because it requires more snapshots to ensure the estimation accuracy of time-frequency transform, compared with the other existing algorithms, the proposed algorithm can perform DOA estimation under multipath or underdetermined conditions.

5. Patents

The authors would like to thank the editor and anonymous reviewers for their careful reading and constructive comments which provide an important guidance for our paper writing and research work. This work was supported by the National Natural Science Foundation of China under Grant 61703433.

Author Contributions: All authors contributed extensively to the study presented in this manuscript. D.Q. designed the main idea, methods and experiments, interpreted the results and wrote the original paper. M.T. and Z.L supervised the main idea, edited the manuscript and provided many valuable suggestions on visualization. S.C. and Y.Z. offered help with the funding.

Funding: This research was funded by National Natural Science Foundation of China, grant number 61703433.

Acknowledgments: The authors would like to thank anonymous reviewers for their valuable comments and suggestions.

Conflicts of Interest: The authors declare no conflict of interest.

References

1. Krim, H.; Viberg, M. Two decades of array signal processing research: The parametric approach. *IEEE Signal Process. Mag.* **1996**, *13*, 67–94. [CrossRef]
2. Zhang, D.; Zhang, Y.; Zheng, G.; Feng, C.; Tang, J. ESPRIT-Like Two-Dimensional DOA Estimation for Monostatic MIMO Radar with Electromagnetic Vector Received Sensors under the Condition of Gain and Phase Uncertainties and Mutual Coupling. *Sensors* **2017**, *17*, 2457. [CrossRef] [PubMed]
3. Liu, Z.; Wang, R.; Zhao, Y. A Bias Compensation Method for Distributed Moving Source Localization Using TDOA and FDOA with Sensor Location Errors. *Sensors* **2018**, *18*, 3747. [CrossRef] [PubMed]
4. Li, W.; Zhang, Y.; Lin, J.; Guo, R.; Chen, Z. Wideband Direction of Arrival Estimation in the Presence of Unknown Mutual Coupling. *Sensors* **2017**, *17*, 230. [CrossRef] [PubMed]
5. Liu, S.; Yang, L.; Yang, S. Robust Joint Calibration of Mutual Coupling and Channel Gain/Phase Inconsistency for Uniform Circular Array. *IEEE Antennas Wirel. Propag. Lett.* **2016**, *15*, 1191–1195. [CrossRef]
6. Hou, Y.; Wen, B.; Tian, Y.; Yang, J. A uniform linear array self-calibration method for UHF river flow detection radar. *IEEE Antennas Wirel. Propag. Lett.* **2017**, *16*, 1899–1902. [CrossRef]
7. Friedlander, B.; Weiss, A.J. Direction finding in the presence of mutual coupling. *IEEE Trans. Antennas Propag.* **1991**, *39*, 273–284. [CrossRef]
8. Xie, J.L.; He, Z.S.; Li, H.Y. A fast DOA estimation algorithm for uniform circular arrays in the presence of unknown mutual coupling. *Prog. Electromagn. Res. C* **2011**, *21*, 257–271. [CrossRef]
9. Wang, M.; Ma, X.C.; Yan, S.F.; Hao, C. An auto-calibration algorithm for uniform circular array with unknown mutual coupling. *IEEE Antennas Wirel. Propag. Lett.* **2015**, *5*, 315–318.
10. Dai, J.S.; Bao, X.; Hu, N.; Chang, C.; Xu, W. A recursive RARE algorithm for DOA estimation with unknown mutual coupling. *IEEE Antennas Wirel. Propag. Lett.* **2014**, *13*, 1593–1596.
11. Wang, H. An Efficient Algorithm for Direction Finding against Unknown Mutual Coupling. *Sensors* **2014**, *14*, 20064–20077. [CrossRef] [PubMed]
12. Wang, B.; Wang, W.; Gu, Y.; Lei, S. Underdetermined DOA Estimation of Quasi-Stationary Signals Using a Partly-Calibrated Array. *Sensors* **2017**, *17*, 702. [CrossRef] [PubMed]

13. Kraus, J.D.; Marhefka, R.J. *Antennas for All Applications*; McGraw: New York, NY, USA, 2002.
14. Wang, B.H.; Hui, H.T.; Leong, M.S. Decoupled 2D Direction of Arrival Estimation Using Compact Uniform Circular Arrays in the Presence of Elevation-Dependent Mutual Coupling. *IEEE Trans. Antennas Propag.* **2010**, *58*, 747–755. [CrossRef]
15. Hui, H.T. Improved compensation for the mutual coupling effect in a dipole array for direction finding. *IEEE Trans. Antennas Propag.* **2003**, *51*, 2498–2503. [CrossRef]
16. Elbir, A. Direction Finding in the Presence of Direction-Dependent Mutual Coupling. *IEEE Antennas Wirel. Propag. Lett.* **2017**, *16*, 1541–1544. [CrossRef]
17. Liu, Y.; Liu, C.; Zhao, Y.; Zhu, J. Wideband array self-calibration and DOA estimation under large position errors. *Digit. Signal Process.* **2018**, *78*, 250–258. [CrossRef]
18. Zhang, C.; Wang, Y.; Jing, F. Underdetermined Blind Source Separation of Synchronous Orthogonal Frequency Hopping Signals Based on Single Source Points Detection. *Sensors* **2017**, *17*, 2074. [CrossRef] [PubMed]

Article

Time Difference of Arrival (TDoA) Localization Combining Weighted Least Squares and Firefly Algorithm

Peng Wu, Shaojing Su, Zhen Zuo *, Xiaojun Guo, Bei Sun and Xudong Wen

College of Intelligence Science and Technology, National University of Defense Technology, Changsha 410073, China; pengwu9510@163.com (P.W.); susj-5@163.com (S.S.); jeanakin@nudt.edu.cn (X.G.); beys1990@163.com (B.S.); wenxudong13@163.com (X.W.)
* Correspondence: z.zuo@nudt.edu.cn

Received: 11 April 2019; Accepted: 2 June 2019; Published: 4 June 2019

Abstract: Time difference of arrival (TDoA) based on a group of sensor nodes with known locations has been widely used to locate targets. Two-step weighted least squares (TSWLS), constrained weighted least squares (CWLS), and Newton–Raphson (NR) iteration are commonly used passive location methods, among which the initial position is needed and the complexity is high. This paper proposes a hybrid firefly algorithm (hybrid-FA) method, combining the weighted least squares (WLS) algorithm and FA, which can reduce computation as well as achieve high accuracy. The WLS algorithm is performed first, the result of which is used to restrict the search region for the FA method. Simulations showed that the hybrid-FA method required far fewer iterations than the FA method alone to achieve the same accuracy. Additionally, two experiments were conducted to compare the results of hybrid-FA with other methods. The findings indicated that the root-mean-square error (RMSE) and mean distance error of the hybrid-FA method were lower than that of the NR, TSWLS, and genetic algorithm (GA). On the whole, the hybrid-FA outperformed the NR, TSWLS, and GA for TDoA measurement.

Keywords: TDoA; weighted least squares; firefly algorithm; hybrid-FA

---

1. Introduction

Target localization based on a group of sensor nodes whose positions are known has been extensively studied in research on signal processing [1–3]. It has been applied widely in military and civil fields, including sensor networks [4], wireless communication [2], radar [5], navigation, and so forth [6–8]. Commonly adopted positioning methods include the signal's time of arrival (ToA) [9], time difference of arrival (TDoA), frequency difference of arrival (FDoA) [10–13], or doppler shift [14]. Compared with FDoA, the TDoA and ToA methods can achieve higher positioning accuracy and require only one channel for each sensor node to perform the measurement, which can minimize the load requirement for a single-sensor node.

For a passive location system based on TDoA, once the measured data are obtained, the range difference between the target and two different sensor nodes can be calculated. In this connection, a set of hyperbolic equations or hyperboloids can be obtained and the solution of the equations is the coordinate of the target. Generally, the solving algorithms commonly adopted include iterative, analytical, and search methods.

The procedure for solving equations from the TDoA method is complex and difficult because the equations are nonlinear, and many studies have been carried out on how to solve this issue. The main idea of the Taylor series method is to expand the first Taylor series of the nonlinear positioning equation at the initial estimation of the target position and then solve the equations by iteration [15].

The advantage of this method is that it can fuse multiple observation data. Yang et al. transformed the equation into a constrained weighted least squares (CWLS) estimation problem by introducing auxiliary variables, and then the Newton iteration method was adopted to solve the problem [16]. In [17], nonconvex TDoA localization was transformed into a convex semidefinite programming (SDP) problem, and the approximate result was taken as the initial value for the Newton iteration method. All of these methods are iterative. Compared with iterative methods, the closed-form method does not need the initial estimation of the target's location and iterative solving is not necessary either. For example, Chan and Ho [18] transformed nonlinear equations into pseudolinear equations by introducing auxiliary variables. Then, the equations were solved by two-step weighted least squares (TSWLS). One downside of this algorithm is that the result is substantially different than the actual position when the signal-to-noise ratio (SNR) is low. Considering this problem, the constrained total least squares (CTLS) method was proposed in [19–21]. While it is not a closed-form method and Newton iterations are needed, the complexity of the CTLS method is much higher than that of the TSWLS method. The approximate maximum likelihood (AML) method [22] was proposed, which can obtain a linear equation from the maximum likelihood function and then the target location can be calculated. The AML method has better positioning performance than the TSWLS method.

In addition to the traditional TSWLS, iteration methods, and so on, many scholars have investigated new methods to enhance positioning accuracy. Two new shrinking-circle methods were proposed (SC-1 and SC-2) to solve a TDoA-based localization problem in a 2-D space [23]. Additionally, a weighted least squares (WLS) algorithm with the cone tangent plane constraint for hyperbolic positioning was proposed, which added the distance between the target and the reference sensor as a new dimension [24]. The theoretical bias of maximum likelihood estimation (MLE) is derived when sensor location errors and positioning measurement noise both exist [25]. Using a rough estimated result by MLE to subtract the theoretical bias can deliver a more accurate source location estimation. Apart from this, research based on certain typical algorithms has been carried out to extract and calculate the TDoA of ultrahigh frequency (UHF) signals [26]. The AML algorithm was proposed for determining a moving target's position and velocity by utilizing TDoA and FDoA measurements [27].

It is also efficient to use a search algorithm to calculate the position of a target. A hybrid genetic algorithm (GA) was proposed to enhance solution accuracy [18]. Nature-inspired algorithms are powerful algorithms for optimization. The firefly algorithm (FA) is one such nature-inspired algorithm, which was proposed in 2008. Using the FA for multimodal optimization applications with high efficiency has been proposed [28,29].

In general, among the methods for solving TDoA equations, analytical and iterative methods both have limitations. Research on algorithms that are robust and have low computational complexity is still worthy of study. Search algorithms for TDoA measurements can provide accurate results, although the efficiency will inevitably decrease when there are many estimated parameters [29]. Therefore, it is necessary to develop a highly efficient search algorithm for TDoA.

This paper is organized as follows: Section 2 introduces the basic model of TDoA measurement. Section 3 formulates the basic principle of WLS. Section 4 provides the main steps of the FA. Section 5 details the hybrid-FA methods proposed in this paper. Section 6 presents the results of simulations and experiments to support the theoretical analysis.

2. Problem Description

In this section, 2-D target localization based on TDoA measurement is presented in the line-of-sight environment. Assume that there are N ($N \geq 3$) sensor nodes, which can also be called basic sensors (BSs), to determine the position of the target. The coordinates of the sensor nodes are known, which are $s_i = (a_i, b_i)^T, i \in \{1, 2, ..., N\}$, where $[\cdot]^T$ denotes the matrix transpose. Assume that the target's coordinate is $p = (x, y)^T$.

As shown in Figure 1a, there are three basic sensors in the 2-D plane to determine the position of the target which form two groups of hyperbolas [24]. The hyperbola has two intersections in the absence of noise and there is one ambiguous position in them. When noise exists, the other two groups of hyperbolas have the other two intersections and both intersections have errors. In order to avoid the ambiguous position, it is advisable to increase the number of the sensors. As demonstrated in Figure 1b, the four basic sensors form three groups of hyperbolas and there is only one intersection without noise, which is the estimated position of the target. When noise exists, it is necessary to follow certain principles to obtain the optimal results.

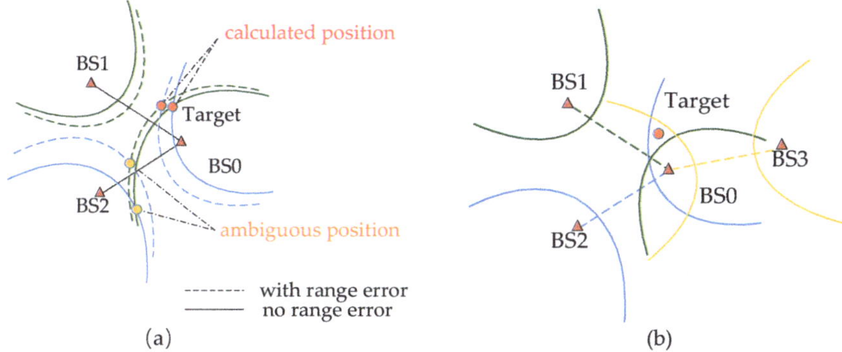

Figure 1. The principle of time difference of arrival (TDoA) measurement. (**a**) The diagram when there are three basic sensors of TDoA. (**b**) Multiple hyperbolas for the optimal position.

Take the first basic sensor BS_1 as a reference sensor and assume that the signal propagates in a straight line between the target and each basic sensor without considering the influence of non-line-of-sight propagation. Assume that the times when the signal arrives at basic sensors BS_1 and BS_i are t_1 and t_i, respectively, and the propagation speed of the signal is c. The range of difference between the target and two basic sensors BS_1 and BS_i is $\{r_{i,1}\}$. This paper assumes that range difference errors $\{n_i\}$ are independent Gaussian random variables with zero mean and known variance σ_i^2, i.e., $\mathbb{N}(0, \sigma_i^2)$. We can obtain

$$r_{i,1} = c|t_1 - t_i| \tag{1}$$

$$r_{i,1} = d_{i,1} + n_{i,1}, i \in \{2,...,N\}. \tag{2}$$

Thus,

$$c|t_1 - t_i| = d_{i,1} + n_{i,1} \tag{3}$$

where $d_{i,1} = d_i - d_1$. Here, distances between the target and the receiver pair BS_1 and BS_i can be expressed as follows:

$$d_1 = \sqrt{(x - a_1)^2 + (y - b_1)^2} \tag{4}$$

$$d_i = \sqrt{(x - a_i)^2 + (y - b_i)^2}, i \in \{2,...,N\}. \tag{5}$$

Actually, the process of obtaining results based on TDoA measurements is the process of solving the $N - 1$ equations as shown in Equation (3) and obtaining the optimal solution.

3. WLS Method

Usually, there are iterative methods, such as those mentioned in Section 1, to solve the equations, for which the computational burden is heavy. In this section, the WLS method is introduced based on TDoA measurements [29]. The sum of squares of residuals is defined as $J_{NLS}(\tilde{x})$:

$$J_{NLS}(\tilde{x}) = min \sum_{i=1}^{N} R_i^2(\tilde{x}) \tag{6}$$

where $\tilde{x}$ represents the optimization variable, and residual $R_i(\tilde{x})$ can be expressed as

$$R_i(\tilde{x}) = \tilde{r}_{i,1} - r_{i,1} \tag{7}$$

where $\tilde{r}_{i,1}$ is the measured value. Therefore, the optimal solution $\hat{p}$ according to the principle of minimum variance is

$$\hat{p} = \underset{\tilde{x} \in R^2}{argmin} J_{NLS}(\tilde{x}). \tag{8}$$

Nonlinear hyperbolic equations can be transformed as follows:

$$r_{i,1} + \sqrt{(x - a_1)^2 + (y - b_1)^2} = \sqrt{(x - a_i)^2 + (y - b_i)^2} + n_{i,1}, \ i \in \{2,...,N\}. \tag{9}$$

After mathematical transformation, we can obtain

$$(x - a_1)(a_i - a_1) + (y - b_1)(b_i - b_1) + r_{i,1}d_1 = \frac{1}{2}\left[(a_i - a_1)^2 + (b_i - b_1)^2 - r_{i,1}^2\right] + d_i n_{i,1}, i \in \{2,...,N\} \tag{10}$$

where the second-order term $n_{i,1}^2$ is ignored and $e_{i,1} = d_i n_{i,1}$. We can obtain

$$AX = \theta + e \tag{11}$$

in which

$$A = \begin{bmatrix} a_2 - a_1 & b_2 - b_1 & r_{2,1} \\ a_3 - a_1 & b_3 - b_1 & r_{3,1} \\ \vdots & \vdots & \vdots \\ a_N - a_1 & b_N - b_1 & r_{N,1} \end{bmatrix} \tag{12}$$

$$X = [x - a_1 \ y - b_1 \ d_1] \tag{13}$$

$$\theta = \frac{1}{2} \begin{bmatrix} (a_2 - a_1)^2 + (b_2 - b_1)^2 - r_{2,1}^2 \\ (a_3 - a_1)^2 + (b_3 - b_1)^2 - r_{3,1}^2 \\ \vdots \\ (a_N - a_1)^2 + (b_N - b_1)^2 - r_{N,1}^2 \end{bmatrix}, \tag{14}$$

$$e = [e_{2,1} \ e_{3,1} \cdots e_{N,1}]^T. \tag{15}$$

Then, the WLS objective function can be expressed as

$$J_{WLS}(X) = (AX - \theta)^T W (AX - \theta) \tag{16}$$

where the weighting matrix is $W = \left(E\{ee^T\}\right)^{-1}$.

Thus,

$$\hat{x}_{WLS} = \left(A^T W A\right)^{-1} A^T W \theta. \tag{17}$$

The WLS method is often adopted because of its simplicity and lower computational burden. As the equations are approximated in the process of simplification, it has low accuracy.

4. Firefly Algorithm

In this section, the principle of the FA is introduced, which is a kind of heuristic algorithm inspired by the flickering behavior of fireflies. The three following idealized rules are needed for the model of this algorithm [30]:

- Each firefly will be attracted by the other fireflies regardless of their sex.
- The higher the brightness of the firefly, the greater the attractiveness of the firefly. In this connection, a less bright firefly will move towards a brighter one.
- The brightness of the fireflies is associated with the objective function.

Through the attraction between the brighter and less bright fireflies, the fireflies will eventually gather around the brightest firefly, which can realize the optimization of the objective function. We can use FA search methods to obtain the optimal result that satisfies Formula (18):

$$J_{NLS}(\tilde{x}) = min \sum_{i=1}^{N} R_i^2(\tilde{x}). \tag{18}$$

In the search space, fireflies move towards brighter fireflies continuously to complete optimization until the preset termination condition of the algorithm is reached.

Assuming that the number of fireflies is N and the dimension is D, the positions of the ith and j^{th} fireflies are $x_i = (x_{i1}, x_{i2}, \cdots, x_{iD}), i = 1, 2, \cdots, N$ and $x_j = (x_{j1}, x_{j2}, \cdots, x_{jD}), j = 1, 2, \cdots, N$. r_{ij} is the distance between the ith and j^{th} fireflies, which can be calculated as follows:

$$r_{ij} = \|x_i - x_j\| = \sqrt{\sum_{d=1}^{D} (x_{id} - x_{jd})^2}. \tag{19}$$

Among them, x_{id} and x_{jd} represent the positions of the ith and j^{th} fireflies, respectively.

The relative brightness of the fireflies is defined as

$$I = I_0 e^{-\gamma r_{ij}^2} \tag{20}$$

where I_0 represents the brightness of the firefly, which is proportional to the value of the objective function. γ is the coefficient of absorbing light intensity, which is usually defined as a constant. r_{ij} denotes the distance of fireflies i and j.

The attractiveness of the fireflies is defined as follows:

$$\beta = \beta_0 e^{-\gamma r_{ij}^2} \tag{21}$$

where β_0 denotes the factor of maximum attraction degree, indicting the attractiveness of the position with the maximum brightness. From Formula (21), we understand that attractiveness decreases with the increase of the distance and the coefficient of absorbing light intensity.

The update of the location is

$$x_{id}(t+1) = x_{id}(t) + \beta \left(x_{jd}(t) - x_{id}(t) \right) + \varepsilon \cdot \alpha_i(t) \tag{22}$$

where $x_{id}(t)$ and $x_{jd}(t)$ are the positions of the ith and j^{th} fireflies after the tth generation. $\alpha_i(t)$ denotes the step factor of the tth generation. The range of the value ε is $[-0.5, 0.5]$, which sequences with uniform distribution.

The process of optimization is as follows. Fireflies with varying degrees of brightness are randomly dispersed in the solution space. The brightness and attractiveness of the fireflies can be calculated according to Equations (20) and (21), respectively. The less bright fireflies will move towards the brighter one. In order to avoid falling into the local optimum, the perturbation term $\varepsilon \cdot \alpha_i(t)$ is added to the process of location updating. Finally, the fireflies will gather around the firefly with highest brightness. The optimal result can thus be obtained. The flowchart for this can be found below.

Step 1. Initialize the parameters in the algorithm. Set the number of fireflies N, the factor of maximum attraction degree β_0, and maximum iteration number or convergence criterion.

Step 2. Initialize the location of the fireflies randomly and calculate the value of the objective function as the original brightness.

Step 3. Calculate the brightness and attractiveness of fireflies referring to Equations (20) and (21), respectively, and determine the moving direction of the fireflies according to their relative brightness.

Step 4. Update the location of the fireflies according to Equation (22) and add the perturbation terms.

Step 5. Recalculate the brightness of the fireflies after updating the location of the fireflies.

Step 6. When the convergence criterion is satisfied or the maximum number of iterations reached, go to the next step; otherwise, go to Step 3.

Step 7. Output the global extremum and optimal value.

5. Hybrid-FA Method

While it usually takes more time to use a search algorithm than iterative methods, this method provides higher accuracy and thus has great potential in practical applications. There are certain reasons for the longer solution time. One significant cause is that it will search the optimal result in the global scope. In this context, if some reasonable regional restrictions are given, a search algorithm, including the FA method, will reduce the computation amount as well as ensure the accuracy of the result.

In this study, the WLS and FA methods were combined for optimal implementation. Since it is easy to obtain the initial result by the WLS method, the initial result can be used to provide the limited area for the FA method.

Assume that the search area is square and the length of a side is l. If the result obtained by the WLS method is (x_{wls}, y_{wls}), which is the initial value, then the constrained region can be given for the FA method as $[x_{wls} \pm \frac{1}{4}l] \times [y_{wls} \pm \frac{1}{4}l]$ to ensure the target falls into the restricted region as much as possible. Additionally, new firefly positions out of the restricted region are ignored in this method.

As shown in Figure 2, the WLS and FA methods are combined for optimal implementation. The initial value is obtained by the WLS method, then the initial result can be used to provide the restrained area for the FA method to search for the optimal result. There are two conditions when the algorithm ends, fulfilling the convergence criterion or implementing maximum iterative times set in advance. As for the former condition, the values of the objective function obtained in the $i - 1^{th}$ and i^{th} iteration are compared. Assuming that the objective function is $\mathbb{F}$, then the ending condition can be expressed as follows:

$$\|X(i^{th}) - X(i - 1^{th})\| \leq \varepsilon_1 \tag{23}$$

$$\|\mathbb{F}(i^{th}) - \mathbb{F}(i - 1^{th})\| \leq \varepsilon_2 \tag{24}$$

where ε_1 and ε_2 are positive predetermined numbers.

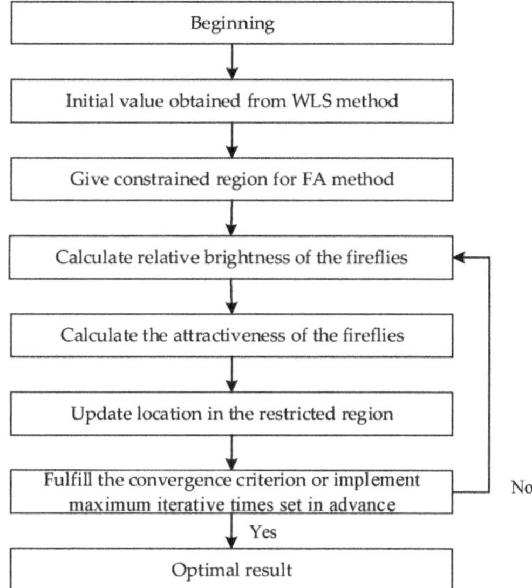

Figure 2. The diagram of hybrid firefly algorithm (hybrid-FA) method.

6. Results

6.1. Preprocessing

Simulations and indoor experiments were conducted and the results of the proposed method and other commonly used methods are presented and compared here. Before the simulations and experiments, the definition of the SNR and the evaluation index are given.

In this paper, the SNR of the signal is defined as

$$SNR = 10 \log \frac{d_{i,1}}{\sigma_i^2} \text{ dB} \tag{25}$$

where $d_{i,1}$ denotes the distance difference between the target and the basic sensors BS_1 and BS_i, and σ_i represents the standard deviation of the noise.

Therefore, when used in practice, if the SNR is known in advance, the variance of noise can be obtained:

$$\sigma_i^2 = \frac{d_{i,1}^2}{10^{SNR/10}}. \tag{26}$$

Otherwise, when the SNR is not known in advance, the variance of noise can be obtained using the approximate method. The performance index of the receiver is usually known according to the specifications or can be measured by testing. Here, the signal's time of arrival was measured by the receiver. Assume that the measurement error targeting at the time of arrival of different receivers is no more than $\Delta \hat{t}_i (i = 1, 2, ..., M)$, where M denotes the number of the receivers. Assume that the time difference of arrival between BS_1 and BS_i is $\Delta t_{1,i}$.

Then, we can get

$$\Delta t_{1,i} = |t_1 - t_i| \tag{27}$$

where t_1 and t_i denote the measured values when the signal arrives at the basic sensors BS_1 and BS_i, respectively.

The variance of the noise of $\Delta t_{1,i}$ can be approximated to

$$\sigma_i^2 = c^2\left(\Delta \hat{t}_1^2 + \Delta \hat{t}_i^2\right) \tag{28}$$

where c represents the propagation velocity of the signal.

The localization performance was evaluated referring to the root-mean-square error (RMSE), which is defined as

$$RMSE = \sqrt{\frac{1}{n}\sum_{i=1}^{n}\left[(\hat{x}_i - x)^2 + (\hat{y}_i - y)^2\right]} \tag{29}$$

where n represents the number of simulation times, (x, y) are the real position coordinates of the target, and $(\hat{x}_i, \hat{y}_i)$ are the estimated positions based on the ith calculation. RMSE was used to measure the average coordinate distance between the estimated target position and the actual target position. The lower it is, the higher the accuracy is.

6.2. Simulation Conditions

The simulation was performed using Matlab 2014a and all the results were obtained on the same computer with a 1.8 GHz CPU and 8 G RAM. Assume that the coordinates of the four basic sensors were $BS1(0,0)$, $BS2(0,10)$, $BS3(10,10)$, and $BS4(10,0)$, where $BS1$ is the reference basic sensor. The layout of the four sensor nodes is shown in Figure 3.

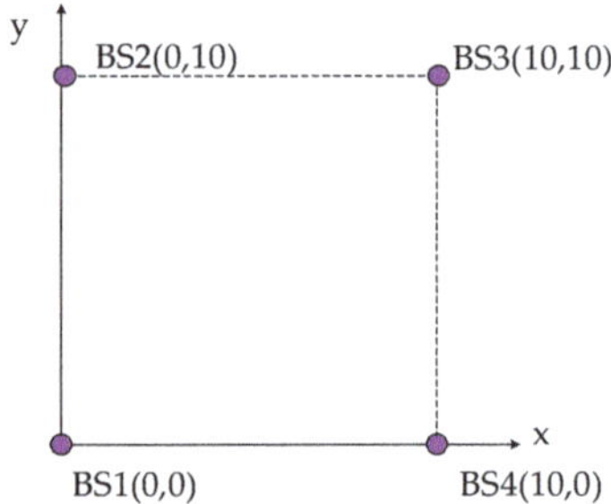

Figure 3. The layout of four sensor nodes in the simulation experiment.

6.3. The Robustness

The results of the simulations were as follows when the SNR = 30 dB. In order to have a better display, the vertical coordinate represents -RMSE. Therefore, the optimal result was the coordinate when the -RMSE reached the maximum.

As illustrated in Figure 4, the FA method searched for the optimal result in the global region ([0,10] × [0,10]). From the figure, we can see that the result obtained by the FA method was even closer to the actual target location than that obtained by the WLS method, while the hybrid-FA method simply searched in the square marked in red. In this connection, it reduced the computational burden and maintained high accuracy.

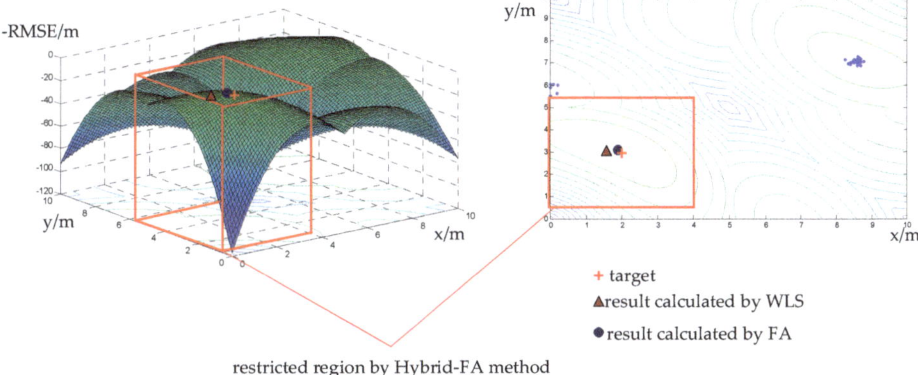

Figure 4. The diagram of hybrid-FA and FA.

Table 1 shows the comparison between FA and hybrid-FA under the same condition when SNR = 30 dB. When the RMSE reached 0.03741 m, the hybrid-FA method only needed to iterate 30 times, while 100 times was required for the FA method by itself. The simulation results demonstrate the efficiency of the hybrid-FA method.

Table 1. The root-mean-square error (RMSE) of the hybrid-FA and FA methods with different numbers of iterations.

	25 Iterations	30 Iterations	50 Iterations	100 Iterations
FA method	0.04334 m	0.04943 m	0.03762 m	0.03741 m
Hybrid-FA method	0.03744 m	0.03741 m	0.03741 m	0.03741 m

In this section, the results of the commonly used algorithms CWLS [31], Newton–Raphson (NR) [29], TSWLS [18], and GA [30], were compared with the hybrid-FA. The GA method is a search algorithm that is commonly used for optimization. For these methods, the number of iterations was 30 and the coordinate of the target was set as (2,3). Figure 5 illustrates the compared results.

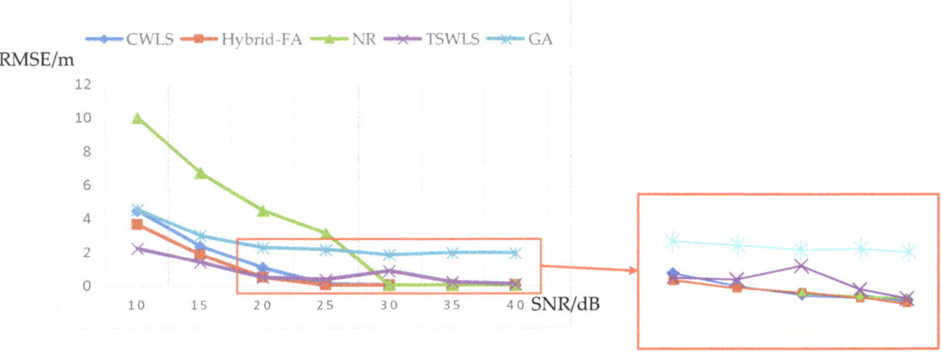

Figure 5. The comparison of the four algorithms for TDoA measurement.

As demonstrated in Figure 5, with the increase of SNR, the RMSE of each algorithm tended to decrease, which means the accuracy of the position had improved. When the SNR = 10 and 15 dB, the RMSE of TSWLS was the lowest, while the RMSE of the other five methods was typically higher than 2 m. It was hard to achieve high accuracy when the SNR was very low, as the acquired data was limited. The data of FDoA or TOA should be combined to get higher accuracy. When the SNR = 20, 25, 35, and 40 dB, the RMSE of the hybrid-FA was the lowest, which was 0.46077, 0.03788, 0.02427, and 0.0177 m, respectively. When SNR = 30 dB, the RMSE of the CWLS method was 0.0001 m lower than that of hybrid-FA method. On the whole, the hybrid-FA method was better than the NR, TSWLS, and GA methods when the SNR ranged from 20 to 40 dB in the simulations. The result of GA is also shown in Figure 5. The RMSE of GA was higher than that of the proposed scheme, which illustrates that the proposed scheme is more appropriate for optimization than GA.

6.4. Experiment

Experiments were carried out to verify the rationality of the algorithms [23,32]. The coordinates of four speakers were *BS1*(0,0)m, *BS2*(0,10)m, *BS3*(10,10)m, and *BS4*(10,0)m, which were used to generate signals. The speakers emitted chirp signals, which were continuous impulse signals of 2.5 kHz. A phone was placed at the same height to receive the sound signal as well as record the receiving time. In this experiment, time-division multiplexing was adopted. The emitting cycle was 1 s, and the speakers emitted 100 ms long signals one after another. The speaker of *BS1* emitted 100 ms long signals at the beginning of every emitting cycle. The speakers of *BS2*, *BS3*, and *BS4* emitted 100 ms long signals at the 250th, 500th, and 750th ms, respectively. Take the speaker of *BS1* as a reference speaker. The receiving time data were saved to a text file and exported to the computer to be processed in Matlab 2014a.

Assume that t_i^j represents the time when the phone receives the signal from the *BSi* speaker at the jth emitting cycle. The time difference of arrival between *BS1* and *BSi* can be expressed as

$$\Delta t = \left| t_i^j - t_1^j - \frac{i-1}{4} T \right| \tag{30}$$

where T denotes the emitting cycle. Thus, the range of difference between the receiver and speakers *BS1* and *BSi* can be expressed as

$$r_{i,1}^o = c^o \Delta t \tag{31}$$

where c^o denotes the propagation velocity of the signal. Then, the equations can be obtained and solved by the localization algorithms.

Firstly, the experiment was conducted to verify the performance of the five localization methods. There were 19 test positions of the receiver and the distribution of receiver occupied the search region as much as possible. At each test position, 50 trials were conducted under the same conditions. In total, 950 trials were carried out in this experiment. The sketch of the experiment is shown in Figure 6.

In Figure 6, the red points represent the position of the receiver and the distance of the two adjacent test points is 2.5 m. For the results, the trials with RMSE greater than 2.5 m were considered as bad results. The result of RMSE is the mean of the results from all trials for each method.

The RMSE of the five methods in this experiment and the amount of bad results are shown in Figure 7. The RMSE of the hybrid-FA method was 0.6778 m, which was lower than that of NR, TSWLS, and GA and 0.0031 m higher than that of CWLS. As for the amount of bad results, for the hybrid-FA method, it was 90, which was less than that of the NR, TSWLS, and GA methods and 3 more than that of CWLS. It can be concluded that the performance of the hybrid-FA method was superior to that of the NR, TSWLS, and GA methods for TDoA measurement.

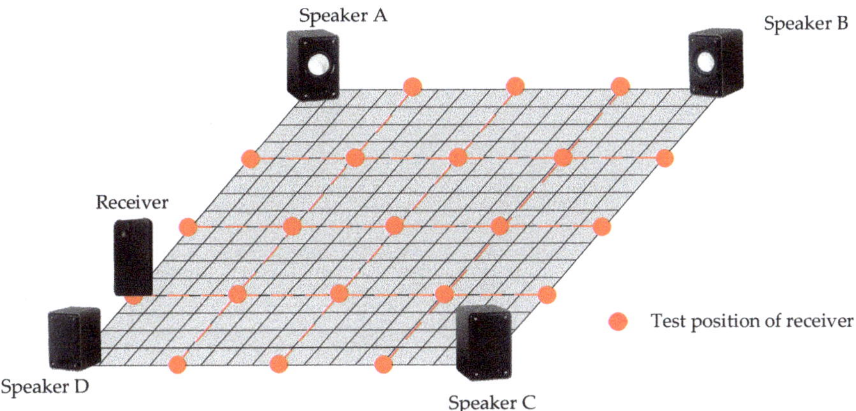

Figure 6. The sketch of the first experiment.

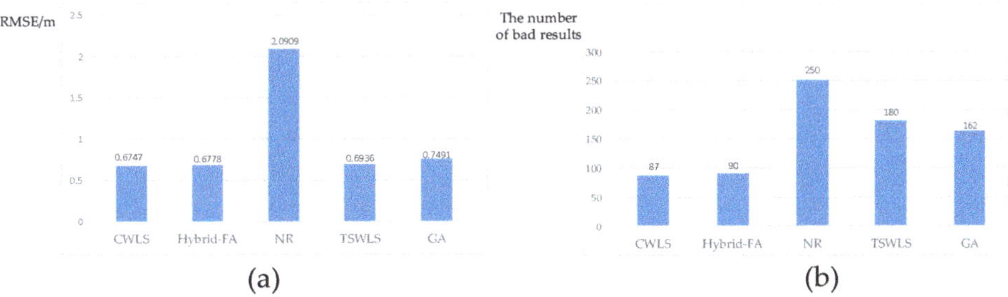

Figure 7. The results of the first experiment. (**a**) The RMSE of the five methods. (**b**) The number of bad results from the five methods.

In the experiment, the phone moved along the red path slowly, as shown in Figure 8. The A series of data was recorded and the results were calculated according to the different methods. The amount of the test position was 19 in the experiment. The discrete position sequence was obtained, then the Kalman filter with the same parameters was used to smooth the motion trail. The final results are shown below.

Figure 9 illustrates the trajectory tracking of the CWLS, hybrid-FA, NR, TSWLS, and GA methods. In Figure 9, the blue point is the discrete position solved by localization algorithms, the black line is the actual trajectory of the target, the red point is the estimated position obtained by smoothing the discrete position sequence using the Kalman filter, and the red line is the smoothed trajectory of the target. It was difficult to arrive at a conclusion solely through observation. For this reason, the mean distance error was introduced to compare the performances of the different methods. Assume that the coordinate of the smoothed position is $\left(x_i^o, y_i^o\right)$ for each method and d_i^o is the distance between the smoothed position and the line y = x in the coordinate system, which is the actual moving path of the receiver. Thus, the mean distance error is defined as follows.

$$mean\ distance\ error\ = \frac{1}{N}\sum_{i=1}^{N} d_i^o \tag{32}$$

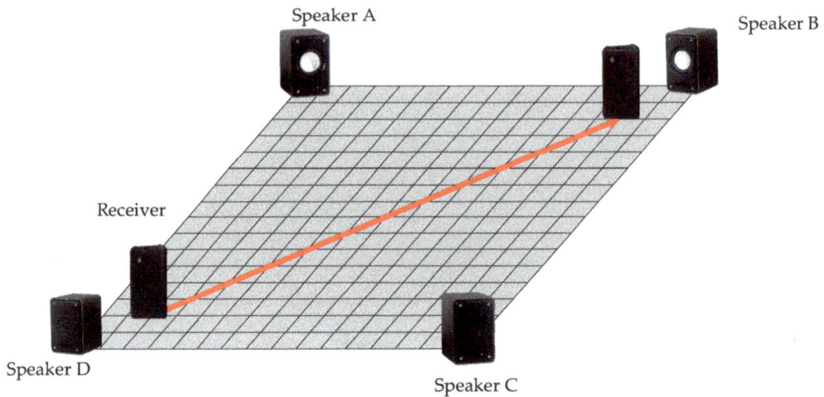

Figure 8. The setting of second experiment.

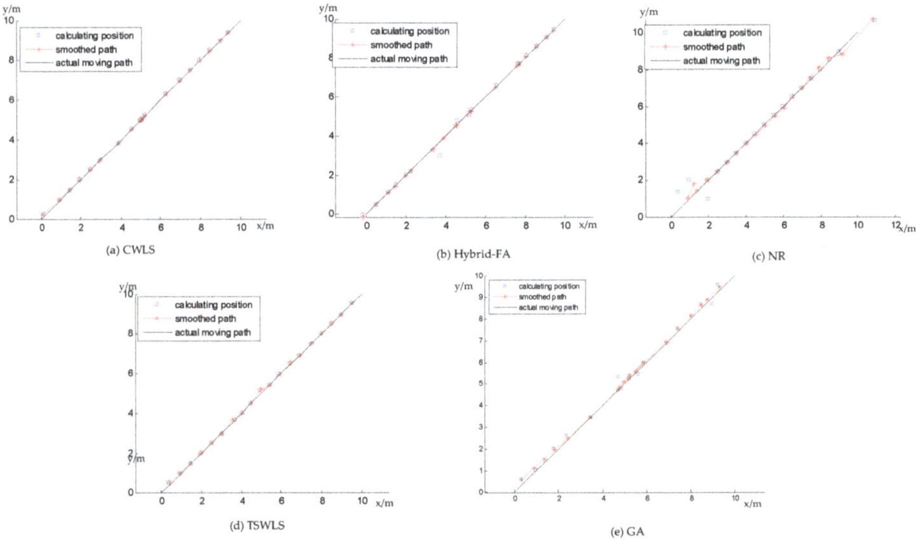

Figure 9. The trajectory tracking of the five methods based on TDoA. (**a**) The trajectory tracking of constrained weighted least squares (CWLS). (**b**) The trajectory tracking of hybrid-FA. (**c**) The trajectory tracking of Newton–Raphson (NR). (**d**) The trajectory tracking of two-step weighted least squares (TSWLS). (**e**) The trajectory tracking of the genetic algorithm (GA).

Table 2 shows the mean distance error of the CWLS, hybrid-FA, NR, TSWLS, and GA methods. The mean distance error of the hybrid-FA method in this experiment was 0.03419 m, which was less than that of the NR, TSWLS, and GA methods and 0.000985 m more than that of CWLS. It can be concluded that the hybrid-FA method outperformed the NR, TSWLS, and GA methods for TDoA measurement.

Table 2. The mean distance error of different methods.

Method	CWLS	Hybrid-FA	NR	TSWLS	GA
Mean distance error (m)	0.033205	0.03419	0.141656	0.062933	0.126473

7. Conclusions

For TDoA measurement, a good algorithm should balance calculation and precision. In this paper, a hybrid-FA method was proposed that combined the WLS and FA methods, which used the result from WLS with low computational burden to provide a reasonable limit to the search region for the FA method. The results of the proposed method were compared with the CWLS, NR, TSWLS, and GA methods using simulations and two experiments, which demonstrated the validity and limitations of the proposed method.

As expected, the hybrid-FA method could cut down the computation of the algorithm with high accuracy compared with using the FA only. Additionally, the hybrid-FA method was compared with the CWLS, NR, TSWLS, and GA methods using simulations and experiments. The RMSE of the hybrid-FA method was lower than that of the NR, TSWLS, and GA methods when the SNR ranged from 20 to 40 dB in the simulations. The result of the first experiment showed that the RMSE of the hybrid-FA method was 0.6778 m, which was lower than that of NR, TSWLS, and GA. The results of the second experiment illustrated that the mean distance error of the hybrid-FA method was 0.03419 m, which was lower than that of NR, TSWLS, and GA. On the whole, the hybrid-FA method outperformed the NR, TSWLS, and GA methods for TDoA measurement.

Author Contributions: Funding acquisition, S.S.; Methodology, P.W.; Resources, X.W.; Software, X.G.; Supervision, B.S.; Validation, Z.Z.

Funding: This work was supported by National Natural Science Foundation of China under Grant No. 51575517, and the Natural Science Foundation of Hunan Province under Grant No. 2018JJ3607.

Conflicts of Interest: The authors declare no conflict of interests.

References

1. Zhou, C.; Huang, G.; Shan, H.; Gao, J. Bias compensation algorithm based on maximum likelihood estimation for passive localization using TDOA and FDOA measurements. *Acta Aeronaut. Astronaut. Sin.* **2015**, *36*, 979–986.
2. Xu, Z.; Qu, C.; Wang, C. Performance analysis for multiple moving observers passive localization in the presence of systematic errors. *Acta Aeronaut. Astronaut. Sin.* **2013**, *34*, 629–635.
3. Weinstein, E. Optimal source localization and tracking from passive array measurements. *IEEE Trans. Acoust. Speech Signal. Process.* **1982**, *30*, 69–76. [CrossRef]
4. Gustafsson, T.; Rao, B.D.; Trivedi, M. Source localization in reverberant environments: Modeling and statistical analysis. *IEEE Trans. Speech Audio Process.* **2003**, *11*, 791–803. [CrossRef]
5. Wang, W.; Wang, X.; Ma, Y. Multi-target localization based on multi-stage Wiener filter for bistatic MIMO radar. *Acta Aeronaut. Astronaut. Sin.* **2012**, *33*, 1281–1288.
6. Stoica, P.; Li, J. Lecture notes-source localization from range-difference measurements. *IEEE Signal Process. Mag.* **2006**, *23*, 63–66. [CrossRef]
7. Wang, C.; Qi, F.; Shi, G.; Ren, J. A linear combination-based weighted least square approach for target localization with noisy range measurements. *Signal. Process.* **2014**, *94*, 202–211. [CrossRef]
8. Griffin, A.; Alexandridis, A.; Pavlidi, D.; Mastorakis, Y.; Mouchtaris, A. Localizing multiple audio sources in a wireless acoustic sensor network. *Signal. Process.* **2015**, *107*, 54–67. [CrossRef]
9. Stansfield, R.G. Statistical theory of DF fixing. *J. IEEE* **1947**, *14*, 762–770.
10. Amar, A.; Weiss, A.J. Localization of Radio Emitters Based on Doppler Frequency Shifts. *IEEE Trans. Signal Process.* **2008**, *56*, 5500–5508. [CrossRef]
11. Chan, Y.; Jardine, F. Target localization and tracking from doppler-shift measurements. *IEEE J. Ocean. Eng.* **1990**, *15*, 251–257. [CrossRef]
12. Qu, X.; Xie, L.; Tan, W. Iterative Constrained Weighted Least Squares Source Localization Using TDOA and FDOA Measurements. *Trans. Signal Process.* **2017**, *65*, 3990–4003. [CrossRef]
13. Yeredor, A.; Angel, E. Joint TDOA and FDOA Estimation: A Conditional Bound and Its Use for Optimally Weighted Localization. *Trans. Signal Process.* **2011**, *59*, 1612–1623. [CrossRef]

14. Ho, K.C.; Lu, X.; Kovavisaruch, L. Source localization using TDOA and FDOA measurements in the presence of receiver location errors: Analysis and solution. *Trans. Signal Process.* **2007**, *55*, 684–696. [CrossRef]

15. Foy, W.H. Position-location solution by taylor-series estimation. *IEEE Trans. Aerosp. Electron. Syst.* **1976**, *12*, 187–194. [CrossRef]

16. Yang, K.; An, J.P.; Bu, X.Y. Constrained total least-squares location algorithm using time-difference-of-arrival measurements. *IEEE Trans. Veh. Technol.* **2010**, *59*, 1558–1562. [CrossRef]

17. Yang, K.H.; Wang, G.; Luo, Z.Q. Efficient convex relaxation methods for robust target localization by a sensor network using time differences of arrivals. *Trans. Signal Process.* **2009**, *57*, 2775–2784. [CrossRef]

18. Chan, Y.T.; Ho, K.C. A simple and efficient estimator for hyperbolic location. *Trans. Signal Process.* **1994**, *42*, 1905–1915. [CrossRef]

19. Yu, H.G.; Huang, G.M.; Gao, J. An efficient con-strained Weighted Least Squares algorithm for moving source location using TDOA and FDOA measure-ments. *IEEE Trans. Wirel. Commun.* **2012**, *11*, 44–47. [CrossRef]

20. Yu, H.G.; Huang, G.M.; Gao, J. Practical constrained least-square algorithm for moving source location using TDOA and FDOA measurements. *J. Syst. Eng. Electron.* **2012**, *23*, 488–494. [CrossRef]

21. Qu, F.Y.; Guo, F.C.; Meng, X.W. Constrained location algorithms based on total least squares method using TDOA and FDOA measurements. In Proceedings of the International Conference on Automatic Control and Artificial Intelligence, Xiamen, China, 3–5 March 2012; IET: London, UK, 2012; pp. 2587–2590.

22. Chan, Y.T.; Hang, H.Y.C.; Ching, P.C. Exact and Approximate Maximum Likelihood localization algorithms. *IEEE Trans. Veh. Technol.* **2006**, *55*, 10–16. [CrossRef]

23. Luo, M.Z.; Chen, X.; Cao, S.; Zhang, X. Two New Shrinking-Circle Methods for Source Localization Based on TDoA Measurements. *Sensors* **2018**, *18*, 1274. [CrossRef] [PubMed]

24. Jin, B.; Xu, X.S.; Zhang, T. Robust Time-Difference-of-Arrival (TDOA) Localization Using Weighted Least Squares with Cone Tangent Plane Constraint. *Sensors* **2018**, *18*, 778.

25. Liu, Z.X.; Wang, R.; Zhao, Y.J. A Bias Compensation Method for Distributed Moving Source Localization Using TDOA and FDOA with Sensor Location Errors. *Sensors* **2018**, *18*, 3747. [CrossRef] [PubMed]

26. Jiang, J.; Wang, K.; Zhang, C.H.; Chen, M.; Zheng, H.; Albarracín, R. Improving the Error of Time Differences of Arrival on Partial Discharges Measurement in Gas-Insulated Switchgear. *Sensors* **2018**, *18*, 4078. [CrossRef] [PubMed]

27. Yu, H.G.; Huang, G.M.; Gao, J.; Wu, X.H. Approximate Maximum Likelihood Algorithm for Moving Source Localization Using TDOA and FDOA Measurements. *Chin. J. Aeronaut.* **2012**, *25*, 593–597. [CrossRef]

28. Yang, X.S. Firefly Algorithms for Multimodal Optimization. In *Stochastic Algorithms: Foundations and Applications*; Watanabe, O., Zeugmann, T., Eds.; Springer: Berlin/Heidelberg, Germany, 2009; pp. 169–178.

29. Rosić, M.; Simić, M.; Pejović, P. Hybrid genetic optimization algorithm for target localization using TDOA measurements. In Proceedings of the IcETRAN 2017, Kladovo, Serbia, 5–8 June 2017.

30. Zhang, L.N.; Liu, L.Q.; Yang, X.S.; Dai, Y.T. A Novel Hybrid Firefly Algorithm for Global Optimization. *PLoS ONE* **2016**, *11*, e0163230. [CrossRef]

31. Cheung, K.W.; So, H.C.; Ma, W.K.; Chan, Y.T. A Constrained Least Squares Approach to Mobile Positioning: Algorith ms and Optimality. *EURASIP J. Adv. Signal. Process.* **2006**, *2006*, 020858. [CrossRef]

32. Moutinho, J.N.; Araújo, R.E.; Freitas, D. Indoor localization with audible sound—Towards practical implementation. *Pervasive Mob. Comput.* **2016**, *29*, 1–16. [CrossRef]

Letter

A Switched-Element System Based Direction of Arrival (DOA) Estimation Method for Un-Cooperative Wideband Orthogonal Frequency Division Multi Linear Frequency Modulation (OFDM-LFM) Radar Signals

Yifei Liu [1,*] , Yuan Zhao [1] , Jun Zhu [1], Jun Wang [2] and Bin Tang [1]

[1] School of Information and Communication Engineering, University of Electronic Science and Technology of China, Chengdu 611731, China; zy_uestc@outlook.com (Y.Z.); uestczhujun@163.com (J.Z.); bint@uestc.edu.cn (B.T.)

[2] School of Electronic Science and Engineering, University of Electronic Science and Technology of China, Chengdu 611731, China; Wangjung@uestc.edu.cn

* Correspondence: flyliu97@foxmail.com; Tel.: +86-159-2874-2900

Received: 4 December 2018; Accepted: 28 December 2018; Published: 2 January 2019

Abstract: This paper proposes a switched-element direction finding (SEDF) system based Direction of Arrival (DOA) estimation method for un-cooperative wideband Orthogonal Frequency Division Multi Linear Frequency Modulation (OFDM-LFM) radar signals. This method is designed to improve the problem that most DOA algorithms occupy numbers of channel and computational resources to handle the direction finding for wideband signals. Then, an iterative spatial parameter estimator is designed through deriving the analytical steering vector of the intercepted OFDM-LFM signal by the SEDF system, which can remarkably mitigate the dispersion effect that is caused by high chirp rate. Finally, the algorithm flow and numerical simulations are given to corroborate the feasibility and validity of our proposed DOA method.

Keywords: wideband OFDM-LFM signal; switched-element system; fractional autocorrelation; DOA estimation

---

1. Introduction

As a novel synthetic aperture radar (SAR) system, the multiple-input multiple-output SAR (MIMO-SAR) utilizes multiple antennas to emit mutually orthogonal waveforms, and employs multiple receiving channels to process the echo signals simultaneously [1–3]. Subject to current technical conditions, wideband Orthogonal Frequency Division Multi Linear Frequency Modulation (OFDM-LFM) modulated waveforms are commonly employed in modern MIMO-SAR systems [1,4], which brings challenge to the passive direction of arrival (DOA) estimation techniques.

Passive DOA estimation techniques have been implemented in electronic warfare equipment. In particular, a review of the most commonly used techniques can be found in literatures [5–7]. However, most of them are derived for narrowband signals, which cannot handle the wideband signal scenario, i.e., the OFDM-LFM signals. In this paper, we focus on the DOA estimation method for un-cooperative wideband OFDM-LFM radar signals. Overview of existing DOA algorithms [8–13] for wideband signals, the common approach is to sample the signals in the frequency domain through the array sensors, then, consider each frequency component into a narrowband signal for processing individually. The broadband beamforming approaches in H. L. Van Trees book [14] utilize arrays with non-uniform element spacing and a time-shift operator to complete decoupling of broadband signals.

Although the mentioned methods can function well, they still suffer from huge cost of hardware and computational resources. Therefore, we exploit the switched-element direction finding (SEDF) system to solve the DOA estimation problem for wideband signals without much cost.

The block diagram of the modified SEDF and the target MIMO radar system are drawn in Figure 1. Its primary advantages include reducing the hardware and storage costs, simplifying the channel calibration process and decreasing the computation load [15–17]. Moreover, SEDF is also suitable for dealing with long-pulse signals, because there is no need to store the entire pulse in each channel. As shown in Figure 1, we consider a SEDF system with two receiving channels whose name are the reference channel (RC) and the switched channel (SC) respectively. When a signal of interest (SOI) is intercepted, the SC starts to switch in a constant period from antenna #1 to antenna #K. Thus, the signal pulse is split into multiple sub-pulses in the SC. Meanwhile, the data are collected via the RC. In formulating the DOA estimation problem for wideband OFDM-LFM signal on this SEDF system, we found that the steering vector is turned into a discrete time LFM-like vector. Hence, we proposed a modified approach to solve this estimation problem, which is inspired by a recently developed parameter estimation algorithm called Fast Iterative Interpolated Digital Fraction Fourier Transform (FII-DFrFT) [18].

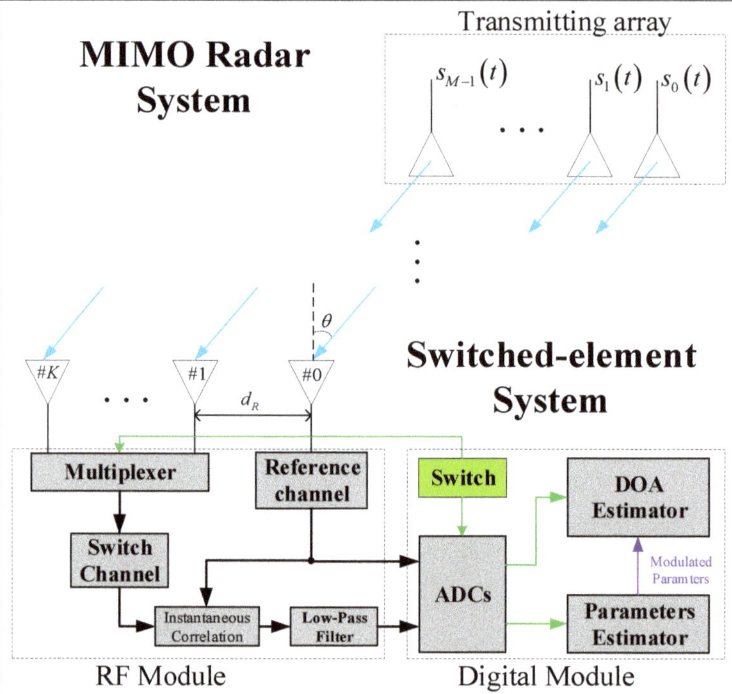

Figure 1. Block diagram of the Switched-Element Direction Finding System (RF is short for the Radio Frequency, ADCs is short for Analog to Digital converters) and Multiple-input multiple-output radar system.

The rest of this paper is organized as follows. In Section 2, we introduce the signal model and the formula derivation for DOA estimation problem. In Section 3, the proposed FII-DFrFT estimator is illustrated in detail. Numerical simulation results are shown in Section 4. Finally, in Section 5 some conclusions are drawn.

2. Problem Formulations

Consider an adversary MIMO-SAR with M transmitters. This radar employs wideband OFDM-LFM waveforms, which were first introduced into the design of an MIMO radar system by F. Cheng [19]. Afterwards, the signal of the mth transmitter is given as:

$$s_m(t) = u_m(t) e^{j2\pi f_0 t}, 0 \le m \le M - 1 \tag{1}$$

$$u_m(t) = e^{j2\pi\left(mf_\Delta t + \frac{1}{2}\gamma_0 t^2\right)} \tag{2}$$

where f_0 denotes the carrier frequency; f_Δ is the frequency step between two adjacent transmitters; γ_0 stands for the chirp rate. Besides, the bandwidth B of the OFDM-LFM signal is defined as $B \triangleq (M-1)f_\Delta + \gamma_0 T_p$, where T_p represents the pulse width of $s_m(t)$.

On the contrary, there are $K + 1$ antennas allocated in the SEDF system with interspace d_R, as shown in Figure 1. Here, we set the intercepted signal via RC as $y_{RC}(t) = s(t - t_0) + n_{RC}(t)$, where t_0 represents the propagation time, and $n_{RC}(t)$ is the additive Gaussian white noise in RC. Since this paper focuses on the DOA, without loss of generality, it is reasonable to set $t_0 = 0$ for the sake of simplicity of derivations. Meanwhile, to avoid redundancy introductions of other scholars' existing work, we assume that the estimation for inner pulse parameters and the radio frequency demodulation have already been accomplished by the techniques and algorithms in References [18,20–22], while using the collected data in the RC. Moreover, we also assume the incident direction θ and the power of the SOI is stable during the switch period T_s. Therefore, the OFDM-LFM signal intercepted via the SC can be written as:

$$
\begin{aligned}
y_{SC}(t) &= A \sum_{m=0}^{M-1} \sum_{k=1}^{K} s_m(t - \tau_k) \, \text{rect}\left(\frac{t-(k-1)T_s}{T_s}\right) + n_{SC}(t) \\
&= A \sum_{k=1}^{K} \sum_{m=0}^{M-1} \exp\left[j2\pi\left(f_m t + \frac{1}{2}\gamma_0 t^2\right) + j2\pi\left(-f_m\tau_k - \gamma_0\tau_k t + \frac{1}{2}\gamma_0\tau_k^2\right)\right] \text{rect}\left(\frac{t-(k-1)T_s}{T_s}\right) + n_{SC}(t) \\
&= A \sum_{k=1}^{K} \sum_{m=0}^{M-1} u_m(t) \, e^{j\varphi_m(\tau_k,t)} \text{rect}\left(\frac{t-(k-1)T_s}{T_s}\right) + n_{SC}(t)
\end{aligned}
\tag{3}
$$

where $\tau_k = [kd_R \sin(\theta)]/c$ is the propagation delay between the #k and #0 antenna, with c represents the speed of light; T_s is the duration for each switch; $n_{SC}(t)$ is the thermal noise in SC; $f_m = f_0 + mf_\Delta$; the phase shift $\varphi_m(\tau_k, t)$ is recast to:

$$\varphi_m(\tau_k, t) = 2\pi\left(-f_m\tau_k - \gamma_0\tau_k t + \frac{1}{2}\gamma_0\tau_k^2\right) \tag{4}$$

which is time related.

Let us consider a common LFM, whose chirp rate has the quantity of 10^{12}Hz/s, while τ_k has the quantity of 10^{-9} s. This means that the third term ($\frac{1}{2}\gamma_0\tau_k^2$) in Equation (4) is almost 0. Thus, we discard this term in the following derivations. Then, ignoring the noise term (its effect will be analyzed in the performance evaluations Section), we can obtain the instantaneous cross correlation between the SC and RC by:

$$
\begin{aligned}
r(t) &= y_{SC}(t) \cdot y_{RC}^*(t) = A^2 \sum_{k=1}^{K} \sum_{m=0}^{M-1} u_m(t) \exp\left[j\varphi_m(\tau_k, t)\right] \sum_{m'=0}^{M-1} u_{m'}^*(t) \\
&= A^2 \sum_{k=1}^{K} \left[\sum_{m=0}^{M-1} \sum_{m'=0}^{M-1} \exp\left[j2\pi(m - m')f_\Delta t + j\varphi_m(\tau_k, t)\right] \text{rect}\left(\frac{t-(k-1)T_s}{T_s}\right)\right]
\end{aligned}
\tag{5}
$$

The above equation reveals that the interested phase shift terms ($\exp\left[j\varphi_m(\tau_k, t)\right]$) are mixed with the cross terms ($\exp\left[j2\pi(m - m')f_\Delta t\right]$), which are caused by the multi-component of the intercepted signal. In order to extract the phase shift term, a low-pass filter $h(t)$ is designed [23] to filter out the cross terms, which ranges from $\pm \exp\left[\pm j2\pi f_\Delta t\right]$ to $\exp\left[\pm j2\pi (M-1) f_\Delta t\right]$. Therefore, we can obtain a new baseband signal $x(t)$ after cross correlation and low-pass filter processing:

$$x(t) = \{y_{SC}(t) \cdot y_{RC}^*(t)\} \otimes h(t) \approx A^2 \sum_{m=0}^{M-1} \sum_{k=1}^{K} \exp\left[j\varphi_m(\tau_k, t)\right] \tag{6}$$

Afterwards, we collect the samples of $x(t)$ every time when the SC switches the antenna, i.e., at $t = 0, T_s, \cdots, (K-1)T_s$. Therefore, the sampled data is given by:

$$\mathbf{x} = \left[x(0)\, x(T_s)\, \cdots\, x((K-1)T_s)\right]_{K\times 1}^{\mathrm{T}} = A^2 \mathbf{a}(\theta) \tag{7}$$

where the steering vector $\mathbf{a}(\theta)$ is expressed as:

$$\mathbf{a}(\theta) = \begin{bmatrix} \sum_{m=0}^{M-1} \exp\left[-j2\pi\left(f_m \frac{d_R \sin(\theta)}{c}\right)\right] \\ \sum_{m=0}^{M-1} \exp\left[-j2\pi\left((f_m + \gamma_0 T_s)\frac{2d_R \sin(\theta)}{c}\right)\right] \\ \vdots \\ \sum_{m=0}^{M-1} \exp\left[-j2\pi\left((f_m + (K-1)\gamma_0 T_s)\frac{K d_R \sin(\theta)}{c}\right)\right] \end{bmatrix}_{K\times 1} \tag{8}$$

For the simplicity of derivations, we define $v \overset{\Delta}{=} d_R \sin(\theta)/c$. Then, the kth entry of $\mathbf{x}$ can be further denoted by:

$$\mathbf{x}[k] = A^2 \sum_{m=0}^{M-1} \exp\left[-j2\pi\left((f_m - \gamma_0 T_s)vk + \gamma_0 T_s v k^2\right)\right] \tag{9}$$

It is interesting to find out that comparing with the traditional narrow band representation, the steering vector of OFDM-LFM signal by SEDF system is also a chirp modulated signal, with respect to k^2. Thus, this spatial signal model brings failure to the regular DOA estimation algorithms such as MUSIC and ESPRIT. Concerning on this, we approach our DOA estimation problem to the parameter estimation for OFDM-LFM signals. Therefore, we define the spatial chirp rate (μ_0) and spatial frequency (ω) as $\mu_0 \overset{\Delta}{=} 2v\gamma_0 T_s$ and $\omega_m = (f_m - T_s\gamma_0)v$ respectively. Then, Equation (9) can be simplified as:

$$\mathbf{x}[k] = A^2 \sum_{m=0}^{M-1} \exp\left[-j2\pi\left(\omega_m k + \frac{\mu_0}{2}k^2\right)\right] \tag{10}$$

To solve this estimation problem, we introduce the fast digital algorithm of FrFT [24] as:

$$X_\alpha\left(\frac{U}{2\Delta x}\right) = \frac{B_\alpha}{2\Delta x} e^{j\pi \tan\left(\frac{\alpha}{2}\right)\left(\frac{U}{2\Delta x}\right)^2} \sum_{k=-K}^{K} e^{j\pi \csc\alpha\left(\frac{U-k}{2\Delta x}\right)^2} e^{j\pi \tan\left(\frac{\alpha}{2}\right)\left(\frac{k}{2\Delta x}\right)^2} x\left(\frac{k}{2\Delta x}\right) \tag{11}$$

where $\Delta x = \sqrt{K}$ and $B_\alpha = \sqrt{(1 - j\cot\alpha)}$.

Substituting Equation (7) into Equation (11) we can obtain:

$$\begin{aligned} X_\alpha\left(\frac{U}{2\Delta x}\right) &= \frac{B_\alpha}{2\Delta x} \sum_{m=0}^{M-1} \sum_{k=-K}^{K} \exp\left[j\pi \frac{\left(\cot\alpha U^2 - 2\csc\alpha Uk + \cot\alpha k^2\right)}{(2\Delta x)^2} - j2\pi\left(\omega_m \frac{k}{2} + \frac{\mu_0}{2}\left(\frac{k}{2}\right)^2\right)\right] \\ &= \frac{B_\alpha}{2\Delta x} \exp\left[j\pi \cot\alpha \left(\frac{U}{2\Delta x}\right)^2\right] \sum_{m=0}^{M-1} \sum_{k=-K}^{K} \exp\left[j\pi\left(-\omega_m - \frac{2U\csc\alpha}{(2\Delta x)^2}\right)k + j\pi\left(-\frac{\mu_0}{4} + \frac{\cot\alpha}{(2\Delta x)^2}\right)k^2\right] \end{aligned} \tag{12}$$

From Equation (12), we can see that $\mathbf{x}$ can be reformulated into multiple (precisely say M) impulses only for a particular $\alpha_0(\cot\alpha_0 = -K\mu_0)$ in the FrFT domain when $K \to \infty$. After peak searching, the peak coordinates (α_B, U_{Bm}) in the FrFT domain can be utilized as an estimator for spatial frequency $v(\theta)$ and DOA θ as:

$$\begin{cases} \hat{v} = -\frac{1}{M-1} \sum_{m=2}^{M} (U_{Bm} - U_{Bm-1}) \frac{\csc \alpha_B}{2Kf_{\Delta}} \\ \hat{\theta} = \arcsin\left(\frac{c\hat{v}}{d_R}\right) \end{cases} \tag{13}$$

However, since the number of antennas K is a limited value, there always some residual terms between the quasi peaks (α_B, U_{Bm}) and real peaks (α_0, U_m). In this paper, we define these residual terms as ϕ_0 and ε_m, where $\alpha_0 = \alpha_B + \phi_0$ and $U_m = U_{Bm} + \varepsilon_m$. Concerning on the influence of these residual terms to the estimation precision of DOA, we propose an iterative high-accuracy method to solve this problem.

3. Proposed Method

3.1. Estimation of Spatial Chirp Rate

As the analytical formulation of $|X_{\alpha_B}(U_{Bm})|$ involves Fresnel integral formula [25], it is difficult to directly construct the estimator for ϕ_0. Thus, we consider utilizing the Fractional Autocorrelation (FA) spectrum of $x(t)$ to form this estimator, which is defined as [26]:

$$\begin{aligned} X_\alpha(\tau) &= \int x\left(t + \frac{\tau}{2}\sin\alpha\right) x^*\left(t - \frac{\tau}{2}\sin\alpha\right) e^{2j\pi t\tau\cos\alpha} dt \\ &= \int \mathrm{rect}\left(\frac{t}{T_K}\right) e^{j2\pi t\tau(\mu_0\sin\alpha+\cos\alpha)} \sum_{m_i=0}^{M-1} \sum_{m_j=0}^{M-1} e^{-j\pi\tau v(\theta)\sin\alpha\left(m_i-m_j\right)f_\Delta} e^{-j2\pi v(\theta)\left(m_i-m_j\right)f_\Delta t} dt \end{aligned} \tag{14}$$

where $T_K \overset{\Delta}{=} KT_s$.

Afterwards, we can calculate the detection statistic [26] interpreted as:

$$L(\alpha) = \int_{-\infty}^{\infty} |X_\alpha(\tau)| d\tau \tag{15}$$

Substituting Equation (14) into Equation (15) yields

$$L(\alpha) = |\Gamma(\alpha)| \int_{-\infty}^{\infty} |T_K \mathrm{Sinc}\left[2\pi T_K(\mu_0\sin\alpha+\cos\alpha)\tau\right]| d\tau \tag{16}$$

where

$$\Gamma(\alpha) = \int_{-\infty}^{\infty} \int_0^{T_K} \sum_{m_i=0}^{M-1} \sum_{m_j=0}^{M-1} e^{-j\pi\tau v f_\Delta\sin\alpha\left(m_i-m_j\right)} e^{-j2\pi t v f_\Delta\left(m_i-m_j\right)} dt d\tau \tag{17}$$

We can ignore the $\Gamma(\alpha)$ in the following derivation as this term does not involve μ_0. Therefore, we can estimate μ_0 by locating the peak of $L(\alpha)$, namely:

$$\hat{\mu}_0 = -\cot\alpha|_{\alpha=\alpha_0} \tag{18}$$

where the coordination of the peak is given by $\alpha_0 = \arg\max\{L(\alpha)\}$.

However, the estimation performance is affected by the grid size of searching, say $\Delta\alpha$, as is demonstrated in Figure 2. To be specific, the actual residual term ϕ_0 between the α_0 and α_B is also defined by $\Delta\alpha$, which is given by:

$$\alpha_0 = \alpha_B + \phi_0 = \alpha_B + \delta_0\Delta\alpha \tag{19}$$

where $\delta_0 \in [-0.5, 0.5]$. Therefore, the fine estimation is now equivalent to obtain an estimate of δ_0. Plugging in Equations (18) and (19), after some trigonometric derivation, we can define the FA coefficient as:

$$L_P = L\left(\alpha_B + P\Delta\alpha\right) = \int_{-\infty}^{\infty} |T_K \mathrm{Sinc}\left[2\pi T_K \csc \alpha_0 \sin\left(\left(P - \delta_0\right)\Delta a\right)\tau\right]| d\tau \tag{20}$$

where $L_p \left(P = \pm 0.5\right)$ calculates the interpolation coefficient at the both edges of α_B. Afterwards, we introduce the error mapping formulation through Algorithm 1 of [27] (see Table I in [7] for more information), which is defined as:

$$h_1\left(\delta\right) = \mathrm{Re}\left\{\frac{L_{0.5} + L_{-0.5}}{L_{0.5} - L_{-0.5}}\right\} \approx \frac{1}{2\delta_0} \tag{21}$$

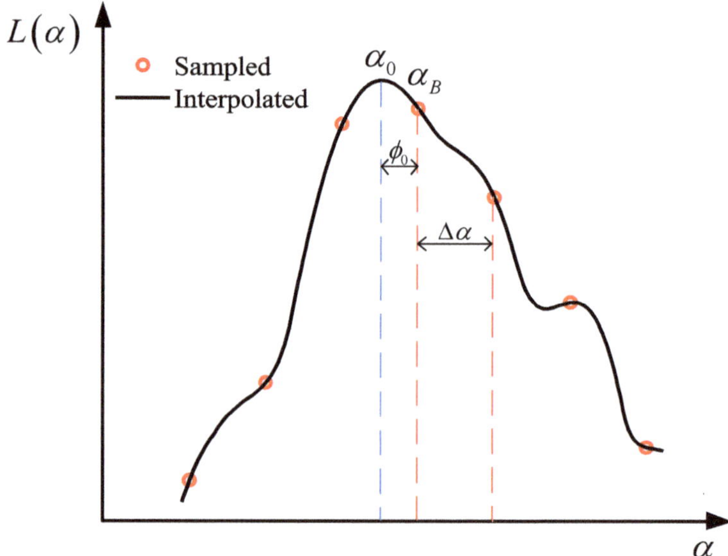

Figure 2. Demonstraction on the effect of the off-grid.

It is worth noting that Equation (21) needs a small enough $\Delta\alpha$, then the following approximations can be utilized: $\sin\left(\delta\Delta\alpha\right) \approx \delta\Delta\alpha$ and $\sin\left[T_K\left(0.5 - \delta\right)\Delta\alpha\pi \csc \tilde{\alpha}\tau\right] \approx \sin\left[T_K\left(0.5 + \delta\right)\Delta\alpha\pi \csc \tilde{\alpha}\tau\right]$. Thus, we can construct the estimator $\hat{\delta}_0 = \frac{1}{2h_1(\delta_0)}$ for δ_0. Then, an iterative process can be combined to improve the estimation accuracy by updating α_B after each iteration, which will be shown in Section 3.3.

3.2. Estimation of Spatial Frequency

Firstly, following Equation (12), we consider one component, say m, of the OFDM-LFM signal with a well estimated spatial chirp rate $-\cot \hat{\alpha}_0 \approx K\mu_0$. Thus, Equation (12) can be rewritten as:

$$X_{\hat{\alpha}_0}\left(\frac{U}{2\Delta x}\right) = \frac{B_{\hat{\alpha}_0}}{2\Delta x} e^{j\pi \cot \hat{\alpha}_0 \left(\frac{U}{2\Delta x}\right)^2} \sum_{k=-K}^{K} e^{j\pi k\left(-\omega_m - \frac{2U \csc \hat{\alpha}_0}{(2\Delta x)^2}\right)} \tag{22}$$

As we analyze in Section 2, the coordination estimated from the discrete searching is bias from the actual value with the finite K. Hence, at the quasi peak $(\hat{\alpha}_0, U_{Bm})$, $X_{\hat{\alpha}_0}\left(\frac{U_{Bm}}{2\Delta x}\right)$ equals:

$$X_{\hat{\alpha}_0}\left(\frac{U_{Bm}}{2\Delta x}\right) = \frac{B_{\hat{\alpha}_0}}{2\Delta x}e^{j\pi\cot\hat{\alpha}_0\left(\frac{U_{Bm}}{2\Delta x}\right)^2}\sum_{k=-K}^{K}e^{j\pi k\left(-\omega_m - \frac{U_{Bm}\csc\hat{\alpha}_0}{2K}\right)} \tag{23}$$

Substituting the real value $\omega_m = -\frac{\csc\hat{\alpha}_0}{2K}U_m$ and $U_m = U_{Bm} + \varepsilon_m$ into Equation (23), we can rewrite it as:

$$X_{\hat{\alpha}_0}\left(\frac{U_{Bm}}{2\Delta x}\right) = \frac{B_{\hat{\alpha}_0}}{2\Delta x}e^{j\pi\cot\hat{\alpha}_0\left(\frac{U_{Bm}}{2\Delta x}\right)^2}\sum_{k=-K}^{K}e^{j\pi k\frac{\varepsilon_m\csc\hat{\alpha}_0}{2K}} \tag{24}$$

Similar to the approach in Section 3.1, we can obtain $X_{\hat{\alpha}_0}\left(\frac{U_{Bm}\pm P}{2\Delta x}\right)$ as:

$$X_{\hat{\alpha}_0}\left(\frac{U_{Bm}\pm P}{2\Delta x}\right) = \Gamma'\left(\hat{\alpha}_0, U_{Bm}\pm P\right)\left[\frac{e^{-j\pi\frac{(\varepsilon_0\mp P)\csc\hat{\alpha}_0}{2}}\left(1 - e^{j\pi(\varepsilon_0\mp P)\csc\hat{\alpha}_0}\right)}{1 - e^{j\pi\frac{(\varepsilon_0\mp P)\csc\hat{\alpha}_0}{2N}}}\right] \tag{25}$$

where $\Gamma'\left(\hat{\alpha}_0, U_{Bm}\pm P\right) = \frac{B_{\hat{\alpha}_0}}{2\Delta x}e^{j\pi\cot\hat{\alpha}_0\left(\frac{U_{Bm}\pm P}{2\Delta x}\right)^2}$.

When $(\varepsilon_m\mp P)\ll N$, we can approximate Equation (25) by using the first order Taylor expansion at $x = 0$ of $1 - e^x \approx x$. Then, similarly to Section 3.1, we could also construct the error mapping through this approximation as:

$$X_{\hat{\alpha}_0}\left(\frac{U_{Bm}\pm P}{2\Delta x}\right) = \Gamma'\left(\hat{\alpha}_0, U_{Bm}\pm P\right)\left[\frac{e^{-j\pi\frac{(\varepsilon_0\mp P)\csc\hat{\alpha}_0}{2}}\left(1 - e^{j\pi(\varepsilon_0\mp P)\csc\hat{\alpha}_0}\right)}{1 - e^{j\pi\frac{(\varepsilon_0\mp P)\csc\hat{\alpha}_0}{2N}}}\right] \tag{26}$$

Hence, we can similarly obtain an estimator $\hat{\delta}_m = 0.5h_2\left(\delta_m\right)$ for the residual term $\hat{\varepsilon}_m$, and combine an iterative process to improve its accuracy.

3.3. Iterative DOA Estimation for OFDM-LFM

In this subsection, the estimators of spatial chirp rate and spatial frequency are combined to estimate the DOA for OFDM-LFM signals. Due to the fact that the FrFT is characterized by linear transformations [28], the major estimation bias between multi-component and mono-component signals through the FrFT based algorithm is caused by the energy leakage from the multi-component. To adapt the above process to the multi-component scenario, we introduce the CLEAN algorithm [27]. Firstly, the noise-free actual fractional coefficient $\tilde{X}_{\hat{\alpha}_0, m}\left(\left(\hat{U}_m\pm P\right)/2\Delta x\right)$ of the mth OFDM component is defined as:

$$\begin{aligned}\tilde{X}_{\hat{\alpha}_0, m}\left(\frac{\hat{U}_m\pm P}{2\Delta x}\right) &= DFRFT_{(\hat{\alpha}_0, \hat{U}_m\pm P)}\left(x\left[k\right]\right)\\ &= X_{\hat{\alpha}_0, m}\left(\frac{\hat{U}_m\pm P}{2\Delta x}\right) + \sum_{l=1,(l\neq m)}^{M}\breve{X}_{\hat{\alpha}_0, l}\left(\frac{\hat{U}_m\pm P}{2\Delta x}\right)\end{aligned} \tag{27}$$

where $\breve{X}_{\hat{\alpha}_0, l}\left(\left(\hat{U}_m\pm P\right)/2\Delta x\right)$ denote the energy leakage from the other $M - 1$ OFDM components, which can be calculated by:

$$\breve{X}_{\hat{\alpha}_0, l}\left(\frac{\hat{U}_m\pm P}{2\Delta x}\right) = A_l DFrFT_{(\hat{\alpha}_0, \hat{U}_m\pm p)}\left(\hat{s}_l\left[n\right]\right) = A_l\frac{B_{\hat{\alpha}_0}}{2\Delta x}e^{j\pi\cot\hat{\alpha}_0\left(\frac{\hat{U}_m\pm P}{2\Delta x}\right)^2}\sum_{n=-N}^{N}e^{j\pi n\left(\frac{(\hat{U}_l - \hat{U}_m\mp P)\csc\hat{\alpha}_0}{2N}\right)} \tag{28}$$

where A_l is the amplitude of the lth $(l = 1, ..., M)$ component.

Then, the target fractional coefficient $\hat{X}_{\hat{\alpha}_0,m}$ can be separated from the mixed term $\tilde{X}_{\hat{\alpha}_0,m}$ by subtracting the leakages as:

$$\hat{X}_{\hat{\alpha}_0,m}\left(\frac{\hat{U}_m \pm P}{2\Delta x}\right) = \tilde{X}_{\hat{\alpha}_0,m}\left(\frac{\hat{U}_m \pm P}{2\Delta x}\right) - \sum_{l=1,(l\neq m)}^{M} \breve{X}_{\hat{\alpha}_0,l}\left(\frac{\hat{U}_m \pm P}{2\Delta x}\right) \tag{29}$$

According to the above derivation, an iteration-based method to accomplish the DOA estimation for OFDM-LFM signal is demonstrated in Algorithm 1.

Algorithm 1: Proposed FII-DFrFT DOA Estimation Method.

Initialization: Set $\hat{\delta}_0 = 0$, $\hat{\varepsilon}_m = 0$, $\hat{A}_m = 0$ and $P = 0.5$, where $m = 1\cdots M$;
               Find $\alpha_B = \arg\operatorname*{Max}_{\alpha}\{L(\alpha)\}$, where $\alpha \in [0 : \Delta\alpha : \pi]$ and $u \in [0 : 1 : N-1]$.

Estimation of Spatial Chirp Rate:

1. Calculate the detection statistic $L(\alpha)$ using Equations (14) and (15),
  find $\alpha_B = \arg\operatorname*{Max}_{\alpha}\{L(\alpha)\}$, where $\alpha \in [0 : \Delta\alpha : \pi]$;

2. for $q = 1 : 1 : Q$ **do**
    | **2.1** Calculate $L(\alpha_B \pm 0.5\Delta\alpha)$ and β using Equations (13) and (16).
    | **2.2** Calculate $\hat{\delta}_q = 1/(2h_1)$, renew $\alpha_B = \alpha_B + \hat{\delta}_q\Delta\alpha$;
end

3. Finally, let $\hat{\alpha}_0 = \alpha_B$.

Estimation of Spatial Frequency:

1. Calculate the $X_{\hat{\alpha}_0}[u] = \text{DFrFT}_{(\hat{\alpha}_0,u)}(x[k])$ using Equation (12).

2. for $m = 1 : 1 : M$ **do**
    | **2.1 if** $q = 1$ **then**
    | | $\tilde{X}_{\hat{\alpha}_0}[u] = X_{\hat{\alpha}_0}[u] - \sum_{l=1,l\neq m}^{M} \hat{A}_l\text{DFrFT}_{(\hat{\alpha}_0,u)}(\hat{s}_l[k])$, $\hat{U}_{Bm} = \arg\operatorname*{Max}_{u}\left(\left|\tilde{X}_{\hat{\alpha}_0}[u]\right|^2\right)$;
    | **end**
    | **2.2** Calculate $\hat{X}_{m,\hat{\alpha}_0}(\hat{U}_m \pm P)$ and $h_2(\delta_m)$ using Equations (26)–(29);
    | **2.3** Calculate $\hat{\varepsilon}_m = 0.5h_2(\delta_m)$, renew $\hat{U}_{Bm} = \hat{U}_{Bm} + \hat{\varepsilon}_m$;
    | **2.4** Calculate $\hat{A}_m = \left|X_{\hat{\alpha}_0}[\hat{U}_m] - \sum_{l=1,l\neq m}^{M} \breve{X}_{\hat{\alpha}_0,l}\left(\frac{\hat{U}_m}{2\Delta x}\right)\right| / \left(\Delta x\left|B_{\hat{\alpha}_0}\right|\right)$
end

3. Calculate $\hat{U}_m = \hat{U}_{Bm} + \hat{\varepsilon}_m$.

Output: $\begin{cases} \hat{v} = -\frac{1}{M-1}\sum_{m=2}^{M}(\hat{U}_m - \hat{U}_{m-1})\frac{\csc\hat{\alpha}_0}{2Kf_\Delta} \\ \hat{\theta} = \arcsin\left(\frac{c\hat{v}}{d_R}\right) \end{cases}$

---

4. Performance Evaluation

In this section, we report our numerical evaluation through a Monte-Carlo simulation. Since the DOA estimation performance is mainly dependent on three factors, which are the signal-to-noise-ratio (SNR), the incidence angle (θ) and the component number (M), we evaluate the estimation performance with respect to these factors in a realistic case.

Consider a coherent MIMO radar (e.g. MIMO-SAR) which employs wideband OFDM-LFM signal. The simulation parameters of this MIMO radar and our SEDF system are listed in Table 1. It is worth noting that we assume the pulse width (T_P) of the OFDM-Signal is greater than KT_s, thus our method can function well. Moreover, we assume the far field sources whose initial phase is uniformly distributed within $[0, 2\pi)$, and we take the thermal noise into consideration, which is modeled as zero-mean Gaussian with variance $\sigma_n^2 = 1$. Additionally, in all simulations, 1000 independent runs are conducted to calculate the Normalized Root Mean Square Error (NRMSE) and Root Mean Square Error (RMSE).

(a) *DOA Estimation versus SNR*

In this simulation, we evaluate the DOA estimation performance with respect to the SNR, while the DOA θ is set as 30 deg. For the sake of comparison, we also simulate the following approaches, Incoherent Signal-subspace Method Conventional Beam Forming (ISM-CBF) [29], Coherent Signal-subspace Method Linearly Constrained Minimum Variance (CSM-LCMV) [11], Rotational Signal Subspace Sparse Asymptotic Minimum Variance (RSS-SAMV) [12] and Sparse Iterative Covariance-based Estimation (SPICE) [9] As these existing approaches are designed for single wideband LFM signal, here, we consider the intercepted signal that received by our switched-element system a mono-component wideband LFM signal ($M = 1$). Then, the above approaches and FII-FrFT are utilized to process the output signal and obtain the DOA estimation results, respectively. These results are collected and organized to NRMSE curves, which are shown in Figure 3. These curves reveal that our FII-FrFT method outperforms most mentioned approaches when SNR is beyond -8 dB. However, the NRMSE curve of FII-FrFT remains stable when SNR is beyond 12 dB and suffers a stable estimation bias, which is caused by the approximations that we employed in the theoretical derivations of Section 3.1 and the off-grid effect. On the other side, although the RSS-SAMV performs best in this simulation, its implementation will consume much more hardware resource (K receiving channels) and computational resource [12].

Table 1. Parameter Settings.

	Number of antennas M	1–4
	Pulse width T_p	20 μs
MIMO Radar Parameters	Carrier frequency f_0	10 GHz
	Chirp rate γ_0	20 MHz/μs
	Frequency step f_Δ	400 MHz
	Number of ULA K	128
	Carrier frequency f_0	10 GHz
	Interspace of ULA d_R	0.015 m
Switched-element System Parameters	Switching interval T_s	0.1 μs
	Searching interval $\Delta\alpha$	0.01
	Iteration number Q	3

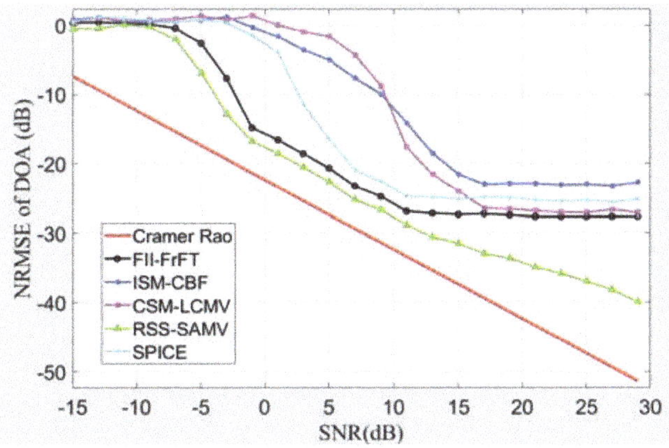

Figure 3. Normalized root mean square error (NRMSE) of DOA versus the signal-to-noise ratio (SNR). Cramer Rao, FII-FrFT, ISM-CBF, CSM-LCMV, RSS-SAMV and SPICE.

(b)   *DOA Estimation versus Real Incident Angle and Component Number M*

In this simulation, we focus on the DOA estimation performance as the function of the real direction θ within $[10, 70]$ degree by the FII-FrFT. We also consider the intercepted OFDM-LFM signals consist different component numbers $M = [2, 3, 4]$. For intuitional comparison with different OFDM-LFM signals, we define a different SNR in this subsection as $\rho = 10 \lg \left(MA^2/\sigma_n^2 \right)$. The root mean square error (RMSE) of DOA estimation results at SNR = 10 dB are given in Figure 4. Firstly, we can see from Figure 4 that our proposed method can handle the OFDM-LFM radar signal well, while its component number affects the RMSE slightly. Secondly, the periodic variation of RMSE curves in Figure 4 reflects the off-grid effect in the fixed searching interval on estimation performance, which is in coincidence with our theoretical analysis in Section 3.2 and the simulation results in Reference [15]. This bias can be reduced by decreasing , i.e., using a denser grid, but it will also lead to the expensive price of computational load. Therefore, our DOA estimation method has to reach a compromise between accuracy and cost.

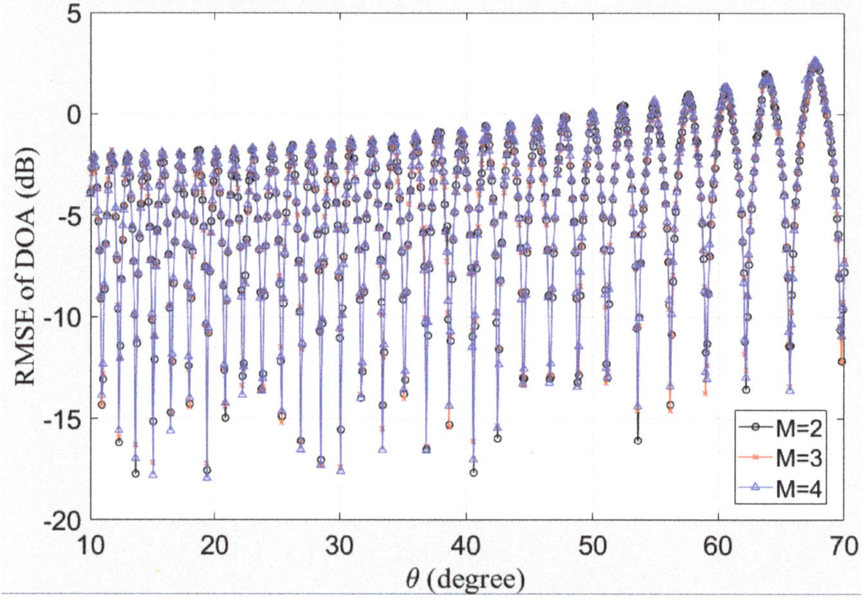

Figure 4. Root mean square error (RMSE) of DOA versus the incident angle.

5. Conclusions

In this paper, a FII-DFrFT based SEDF system was introduced to improve the DOA estimation performance considering the wideband OFDM-LFM signals. The steering vector was reformulated followed by the iterative interpolation in both FA and DFrFT spectrum. Numerical simulations illustrated the validity and superiority of our algorithm compared with some other wideband DOA estimation approaches like ISM-CBF, CSM-LCMV, RSS-SAMV and SPICE. On the other hand, in the practice scenario, the modulated parameters of un-cooperative MIMO radar are generally unknown. This will cause the DOA estimation to be possibly ambiguous. Fortunately, taking advantage of a flexible switching interval, we can design a multi-interval SEDF system to resolve this ambiguity. Finally, the estimation bias caused by the off-grid effect and approximation are also of interest and will be the subject of our further investigation.

Author Contributions: Y.L. designed and wrote this paper under the supervision of B.T. Y.Z. proposed the main idea and assisted with the methodology with formal analysis. J.Z. and J.W. provided the support of the entire study. Y.Z. and B.T. reviewed and edited the manuscript.

Funding: This research was funded by the National Natural Science Foundation of China (Grant No. 61571088).

Acknowledgments: Thanks to the University of Electronic Science and Technology of China. Thanks to my supervisor Bin Tang. Thank you to my dear love Silei Gui and all the people who support us.

Conflicts of Interest: The authors declare no conflict of interest.

References

1. Han, K.; Wang, Y.; Peng, X.; Hong, W. Modulating multicarriers with chirp for MIMO-SAR waveform diversity design. In Proceedings of the 2013 IEEE International Conference on Signal Processing, Communication and Computing (ICSPCC 2013), KunMing, China, 5–8 August 2013; pp. 1–4.
2. Gu, F.F.; Zhang, Q.; Chi, L.; Chen, Y.A.; Li, S. A Novel Motion Compensating Method for MIMO-SAR Imaging Based on Compressed Sensing. *IEEE Sens. J.* **2015**, *15*, 2157–2165. [CrossRef]
3. Hu, C.; Wang, J.Y.; Tian, W.M.; Zeng, T.; Wang, R. Design and Imaging of Ground-Based Multiple-Input Multiple-Output Synthetic Aperture Radar (MIMO SAR) with Non-Collinear Arrays. *Sensors* **2017**, *17*, 598. [CrossRef] [PubMed]
4. Zhuge, X.; Yarovoy, A.G. A Sparse Aperture MIMO-SAR-Based UWB Imaging System for Concealed Weapon Detection. *IEEE Trans. Geosci. Remote Sens.* **2011**, *49*, 509–518. [CrossRef]
5. Krim, H.; Viberg, M. Two decades of array signal processing research—The parametric approach. *IEEE Signal Process. Mag.* **1996**, *13*, 67–94. [CrossRef]
6. Farina, A.; Gini, F.; Greco, M. DOA estimation by exploiting the amplitude modulation induced by antenna scanning. *IEEE Trans. Aerosp. Electron. Syst.* **2002**, *38*, 1276–1286. [CrossRef]
7. Aboutanios, E.; Hassanien, A.; Amin, M.G.; Zoubir, A.M. Fast Iterative Interpolated Beamforming for Accurate Single-Snapshot DOA Estimation. *IEEE Geosci. Remote Sens. Lett.* **2017**, *14*, 574–578. [CrossRef]
8. Malioutov, D.; Cetin, M.; Willsky, A.S. A sparse signal reconstruction perspective for source localization with sensor arrays. *IEEE Trans. Signal Process.* **2005**, *53*, 3010–3022. doi:10.1109/TSP.2005.850882. [CrossRef]
9. Stoica, P.; Babu, P.; Li, J. SPICE: A Sparse Covariance-Based Estimation Method for Array Processing. *IEEE Trans. Signal Process.* **2011**, *59*, 629–638. [CrossRef]
10. Abeida, H.; Zhang, Q.; Li, J.; Merabtine, N. Iterative Sparse Asymptotic Minimum Variance Based Approaches for Array Processing. *IEEE Trans. Signal Process.* **2013**, *61*, 933–944. [CrossRef]
11. Valaee, S.; Kabal, P. Wideband array processing using a two-sided correlation transformation. *IEEE Trans. Signal Process.* **1995**, *43*, 160–172. doi:10.1109/78.365295. [CrossRef]
12. Zhang, X.; Sun, J.; Cao, X.; Wu, J.; Chen, Y. Wideband Signal DOA Estimation Based on Sparse Asymptotic Minimum Variance. *Mod. Radar* **2018**, 30–35. doi:10.16592/j.cnki.1004-7859.2018.01.007. [CrossRef]
13. Mishra, K.V.; Kahane, I.; Kaufmann, A.; Eldar, Y.C. High spatial resolution radar using thinned arrays. In Proceedings of the 2017 IEEE Radar Conference (RadarConf), Seattle, WA, USA, 8–12 May 2017; pp. 1119–1124. doi:10.1109/RADAR.2017.7944372. [CrossRef]
14. Trees, H.L.V. *Optimum Array Processing*; John Wiley and Sons Inc.: New York, NY, USA, 2002; p. 1433.
15. Tennant, A.; Chambers, B. Direction finding using a four-element time-switched array system. In Proceedings of the 2008 Loughborough Antennas and Propagation Conference, Loughborough, UK, 17–18 March 2008; pp. 181–184. doi:10.1109/LAPC.2008.4516896. [CrossRef]
16. Wu, W.; Cooper, C.C.; Goodman, N.A. Switched-Element Direction Finding. *IEEE Trans. Aerosp. Electron. Syst.* **2009**, *45*, 1209–1217. doi:10.1109/TAES.2009.5259194. [CrossRef]
17. Zhao, Y.; Giniy, F.; Grecoy, M.; Liu, Y.; Zhu, J. Iterative Interpolar based Switched Element Direction Finding for Wideband Linear Frequency Modulated Signals. In Proceedings of the 2019 IEEE Radar Conference, Boston, MA, USA, 22–26 April 2019; submitted.
18. Liu, Y.; Zhao, Y.; Zhu, J.; Xiong, Y.; Tang, B. Iterative High-Accuracy Parameter Estimation of Uncooperative OFDM-LFM Radar Signals Based on FrFT and Fractional Autocorrelation Interpolation. *Sensors* **2018**, *18*, 3550. [CrossRef] [PubMed]

19. Fang, C.; Zishu, H.; Liu, H.M.; Jun, L. The Parameter Setting Problem of Signal OFDM-LFM for MIMO Radar. In Proceedings of the 2008 International Conference on Communications, Circuits and Systems, Fujian, China, 25–27 May 2008; pp. 981–985.

20. Jayaprakash, A.; Reddy, G.R. Robust Blind Carrier Frequency Offset Estimation Algorithm for OFDM Systems. *Wirel. Pers. Commun.* **2017**, *94*, 777–791. [CrossRef]

21. Howard, S.; Sirianunpiboon, S.; Cochran, D. Detection and characterization of MIMO radar signals. In Proceedings of the 2013 International Conference on Radar, Adelaide, SA, Australia, 9–12 September 2013; pp. 330–334. [CrossRef]

22. Li, Y.H.; Tang, B. Parameters estimation and detection of MIMO-LFM signals using MWHT. *Int. J. Electron.* **2016**, *103*, 439–454. [CrossRef]

23. Yih-Min, C.; Kuo, I.Y. Design of lowpass filter for digital down converter in OFDM receivers. In Proceedings of the 2005 International Conference on Wireless Networks, Communications and Mobile Computing, Maui, HI, USA, 13–16 June 2005; Volume 2, pp. 1094–1099. [CrossRef]

24. Ozaktas, H.M.; Ankan, O.; Kutay, M.A.; Bozdagi, G. Digital computation of the fractional Fourier transform. *IEEE Trans. Signal Process.* **1996**, *44*, 2141–2150. [CrossRef]

25. Li, C.P.; Dao, X.H.; Guo, P. Fractional derivatives in complex planes. *Nonlinear Anal. Theory Methods Appl.* **2009**, *71*, 1857–1869. [CrossRef]

26. Akay, O.; Boudreaux-Bartels, G.F. Fractional autocorrelation and its application to detection and estimation of linear FM signals. In Proceedings of the IEEE-SP International Symposium on Time-Frequency and Time-Scale Analysis (Cat. No. 98TH8380), Pittsburgh, PA, USA, 9 October 1998; pp. 213–216.

27. Ye, S.L.; Aboutanios, E. An Algorithm for the Parameter Estimation of Multiple Superimposed Exponentials in Noise. In Proceedings of the 2015 IEEE International Conference on Acoustics, Speech and Signal Processing (ICASSP), Brisbane, QLD, Australia, 19–24 April 2015; pp. 3457–3461.

28. Ozaktas, H.M.; Mendlovic, D. Fractional Fourier Optics. *J. Opt. Soc. Am. A Opt. Image Sci. Vis.* **1995**, *12*, 743–751. [CrossRef]

29. Velni, J.M.; Khorasani, K. Localization of wideband sources in colored noise VIA generalized least squares (GLS). In Proceedings of the IEEE/SP 13th Workshop on Statistical Signal Processing, Bordeaux, France, 17–20 July 2005; pp. 525–530. [CrossRef]

Article

Improved Bound Fit Algorithm for Fine Delay Scheduling in a Multi-Group Scan of Ultrasonic Phased Arrays

Yuzhong Li [1,2], Wenming Tang [1] and Guixiong Liu [1,*]

[1] School of Mechanical & Automotive Engineering, South China University of Technology, Guangzhou 510641, China; melyz@mail.scut.edu.cn (Y.L.); tang.wm@mail.scut.edu.cn (W.T.)

[2] School of Information Engineering, Huizhou Economic and Polytechnic College, Huizhou 516057, China

* Correspondence: megxliu@scut.edu.cn; Tel.: +86-020-8711-0568

Received: 19 December 2018; Accepted: 16 February 2019; Published: 21 February 2019

Abstract: Multi-group scanning of ultrasonic phased arrays (UPAs) is a research field in distributed sensor technology. Interpolation filters intended for fine delay modules can provide high-accuracy time delays during the multi-group scanning of large-number-array elements in UPA instruments. However, increasing focus precision requires a large increase in the number of fine delay modules. In this paper, an architecture with fine delay modules for time division scheduling is explained in detail. An improved bound fit (IBF) algorithm is proposed, and an analysis of its mathematical model and time complexity is provided. The IBF algorithm was verified by experiment, wherein the performances of list, longest processing time, bound fit, and IBF algorithms were compared in terms of frame data scheduling in the multi-group scan. The experimental results prove that the scheduling algorithm decreased the makespan by 8.76–21.48%, and achieved the frame rate at 78 fps. The architecture reduced resource consumption by 30–40%. Therefore, the proposed architecture, model, and algorithm can reduce makespan, improve real-time performance, and decrease resource consumption.

Keywords: ultrasonic phased array; scheduling algorithm; multi-group sensors; FPGA

---

1. Introduction

Ultrasonic phased array (UPA) technology is an important nondestructive testing method that is widely used in aerospace, shipbuilding, port machinery, and nuclear energy. With its multiple-group scanning functionality and a large number of other elements, the multi-group scan UPA system can provide extended scanning flexibility and image contrast, increased focal law diversification, and high signal-to-noise ratio (SNR). Within the system, a number of filters in a given module determine the precision of fine delay. The higher the precision, the better the image resolution. Classical all-parallel fine delay modules require a lot of hardware resources, i.e., a multiplier, look-up table (LUT), register (Reg), and an in field programmable gate array (FPGA). Synchronization and integration difficulty need to be considered in the use of multi-chip schemes, while hardware resources are limited in single chip schemes. Therefore, an architecture with time-division multiplexing is used to schedule frame tasks between fine delay modules in a single chip. This method can significantly improve resource utilization and reduce the number of resources used. However, when the sampling depth or the value of the focal law is large, the frame rate (frames per second, fps) decreases, leading to worse real-time performance of the distributed UPA instrument and a greatly reduced application scope. Therefore, it is necessary to coordinate fine modules and frame tasks for multi-group scanning through algorithm schedules, minimize idle time slots of resources in the fine delay modules, and reduce the makespan of all frame tasks to improve time performance.

In order to reduce the trade-off between the real-time processing of big data and system complexity, many studies have been conducted on high-performance hardware architecture and corresponding algorithms. Thus, various resource optimizations have been proposed. Holmes et al. [1] proposed a UPA system called the full matrix capture and total focus method (FMC-TFM), which requires the processing of large focus and delay data, and has been the mainstream architecture for research in recent years. Njiki el al. [2] proposed a hardware architecture for big data processing based on FMC and a large-scale phased array instrument, applied in the M2M NDT (nondestructive testing) (Eddyfi Technologies, Québec, QC, Canada) UPA system. The proposed FMC-TFM architecture can achieve a frame rate of 73.6 fps and 128 × 128 pixels in the region of interest. Shao and Yuan [3] proposed a method based on the compute unified device architecture (CUDA) interface, a parallel graphics processing unit (GPU), and whole parallel echo signal processing, wherein parallel GPUs accelerate the method 2–3-fold compared to MATLAB (Mathworks, Co., Ltd., Natick, MA, USA), while the multithreading CPU provides four times higher acceleration than a single thread. Guo et al. [4] improved a TFM imaging system and proposed an algorithm based on read-only memory (ROM). Zhang et al. [5] used a state machine, operation unit, and a large data storage unit to form the TFM algorithm imaging system, which achieved good performance. Tang et al. [6] proposed a data transmission algorithm for UPAs, but it does not work with delay and focus. Liu et al. [7] proposed an improved 8× interpolation cascaded integrator-comb (CIC) filter parallel algorithm, which reduced 12.5% of addition and 29.2% of multiplication and yielded a time delay accuracy of 1 ns at 125 MHz. Su et al. [8] proposed a parallel delay multiply and sum beamforming (PDMAS) algorithm, based on a graphics processing unit (GPU) that improved the parallelism and stability of the beamformer with a frame rate of 83 fps. However, these papers only focus on the performance of the delay and focus module, and not the multi-group scan and its frame task scheduling.

Although a multi-core CPU with single instruction multiple data (SIMD) and the GPU programmed by CUDA also realizes the beamform function (delay and focus), Asano et al. [9] found that a GPU was slower than a CPU for complex algorithms. Furthermore, they also found that a GPU only has potential for naïve computation methods, due to its small local memory and the memory access limitation in the architecture. The performance of a quad CPU is 1/12 to 1/7 that of a field programmable gate array (FPGA). The performance of an FPGA is only limited by its size and bandwidth. FPGA is the mainstream solution for portable UPA instruments, and is supported by manufacturers. Moreover, it is convenient for design and verification of the UPA system's integrated circuits. Therefore, this paper uses FPGA to implement the algorithm and architecture of the multi-group scan UPA system.

The fine delay scheduling problem in the multi-group scanning of UPA systems, which we address here, can be considered as a parallel machine scheduling problem. The aim is to decrease the makespan, which can be represented as $P_m \mid \mid C_{max}$. It is a non-deterministic polynomial-time hard (NP-hard) problem [10], which cannot be solved using polynomial algorithms. Heuristic algorithms are a simple and effective method used to address NP-hard problems at present.

The most commonly used heuristic algorithms are the longest processing time algorithm (LPT) [11] and the MULTIFIT algorithm. The MULTIFIT algorithm proposed by Coffman et al. [12] is based on the first fit decreasing (FFD) iteration algorithm, which is used in bin-packing problems. However, the MULTIFIT algorithm has much better performance than the LPT algorithm. Freisen et al. [13] studied the absolute performance and time complexity of the MULTIFIT algorithm. Lee et al. [14] used a combination of the LPT and MULTIFIT algorithms. Kang et al. [15] simplified the MULTIFIT algorithm and combined it with the prepare algorithm (PA), in order to form the bound fit (BF) algorithm. Li et al. [16] proposed the QUICKFIT algorithm, which is an improved BF algorithm in the iteration stage. Based on the advantages of the LPT and BF algorithms, the improved bound fit (IBF) algorithm is proposed here.

In this paper, a fine delay scheduling architecture was also analyzed considering multi-group-scan echo data diversity, using a non-preempt model for the scheduling problem and proposing the IBF algorithm for optimization.

The paper is organized as follows. In Section 2, the architecture of the fine delay module scheduling for the multi-group scanning of UPA systems is presented, and the multi-group scan problem is explained. In Section 3, the IBF algorithm is proposed and an analysis of its performance and time complexity is provided. LIST, LPT, BF, and IBF algorithms are compared in Section 4. Finally, a conclusion is provided in Section 5.

2. Fine Delay Module for Multi-Group Scanning of UPAs

2.1. Fine Delay Scheduling Principle

The delay method and focus scheduling based on different UPA instrument focal parameters (e.g., number of apertures, sending and receiving time, and data amount), which control the pulse repetition frequency (PRF) and frame formation, are used for scheduling in multi-group scans. The delay precision is 1.25 ns. Due to the limitation of the resources of the FPGA in our experiments, the system architecture is designed as four groups and two fine delay modules. Each group has eight channels, and each channel has 10-bit analog-digital converter (ADC). Sampling depth is 2–8 K, the number of focal law ≤128, and read parameter length is 1024 in each group. The design frame rate is not less than 24 fps, which meets the requirements of real-time display.

A diagram of the for mutli-group scanning is shown in Figure 1, labels ①–⑤ in Figure 1 are described below.

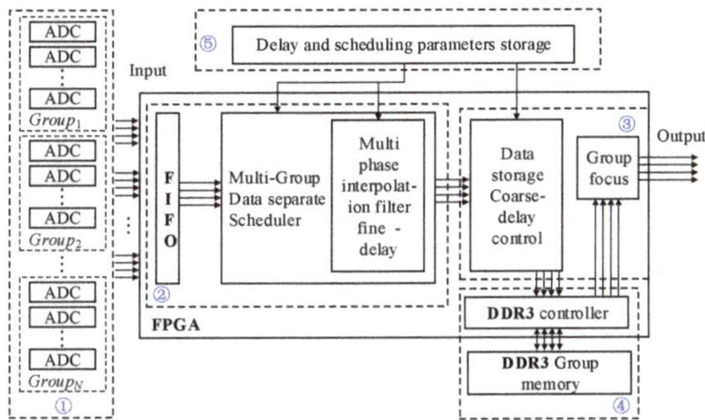

Figure 1. Diagram of the fine delay module for multi-group scanning.

The presented block diagram includes the following parts:

(1) High speed multi-channel ADC module (HADC): Ultrasonic echo signals are subjected to high-speed multi-channel ADC acquisition, conditioning conversion, and transformation into low-voltage differential signaling (LVDS) serial signals. They are then fed to the FPGA for further processing. ADCs are divided into groups according to the probe socket and multi-group scan.

(2) Fine delay scheduling module (FDS): The LVDS serial signal is first converted into a parallel signal, then the parallel signal generated by the IP core is sent to the multi-channel first-in first-out memory (FIFO), which is used for buffering and scheduling. The scheduling module consists of several fine delay modules. The signal buffered in the FIFO is then fed to the scheduling module, where it is forwarded to different fine delay modules. Thus, time division multiplexing is achieved.

The fine-delay module used in this study contains the multi-level half-band filter that was proposed by Liu and Tang [17]. A diagram of the multi-level half-band fine delay filter is presented in Figure 2, whereas its simulation diagram created in ModelSim (Mentor Co., Ltd., Wilsonville, OR, USA) is shown in Figure 3.

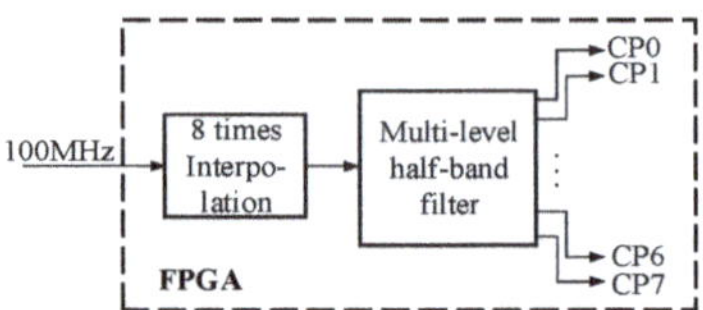

Figure 2. Diagram of fine delay module.

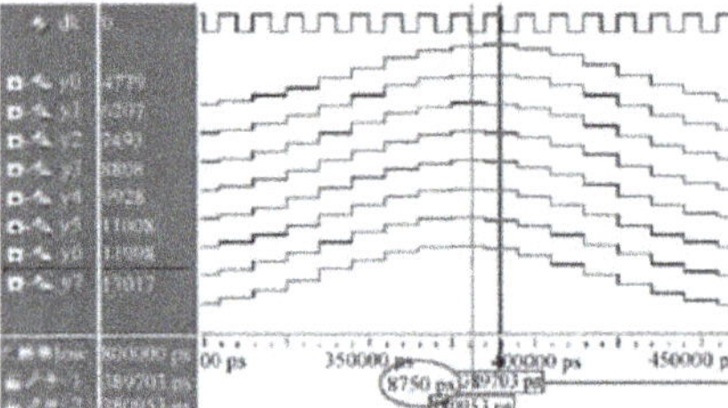

Figure 3. ModelSim simulation diagram of the multi-level half-band fine delay filter.

The multi-level half-band fine delay filter uses the interpolation method with eight time intervals to design a half-band filter. The implementation of synthetic technology in the multi-level half-band interpolation filter results in filter decomposition into eight sub-filters. Simultaneously, interpolation with poly-phase decomposition is achieved. The eight filters delay the original signal for 0, 1.25, 2.5, 3.75, 5, 6.25, 7.5, and 8.75 ns. The data samples have a 10-bit length, and thus two 9-bit multipliers are needed for multiplications. However, the multi-level half-band filter uses six 9-bit multipliers. In addition, each channel has eight fine delay channels, so there are 96 (i.e., $6 \times 2 \times 8 = 96$) 9-bit multipliers. If all parallel delay is used in a 256-element UPA system, then 24,576 multipliers would be needed. Given such large resource consumption, the integration of a single FPGA in the multi-group scan module of a UPA system would be difficult.

(3)   Coarse delay and sum module (CDS): Coarse delay is based on counter clock delay technology. All the relative delay parameters of focal laws, calculated by a PC, can be loaded from the "delay and scheduling parameters storage" block in Figure 1. The double data rate 3 (DDR3) synchronous dynamic random access memory input signal addresses the corresponding coarse delay parameter counted by the clock, and thus fixed integer coarse delay is achieved. The sum module merges signals processed by fine delay and coarse delay blocks in an ultrasonic digital beam, which represents the complete beamform of the focal laws. All signals of the ultrasonic digital beam are stored in memory, and all signal groups form a corresponding beamform. In other words, each focal law forms a digital beamform, and all the beamforms of the same group generate the initial image information of that group.

(4)   External DDR3: Since the internal RAM capacity of the FPGA is insufficient, a DDR3 controller with two DDR3 memories is used for coarse delay data storage. DDR3 memory has a coarse delay and reads the group focus module according to the group.

(5)   Delay and scheduling parameters storage (DSPS): Delay and scheduling parameters storage is a large-scale storage block in the FPGA. The delay and scheduling parameters are calculated using a focal law calculator in the PC, corresponding to the input data entered by the user. DSPS contains a scheduling table, the pulse repetition frequency of each group, and the time delay parameter for both fine and coarse delays according to focal laws. It also includes algorithmic

control for scheduling Mux and Demux based on the above parameters. A fine delay scheduling model diagram in the multi-scan group is presented in Figure 4.

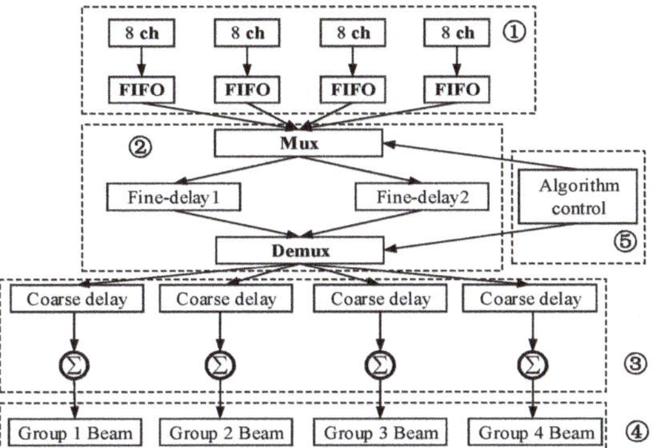

Figure 4. Fine delay scheduling model diagram in the multi-scan group.

2.2. Fine Delay Scheduling Problem in Multi-Group Scanning

The parameters of the fine delay module for multi-group scanning of UPAs are presented in Table 1. Here, we represent the symbols used in the scheduling problems with brackets.

Table 1. Parameters of the fine delay module for multi-group scanning of a ultrasonic phased array (UPA) system.

Symbol	Parameter
N_{Group}	Number of groups (n) [1]
N^i_{FocalLaw}	Number of focal laws in the ith group
D^i_{Sample}	Sample depth of the ith group
N_{FDModule}	Number of fine delay modules (m) [1]
T^i_{RP}	Read parameter time of focal law
$T_{\text{clock-cycle}}$	Clock period in FPGA
t^i_p	Processing time in the ith group (p_i) [1]

[1] Symbols in brackets are those used in the scheduling problem.

Fine-delay scheduling for multi-group scanning of UPAs must satisfy four conditions:

(1) Each focal law must be separately processed in fine delay modules. In other words, one fine delay module must process only one focal law datum.

(2) The process cannot be interrupted or preemptive, i.e., a no-interrupt non-preemptive (NINP) model is adopted.

(3) There is no time gap between the start time of focal law and the start time of the pulse repetition period.

(4) The sample depth is less than the pulse repetition period.

Condition (1) avoids timing confusion, condition (2) avoids interruption of the fine delay signal processing, and condition (3) compacts the frame task for scheduling and decreases the time slot waste. Condition (4) ensures that the fine delay processing will not exceed its abilities, leading to echo data overlap.

Before a description of the fine delay scheduling problem is presented, some parameters must be defined:

Definition 1. *Frame task.*

If it is assumed that the *i*th scan has focal law frame N^i_{FocalLaw} and sample depth D^i_{Sample}, then the frame task is the time needed to complete all beamforms (or focal laws) of the image.

Definition 2. *Frame task deadline.*

The frame task deadline represents the time the system needs to generate a complete image for all groups, and it must be less than 1/24 s for real-time applications.

Schematic diagrams of the frame task and frame task deadline are presented in Figure 5a,b, respectively.

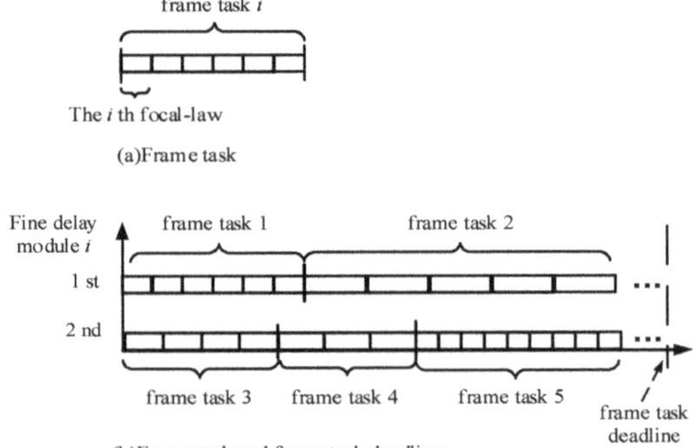

Figure 5. Schematic diagram of: (**a**) Frame task and (**b**) Frame task and frame task deadline.

The time parameters used in the proposed algorithm are defined as follows.

Start time, t^i_s, is defined by:

$$t^i_s = 0 \quad i = 1, 2, \ldots, N_{\text{Group}} \tag{1}$$

Processing time, t^i_p, is defined by:

$$t^i_p = (D^i_{\text{sample}} \times T_{\text{clock-cycle}} + T^i_{\text{RP}}) \times N^i_{\text{FocalLaw}} \quad i = 1, 2, \ldots, N_{\text{Group}} \tag{2}$$

End time, t^i_d, is defined by:

$$t^i_d = 1/24 \ s \quad i = 1, 2, \ldots, N_{\text{Group}} \tag{3}$$

Therefore, the question can be set as $P_m \| C_{\max}$, and the scheduling model is defined by:

$$Min\ z = Max\left(\sum_{j=1}^{n} t^j_p x_{ij}\right) \quad i = 1, 2, \ldots, m \tag{4}$$

subject to:

$$\sum_{j=1}^{n} t^j_p x_{ij} \le t_d y_i \quad i = 1, 2, \ldots, m \quad j = 1, 2, \ldots, n \tag{5}$$

$$\sum_{i=1}^{m} x_{ij} = 1 \quad i = 1, 2, \ldots, m \quad j = 1, 2, \ldots, n \tag{6}$$

$$x_{ij} \in \{0, 1\} \tag{7}$$

$$t_d \le 1/24 \tag{8}$$

Equation (4) refers to the scheduling goal of minimizing the project's maximum completion time, which represents the time needed for the completion of all project tasks. In this paper, we consider the frame task as the job or task of the scheduling problem. According to Equation (5), the time allocation of each fine delay module cannot be greater than t_d. Equations (6) and (7) show that any task can be assigned only to one processor, and x_{ij} is an assigned variable that is equal to zero or one. Equation (8) represents all fame tasks that must be finished before the frame task deadline.

3. IBF Algorithm

Since there is no dependency between tasks, the fine delay scheduling problem in multi-group scanning can be considered as an independent, parallel processor scheduling task.

The IBF algorithm parameters are defined as follows. Input is the set of tasks $T = \{t_i, i = 1,2, \ldots ,n\}$, the number of fine-delay modules is m, and the number of tasks is n. Output is the maximal processing time, $C_{\max}^{\text{IBF}}$.

The IBF algorithm steps are as follows:

Step 1. Sort tasks T in descending order according to the task processing time: $p_i, i = 1,2, \ldots ,n$;

Step 2. Assume that $A = \frac{1}{m} \sum_{i=1}^{n} p_i$ and $L_j, j = 1,2,\ldots,m$ are the focus and delay module pointers, respectively;

Step 3. Use the LPT algorithm to obtain the maximal processing time $C_{\max}^{\text{LPT}}$. Let $l = 1$ and $B(1) = C_{\max}^{\text{LPT}}$;

Step 4. If $A < \max(L_j) < B(l)$, go to step 5; otherwise, go to step 8;

Step 5. Let $l = l + 1$, $i = 1$, $B(l) = \min(\max(L_j), B(l-1) - 1)$;

Step 6. If there is at least one j that satisfies the condition $L_j + p_i \leq B(l)$, then allocate task t_i to the focus and delay module, which satisfies condition $L_j + p_i \leq B(l)$. Otherwise, allocate the task to the focus and delay module, which provides the minimal value of $L_j + p_i$;

Step 7. Set $i = i + 1$, and if $i \leq n$, go back to step 6; otherwise, go back to step 4;

Step 8. $C_{\max}^{\text{IBF}} = \min(B(1), B(2), B(l-1))$.

In step 3, the LPT algorithm is used to calculate the initial processing time in order to better approximate the initial conditions. Steps 4–8 represent the prepare algorithm (PA). Thus, the IBF algorithm is a combination of LPT and PA that improves the boundary and convergence of iteration, and achieves better performance in terms of local search and iterative progression. The IBF flowchart is shown in Figure 6.

The IBF algorithm analysis is obtained for $B(1) = C_{\max}^{\text{LPT}}$. In the case the iteration stops at $l = 2$, then the output algorithm result will be $C_{\max}^{\text{IBF}} = C_{\max}^{\text{LPT}}$. If the iteration stops at $l = 3$, then the output result will be $C_{\max}^{\text{IBF}} = C_{\max}^{\text{PA}(B(0))}$, and that wil be the makespan. If the iteration stops at $l > 3$, then $C_{\max}^{\text{IBF}} = C_{\max}^{\text{PA}(B(l-1))}$.

From $B(l) = \min(\max(L_j), B(l-1) - 1)$, we obtain $B(l) \leq B(l-1) - 1$. Thus, $B(2) \leq B(1) - 1$, $B(3) \leq B(2) - 1 \leq (B(1) - 1) - 1 = B(1) - 2$.

After induction $B(l) \leq B(1) - (l - 1)$. Therefore, the absolute performance of the IBF algorithm is defined by:

$$C_{\max}^{\text{IBF}} \leq \left(\frac{4}{3} - \frac{1}{3m} \right) C_{\max}^{\text{OPT}} - (l-1) \tag{9}$$

If the iteration number is equal to one, the IBF time complexity is defined by:

$$O(n \log n + nl \log m) \tag{10}$$

If the number of iterations is greater than one, IBF employs the PA, which represents the FFD algorithm used in the bin-packing problem.

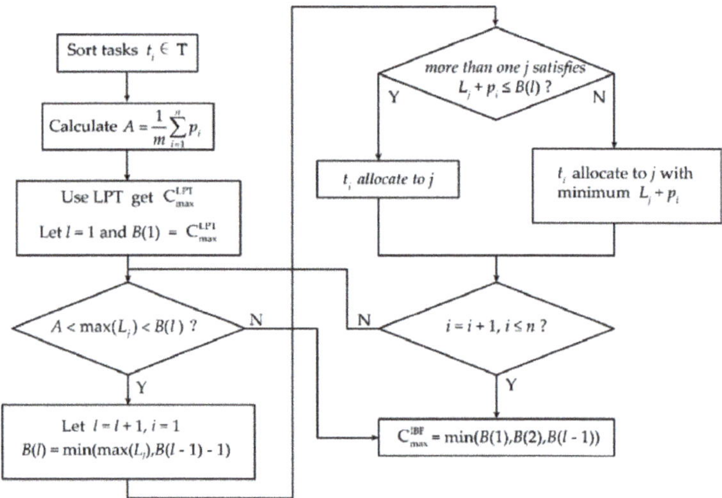

Figure 6. Improved bound fit algorithm (IBF) flowchart. LPT: longest processing time algorithm.

4. Experimental Results

4.1. Time Performance

In order to determine the real-time performance of the IBF algorithm, a randomly generated set of tasks was used. The set and real-time deadline were used to simulate a UPA multi-group fine delay scheduling problem. The specific task generation process was as follows. First, m time blocks were generated. The length of each time block was as long as the deadline t_d. Then, each task block was divided into $h = \lceil n/m \rceil + 1$ parts, and thus $h \times m$ tasks were obtained in m time blocks. Afterward, n tasks from $h \times m$ tasks that were generated from the previous step were chosen to create a set of tasks, and all task lengths were multiplied by 0.99. Thus, a random generation of a set of tasks was produced. The whole experiment ran in I7-4850HQ (Intel Corporation, Santa Clara, CA, USA) 8 GB RAM with MATLAB 2016a.

This process was conducted to ensure that the processing time of each generated task was not greater than the real-time deadline. All generated tasks did not exceed the calculating ability of the fine-delay module. In other words, a feasible solution always existed for a given scheduling in terms of the number of modules that satisfied the required conditions. The generated set was subjected to a random uniform distribution, and a variety of large scopes were covered.

Five tests were conducted with the following parameters: the number of fine-delay modules m, the ratio of number of tasks and fine delay modules $k = n/m$, the real-time deadline d, the number of iterations K, and makespan C_{max}. Each test was generated 100 times, and the average result was calculated. The LIST, LPT, BF, and IBF algorithms were compared.

Test 1 compared LPT, BF, and IBF algorithms in terms of makespan. In Figure 7a, the parameter settings were: $m = 4$, $k = 2$–10, and $d = 1000$. Note that each curve had a peak value at $k = 3$, because when $k = 3$, the method generating the problem reduced the number of tasks and increased the length. Under this condition, the problem was difficult to schedule. With gradually increasing k, all curves gradually declined. IBF had the smallest makespan at $k < 8$, and when $k \ge 8$, IBF and BF almost had the same makespan performance. This is because with the increase in k, the problem produced more tasks and the length decreased. That is, the smaller the granularity of the tasks, the greater the role of the scheduling algorithm. In Figure 7b, the parameter settings were: $m = 2$–10, $k = 4$, and $d = 1000$. We can see that the IBF algorithm still had the smallest makespan, but with the increase in m, the gap between BF and IBF continued to narrow. Although k was unchanged, the larger the value of m, the greater the permutations and combinations of the scheduling algorithm were. In makespan comparisons, IBF always had the best performance, but, as parameters k and m increased, the performance of BF and IBF gradually approached each other.

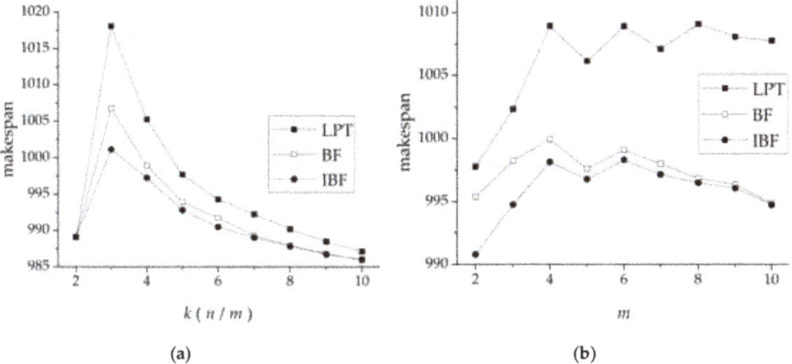

Figure 7. Comparison of LPT, bound fit (BF), and IBF in terms of makespan with (**a**) variable k (ratio of the number of tasks n and the number of fine delay modules m) and (**b**) variable number of fine delay modules m.

Test 2 compared LPT, BF, and IBF in terms of the missed deadline rate (MDR) with variables k and m. The parameter settings in Figure 8a were the same as in Figure 7a, and those in Figure 7b were applied to Figure 8b. The MDR is defined as the number of times a deadline was missed when a scheduling problem was generated randomly 100 times. Figure 8a shows that all curves had a peak value at $k = 3$, and then gradually decreased with increasing k. The reason is similar to test 1. Note that in Figure 8b, IBF had the smallest makespan, but when $m > 9$, the values of BF and IBF were basically the same. IBF was still the best in MDR performance, and with the increase in k, the scheduling performance improved as well. When $k > 8$, IBF was not significantly superior to BF.

Test 3 compared LPT, BF, and IBF using statistical plots. Parameter settings were $m = 4$, $k = 4$, and calculation was run 100 times to obtain the makespan. Figure 9a shows the box plot. Note that the IBF algorithm had the lowest median and upper limits and the narrowest interquartile range (IQR). This shows that IBF scheduling had the best overall performance and the most centralized data. In the 95% confidence interval (CI) plot in Figure 9b, IBF had the lowest mean and the narrowest 95% CI. The IBF algorithm outperformed the BF and LPT algorithms in terms of statistical performance.

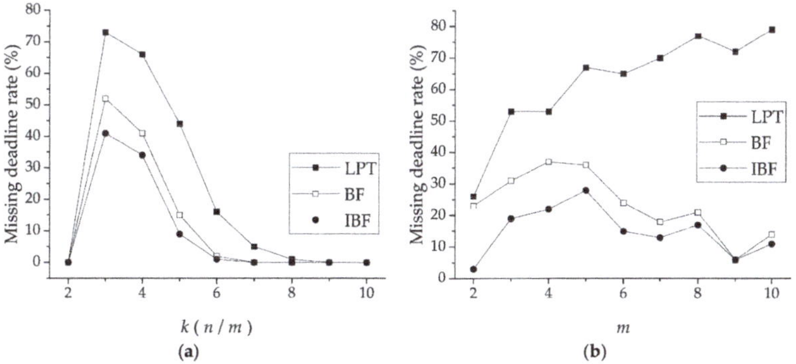

Figure 8. Comparison of LPT, BF, and IBF in terms of missed deadline rate (MDR) with (**a**) k (the ratio of the number of tasks n and the number of fine-delay modules m) and (**b**) variable number of fine delay modules m.

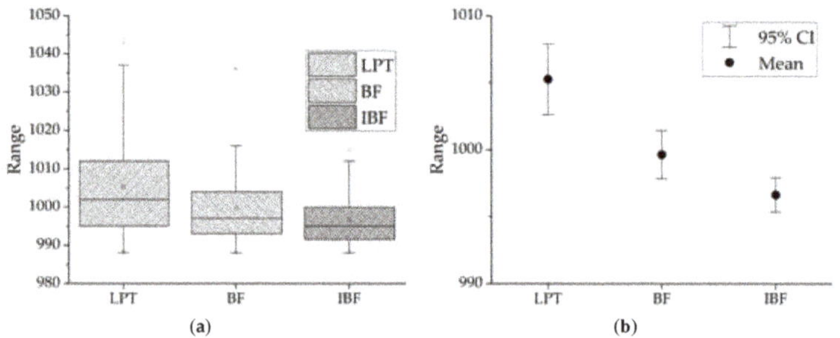

Figure 9. Comparison of LIST, LPT, and IBF algorithms in (**a**) boxplot and (**b**) 95% confidence interval (CI) plot.

Test 4 compared the performance of LIST, LPT, BF, and IBF algorithms (Table 2). The test parameter settings were $m = 4$, $k = 4$, $d = 1000$, and the average of 100 runs was taken. The LIST algorithm had the worst performance, which affected the display of the figures. In order to clearly compare BF and IBF, which was not mentioned in the previous experiments, $R_{\text{IBF/LIST}}$ was defined as follows:

$$R_{\text{IBF/LIST}} = \frac{\overline{C_{\max}^{\text{LIST}}} - \overline{C_{\max}^{\text{IBF}}}}{\overline{C_{\max}^{\text{LIST}}}} \times 100\% \tag{11}$$

where $\overline{C_{\max}^{\text{LIST}}}$, $\overline{C_{\max}^{\text{LPT}}}$, $\overline{C_{\max}^{\text{BF}}}$, and $\overline{C_{\max}^{\text{IBF}}}$ represent the average makespans of LIST, LPT, BF, and IBF obtained from 100 runs, respectively. In addition, $\overline{K^{\text{BF}}}$ and $\overline{K^{\text{IBF}}}$ represent the average number of iterations for BF and IBF. As shown in Table 2, IBF had the lowest average makespan, but its average number of iterations was slightly greater than that of the BF algorithm. This was also reflected in the elapsed time. In the worst case of our experiment, the average elapsed times at $m = 10$, $k = 4$ for LIST, LPT, BF, and IBF algorithms were 2.70, 2.63, 40.61, and 55.21 ms, respectively. The elapsed time of IBF was greater than BF by about 35.95%. However, as shown in the last column of Table 2, IBF improved performance by 8.76–21.48% compared to the LIST algorithm.

Table 2. Comparison of LIST, LPT, BF, and IBF performance.

m	LIST	LPT	BF		IBF		$R_{\text{IBF/LIST}}$
	$\overline{C_{\max}^{\text{LIST}}}$	$\overline{C_{\max}^{\text{LPT}}}$	$\overline{C_{\max}^{\text{BF}}}$	$\overline{K^{\text{BF}}}$	$\overline{C_{\max}^{\text{IBF}}}$	$\overline{K^{\text{IBF}}}$	
2	1092.39	996.67	995.96	3.02	991.17	2.64	8.76%
4	1174.41	1005.26	999.65	4.01	996.65	5.49	14.40%
6	1264.74	1007.05	997.98	5.28	996.91	8.11	20.37%
8	1258.18	1008.45	997.63	6.86	997.16	10.1	19.85%
10	1282.96	1007.41	996.22	8.01	996.06	11.61	21.48%

Test 5 was used to examine the relationship of IBF with the number of iterations. In Figure 10a, all curves had a peak value at $k = 3$–5, and then slowly declined. This occurred because when $k = 35$, the generated tasks had large granularity, which facilitated iteration without satisfying the conditions, so the number of iterations was greater. The number of iterations with larger m was greater than that with smaller m, because a large m leads to more permutations and combinations. When $k > 8$, the number of iterations decreased gradually and tended to be the same. Due to the small size of the task, the initial LPT algorithm was more effective, so the number of iterations decreased. In Figure 10b, except for the case of $k = 2$, the other curves increased gradually, and the larger the value of k, the smaller the number of iterations. Therefore, the greater the task granularity, the greater the value of m and the greater the number of iterations.

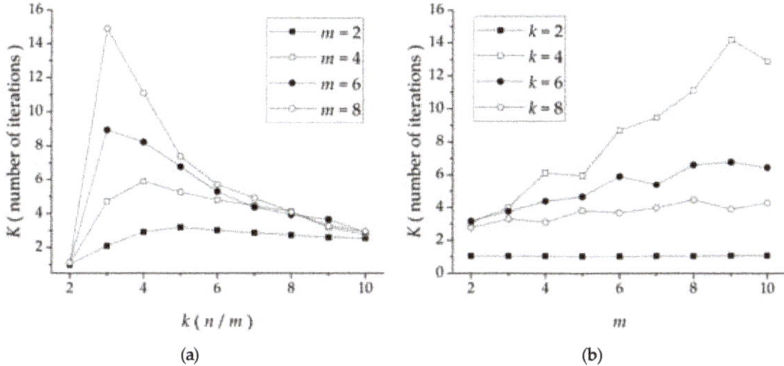

Figure 10. IBF number of iterations for: (**a**) k = 2–10 and (**b**) m = 2–10.

4.2. Resource Consumption

In the experiment, an Altera Cyclone VI EP4CE115F29C8 and Quartus II 13.0 (Intel Corporation, Santa Clara, CA, USA) were used to compare all parallel and 1/2 scheduling for 32-channel and 64-channel architectures. Then, the TimeQuest Timing Analyzer in Quartus II was used to determine the maximal clock frequency for the listed architectures. The clock frequency was set to 100 MHz. The obtained resource consumption and maximal frequencies of all architectures are presented in Table 3, wherein "number of groups" represents the number of scan groups in the multi-group UPA system; "number of modules" represents the number of fine delay modules in the system; "Total LUT" (LUT: look up table), "Total Reg.", and "Total 9-bit Mult." refer to the consumption of total logic unit, total register, and total 9-bit multiplier, respectively; and Fmax represents the maximum clock frequency. Percentages with brackets in the Total LUT and Total 9-bit Mult. columns represent their share of all the same resources in the entire FPGA.

Table 3. Resource consumption and max frequency of all parallel and 1/2 scheduling for 32-channel and 64-channel architectures.

	Number of Groups	Number of Modules	Total LUT	Total Reg.	Total 9-bit Mult.	Fmax (MHz)
All-par. 32 ch.	4	4	5086 (4.44%)	3977	320 (60%)	137.74
1/2 Sch. 32 ch.	4	2	2902 (2.53%)	2445	160 (30%)	146.99
All-par. 64 ch.	8	8	14,340 (12.53%)	8569	532 (100%) [1]	113.77
1/2 Sch. 64 ch.	8	4	5902 (5.16%)	4857	320 (60%)	126.53

[1] Due to resource limitations, the total 9-bit multiplier in the FPGA was 532.

Table 3 shows that all parallel architectures demand more resources and have lower maximal frequencies than 1/2 scheduling architectures. The 1/2 scheduling architecture could save about 57.06–58.84% in LUT and 30–40% in 9-bit multipliers. Table 3 also demonstrates that maximum frequency decreased as the number of channels increased. The bold text in column Fmax are the best Fmax in same number of channels, respectively. Therefore, based on the premise of guaranteeing real-time performance, the proposed architecture and IBF algorithm can reduce resource consumption, shorten timing, and increase the maximum clock frequency.

4.3. Real-Time Verification

Figure 11 displays the results of the pre-synthesis simulation in four groups of two fine delay modules, using ModelSim 10.2 SE electronics design automation tools (Mentor Co., Ltd., Wilsonville, OR, USA). The other experimental conditions are described in the previous section, and the experimental parameters are shown in Table 4. The delay caused by fine-delay filters with eight clock-cycles has been taken into account and combined into time of read parameter. Units are clock cycles of the FPGA in Table 4 columns 2–4.

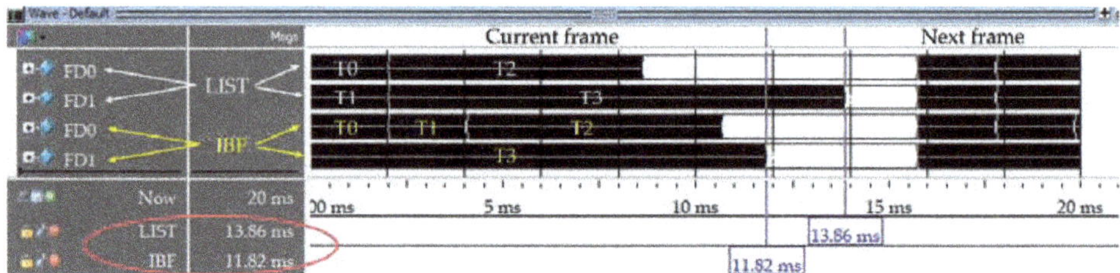

Figure 11. Four groups scheduled in two fine delay modules' simulation by ModelSim.

Table 4. Four groups of two fine delay modules simulation parameters.

Group	Number of Focal Laws $(N^i_{FocalLaw})$	Sample Depth (D^i_{Sample})	Processing Time [1] (p_i)
0	64	2048	196,608
1	64	2048	196,608
2	128	4096	655,360
3	128	8192	1,179,648

[1] Time of read parameters $T^i_{RP} = 1024$.

In Figure 11, the tasks were T0–T3, corresponding to frame tasks of Group 0–3, and FD0 and FD1 are fine delay modules. The upper FD0 and FD1 were scheduled by LIST, and the lower FD0 and FD1 were scheduled by IBF. In the case of maximum 8 K sampling depth, 128 focal laws (Group 3), the makespan of LIST was 13.86 ms, whereas the makespan of IBF was 11.82 ms, so IBF is superior to LIST. At a waiting time of more than 1 ms between frames, the frame periods of LIST and IBF were 14.86 and 12.82 ms, respectively, which correspond to frame rates of 67 and 78 fps, respectively. Therefore, the IBF algorithm generally reduced the makespan of the frame tasks, increased the frame rate, and improved real-time performance of the multi-group scan UPA instrument.

5. Conclusions

In this paper, a fine delay scheduling architecture in the multi-group scanning of a UPA system was presented. The diversity of echo data in multi-group scanning and the number of focal laws were considered, and the multi-group scan problem was modelled by a linear equation. The IBF algorithm was proposed, and its time complexity and absolute performance were analyzed. The experimental results showed that compared to LIST, LPT, and BF algorithms, the IBF algorithm decreased the makespan by 8.76–21.48%, while the frame rate reached 78 fps, and the architecture reduced FPGA resources by 30–40%. The IBF algorithm was superior to BF in terms of its small task-to-module ratio. The proposed algorithm and mathematical model was applied to a UPA. uUsing the proposed architectures effectively improved integration, increased maximum frequency, improved real-time performance, and finally, decreased resource consumption. Therefore, the instrument's flexibility and performance was improved. The next step is to study another processing module scheduling and multi-FPGA situation, integrated in a distributed environment.

Author Contributions: Y.L., W.T. and G.L. conceived the idea of the paper; Y.L. performed the experiments, and Y.L. and W.T. carried out the system model; Y.L. wrote the paper.

Funding: This work was financially supported by the National Key Foundation for Exploring Scientific Instrument (2013YQ230575) and Guangzhou Science and Technology Plan Project (201509010008).

Conflicts of Interest: The authors declare no conflict of interest.

References

1. Holmes, C.; Drinkwater, B.W.; Wilcox, P.D. Post-processing of the full matrix of ultrasonic transmit–receive array data for non-destructive evaluation. *Ndt E Int.* **2005**, *38*, 701–711. [CrossRef]
2. Njiki, M.; Bouaziz, S.; Elouardi, A.; Casula, O.; Roy, O. A multi-FPGA implementation of real-time reconstruction using Total Focusing Method. In Proceedings of the 2013 IEEE International Conference on Cyber Technology in Automation, Control and Intelligent Systems, Nanjing, China, 26–29 May 2013; pp. 468–473.
3. Shao, Z. Research on a GPU-based Real-Time Ultrasound Imaging System. Ph.D. Thesis, Nanjing University, Nanjing, China, 2014.
4. Guo, J.Q.; Li, X.; Gao, X.; Wang, Z.; Zhao, Q. Implementation of total focusing method for phased array ultrasonic imaging on FPGA. In Proceedings of the International Symposium on Precision Engineering Measurement & Instrumentation, Changsha, China, 8–11 August 2014.
5. Zhang, X.; Guo, J.; Luo, X.; Gao, X.; Wang, Z.; Zhao, Q.; Zheng, B. Defect detection study on total focus method of sound field imaging based on parallel processing. In Proceedings of the 2014 IEEE Far East Forum on Nondestructive Evaluation/Testing (FENDT), Chengdu, China, 20–23 June 2014; pp. 112–116.
6. Tang, W.; Liu, G.; Li, Y.; Tan, D. An improved scheduling algorithm for data transmission in ultrasonic phased arrays with multi-group ultrasonic sensors. *Sensors* **2017**, *17*, 2355. [CrossRef] [PubMed]
7. Liu, P.; Li, X.; Li, H.; Su, Z.; Zhang, H. Implementation of high time delay accuracy of ultrasonic phased array based on interpolation CIC filter. *Sensors* **2017**, *17*, 2322. [CrossRef] [PubMed]
8. Su, T.; Yao, D.J.; Li, D.Y.; Zhang, S. Ultrasound parallel delay multiply and sum beamforming algorithm based on GPU. In Proceedings of the 2nd IET International Conference on Biomedical Image and Signal Processing (ICBISP 2017), Wuhan, China, 13–14 May 2017.
9. Asano, S.; Maruyama, T.; Yamaguchi, Y. Performance comparison of FPGA, GPU and CPU in image processing. In Proceedings of the International Conference on Field Programmable Logic & Applications, Prague, Czech Republic, 31 August–2 September 2009.
10. Ullman, S.D. *Complexity of Sequencing Problems. Computers and Job-Shop Scheduling*; John Wiley: New York, NY, USA, 1976.
11. Graham, R.L. Bounds on multiprocessing timing anomalies. *SIAM J. Appl. Math.* **1969**, *17*, 416–429. [CrossRef]
12. Coffman, F.G., Jr.; Garey, M.R.; Johnson, D.S. An application of bin-packing to multiprocessor scheduling. *SIAM J. Compt.* **1978**, *7*, 1–17. [CrossRef]
13. Friesen, D.K. Tighter bounds for the multifit processor scheduling algorithm. *SIAM J. Comput.* **1984**, *13*, 170–181. [CrossRef]
14. Lee, C.Y.; Massey, J.D. Multiprocessor scheduling: Combining LPT and MULTIFIT. *Discret. Appl. Math.* **1988**, *20*, 233–242. [CrossRef]
15. Kang, Y.; Zheng, Y. Independent tasks scheduling on identical parallel processors. *Acta Autom. Sin.* **1997**, *23*, 81–84.
16. Li, X.P.; Xu, X.F.; Zhan, D.C. A Quick Algorithm for Independent Tasks Scheduling on Identical Parallel Processors. *J. Softw.* **2002**, *13*, 812–814.
17. Liu, G.; Tang, W.; Tan, D. Focusing time delay of ultrasonic phased array based on multistage half-band filter. *Opt. Precis. Eng.* **2014**, *22*, 1571–1576.

Letter

Iterative High-Accuracy Parameter Estimation of Uncooperative OFDM-LFM Radar Signals Based on FrFT and Fractional Autocorrelation Interpolation

Yifei Liu *, **Yuan Zhao**, **Jun Zhu, Ying Xiong and Bin Tang**

School of information and Communication Engineering, University of Electronic Science and Technology of China, Chengdu 611731, China; zy_uestc@outlook.com (Y.Z.); uestczhujun@163.com (J.Z.); Xiongy@uestc.edu.cn (Y.X.); BinT@uestc.edu.cn (B.T.)
* Correspondence: flyliu97@foxmail.com; Tel.: +86-159-2874-2900

Received: 3 September 2018; Accepted: 17 October 2018 ; Published: 19 October 2018

Abstract: To improve the parameter estimation performance of uncooperative Orthogonal Frequency Division Multi- (OFDM) Linear Frequency Modulation (LFM) radar signals, this paper proposes an iterative high-accuracy method, which is based on Fractional Fourier Transform (FrFT) and Fractional Autocorrelation (FA) interpolation. Two iterative estimators for rotation angle and center frequencies are derived from the analytical formulations of the OFDM-LFM signal. Both estimators are designed by measuring the residual terms between the quasi peak and the real peak in the fractional spectrum, which were obtained from the finite sampling data. Successful elimination of spectral leakage caused by multiple components of the OFDM-LFM signal is also proposed by a sequential removal of the strong coefficient in the fractional spectrum through an iterative process. The method flow is given and its superior performance is demonstrated by the simulation results.

Keywords: uncooperative sensor signal processing; MIMO radar; fractional Fourier transform; fractional autocorrelation interpolation

---

1. Introduction

As a novel radar system, the Multiple-Input Multiple-Output (MIMO) radar employs multiple transmitting antennas to emit mutually orthogonal waveforms and uses multiple receiving antennas to process the echo signals simultaneously [1]. Subject to current technical conditions, the coherent MIMO radar technique is commonly used in modern MIMO radar systems [2].

In this paper, we focus on the high-accuracy parameter estimation of uncooperative Orthogonal Frequency Division Multi- (OFDM) Linear Frequency Modulation (LFM) signals, which have been widely used in coherent MIMO radar systems. In the past decades, much research has been conducted on OFDM-LFM waveform design [1–3]. However, only a few studies have discussed parameter estimation for uncooperative OFDM-LFM signals in electronic warfare systems. The signal model of the intercepted MIMO signals based on a single-channel receiver has been analyzed in the literature [4,5]. Moreover, an improved Multiple Wigner–Hough Transform (MWHT) [6] was proposed to enhance the performance of signal detection and parameter estimation for OFDM-LFM signals. Other estimation algorithms based on likelihood estimators or optimization methods for multicomponent LFM signals were applied in [7–9]. However, these earlier algorithms have limitations on the estimation accuracy and efficiency due to the cross-term and picket fence effects. Besides, most of these algorithms also lack computational efficiency, making them more difficult and expensive to realize [10].

Based on the fast algorithm of Fractional Fourier Transform (FrFT) that was proposed by Ozaktas [11], related research works [12–17] have brought the application of digital FrFT (DFrFT) to maturity. On the one hand, fractional derivatives and calculus in a complex plane were studied in [13–16], which are beneficial in establishing fractional models in engineering. Furthermore, the contributions in [17] proposed the fractional geometric calculus and extended the fractional calculus to any dimension. On the other hand, the analytical FrFT formulations of multicomponent LFM signals were introduced in [10]. Conventional DFrFT utilizes a coarse-fine search strategy to improve the estimation accuracy. The coarse-fine strategy firstly obtains a crude estimation by searching the maximum FrFT coefficient of the received data, and then, the result is refined by modified methods such as Newton-type methods [8] and interpolation methods [10,18]. However, these methods require numerous extra calculations and make it difficult to handle the OFDM-LFM signals.

Inspired by the recently-developed fast iterative interpolated beamforming estimation method [19], we propose a fast and high accuracy estimator for uncooperative OFDM-LFM signals based on DFrFT and Fractional Autocorrelation (FA). We refer to the proposed method as Fast Iterative Interpolated DFrFT (FII-DFrFT), which exhibits desirable convergence properties and the same order computational complexity as Digital Fourier Transform (DFT).

The rest of this paper is organized as follows. In Section 2, the signal model of the intercepted uncooperative OFDM-LFM signal and the analytical formulations of DFrFT for this signal are given. In Section 3, the proposed FII-DFrFT is described. Section 4 gives the numerical simulation of the proposed algorithm, and some conclusions are drawn in Section 5.

2. Signal Model and FrFT

Let us consider a single-channel reconnaissance receiver and an adversary MIMO radar system with M transmitters. Assume that this radar employs OFDM-LFM waveforms, which were firstly introduced into the design of an MIMO radar system by F. Cheng [1]. Afterwards, the signal of the m-th transmitter is given as [1]:

$$s_m(t) = u_m(t) e^{j2\pi f_0 t}, 1 \leq m \leq M \tag{1}$$

$$u_m(t) = \frac{1}{\sqrt{T_p}} \mathrm{rect}\left(\frac{t}{T_p}\right) e^{j2\pi\left(mf_\Delta t + \frac{1}{2}\mu_0 t^2\right)} e^{j\phi_m}, 1 \leq m \leq M \tag{2}$$

where f_0 denotes the carrier frequency of the victim radar system, T_p is the pulse duration, f_Δ is the frequency step between two adjacent transmitters, μ_0 is the chirp rate and ϕ_m is the initial phase of the m-th transmitting signal. Here, we also assume $\mu_0 T_p \ll f_0$ [1].

Therefore, the MIMO radar signal intercepted by the reconnaissance receiver can be written as:

$$x(t) = A_m \sum_{m=0}^{M-1} s_m(t) + \omega(t) \tag{3}$$

where A_m is the complex constant amplitude of the m-th subpulse and $\omega(t)$ represents zero-mean white Gaussian noise with variance σ^2. The receiver first detects the observed signal energy and estimates the carrier frequency. Here, it is assumed that the above steps have been accomplished [20,21]. Then, these detected pulse observations are demodulated into intermediate frequency and sampled at an appropriate frequency, f_s, which satisfies the bandpass sampling theorem. Thus, we can collect N successive samples of the signal pulse represented as:

$$x[n] = A_m \sum_{m=0}^{M-1} e^{j2\pi\left[f_m n T_s + \frac{1}{2}\mu_0 (nT_s)^2\right]} e^{j\phi_m} + \omega[n] \tag{4}$$

where $T_s = 1/f_s$, $f_m = f_I + mf_\Delta$, f_I denotes the demodulated intermediate frequency and $n = 0, 1, \cdots, N-1 \left(N = T_p f_s\right)$. The classical definition of FrFT [11] is:

$$X_\alpha(u) = \int_{-\infty}^{\infty} K_\alpha(t, u) x(t) \, dt \tag{5}$$

where $K_\alpha(t, u)$ is the kernel function with:

$$K_\alpha(t, u) = \begin{cases} B_\alpha exp\left[j\pi\left((u^2 + t^2)\cot\alpha - 2ut\csc\alpha \right) \right], & \alpha \neq k\pi \\ \delta(t - u) & \alpha = 2k\pi \\ \delta(t + u) & \alpha = 2(k+1)\pi \end{cases} \tag{6}$$

and $B_\alpha = \sqrt{(1 - j\cot\alpha)}$. $\alpha = p\pi/2$ is called the rotation angle, while p is the order of FrFT. u is a spectral parameter. We employed the fast digital algorithm of FrFT [11], which is represented as:

$$X_\alpha\left(\frac{U}{2\Delta x} \right) = \frac{B_\alpha}{2\Delta x} e^{j\pi\tan\left(\frac{\alpha}{2}\right)\left(\frac{U}{2\Delta x}\right)^2} \sum_{n=-N}^{N} e^{j\pi\csc\alpha\left(\frac{U-n}{2\Delta x}\right)^2} e^{j\pi\tan\left(\frac{\alpha}{2}\right)\left(\frac{n}{2\Delta x}\right)^2} x\left(\frac{n}{2\Delta x} \right) \tag{7}$$

where $U = u2\Delta x$ and $\Delta x = \sqrt{N}$.

The OFDM-LFM signal is reformulated into multiple impulses only for a particular p in the FrFT domain, while the Gaussian white noise term is distributed evenly in the (α, U) plane. After peak searching, the estimated coordinates $(\hat{\alpha}_0, \hat{U}_m)$ can be used to obtain the estimators for OFDM-LFM signal parameters as [10]:

$$\begin{cases} \hat{\mu}_0 = -\cot(\hat{\alpha}_0)\frac{f_s^2}{N} \\ \hat{f}_m = \hat{U}_m\csc(\hat{\alpha}_0)\frac{f_s}{N} \\ \hat{f}_\Delta = \frac{1}{M-1}\sum_{m=2}^{M-1}(f_m - f_{m-1}) \\ \hat{A}_m = \frac{|X_{\hat{\alpha}_0}(\hat{U}_m)|}{\Delta x |B_{\hat{\alpha}_0}|} \end{cases} \tag{8}$$

However, this estimation performance depends on the grid size used for searching, while the ideal impulses require that Equation (7) is computed on an infinite number of grid points. In practice, due to the finite sampling data and leakage of other components' energy, there always exists some residual terms between the estimated quasi peaks (α_B, U_{Bm}) and real peaks (α_0, U_m), where α_B and U_{Bm} represent the bias estimations. Here, we set the residual terms as δ_0 and ε_m, where $\alpha_0 = \alpha_B + \varphi_0$ and $U_m = U_{Bm} + \varepsilon_m$. Furthermore, we set $\varphi_0 = \delta_0\Delta\alpha$, where $\Delta\alpha$ is the coarse searching interval of rotation α. In addition, it is reasonable to assume that $\delta_0, \varepsilon_m \in [-0.5, 0.5]$. Therefore, the residual term is the decisive point affecting the parameter estimation precision in Equation (8). Through conventional algorithms such as Newton-type [8] and interpolation [18], the residual term can be estimated. However, the first method suffers from a huge computational cost, and the second is only developed for monocomponent signals.

3. The Proposed Method

Our proposed estimation method is inspired by the multiple component estimator in [19], which was designed for direction of arrival estimation and implemented by DFT. However, if we want to use that idea in OFDM-LFM radar signal parameter estimation, some improvements on DFrFT should be conferred.

Substituting Equation (3) into Equation (5) and ignoring the noise term, the energy of the OFDM-LFM signal concentrates in the DFrFT domain:

$$X_{\alpha_0}(U) = A_m B_{\alpha_0} e^{j\pi U^2(-\mu)} \sum_{m=0}^{M-1} \left\{ e^{j\phi_m} \delta\left[2\pi\left(mf_\Delta - U\csc\alpha_0 \right) \right] \right\} \tag{9}$$

After peak searching at a sufficient grid interval, $\Delta\alpha$, we can obtain M peak coordinates (α_B, U_{Bm}). Then, the true chirp rate and true center frequency of the m-th component is given by:

$$\begin{cases} \mu_0 = -\cot(\alpha_B + \delta_0\Delta\alpha)\frac{f_s^2}{N} \\ f_m = mf_\Delta = (U_{Bm} + \varepsilon_m)\csc(\hat{\alpha}_0 + \delta_0\Delta\alpha)\frac{f_s}{N} \end{cases} \tag{10}$$

In the following subsections, we will derive the estimator for chirp rate μ_0 based on the δ_0 and the estimator for center frequency f_m based on the ε_m.

3.1. Estimator for Chirp Rate

Due to the fact that the analytical formulation of the quasi-peak amplitude $|X_{\alpha_B}(U_{Bm})|$ in the FrFT domain involves the Fresnel integral formula [10], it is hard to construct the estimator for δ_0 directly through the iterative method. Therefore, we introduce the FA algorithm to remove the Fresnel term, which is given as [22]:

$$(x_\alpha^* x)(\tau) = \int x\left(t + \frac{\tau}{2}\sin\alpha\right)x^*\left(t - \frac{\tau}{2}\sin\alpha\right)e^{2j\pi t\tau\cos\alpha}dt \tag{11}$$

where τ represents the delay factor. Then, the FA envelope statistic is also given as:

$$L(\alpha) = \int_{-\infty}^{\infty}|(x_\alpha^* x)(\tau)|d\tau \tag{12}$$

Substituting Equation (3) into Equations (11) and (12) and ignoring the noise term, we can derive the FA envelope of the OFDM-LFM signal:

$$\begin{aligned}(x_\alpha^* x)(\tau) &= \int_{-\infty}^{\infty}\sum_{m=0}^{M-1}A_m s_m\left(t + \frac{\tau}{2}\sin\alpha\right)\sum_{m=0}^{M-1}A_m s_m\left(t - \frac{\tau}{2}\sin\alpha\right)e^{2j\pi t\tau\cos\alpha}dt \\ &= \int_{-\infty}^{\infty}\gamma(t)e^{j2\pi t\tau(\mu_0\sin\alpha+\cos\alpha)}\sum_{m_i=0}^{M-1}\sum_{m_j=0}^{M-1}A_{m_i}A_{m_j}e^{j\pi\tau f_\Delta\sin\alpha(m_i-m_j)}e^{j2\pi tf_\Delta(m_i-m_j)}dt\end{aligned} \tag{13}$$

$$L(\alpha) = \int_{-\infty}^{\infty}|(x_\alpha^* x)(\tau)|d\tau \tag{14}$$

where $\gamma(t) = 1/\sqrt{T_p}\,\mathrm{rect}\left(t/\sqrt{T_p}\right)$ and:

$$\Gamma(\alpha) = \int_{-\infty}^{\infty}\int_{-\frac{T_p}{2}}^{\frac{T_p}{2}}\sum_{m_i=0}^{M-1}\sum_{m_j=0}^{M-1}A_{m_i}A_{m_j}e^{j\pi\tau f_\Delta\sin\alpha(m_i-m_j)}e^{j2\pi tf_\Delta(m_i-m_j)}dtd\tau \tag{15}$$

It is noticed that $\Gamma(\alpha)$ does not involve μ_0; therefore, we can ignore it in the following analysis of this subsection. The real peak of $L(\alpha_0)$ satisfies $\mu_0 = -\cot\alpha_0$. Substituting $\alpha_B = \alpha_0 - \delta_0\Delta\alpha$ into Equation (14), we can obtain:

$$\begin{aligned}L(\alpha_B) &= \int_{-\infty}^{\infty}|T_p\mathrm{Sinc}\left[\pi\tau\left(-\cot\alpha_0\sin(\alpha_0 - \delta_0\Delta\alpha) + \cos(\alpha_0 - \delta_0\Delta\alpha)\right)\right]|d\tau \\ &= \int_{-\infty}^{\infty}|T_p\mathrm{sinc}\left[\pi\tau\csc\alpha_0\sin(\delta_0\Delta\alpha)\right]|d\tau\end{aligned} \tag{16}$$

When the searching interval $\Delta\alpha$ is small enough, it is reasonable to use the approximations $\sin(\delta_0\Delta\alpha) \approx \delta_0\Delta\alpha$ and $\sin\left[\pi\tau\csc\alpha_0(0.5 - \delta_0)\Delta\alpha\right] \approx \sin\left[\pi\tau\csc\alpha_0(0.5 + \delta_0)\Delta\alpha\right]$ in Equation (16). Then, we can construct the error mapping as:

$$\beta = \frac{L(\alpha_B + 0.5\Delta\alpha) + L(\alpha_B - 0.5\Delta\alpha)}{L(\alpha_B + 0.5\Delta\alpha) - L(\alpha_B - 0.5\Delta\alpha)} \approx \int_{-\infty}^{\infty}\frac{\left|\frac{1}{\pi\csc\alpha_0(0.5-\delta_0)\alpha_s\tau}\right| - \left|\frac{1}{\pi\csc\alpha_0(0.5+\delta_0)\alpha_s\tau}\right|}{\left|\frac{1}{\pi\csc\alpha_0(0.5-\delta_0)\alpha_s\tau}\right| + \left|\frac{1}{\pi\csc\alpha_0(0.5+\delta_0)\alpha_s\tau}\right|}d\tau = \frac{1}{2\delta_0} \tag{17}$$

Hence, $\hat{\delta}_0 = 1/2\beta$ can be used as an estimator for δ_0.

Finally, the new estimation of rotation angle α_0 is presented as $\hat{\alpha}_0 = \alpha_B + \hat{\delta}_0 \Delta\alpha$. Then, by substituting $\alpha_B = \hat{\alpha}_0$ and renewing $\hat{\alpha}_0$, an iterative method can be combined to improve the estimation accuracy.

3.2. Estimator for Center Frequency

Firstly, we consider the DFrFT for a monocomponent LFM signal. Substituting Equation (2) into Equation (7), we can obtain:

$$X_\alpha \left(\frac{U}{2\Delta x} \right) = \frac{B_\alpha}{2\Delta x} e^{j\pi \cot \alpha \left(\frac{U}{2\Delta x} \right)^2} \sum_{n=-N}^{N} e^{j\pi 2n \left(\frac{f_m}{2f_s} - \frac{U \csc \alpha}{(2\Delta x)^2} \right) + j\pi n^2 \left(\frac{k_0}{(2f_s)^2} + \frac{\cot \alpha}{(2\Delta x)^2} \right)} \tag{18}$$

According to the analysis in Section 3.1, we assume that $\hat{\alpha}_0 \approx \alpha_0$. Hence, at the quasi peak $(\hat{\alpha}_0, U_{Bm})$, Equation (18) can be approximated by:

$$X_{\hat{\alpha}_0} \left(\frac{U_{Bm}}{2\Delta x} \right) = \frac{B_{\hat{\alpha}_0}}{2\Delta x} e^{j\pi \cot \hat{\alpha}_0 \left(\frac{U_{Bm}}{2\Delta x} \right)^2} \sum_{n=-N}^{N} e^{j\pi n \left(\frac{f_m}{f_s} - \frac{U_{Bm} \csc \hat{\alpha}_0}{2N} \right)} \tag{19}$$

Using $f_m = U_m f_s \csc \alpha_0 / 2N$ and $U_m = U_{Bm} + \varepsilon_0$, we can rewrite Equation (19) as:

$$X_{\hat{\alpha}_0} \left(\frac{U_{Bm}}{2\Delta x} \right) = \frac{B_{\hat{\alpha}_0}}{2\Delta x} e^{j\pi \cot \hat{\alpha}_0 \left(\frac{U_{Bm}}{2\Delta x} \right)^2} \sum_{n=-N}^{N} e^{j\pi n \frac{\varepsilon_0 \csc \hat{\alpha}_0}{2N}} \tag{20}$$

Similar to the approach in Section 3.1, we can obtain $X_{\hat{\alpha}_0} \left(\frac{U_{Bm} \pm P}{2\Delta x} \right)$ as:

$$X_{\hat{\alpha}_0} \left(\frac{U_{Bm} \pm P}{2\Delta x} \right) = \Gamma' (\hat{\alpha}_0, U_{Bm} \pm P) \left[\frac{e^{-j\pi \frac{(\varepsilon_0 \mp P) \csc \hat{\alpha}_0}{2}} \left(1 - e^{j\pi (\varepsilon_0 \mp P) \csc \hat{\alpha}_0} \right)}{1 - e^{j\pi \frac{(\varepsilon_0 \mp P) \csc \hat{\alpha}_0}{2N}}} \right] \tag{21}$$

where:

$$\Gamma' (\hat{\alpha}_0, U_{Bm} \pm P) = \frac{B_{\hat{\alpha}_0}}{2\Delta x} e^{j\pi \cot \hat{\alpha}_0 \left(\frac{U_{Bm} \pm P}{2\Delta x} \right)^2} \tag{22}$$

When $(\varepsilon_0 \mp P) \ll N$, it is reasonable to use the approximation $1 - e^x \approx x (x \to 0)$ in Equation (21). Then, by setting $P = 1 / \csc \hat{\alpha}_0$, we can construct the error mapping as:

$$h = \frac{\left| X_{\hat{\alpha}_0} \left(\frac{U_{Bm} + P}{2\Delta x} \right) \right| + \left| X_{\hat{\alpha}_0} \left(\frac{U_{Bm} - P}{2\Delta x} \right) \right|}{\left| X_{\hat{\alpha}_0} \left(\frac{U_{Bm} + P}{2\Delta x} \right) \right| - \left| X_{\hat{\alpha}_0} \left(\frac{U_{Bm} - P}{2\Delta x} \right) \right|} = \frac{\varepsilon_0}{\csc \alpha_0} \tag{23}$$

Hence, $\hat{\varepsilon}_0 = h \csc \hat{\alpha}_0$ can be used as an estimator for $\hat{\varepsilon}_0$. The fine estimation of U_m is presented as $\hat{U}_m = U_{Bm} + \hat{\varepsilon}_0$. Then, by substituting $U_B = \hat{U}_0$ and renewing $\hat{U}_0$, an iterative method can also be combined to improve the estimation accuracy.

3.3. Iterative Estimation for OFDM-LFM

In this subsection, we extend the proposed center frequency estimation method to the OFDM-LFM signals. The major difference between multiple center frequency estimation and single center frequency estimation is the estimation error that is caused by the leakage of multiple components in the OFDM-LFM signal. This error will lead to a bias in the interpolated DFrFT coefficients, deviating from their true values. Assuming the noise-free actual coefficients $\tilde{X}_{\hat{\alpha}_0,m}\left(\left(\hat{U}_m \pm P\right)/2\Delta x\right)$ of the m-th component, we obtain:

$$\tilde{X}_{\hat{\alpha}_0,m}\left(\frac{\hat{U}_m \pm P}{2\Delta x}\right) = DFRFT_{(\hat{\alpha}_0,\hat{U}_m \pm p)}\left(x\left[n\right]\right) = X_{\hat{\alpha}_0,m}\left(\frac{\hat{U}_m \pm P}{2\Delta x}\right) + \sum_{l=1,(l\neq m)}^{M} \breve{X}_{\hat{\alpha}_0,l}\left(\frac{\hat{U}_m \pm P}{2\Delta x}\right) \tag{24}$$

where $\breve{X}_{\hat{\alpha}_0,l}\left(\left(\hat{U}_m \pm P\right)/2\Delta x\right)$ represent the leakage terms introduced by the other $M-1$ OFDM-LFM components, which can be calculated by:

$$\breve{X}_{\hat{\alpha}_0,l}\left(\frac{\hat{U}_m \pm P}{2\Delta x}\right) = A_l DFrFT_{(\hat{\alpha}_0,\hat{U}_m \pm p)}\left(\hat{s}_l\left[n\right]\right) = A_l \frac{B_{\hat{\alpha}_0}}{2\Delta x} e^{j\pi \cot \hat{\alpha}_0 \left(\frac{\hat{U}_m \pm P}{2\Delta x}\right)^2} \sum_{n=-N}^{N} e^{j\pi n \left(\frac{(\hat{U}_l - \hat{U}_m \mp P)\csc \hat{\alpha}_0}{2N}\right)} \tag{25}$$

where A_l represents the complex amplitude of the l-th $(l=1,\dots,M)$ component in the fractional domain. Therefore, the estimation error of interpolated coefficients can be reduced by subtracting the sum of the leakage from other components, which is:

$$\hat{X}_{\hat{\alpha}_0,m}\left(\frac{\hat{U}_m \pm P}{2\Delta x}\right) = \tilde{X}_{\hat{\alpha}_0,m}\left(\frac{\hat{U}_m \pm P}{2\Delta x}\right) - \sum_{l=1,(l\neq m)}^{M} \breve{X}_{\hat{\alpha}_0,l}\left(\frac{\hat{U}_m \pm P}{2\Delta x}\right) \tag{26}$$

Based on the above analysis, we propose an iteration-based algorithm to accomplish the parameter estimation of the OFDM-LFM signal, which is given in Algorithm 1.

Algorithm 1: Proposed fast iterative interpolated digital fractional Fourier transform method

---

Initialization: Set $q=0$, $\hat{\delta}_0=0$, $\hat{\varepsilon}_m=0$ and $\hat{A}_m=0$ $(m=1\cdots M)$.

Calculate $L\left(\alpha\right) = \int_{-T_p}^{T_p} \left|\left(x_\alpha^* x\right)\left(\tau\right)\right| d\tau$.

Find $\alpha_B = \arg \underset{\alpha}{\text{Max}} \left\{L\left(\alpha\right)\right\}$,

where $\alpha \in [0:\Delta\alpha:\pi]$ and $u \in [0:1:N-1]$.

1 **Repeat**

2   Calculate $L\left(\alpha_B \pm 0.5\Delta\alpha\right)$ and β using Equations (14) and (17).

3   Renew $\alpha_B = \alpha_B + \hat{\delta}_0\Delta\alpha$, where $\hat{\delta}_0 = 0.5\beta$.

4 **Until** $q=Q$;

5 Let $X_{\hat{\alpha}_0}\left[u\right] = DFrFT_{(\hat{\alpha}_0,u)}\left(x\left[n\right]\right)$, $P = \left(\csc\hat{\alpha}_0\right)^{-1}$

6 **Repeat**

7   **for** $m=1$ **to** M **do**

8   **if** $(q==1)$ **then**

9               $\tilde{X}_{\hat{\alpha}_0}\left[u\right] = X_{\hat{\alpha}_0}\left[u\right] - \sum_{l=1,l\neq m}^{M} \hat{A}_l DFrFT_{(\hat{\alpha}_0,u)}\left(\hat{s}_l\left[n\right]\right)$

10              $\hat{U}_{Bm} = \arg \underset{u}{\text{Max}} \left(\left|\tilde{X}_{\hat{\alpha}_0}\left[u\right]\right|^2\right)$

11   **end**

12   Calculate $\hat{X}_{m,\hat{\alpha}_0}\left(\hat{U}_m \pm P\right)$ and h by Equations (23), (25) and (26).

13   Renew $\hat{U}_{Bm} = \hat{U}_{Bm} + \hat{\varepsilon}_m$, where $\hat{\varepsilon}_m = h\csc\hat{\alpha}_0$.

14          $\hat{A}_m = \left|X_{\hat{\alpha}_0}\left[\hat{U}_m\right] - \sum_{l=1,l\neq m}^{M} \breve{X}_{\hat{\alpha}_0,l}\left(\frac{\hat{U}_m}{2\Delta x}\right)\right| / \left(\Delta x \left|B_{\hat{\alpha}_0}\right|\right)$

15   **end**

16 **Until** $q=Q$;

Result: $\hat{U}_m = \hat{U}_{Bm} + \hat{\varepsilon}_m$, $\hat{\alpha}_0 = \alpha_B + \hat{\delta}_0\Delta\alpha$. Calculate $\hat{\mu}_0$, $\hat{f}_m$, $\hat{f}_\Delta$ by Equation (8).

Next, we discuss the computational complexity of this method. The proposed method consists of two parts, which are the estimation process of α_0 and the estimation process of U_m. The major computational load in the first part is due to the FA, which is about $O\left[G\left(2N\log N + N\right)\right]$ [22], where $G = \lfloor \pi/\Delta\alpha \rfloor$ ($\lfloor \bullet \rfloor$ indicates the floor operator). The major computational load in the second part is due to the DFrFT during coarse searching, which is about $O\left(N\log N\right)$ [11]. In addition, there are M-times DFrFT coefficient calculations, whose computation complexity is about $O\left(2MN\right)$ during each iteration. Consequently, the overall complexity of the above-mentioned method can be expressed as approximately $O\left[GN\log N + N\log N + MN\right]$, which is more efficient than the methods in [6] (requires $O\left(GN^2\log N\right)$), [7] (requires $O\left(N^3\right)$) and [10] (requires $O\left(GN^2\log N\right)$), but less efficient than the method in [18] (requires $O\left(8N + N\log N\right)$).

4. Simulations

The goal of this section is to evaluate the estimation performance of the proposed method through Monte Carlo simulations. Consider two kinds of OFDM-LFM radar waveforms that are applied in different radar modes (searching and tracking modes). The employed simulation parameters are listed in Table 1, which are consistent with the simulation settings in [1]. As defined in [1], we consider that the signal amplitude A_m of each subpulse is equal to A_0. As analyzed in Section 2, it is assumed that the signal detection and carrier frequency estimation have been accomplished. Here, we ignore the influence of signal detection probability and the accuracy of the carrier frequency estimation for the results. Then, the demodulated baseband pulse observations are sampled at the frequency $f_s = 50$ MHz. In this context, the Normalized Mean Squared Error (NMSE) is used to evaluate the estimation accuracy. Furthermore, we define the Signal-to-Noise Ratio (SNR) as $\rho = 10\lg\left(A_1^2/\sigma^2\right)$ and set the searching interval of rotation to be $\Delta\alpha = 0.001$. Some other algorithms reported in [6,7,10,18] and Cramer–Rao lower Bounds (CRB) [23] are also reviewed for comparison.

Table 1. Parameters of Orthogonal Frequency Division Multi-Linear Frequency Modulation (OFDM-LFM) signals.

	Ω_1	Ω_2
Radar operation mode	Searching	Tracking
Number of antennas M	4	4
Pulse duration T_p	20 µs	20 µs
Chirp rate μ_0	0.15 MHz/µs	0.15 MHz/µs
Bandwidth B_0	3 MHz	3 MHz
Frequency step f_Δ	5 MHz	1.5 MHz

Figure 1 gives the NMSE of the chirp rate estimation, $\hat{\mu}_0$, versus different SNRs. In this simulation, Monte Carlo experiments were repeated 500 times for each SNR from -18 dB to 2 dB. It is obvious from Figure 1 that most NMSE curves of estimation algorithms approach or achieve the CRB at specific SNRs. Among them, the performance of the proposed method coincides with the CRB at the lowest SNR, which is -11 dB. Moreover, the simulation result from Figure 1 confirms that the estimation performance of the proposed method slightly outperforms the other algorithms at all SNRs. Here, the iteration number is set to $Q = 3$, which is demonstrated in Figure 2.

In Figure 2, we study the effect of the iteration number, Q, on the convergence characteristics of the proposed method. In this simulation, the NMSE curves of frequency step (f_Δ) estimation versus the iteration number (Q) when the SNR is set to $[-10, -7, -4, -1]$ dB are depicted. Here, both signals with parameters from Ω_1 and Ω_2 are used for the simulation. As can be seen, the parameter estimation performance converges after three iterations for almost all of the SNRs. Therefore, through the simulation in Figure 2, the iteration number Q is suggested to be chosen as three.

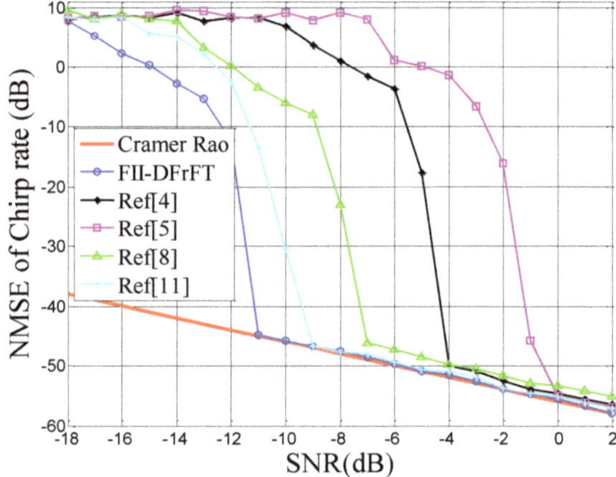

Figure 1. Normalized Mean Squared Error (NMSE) of μ_0 for signal Ω_1 versus the signal-to-noise ratio. FII, Fast Iterative Interpolated.

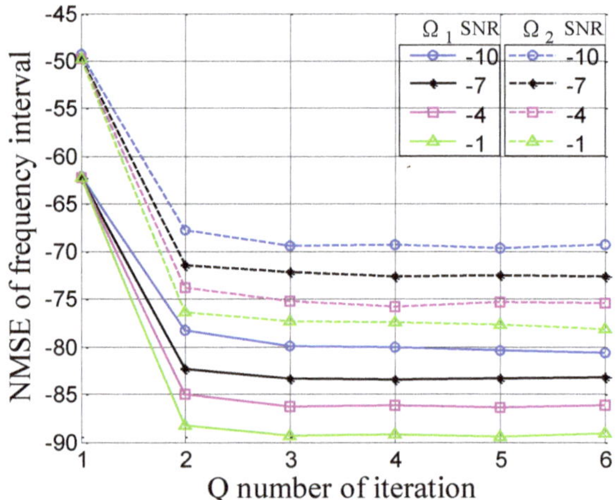

Figure 2. NMSE of f_Δ for signals Ω_1 and Ω_2 versus the number of iterations.

5. Conclusions

In this study, we have derived the analytical AF and DFrFT approximation of OFDM-LFM radar signals. A new method called FII-DFrFT was proposed for uncooperative OFDM-LFM parameter estimation, which was formulated by locating the bottleneck issue that affects the estimation performance. The analytical formulas were hence derived, as well as their performance evaluation. Numerical simulations showed the validity and superiority of the proposed method, through comparisons with some existing algorithms at different SNRs. Nevertheless, as an uncooperative facility, especially for hostile MIMO radars, the estimation performance in the presence of clutter and other structure interferences is still a challenge for most cases. Hence, in future research, we would like to focus on the derivation and evaluation, taking into consideration the keen factors' uncertainty, as well as the clutter background, before the proposed scheme is employed for practical applications.

Author Contributions: Y.L. designed and wrote this paper under the supervision of B.T., Y.Z. assisted with the methodology and formal analysis. J.Z. and Y.X. provided the support of the entire study. Y.Z. and B.T. reviewed and edited the manuscript.

Funding: This research was funded by the National Natural Science Foundation of China (Grant No. 61571088).

Acknowledgments: Thanks to the University of Electronic Science and Technology of China. Thanks to mysupervisor Bin Tang. Thank you to mydear love Leilei and all the people who support us.

Conflicts of Interest: The authors declare no conflict of interest.

Abbreviations

The following abbreviations are used in this manuscript:

OFDM	Orthogonal Frequency Division Multi-
LFM	Linear Frequency Modulation
FrFT	Fractional Fourier Transform
FA	Fractional Autocorrelation
MWHT	Multiple Wigner–Hough Transform
DFrFT	Digital FrFT
DFT	Digital Fourier Transform
NMSE	Normalized Mean Squared Error
SNR	Signal-to-Noise Ratio
CRB	Cramer–Rao lower Bound

References

1. Cheng, F.; He, Z.; Liu, H.M.; Li, J. The Parameter Setting Problem of Signal OFDM-LFM for MIMO Radar. In Proceedings of the International Conference on Communications, Circuits and Systems, Fujian, China, 25–27 May 2008; pp. 981–985.
2. Li, J.; Stoica, P. *MIMO Radar Signal Processing*; Wiley-IEEE Press: Hoboken, NJ, USA, 2009.
3. Yao, Y.; Zhao, J.; Wu, L. Adaptive Waveform Design for MIMO Radar-Communication Transceiver. *Sensors* **2018**, *18*, 1957. [CrossRef] [PubMed]
4. Li, F.; He, F.; Dong, Z.; Wu, M.; Zhang, Y. General Signal Model for Multiple-Input Multiple-Output GMTI Radar. *Sensors* **2018**, *18*, 2576. [CrossRef] [PubMed]
5. Li, Y.H.; Wang, J.; He, X.D.; Tang, B. A method for PRI estimation of multicomponent LFM signals from MIMO radars. In Proceedings of the IEEE 17th International Conference on Computational Science and Engineering (CSE), Chengdu, China, 19–21 December 2014; pp. 1034–1038. [CrossRef]
6. Li, Y.H.; Tang, B. Parameters estimation and detection of MIMO-LFM signals using MWHT. *Int. J. Electron.* **2016**, *103*, 439–454. [CrossRef]
7. Lin, Y.; Peng, Y.N.; Wang, X.T. Maximum likelihood parameter estimation of multiple chirp signals by a new Markov chain Monte Carlo approach. In Proceedings of the IEEE 2004 Radar Conference, Philadelphia, PA, USA, 29 April 2004; pp. 559–562. [CrossRef]
8. Yang, P.; Liu, Z.; Jiang, W.L. Parameter estimation of multi-component chirp signals based on discrete chirp Fourier transform and population Monte Carlo. *Signal Image Video Process.* **2015**, *9*, 1137–1149. [CrossRef]
9. Wang, J.Z.; Su, S.Y.; Chen, Z.P. Parameter estimation of chirp signal under low SNR. *Sci. China-Inf. Sci.* **2015**, *58*, 1–3. [CrossRef]
10. Serbes, A. On the Estimation of LFM Signal Parameters: Analytical Formulation. *IEEE Trans. Aerosp. Electron. Syst.* **2018**, *54*, 848–860. [CrossRef]
11. Ozaktas, H.M.; Ankan, O.; Kutay, M.A.; Bozdagi, G. Digital computation of the fractional Fourier transform. *IEEE Trans. Signal Process.* **1996**, *44*, 2141–2150. [CrossRef]
12. Mei, L.; Zhang, Q.Y.; Sha, X.J.; Zhang, N.T. Digital computation of the weighted-type fractional Fourier transform. *Sci. China-Inf. Sci.* **2013**, *56*, 1–2. [CrossRef]
13. Li, C.P.; Dao, X.H.; Guo, P. Fractional derivatives in complex planes. *Nonlinear Anal.-Theory Methods Appl.* **2009**, *71*, 1857–1869. [CrossRef]
14. Ortigueira, M.D. A coherent approach to non-integer order derivatives. *Signal Process.* **2006**, *86*, 2505–2515. [CrossRef]
15. Guariglia, E. *Fractional Derivative of the Riemann Zeta Function*; Walter de Gruyter GmbH & Co KG.: Berlin, Germany, 2015.

16. Guariglia, E.; Silvestrov, S. A Functional Equation for the Riemann Zeta Fractional Derivative. *AIP Conf. Proc.* **2017**, *1798*. [CrossRef]

17. Wang, X. Fractional Geometric Calculus: Toward A Unified Mathematical Language for Physics and Engineering. In Proceedings of the Fifth Symposium on Fractional Differentiation and Its Applications, Nanjing, China, 14–17 May 2012.

18. Song, J.; Wang, Y.X.; Liu, Y.F. Iterative Interpolation for Parameter Estimation of LFM Signal Based on Fractional Fourier Transform. *Circuits Syst. Signal Process.* **2013**, *32*, 1489–1499. [CrossRef]

19. Aboutanios, E.; Hassanien, A.; Amin, M.G.; Zoubir, A.M. Fast Iterative Interpolated Beamforming for Accurate Single-Snapshot DOA Estimation. *IEEE Geosci. Remote Sens. Lett.* **2017**, *14*, 574–578. [CrossRef]

20. Jin, M.; Guo, Q.H.; Li, Y.M.; Xi, J.T.; Yu, Y.G. Energy Detection With Random Arrival and Departure of Primary Signals: New Detector and Performance Analysis. *IEEE Trans. Veh. Technol.* **2017**, *66*, 10092–10101. [CrossRef]

21. Jayaprakash, A.; Reddy, G.R. Robust Blind Carrier Frequency Offset Estimation Algorithm for OFDM Systems. *Wirel. Pers. Commun.* **2017**, *94*, 777–791. [CrossRef]

22. Akay, O.; Boudreaux-Bartels, G.F. Fractional autocorrelation and its application to detection and estimation of linear FM signals. In Proceedings of the IEEE-SP International Symposium on Time-Frequency and Time-Scale Analysis (Cat. No.98TH8380), Pittsburgh, PA, USA, 9 October 1998; pp. 213–216. [CrossRef]

23. Ristic, B.; Boashash, B. Comments on "The Cramer-Rao lower bounds for signals with constant amplitude and polynomial phase". *IEEE Trans. Signal Process.* **1998**, *46*, 1708–1709. [CrossRef]

Article

Differential Run-Length Encryption in Sensor Networks

Chiratheep Chianphatthanakit, Anuparp Boonsongsrikul * and Somjet Suppharangsan

Department of Electrical Engineering, Faculty of Engineering, Burapha University Chonburi Campus, Chonburi 20131, Thailand
* Correspondence: anuparp@eng.buu.ac.th; Tel.: +66-3810-2222 (ext. 3380)

Received: 28 May 2019; Accepted: 17 July 2019; Published: 19 July 2019

Abstract: Energy is a main concern in the design and deployment of Wireless Sensor Networks because sensor nodes are constrained by limitations of battery, memory, and a processing unit. A number of techniques have been presented to solve this power problem. Among the proposed solutions, the data compression scheme is one that can be used to reduce the volume of data for transmission. This article presents a data compression algorithm called Differential Run Length Encryption (D-RLE) consisting of three steps. First, reading values are divided into groups by using a threshold of Chauvenet's criterion. Second, each group is subdivided into subgroups whose consecutive member values are determined by a subtraction scheme under a K-RLE based threshold. Third, the member values are then encoded to binary based on our ad hoc scheme to compress the data. The experimental results show that the data rate savings by D-RLE can be up to 90% and energy usage can be saved more than 90% compared to data transmission without compression.

Keywords: data compression; wireless sensor networks; energy consumption

1. Introduction

Wireless sensor networks (WSNs) consist of smart wireless sensors working together to monitor areas and to collect data such as temperature and humidity from the environment. However, sensor nodes are faced with resource constraints in terms of energy, memory, and a processing unit [1]. Many works [2–4] proposed a variety of solutions to overcome the restrictions. A challenge is how to prolong sensor life time during data delivery from sensor nodes to a base station. Energy consumption is a hard problem in the design and deployment in WSNs because sensor nodes may be deployed in harsh environments where it is not easy to replace the batteries [5]. Energy consumption mostly occurs in either data computing or data transmission. Wang et al. [6] reported that the ratio between computing and communication incurred energy consumption is about 1:3000; therefore, sensor nodes should focus on effective data communication. If sensor nodes reduce the number of data transmissions, this can obviously save energy consumption in the entire network. The end-to-end energy cost and network lifetime are greatly restricted if the cooperative transmission model is not designed properly [7]. The most common technique for saving energy is the use of a sleep-wake scheduling scheme [8] in which a significant part of the sensor's transceiver is switched off. However, the solution induces a problem of time synchronization [9] and the possibility of retransmitting data. Sensor network topologies also have a massive impact on energy usage in data transmission. In tree-based topologies [10], data aggregation approaches are often mentioned in order to reduce data redundancy and resulted in decreasing the number of data transmissions. However, it is merely given an approximate data value in a local area [11]. In cluster-based topologies [7,12], a cluster head plays an important role that collects and forwards all data from neighboring nodes to the base station. The cluster head consumes higher energy than other neighboring nodes and results in failures if it is out of energy more quickly. If the number of failures exceeds the tolerance level, a system may collapse [13]. To disregard this problem, a cluster head [14] is assumed as a special node having more sufficient energy than its neighboring nodes. On the other hand, our proposed solution does not require a special node. It simply can be applied for all sensor nodes including a cluster head that can be exhausted when it works hard. Residual energy is a criterion in selection of a cluster head [15].

In this paper, data compression is a proposed solution that can reduce a data packet size and amount of data transmission and result in prolonging the battery life of wireless sensor nodes. The proposed concept shown in Figure 1 can apply either lossless or lossy data compression. Furthermore, the proposed data compression does not require extra RAM. Data compression can be divided into lossless and lossy algorithms. Lossless compression provides data accuracy but normally requires extensive use of memory for making a lookup table. A sensor LZW (S-LZW) algorithm [16] is an extension of a lossless data compression algorithm created by Abraham Lempel, Jacob Ziv, and Terry Welch (LZW) [17,18]. Capo-Chichi et al. [19] and Roy et al. [20] reported a concept of S-LZW that is a dictionary-based algorithm that is initialized by all standard characters of 255 ASCII codes. However, a new string in the input stream creates a new entry and results in the limitation of memory in a sensor node. In [21–24], their schemes are based on lossless entropy compression (LEC) for data compression by using the Huffman variable length codes. The data difference is an input to an entropy encoder. LEC is one of the efficient schemes in data compression, therefore LEC is applied for reliable data transmission to monitor a structural health in wireless sensor networks [25]. On the other hand, lossy compression [26] is data compression that is appropriate for sending approximate data or repeated data [27]. The *K*-Run-Length Encoding (*K*-RLE) algorithm [28] is a lossy compression that is an adaptation of RLE [17]. *K*-RLE's data accuracy and compression ratio depend on the K-precision.

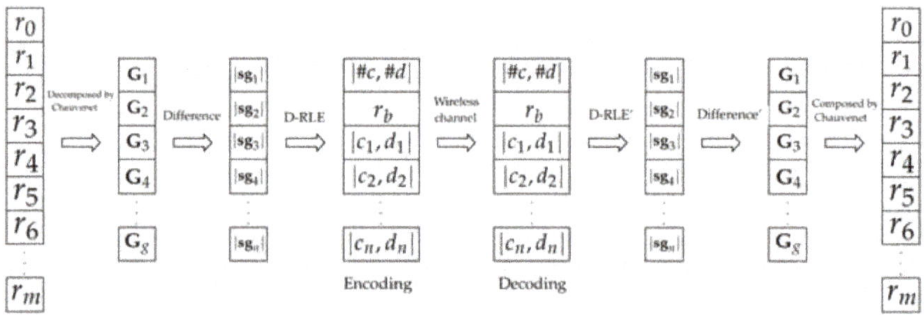

Figure 1. Data encoding and decoding process.

2. Related Work

In [26], researchers presented a comparison of data compression schemes with different sensor data types and sensor data sets in WSNs. In [24], data types in compression can be considered and divided into smooth temperature and relative humidity data and dynamic volcanic data that exhibit dramatic different characteristics. Since power consumption is one of the main concerns, Koc et al. [29] studied and measured power consumption during data compression by using the MSP432 family of microcontrollers. With the same fixed parameters of wireless environments, the energy usage for a fixed size packet would be the same on delivering the packet. Reducing the number of data packets would help reduce the energy consumption. Therefore, data compression has played a significant role in WSNs. Two types of data compression are generally categorized and referred to as lossy and lossless compression. The former compression permanently removes a certain amount of data, reducing the size of data to much smaller than the original ones, but it degrades the quality of data. While the latter compression reduces the data size without any data quality loss, its compression rate is lower than the former compression. In our experiments, we compared our results with three lossless algorithms that is LEC [21], Lempel-Ziv-welch (LZW) [17,18] and run-length encoding (RLE), whereas we compared our results with K-RLE for the lossy scheme. LEC is a lossless algorithm based on the Huffman concept in which the entropy is used for defining the Huffman codes. Table 1 shows the prefix and suffix codes used in LEC. LEC computes the difference data values and then replaces the difference by the corresponding codes from Table 1. LEC Algorithm is shown in Algorithm 1.

Table 1. LEC codes.

Level (n_i)	Bits	Prefix (s_i)	Suffix Range (a_i)	Value (d_i)
0	2	00	-	0
1	4	010	0...1	$-1, 1$
2	5	011	00...11	$-3, -2, 2, 3$
3	6	100	000...111	$-7, ..., -4, 4, ..., 7$
4	7	101	0000...1111	$-15, ..., -8, 8, ..., 15$
5	8	110	00000...11111	$-31, ..., -16, 16, ..., 31$
6	10	1110	000000...111111	$-63, ..., -32, 32, ..., 63$
7	12	11110	0000000...1111111	$-127, ..., -64, 64, ..., 127$

Algorithm 1 LEC Pseudocode

Require: d_i, Table of LEC codes
Ensure: bs_i
  if ($d_i == 0$) **then**

    $n_i = 0$
  else

    $n_i = \lceil \log_2(|d_i|) \rceil + 1$
  end if
  $s_i = \text{Table}(n_i)$ ▷ extract s_i from LEC Table
  if ($n_i == 0$) **then**

    $bs_i = s_i$
    return bs_i
  end if
  if ($d_i > 0$) **then**

    $a_i = (d_i)|_{n_i}$ ▷$(v)|_{n_i}$ is the n_i low-order bits of v.
  else

    $a_i = (d_i - 1)|_{n_i}$
  end if
  $bs_i = (s_i, a_i)$
  return bs_i

When d_i is negative, low-order bits of the two's complement representation of $(d - 1)$ are used for the suffix code. For example, suppose we have a data set: <19, 18, 20, 21>. Starting with the first data 19, we then compute the value difference between a pair of consecutive data, resulting in <19, −1, 2, 1>. By using Table 1 and Algorithm 1 above, we obtain the following encoded bs_i: (0001 0011), (010,0) (011,10) (010,1).

LZW is a dictionary-based lossless compression. LZW used in the experiments begins with the value of 256 onwards to avoid repeating the value of the first 256 ASCII codes. The algorithm repeatedly reads a symbol input to form a string and checks if the string is not in the dictionary. Once such a string is found, the corresponding output code for the string without the last symbol that is the longest string in the dictionary is sent out, and the new found string is added to the dictionary with the next available output code. Table 2 shows an example of data input string AAAABAAAABCC. Applying Algorithm 2, the seven new codes (code from 256 to 262) are added into the dictionary and the output strings are <A, AA, A, B, AAA, B, C, C>. The output codes for those output strings are <65,

256, 65, 66, 257, 66, 67, 67> where each output code uses nine bits, so in total the encoding output uses 72 bits compared to original input 96 bits. Nevertheless, a larger dictionary requires larger memory.

Table 2. LZW codes.

String	Output	Dictionary	Total Bits
A	65	256 = AA	9
AA	256	257 = AAA	18
A	65	258 = AB	27
B	66	259 = BA	36
AAA	257	260 = AAAB	45
B	66	261 = BC	54
C	67	262 = CC	63
C	67		72

Algorithm 2 LZW Pseudocode

---

initialize Dictionary[0-255] = first 256 ASCII codes
STRING ← get input symbol
while there are still input symbols **do**

    SYMBOL ← get input symbol
    if (STRING+SYMBOL is in Dictonary) **then**

        STRING = STRING+SYMBOL
    else

        output the code for STRING
        add STRING+SYMBOL to Dictionary
        STRING = SYMBOL
    end if
end while
output the code for STRING

RLE is the simplest compression, working by counting the amount of repeating consecutive identical data. The amount of consecutive identical data followed by the data symbol is replaced for the original repeating data. For example, the data of AAABBCEEFFFFFFFFAA are compressed to 3A2B1C2E8F2A, which implies that there are 3 A's, 2 B's, C, 2 E's, 8 F's and 2 A's next to each other in series. RLE pseudocode is shown in Algorithm 3.

Algorithm 3 RLE Pseudocode

---

while there are still input symbols **do**

    count = 0
    repeat

        get input symbol
        count = *count* + 1
    until symbol unequal to next symbol
    output count and symbol
end while

K-RLE is based on a RLE algorithm allowing quality loss to such an extent. The value of K indicates the range of different data values. If K=1, for example, the data of <19, 18, 20, 21> will be encoded as <(3, 19), (1, 21)> because the first three pieces of data are in the range of 19 ± 1 and the last two pieces of data are in the range of 20 ± 1. The encoded data <(3, 19), (1, 21)> then will be decoded as <19, 19, 19, 21>. Obviously, the decoded data are different from the original data due to the lossy scheme. *K*-RLE pseudocode is shown in Algorithm 4.

Algorithm 4 *K*-RLE Pseudocode

$v1 \leftarrow$ read input value
$count = 1$
while there are still input values **do**

    $v2 \leftarrow$ read next input value
    if $(|v1 - v2| \leq K)$ **then**

        $count = count + 1$
    else

        output $(count, v1)$
        $v1 \leftarrow v2$
        $count = 1$
    end if
end while
output $(count, v1)$

LZW compresses the same data with the same encoding though these data are at different positions in the data input stream. In contrast, RLE requires that the same data must stand next to each other in a row. Both LEC and LZW apply a similar concept in terms of the prefix codes. However, LEC also has suffix codes addressing the different value; hence, each encoding of difference value in the input stream consists of prefix and suffix codes. If the input stream has many pieces of consecutive identical data, RLE performs very well. LZW would be preferred to RLE if the input stream consisted of many repeating data with shorter output codes. LEC works even better if the input stream has many of the same levels of the difference values with shorter prefix and suffix codes. Aforementioned algorithms have the linear time complexity $\mathcal{O}(n)$ and perform best in their own characteristics, not for general datasets. This gap stimulates how we can combine each strong point of those algorithms to compress the data. To this end, we have developed Differential Run Length Encryption (D-RLE), which also has the linear time complexity $\mathcal{O}(n)$, and will explain its concept in the next section.

3. Differential Run Length Encryption

This section presents the proposed algorithm called Differential Run Length Encryption or D-RLE, which consists of three steps. First, raw data are collected and divided into several groups by using a threshold of Chauvenet's criterion. Second, consecutive data are subtracted and arranged into multiple subgroups of differential values based on a threshold of *K*-RLE. Third, our adaptive data compression, which is the DSC-based scheme [30], is employed to each piece of subgroup data. Data formats in D-RLE are shown in Table 3. As a result, D-RLE significantly reduces the amount of data delivery, saves energy consumption and prolongs the battery life of the sensor nodes.

Table 3. Data formats in D-RLE.

Process	Format
raw data	$< r_0, r_1, ..., r_m >$
group division	$< \mathbf{G}_1, \mathbf{G}_2, ..., \mathbf{G}_g >$, where $\mathbf{G}_1 =< r_0, r_1, ..., r_{j-1} >$, $\mathbf{G}_2 =< r_j, r_{j+1}, ..., r_{k-1} >, ..., \mathbf{G}_g =< r_l, r_{l+1}, ..., r_m >$
subgroup division	for each $\mathbf{G}_i = \mathbf{g}_{be} =< r_b, r_{b+1}, r_{b+2}, ..., r_e >$ $\Rightarrow\ < r_b, \|sg_1\|, \|sg_2\|, ..., \|sg_n\| >$ $\Rightarrow\ < r_b, \|c_1, r_{b+1}, ..., r_{b+c_1}\|, \|c_2, r_{b+c_1+1}, ..., r_{b+c_1+c_2}\|, ..., \|c_n, r_{b+c_1+c_2+...+c_{n-1}+1}, ..., r_e\| >$ $\Rightarrow\ < r_b, \|c_1, (r_{b+c_1} - r_b)\|, \|c_2, (r_{b+c_1+c_2} - r_{b+c_1})\|, ..., \|c_n, (r_e - r_{b+c_1+c_2+...+c_{n-1}})\| >$ $\Rightarrow\ < r_b, \|c_1, d_1\|, \|c_2, d_2\|, ..., \|c_n, d_n\| >$
encoded data	$< \|\#c, \#d\|, r_b, \|c_1, d_1\|, \|c_2, d_2\|, ..., \|c_n, d_n\| >$

3.1. Group Division by Chauvenet's Criterion

The raw data can be considered as a random sample $<r_0, r_1, \ldots, r_m>$ and they are grouped by using a pre-defined Chauvenet's criterion [31,32] D_{max}, which is set to 1.96 according to the significance level of 0.05 from statistics. The data r_i is passed to Equation (1) for calculating D_i in which μ and σ are mean and standard deviation, respectively. The value of D_i is then compared with D_{max} to consider if r_i should belong to the same group or not. We maintain r_i to the same group if D_{max} is greater than or equal to D_i; otherwise, we split r_i into the next group. For example, suppose we have the following raw data $<21,25,28,30,31,35,37,42,47,49,50,55,62,76,82,95,105,103,92,86,71,63,59,52,41,34,30,26,25,21>$. As the raw data coming into Equation (1), we found that $D_{15} = |95 - 52.43|/25.34 = 1.68$, whereas $D_{16} = |105 - 52.43|/25.34 = 2.07$. Since $D_{max} = 1.96$ is less than $D_{16} = 2.07$, data r_{16} is split into a different group. Therefore, in this data sample set, there are two groups where the first group is $<21,25,28,30,31,35,37,42,47,49,50,55,62,76,82,95>$, and the other group is $<105,103,92,86,71,63,59,52,41,34,30,26,25,21>$.

$$D_i = \frac{|r_i - \mu|}{\sigma} \tag{1}$$

Once a group is split, the mean and standard deviation are recalculated and updated with the remaining data and used in Equation (1) for the next round of group divisions. Algorithm 5 shows the algorithm for the group division step.

Algorithm 5 Group Division

Require: $r[m+1] = \{r_0, r_1, \ldots, r_m\}$
Ensure: $\mathbf{G} = \{\mathbf{G}_1, \mathbf{G}_2, \ldots, \mathbf{G}_g\}$
   initialize $N = m + 1, D_{max} = 1.96, g = 1$
   initialize arrays $D_i[N]$
   compute μ and σ
   add r_0 into $\mathbf{G}_1$
   for $(i = 1$ to $m)$ **do**
      $D_i[i] = \frac{|r_i - \mu|}{\sigma}$
      if $(D_{max} < D_i[i])$ **then**
         $g = g + 1$
         add r_i into $\mathbf{G}_g$
         update μ, σ
      else
         add r_i into $\mathbf{G}_g$
      end if
   end for

3.2. Subgroups Division

Each group from the first step will be sub-divided based on the K-RLE scheme. The value of K implies the degree of tolerance. It is lossless if K equals zero. The result of the subgroup division is written in the compact format: $<r_b, |c_1, d_1|_1, |c_2, d_2|_2, \ldots, |c_n, d_n|_n>$, where the symbol $|c_i, d_i|_i$ is used to separate ith subgroup and r_b is the first value of the group. The number of elements in ith subgroup is denoted by c_i and the value difference between the last raw data of the present subgroup ith and the last raw data in the previous subgroup $(i-1)$th is denoted by d_i. The algorithm of discovering the c_i and d_i is shown in Algorithm 6. As we start off the index from zero, $m + 1$ is the number of data members in the group. The index n is the number of subgroups that is also the number of members in set C_i and D_i. The relationship between m and c_i is shown by Equation (2):

$$m = \sum_{i=1}^{n} c_i. \tag{2}$$

Algorithm 6 Computing $|c_i, d_i|$

Require: $K, r = \{r_0, r_1, \ldots, r_m\}$
Ensure: $C_i = \{c_1, c_2, \ldots, c_n\}, D_i = \{d_1, d_2, \ldots, d_n\}$
  initialize $i = 1, j = 1, count = 1, f = 1, s = 0$
  while $j < (m + 1)$ **do**

    $s = r[f] - r[j + 1]$
    if $|s| \leq K$ **then**

      $count = count + 1$
    else

      $C[i] = count$
      $D[i] = r[j] - r[f - 1]$
      $i = i + 1$
      $f = j + 1$
      $count = 1$
    end if
    $j = j + 1$
  end while

For the first group $<21, 25, 28, 30, 31, 35, 37, 42, 47, 49, 50, 55, 62, 76, 82, 95>$, we ran the algorithm from Algorithm 6 above and had the following subgroup division result: $<21, |1,4|, |1,3|, |2,3|, |1,4|, |1,2|, |1,5|, |1,5|, |2,3|, |1,5|, |1,7|, |1,14|, |1,6|, |1,13|>$. Value 21 is the first raw data r_b of the group, followed by $|c_i, d_i|_{i=1 \, to \, 13}$. The variable c_i is just a counter, starting from one, indicating how many pairs that the absolute difference between the first and next raw data in the same subgroup do not differ more than the pre-defined K value. The variable d_i roughly dictates to us how different the data are between the present and previous subgroups. The following subgroup division result: $<105, |1,-2|, |1,-11|, |1,-6|, |1,-15|, |1,-8|, |1,-4|, |1,-7|, |1,-11|, |1,-7|, |1,-4|, |2,-5|, |1,-4|>$ is obtained for the second group from the first step.

3.3. Adaptive Data Encoding

The last step is the process of adaptively encoding each subgroup division result. Shortened opcodes for c_i and d_i are used to compress raw data via the encoding process. The number of bits to represent c_i and d_i is determined by set $C_i = \{c_1, c_2, \ldots, c_n\}$ and set $D_i = \{d_1, d_2, \ldots, d_n\}$, respectively. We shall use the first subgroup division result, $<21, |1,4|, |1,3|, |2,3|, |1,4|, |1,2|, |1,5|, |1,5|, |2,3|, |1,5|, |1,7|, |1,14|, |1,6|, |1,13|>$, as an illustration of the encoding process. To begin with, set C_i is $\{1,1,2,1,1,1,1,2,1,1,1,1,1\}$ and set D_i is $\{4,3,3,4,2,5,5,3,5,7,14,6,13\}$. Then, Equations (3)–(5) are used to compute the number of bits for c_i and d_i, respectively:

$$\#c = \left\lceil \log_2 \left(\max_{i=1}^{n}(c_i) \right) \right\rceil, \tag{3}$$

$$d_L = d_i \text{ where } i = \underset{i}{\text{argmax}} \, |D_i|, \tag{4}$$

$$\#d = \begin{cases} k+1 & \text{if } -(2^k) + 1 \leq d_L \leq 2^k, \\ k+2 & \text{if } d_L = -(2^k), \end{cases} \text{ where } k \in I^+. \tag{5}$$

Because the maximum value of C is 2, the number of bits for c_i equals $\lceil (\log_2 2) \rceil = 1$ bit. The argument of the maximum for index i of absolute value of D_i is $i = 11$; hence, $d_L = d_{11} = 14$ and $14 \leq 2^{k=4}$; then, the number of bits for d_i equals $k + 1 = 4 + 1 = 5$ bits. In binary code, 1 and 5 are represented by 0001 and 0101, respectively. We combine 0001 with 0101 to form a byte as 0001 0101 referred to as $<|\#c, \#d|>$ and this byte will be our length field to notify the decoder of the bit sizes of c_i and d_i. We use these bit sizes to limit the bit length used for binary codes of c_i and d_i . In addition, the value of each c_i before changing to binary code is decreased by one as doing so will help reduce bit sizes for c_i. For example, the binary code $c_i = 8$ is 1000 (4 bits), but we could obtain the shortened binary code 111 (3 bits) if we decrease 8 to 7. We could not do the same decrease for d_i as d_i can be either positive or negative values, whereas c_i is only positive integers. Subsequently, the data field is set to the format $<r_b, |c_1, d_1|_1, |c_2, d_2|_2, \ldots, |c_n, d_n|_n>$ in which r_b is represented by its corresponding 8-bit binary code, and $|c_i, d_i|_{i=1 \text{ to } n}$ are opcodes for c_i and d_i denoted by shorten binary codes where $c_i = c_i - 1$. Two's complement is used for negative values of d_i. The length and data fields join together to make a data payload. Finally, the data payload $<|\#c, \#d|, r_b, |c_1, d_1|_1, |c_2, d_2|_2, \ldots, |c_n, d_n|_n>$ of our example is $<|1,5|, 21, |0,4|_1, |0,3|_2, |1,3|_3, |0,4|_4, |0,2|_5, |0,5|_6, |0,5|_7, |1,3|_8, |0,5|_9, |0,7|_{10}, |0,14|_{11}, |0,6|_{12}, |0,13|_{13}>$ and will be encoded as $<|0001,0101|, |0001\,0101|, |0,00100|_1, |0,00011|_2, |1,00011|_3, |0,00100|_4, |0,00010|_5,$ $|0,00101|_6, |0,00101|_7, |1,00011|_8, |0,00101|_9, |0,00111|_{10}, |0,01110|_{11}, |0,00110|_{12}, |0,001101|_{13}>$. The total number of encoded bits for each subgroup division result can be found by Equation (6). In our example case $K = 1$, the total number of encoded bits is $16 + [13 * (1 + 5)] = 94$ bits compared to $2 * 16 * 8 = 256$ bits without compression. This means we have saved approximately 63.28% of data delivery:

$$Size_{bit} = 16 + [n \times (\#c + \#d)]. \tag{6}$$

For the second subgroup division result, the number of bits for c_i and d_i equal to 1 and 5 bits, respectively. Therefore, the data payload of the second subgroup is $<|0,5|, 105, |0,-2|, |0,-11|, |0,-6|, |0,-15|, |0,-8|, |0,-4|, |0,-7|, |0,-11|, |0,-7|, |0,-4|, |1,-5|, |0,-4|>$ and encoded as $<|0000,0101|, |0110\,1001|, |0,11110|_1, |0,10101|_2, |0,11010|_3, |0,10001|_4, |0,11000|_5,$ $|0,11100|_6, |0,11001|_7, |0,10101|_8, |0,11001|_9, |0,11100|_{10}, |1,11011|_{11}, |0,11100|_{12}>$. The total number of encoded bits is $16 + [12 * (1 + 5)] = 88$ bits compared to 240 bits without compression for the second subgroup. This means we have saved approximately 63.33% of data delivery.

This example shows that, from the raw data size of $256 + 240 = 496$ bits, our D-RLE has compressed the raw data to $94 + 88 = 182$ bits, which means we have saved bits sent up to 63.31% of the raw data. Note that we have demonstrated only 30 pieces of raw data for illustration purposes, but, in practice, each sensor node would collect more data in the long run before delivering the data. In the next section, we will show that D-RLE impressively improves accomplishment for the longer data delivery.

4. Performance Evaluation

In our testbed, we assumed that data sets are already stored in a sensor node. The sensor node could be a cluster head that collects data either from itself or from neighboring nodes. Later, the sensor node performs data compression and wirelessly sends compressed data to a receiver. This section analyzes the performance of our algorithm and compares the performance with four benchmark algorithms on five datasets.

4.1. Effect of K Value

We have evaluated the performance of our algorithm in terms of data rate savings (DRS) [30] as shown in Equation (7). Particularly, the case of lossless compression ($K = 0$) and the case of lossy compression ($K = 1$) were evaluated. The lossy case for $K = 1$ has been shown in the previous section and its DRS is 63.31%. We did repeat the same procedure except for $K = 0$ in the lossless case on the same raw data $<21, 25, 28, 30, 31, 35, 37, 42, 47, 49, 50, 55, 62, 76, 82, 95, 105, 103, 92, 86, 71, 63, 59, 52, 41, 34, 30, 26, 25, 21>$, and received the corresponding encoded data payload $<|0,5|, r_b, d_1, d_2, \ldots, d_n>$. Notice that the number of bits for the opcode c_i is 0; hence, there is no need for sending c_i of each subgroup. The total number of encoded bits used is 172. Therefore, the DRS of the lossless case is $172/496 = 65.32\%$, compared to 63.31% of the above lossy case:

$$DRS = \left(1 - \frac{\text{size of compressed data}}{\text{size of raw data}}\right) \times 100\%. \tag{7}$$

To achieve more data rate savings, we could allow more data lost or distorted to some extent. It depends on applications how much the quality loss is acceptable and this is done via the value of K in the algorithm. Table 4 shows the effect of varying K. We obtained more data rate savings when $K \geq 4$ for the same raw data previously illustrated.

Table 4. Comparison results of varying K values.

K	Uncompression (Bits) (G_1, G_2)	Compression (Bits) (G_1, G_2)	DRS (%) (G_1, G_2, G)
0	256, 240	91, 81	64.45, 66.25, 65.32
1	256, 240	94, 88	63.28, 63.33, 63.31
2	256, 240	88, 88	65.63, 63.33, 64.52
3	256, 240	86, 88	66.41, 63.33, 64.92
4	256, 240	86, 76	66.41, 68.33, 67.34
5	256, 240	79, 79	69.14, 67.08, 68.15

4.2. Evaluation Results

We have split our experiments into two sets. The dataset we used in the first set is varied in size—roughly speaking as small, medium, and large sizes—while, for the second set, we have fixed each dataset with the same size. Subsequently, benchmark algorithms as well as our proposed D-RLE algorithm were applied to compress those datasets and then the energy consumption for data compression and transmission would be measured.

Starting with the first set, we have simulated a 100-byte temperature dataset shown in Figure 2c and its shape resembles Figure 2d, which is a real collected dataset by [33]. In addition to these datasets, three more datasets referred to as sine-like, chaotic, and temperatureMin datasets as illustrated in Figure 2 were used for evaluating effectiveness of compression algorithms in our experiments. The simulated temperature dataset was created by Algorithm 7 while the temperatureHr dataset was the actual hourly recorded temperature data for 48 h. The temperatureMin dataset was retrieved from the same source of the temperatureHr dataset, but minutely recorded data. For the sine-like dataset, the minimum and maximum data values are 2 and 19, respectively. The neighboring data value next to 2 is 3 and then the data value is increased by 3 until reaching the maximum value. The data value after the maximum was set to 18 then decreased by 3 until touching the minimum value. By doing so for two cycles, the sine-like dataset has 30 pieces of data. We have fixed data ranging from 2 to 19 for the sine-like dataset, whereas we have randomly selected data ranging from 0 to 20 for the chaotic dataset. For chaotic and simulated temperature datasets, each dataset consists of 100 pieces of raw data. The temperatureHr dataset only has 48 pieces of data as the data were hourly recorded for 48 h, whereas the temperatureMin dataset has 2880 pieces of data since the data were minutely recorded for the same 48 h period. The raw data in each dataset were recorded as a series of strings; hence, bit sizes of data 2, 12 , and 102 are 8, 16, and 24 bits, respectively.

Algorithm 7 Creating simulated temperature data

Require: $g, uppertemp, lowertemp$
Ensure: $t[g] = \{t_1, t_2, \ldots, t_g\}$
  initialize $min = 0, max = 10, temp = 20, i = 1, up = true$
  while $i \leq g$ **do**
    while $(temp \leq uppertemp)$ AND $(up$ is true$)$ **do**
      $temp = temp + Random(min, max)$
      if $(uppertemp < temp)$ **then**
        $up = false$
        $temp = uppertemp$
      end if
      $t[i] = temp$
      $i = i + 1$
    end while
    while $(lowertemp \leq temp)$ AND $(up$ is false$)$ **do**
      $temp = temp - Random(min, max)$
      if $(temp < lowertemp)$ **then**
        $up = true$
        $temp = lowertemp$
      end if
      $t[i] = temp$
      $i = i + 1$
    end while
  end while

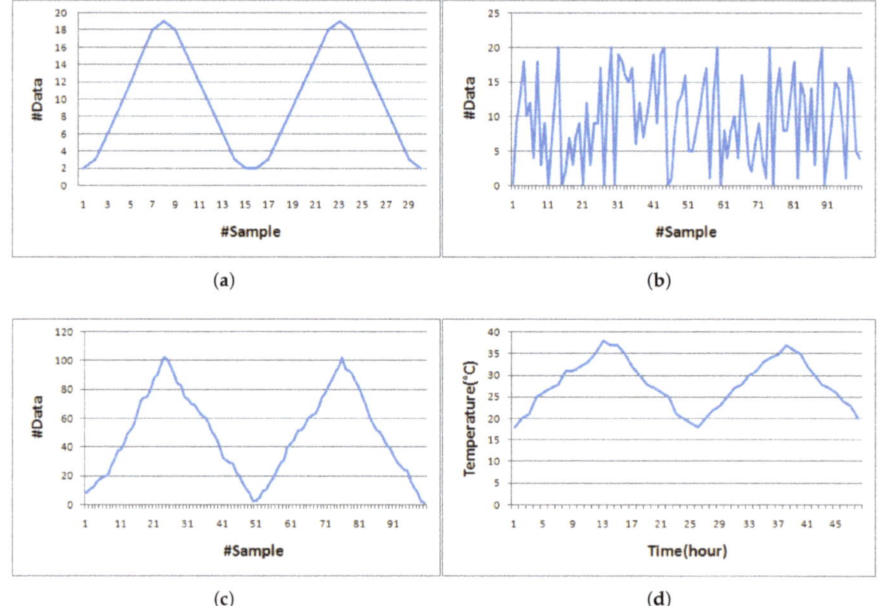

(a)   (b)

(c)   (d)

Figure 2. *Cont.*

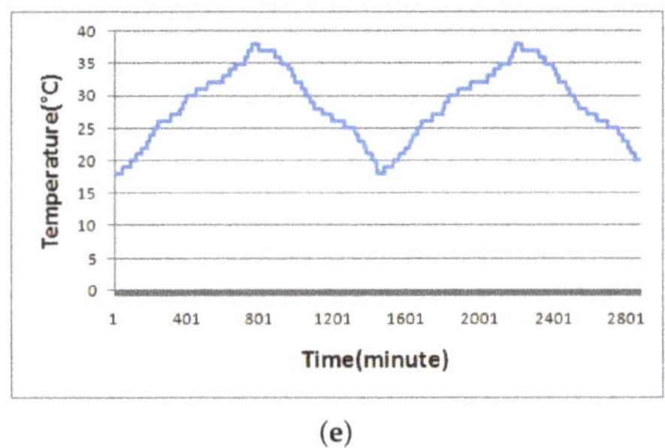

(**e**)

Figure 2. Five datasets used in the experiment. (**a**) sine-like; (**b**) chaotic; (**c**) simulated temperature; (**d**) temperatureHr; (**e**) temperatureMin.

For the second set of experiments, we extended the size of each dataset except for temperatureMin dataset to 46,080 bits. The reason to do this experiment is to investigate the energy usage for lengthy data transmission in one shot compared to multishot transmission of a small amount of data. We expected that the one shot delivery should have more efficient energy usage than the multishot since the sensor node in the single shot would be in a silent or power saving mode longer, whereas the multishot could wake up the sensor node more frequently. For both sets of experiments, four selected benchmark compression algorithms which are RLE, K-RLE, LEC, and LZW were used and compared to our D-RLE algorithm. For K-RLE, we set $K = 1$. A sensor node has been used in the experiments and run these algorithms to compress the datasets before transmitting data to a base station. The DRS obtained and energy consumed by each algorithm then were recorded for comparisons. The algorithms were implemented into a LAUNCHXL-CC1310 board [34] acting as the sensor node equipped with an RF module. To measure the power and energy usage in data compression and transmission, an MSP430FR5969 board [35] and code composer studio (CCS) program (version 8.0.0, Texas Instrument Inc., Dallas, Texas, USA) [36] were used. The MSP430FR5969 board was connected to the LAUNCHXL-CC1310 board as shown in Figure 3, and the amount of energy used was then measured by the EnergyTrace() function in CCS software. We took off the jumper connecting between the microcontroller and the debug parts of the CC1310 board to ensure that the power source came from the MSP430FR5969 board. The corresponding 3.3V Vcc and ground pins between the two boards were wired up as illustrated by black and white lines in Figure 3.

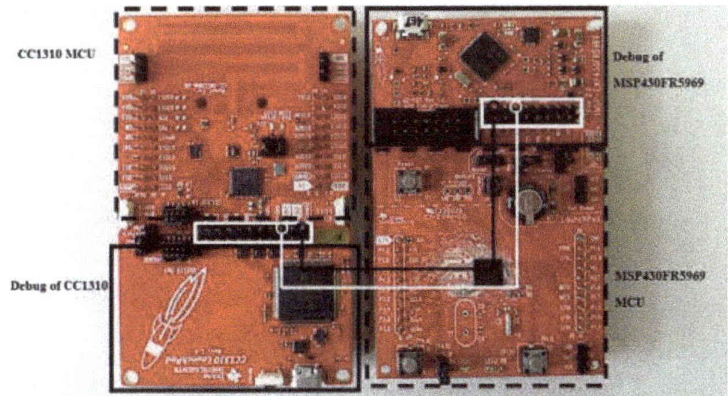

Figure 3. Board configuration for measuring energy use.

Tables 5 and 6 show the comparison results and total energy consumption on the datasets with different sizes while Tables 7 and 8 show the comparison results and total energy consumption on lengthy datasets with the same size of 46,080 bits, respectively. In the matter of DRS, the sine-like data gradually change values and there are no repeating values in adjacent data, so it is obvious that

RLE and *K*-RLE poorly perform while D-RLE works much better than others. On the other hand, the temperatureMin dataset has many repeating values and this characteristic does help RLE, *K*-RLE, and D-RLE to have higher DRS than LEC and LZW. The temperatureMin dataset has many nearby identical repeating data making it possible for LZW to create a dictionary with shorter encoding bits than LEC, and hence LZW has compressed data better than LEC. For a simulated temperature dataset, LZW, RLE and *K*-RLE perform worse than LEC and D-RLE since there are no repeating data values. LEC and D-RLE share a similar concept in the way of encoding the difference values. While LEC considers the all of the data as one group for data encoding, D-RLE divides the data into several subgroups in which the members within the same subgroup are not much different, leading to smaller encoding bits. For the temperatureHr dataset, D-RLE performs very well while among *K*-RLE, LEC and LZW work comparably, but RLE is the worst. RLE is the worst algorithm for the datasets that there are no repeating values. The negative DRS of RLE means RLE could not compress data at all, and it also adds extra overheads into the raw data. For the chaotic dataset, LEC, LZW and D-RLE perform better than the RLE family due to the data fluctuation and the dataset rarely has repeating data. We found that, in terms of DRS on those datasets, our D-RLE is the winner on both single shot and multi-shot patterns. Though the compression time by D-RLE is longer than others except for LZW, the compression energy used by D-RLE is not much different from others in multi-shot patterns. For the single shot pattern, D-RLE spends compression energy similar to LEC and consumes more compression energy than RLE and *K*-RLE, but less than LZW. We would suggest using D-RLE for a dataset that has a long sequence of data and it works best for repeating data or a gradual change in data values.

Table 5. Comparison between compression and transmission steps for datasets with different size.

Dataset	Algorithm	Compression Step				Transmission Step		
		#Bits	Time (s)	DRS (%)	Energy (mJ)	Time (s)	#Packets	Energy (mJ)
sine-like (352 bits)	RLE	464	0.023	−31.82	0.275	0.015	0.453	0.781
	K-RLE	320	0.023	9.09	0.303	0.011	0.313	0.473
	LEC	149	0.037	57.67	0.313	0.005	0.146	0.222
	LZW	207	0.110	41.19	0.396	0.008	0.202	0.309
	D-RLE	132	0.075	62.50	0.327	0.004	0.129	0.203
chaotic (1176 bits)	RLE	1504	0.076	−27.89	0.919	0.050	1.469	2.530
	K-RLE	1168	0.077	0.68	1.014	0.041	1.141	1.728
	LEC	733	0.122	37.67	1.047	0.024	0.716	1.092
	LZW	648	0.367	44.90	1.323	0.025	0.633	0.967
	D-RLE	616	0.249	47.62	1.092	0.019	0.602	0.947
simulated temperature (1600 bits)	RLE	1600	0.103	0.00	1.250	0.053	1.563	2.692
	K-RLE	1280	0.104	20.00	1.379	0.045	1.250	1.893
	LEC	651	0.166	59.31	1.424	0.021	0.636	0.970
	LZW	1584	0.500	1.00	1.800	0.061	1.547	2.363
	D-RLE	416	0.339	74.00	1.486	0.013	0.406	0.640
temperatureHr (768 bits)	RLE	752	0.049	2.08	0.600	0.025	0.734	1.265
	K-RLE	512	0.050	33.33	0.662	0.018	0.500	0.757
	LEC	528	0.080	31.25	0.684	0.017	0.516	0.787
	LZW	504	0.240	34.38	0.864	0.020	0.492	0.752
	D-RLE	204	0.163	73.44	0.713	0.006	0.199	0.314
temperatureMin (46,080 bits)	RLE	984	2.963	97.87	36.005	0.033	0.961	1.655
	K-RLE	656	3.009	98.58	39.713	0.023	0.641	0.970
	LEC	6240	4.780	86.46	41.018	0.202	6.094	9.299
	LZW	4230	14.396	90.82	51.847	0.164	4.131	6.310
	D-RLE	467	9.776	98.99	42.802	0.015	0.456	0.718

Table 6. Total energy use for the datasets with different sizes.

Dataset	Total Energy Use (mJ)				
	RLE	***K*-RLE**	**LEC**	**LZW**	**D-RLE**
sine-like	1.056	0.777	0.535	0.705	0.530
chaotic	3.449	2.741	2.139	2.290	2.040
simulated temperature	3.942	3.272	2.394	4.163	2.126
temperatureHr	1.865	1.419	1.470	1.616	1.027
temperatureMin	37.660	40.683	50.317	58.157	43.520

Table 7. Comparison between compression and transmission steps for datasets with the same size of 46,080 bits.

Dataset	Algorithm	Compression Step				Transmission Step		
		#Bits	**Time (s)**	**DRS (%)**	**Energy (mJ)**	**Time (s)**	**#Packets**	**Energy (mJ)**
sine-like	RLE	20,352	3.044	55.83	34.903	0.680	19.875	34.239
	K-RLE	16,056	3.060	65.16	37.493	0.570	15.680	23.749
	LEC	6876	4.862	85.08	40.778	0.223	6.715	10.246
	LZW	6588	14.156	85.70	50.407	0.256	6.434	9.828
	D-RLE	4476	10.087	90.29	42.703	0.140	4.371	6.883
chaotic	RLE	21,888	2.991	52.50	34.651	0.731	21.375	36.823
	K-RLE	18,816	3.126	59.17	35.928	0.668	18.375	27.832
	LEC	8796	4.762	80.91	40.538	0.285	8.590	13.107
	LZW	7776	14.052	83.13	49.407	0.302	7.594	11.600
	D-RLE	7332	9.367	84.09	41.047	0.230	7.160	11.275
simulated temperature	RLE	19,200	3.023	58.33	34.918	0.641	18.750	32.301
	K-RLE	16,560	3.005	64.06	36.986	0.588	16.172	24.495
	LEC	7812	4.836	83.05	40.596	0.253	7.629	11.641
	LZW	19,008	13.346	58.75	49.447	0.737	18.563	28.355
	D-RLE	7332	9.967	84.09	41.630	0.230	7.160	11.275
temperatureHr	RLE	45,120	2.960	2.08	34.577	1.506	44.063	75.906
	K-RLE	30,720	3.006	33.33	35.976	1.091	30.000	45.439
	LEC	31,680	4.755	31.25	39.098	1.028	30.938	47.208
	LZW	30,240	13.368	34.38	47.657	1.173	29.531	45.111
	D-RLE	12,240	9.420	73.44	41.023	0.384	11.953	18.823
temperatureMin	RLE	984	2.963	97.87	36.005	0.033	0.961	1.655
	K-RLE	656	3.009	98.58	39.713	0.023	0.641	0.970
	LEC	6240	4.780	86.46	41.018	0.202	6.094	9.299
	LZW	4230	14.396	90.82	51.847	0.164	4.131	6.310
	D-RLE	467	9.776	98.99	42.802	0.015	0.456	0.718

Table 8. Total energy use for the datasets with the same size of 46,080 bits.

Dataset	Total Energy Use (mJ)				
	RLE	***K*-RLE**	**LEC**	**LZW**	**D-RLE**
sine-like	69.142	61.242	51.025	60.235	49.586
chaotic	71.474	63.760	53.646	61.007	52.322
simulated temperature	67.218	61.481	52.237	77.803	52.906
temperatureHr	110.483	81.415	86.306	92.768	59.846
temperatureMin	37.660	40.683	50.317	58.157	43.520

As a result of highest DRS performance, the number of packets for data delivery by D-RLE is smaller than the number of packets by other algorithms, leading to less transmission energy. In terms of energy use, D-RLE uses the least total energy compared to other algorithms on most of the datasets as shown in Tables 6 and 8. For example, in temperatureHr dataset of 46,080 bits, D-RLE approximately sends only 12 packets with the total energy use of 18.82 mJ, while others use more than 30 packets with the total of energy greater than 45 mJ. The total power use, P_{total}, consists of two parts from compression and transmission steps, which is referred to as compression power, P_c, and transmission power, P_t, respectively. P_{total} is determined by Equation (8) in which the subscript i indicates the ith group number when we compress the data, and j expresses the jth payload or packet number that we deliver. The values of g and p are the number of groups and the number of packets. To calculate corresponding energy used in each step, we use the relationships between energy and power from Equations (9) and (10), where T_{c_i} and T_{t_j} are the compression time spending for compressing ith group and transmission time spending for delivering jth packet, respectively. Lastly, total energy consumption, E_{total}, is computed by Equation (11), adding transmission and compression energy. The experimental results show that D-RLE takes minimal transmission energy in exchange for slightly more compression energy, but it is worthy of being considered, as D-RLE significantly reduces total energy use while other benchmark algorithms consume much higher total energy level on the same datasets:

$$P_{total} = \sum_{i=1}^{g} P_{c_i} + \sum_{j=1}^{p} P_{t_j}, \tag{8}$$

$$E_c = \sum_{i=1}^{g} P_{c_i} \times \Delta T_{c_i}, \tag{9}$$

$$E_t = \sum_{j=1}^{p} P_{t_j} \times \Delta T_{t_j}, \tag{10}$$

$$E_{total} = E_t + E_c. \tag{11}$$

Figure 4 compares power and energy consumed by D-RLE during data compression and transmission between single shot and multishot cases on the TemperatureMin dataset, respectively. Both cases have the same data length of 46,080 bits. The single shot receives all of the data before starting compression and transmission while the multishot would receive several data portions in which each portion has the same data length. According to the graph, it is clearly seen that the transmission period demands higher power consumption than the compression period. However, the energy consumption during the transmission indicated as T in the graph is less than the energy used during compression indicated as C in the graph. The smaller size of compressed data gives the shorter period of data transmission time for the single shot case. On the other hand, the multishot case has to repeat many compression and transmission cycles and take more time. In each cycle of the multishot, the compression step is reinitiated, which gives us lower compression efficiency and hence the accumulated energy used by the multishot is greater than the accumulated energy used by the single shot. Therefore, the single shot is more efficient and energy saving compared to the multishot. Other datasets have the graphs in the same manner as the Temperature dataset.

4.3. Performance Visualization

We have plotted radar charts as shown in Figure 5 according to five categories for making a simple way to visualize performance comparison among the algorithms. The first three categories are DRS and data accuracy (in the sense of how much difference there is between the decoded data and its original data). The last three ad hoc categories are called compression time efficiency (CTE), compression energy efficiency (CEE), and transmission energy efficiency (TEE) in which they are defined by Equations (12)–(14), respectively. The parameter A in those equations is the number of algorithms we used in the experiments , i.e., $A = 5$. T_{c_i}, E_{c_i} and E_{t_i} are the compression time, compression energy, and transmission energy of $i$thalgorithm, respectively. Each category has a score from 0 to 100; the higher the score, the better the performance. The left panel of Figure 5 shows comparisons among RLE, K-RLE and D-RLE while the right panel of Figure 5 shows comparison among LEC, LZW and D-RLE. For the left panel, D-RLE performs better than the others on most categories except for CTE given the fact that D-RLE takes more time in the compression step. For the right panel, D-RLE performs equally or slightly better than the others in terms of CEE, TEE, and accuracy. D-RLE is located between LEC and LZW on CTE, whereas D-RLE mostly achieves better DRS (DRS results on the radar charts Figure 5a–c might not be clearly seen as shown as the number in Table 7). On average, D-RLE gets a high score and is well balanced, reaching the vertex of pentagon in the graph when compared to other algorithms in each category.

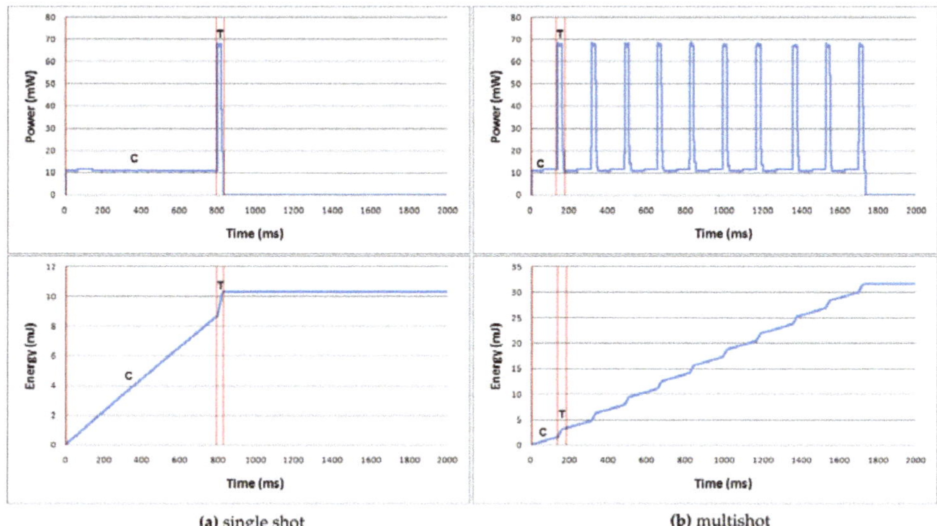

(a) single shot **(b)** multishot

Figure 4. Power and energy comparison between single shot and multishot on the TemperatureMin dataset.

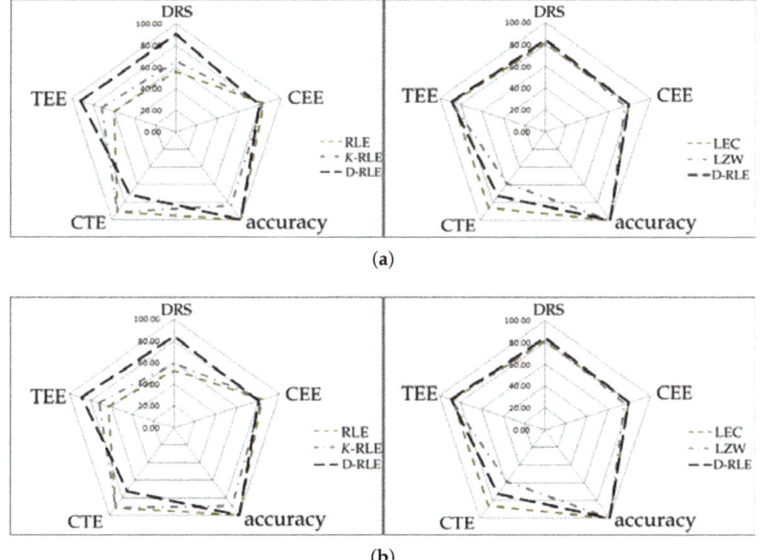

(a)

(b)

Figure 5. *Cont.*

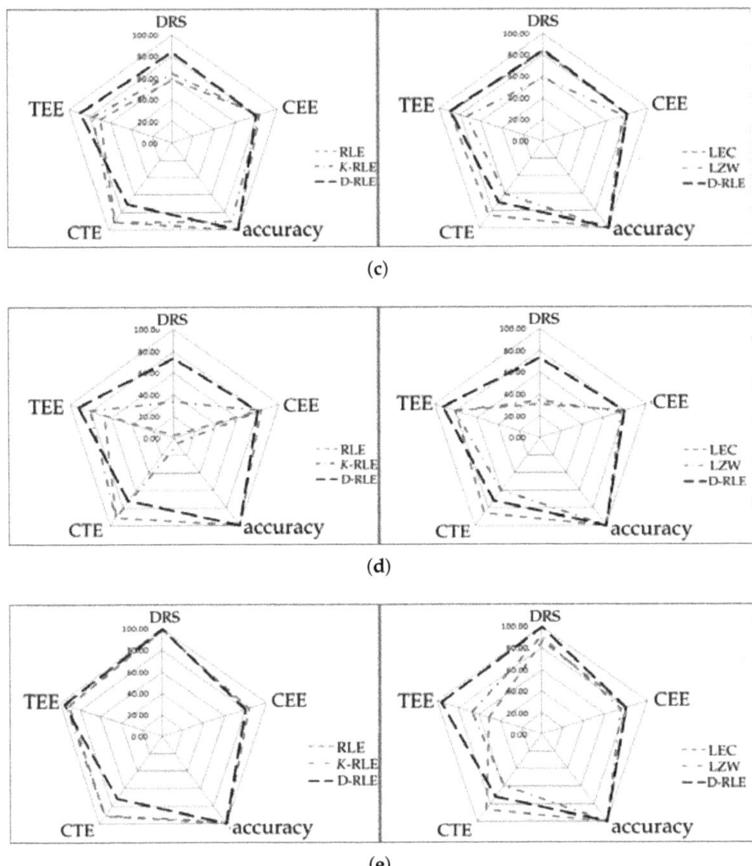

Figure 5. Radar chart comparison. (**a**) sine-like; (**b**) chaotic; (**c**) simulated temperature; (**d**) temperatureHr; (**e**) temperatureMin.

$$CTE_i = (1 - \frac{T_{c_i}}{\sum_{a=1}^{A} T_{c_a}}) \times 100 \tag{12}$$

$$CEE_i = (1 - \frac{E_{c_i}}{\sum_{a=1}^{A} E_{c_a}}) \times 100 \tag{13}$$

$$TEE_i = (1 - \frac{E_{t_i}}{\sum_{a=1}^{A} E_{t_a}}) \times 100 \tag{14}$$

5. Conclusions

We have presented a compression algorithm called D-RLE applied to the domain of wireless sensor nodes in which energy use is one of the most important aspects. It starts with dividing the data into many groups based on Chauvenet's criterion and then each group further forms subgroups to which an adaptive encoding is applied. According to the experimental results, D-RLE have demonstrated that it performs very well, gives the highest data savings rate and spends less energy compared to other benchmark algorithms. In particular, D-RLE is suitable for big amounts of data with repeating or gradually changed values and for a single shot delivery mode. Due to its highest compression rate, the amount of data transmission is significantly reduced and hence less energy is demanded. This prolongs the battery life of the sensor nodes. This work is an alternative way to increase the performance of the sensor node concerning the energy.

Author Contributions: C.C. designed the core architecture and performed the hardware/software implementation and experiments; A.B. provided supervision to the project and has the responsibility as the main corresponding author; S.S. co-supervised the project and contributed to the experimental design, partial software development and result analysis.

Funding: This research was supported by the Faculty of Engineering, Burapha University, Thailand under the contract number 15/2555. The authors gratefully acknowledge the contributions of all committee members for their valuable suggestions.

Conflicts of Interest: The authors declare no conflict of interest. The founding sponsors had no role in the design of the study; in the collection, analyses, or interpretation of data; in the writing of the manuscript, and in the decision to publish the results.

References

1. Sheng, Z.; Wang, H.; Yin, C.; Hu, X.; Yang, S.; Leung, V.C.M. Lightweight Management of Resource-Constrained Sensor Devices in Internet of Things. *IEEE Internet Things J.* **2015**, *2*, 402–411. [CrossRef]
2. He, S.; Xie, K.; Chen, W.; Zhang, D.; Wen, J. Energy-Aware Routing for SWIPT in Multi-Hop Energy-Constrained Wireless Network. *IEEE Access* **2018**, *6*, 17996–18008. [CrossRef]
3. Alvi, S.A.; Zhou, X.; Durrani, S. Optimal Compression and Transmission Rate Control for Node-Lifetime Maximization. *IEEE Trans. Wirel. Commun.* **2018**, *17*, 7774–7788. [CrossRef]
4. Gana Kolo, J.; Anandan Shanmugam, S.; Wee Gin Lim, D.; Ang, L.M.; Seng, K. An Adaptive Lossless Data Compression Scheme for Wireless Sensor Networks. *J. Sens.* **2012**, *2012*, 539638.
5. Wang, J.; Gao, Y.; Liu, W.; Sangaiah, A.K.; Kim, H.J. Energy Efficient Routing Algorithm with Mobile Sink Support for Wireless Sensor Networks. *Sensors* **2019**, *19*, 1494. [CrossRef] [PubMed]
6. Wang, J.; Tawose, O.T.; Jiang, L.; Zhao, D. A New Data Fusion Algorithm for Wireless Sensor Networks Inspired by Hesitant Fuzzy Entropy. *Sensors* **2019**, *19*, 784. [CrossRef] [PubMed]
7. Cheng, J.; Gao, Y.; Zhang, N.; Yang, H. An Energy-Efficient Two-Stage Cooperative Routing Scheme in Wireless Multi-Hop Networks. *Sensors* **2019**, *19*, 1002. [CrossRef] [PubMed]
8. Kim, J.; Lin, X.; Shroff, N.B. Optimal Anycast Technique for Delay-Sensitive Energy-Constrained Asynchronous Sensor Networks. *IEEE/ACM Trans. Netw.* **2011**, *19*, 484–497. [CrossRef]
9. Wang, H.; Zeng, H.; Wang, P. Linear Estimation of Clock Frequency Offset for Time Synchronization Based on Overhearing in Wireless Sensor Networks. *IEEE Commun. Lett.* **2016**, *20*, 288–291. [CrossRef]
10. Liu, X. Atypical Hierarchical Routing Protocols for Wireless Sensor Networks: A Review. *IEEE Sens. J.* **2015**, *15*, 5372–5383. [CrossRef]
11. Boubiche, S.; Boubiche, D.E.; Bilami, A.; Toral-Cruz, H. Big Data Challenges and Data Aggregation Strategies in Wireless Sensor Networks. *IEEE Access* **2018**, *6*, 20558–20571. [CrossRef]
12. Lin, H.; Üster, H. Exact and Heuristic Algorithms for Data-Gathering Cluster-Based Wireless Sensor Network Design Problem. *IEEE/ACM Trans. Netw.* **2014**, *22*, 903–916. [CrossRef]
13. Mahapatro, A.; Khilar, P.M. Fault Diagnosis in Wireless Sensor Networks: A Survey. *IEEE Commun. Surv. Tutor.* **2013**, *15*, 2000–2026. [CrossRef]
14. Wang, J.; Al-Mamun, A.; Li, T.; Jiang, L.; Zhao, D. Toward Performant and Energy-efficient Queries in Three-tier Wireless Sensor Networks. In Proceedings of the 47th International Conference on Parallel Processing, ICPP 2018, Eugene, OR, USA, 13–16 August 2018; ACM: New York, NY, USA, 2018; pp. 42:1–42:10.
15. Wang, X.; Liu, X.; Wang, M.; Nie, Y.; Bian, Y. Energy-Efficient Spatial Query-Centric Geographic Routing Protocol in Wireless Sensor Networks. *Sensors* **2019**, *19*, 2363. [CrossRef] [PubMed]
16. Sadler, C.M.; Martonosi, M. Data Compression Algorithms for Energy-constrained Devices in Delay Tolerant Networks. In Proceedings of the 4th International Conference on Embedded Networked Sensor Systems, SenSys '06, Boulder, CO, USA, 31 October–3 November 2006; ACM: New York, NY, USA, 2006; pp. 265–278.

17. Salomon, D. *Data Compression: The Complete Reference*; Springer: Berlin/Heidelberg, Germany, 2006.

18. Welch, T.A. Technique for High-Performance Data Compression. *Computer* **1984**, *17*, 8–19. [CrossRef]

19. Capo-Chichi, E.P.; Guyennet, H.; Friedt, J. K-RLE: A New Data Compression Algorithm for Wireless Sensor Network. In Proceedings of the 2009 Third International Conference on Sensor Technologies and Applications, Athens, Greece, 18–23 June 2009; pp. 502–507.

20. Roy, S.; Panja, S.C.; Patra, S.N. DMBRLE: A lossless compression algorithm for solar irradiance data acquisition. In Proceedings of the 2015 IEEE 2nd International Conference on Recent Trends in Information Systems (ReTIS), Kolkata, India, 9–11 July 2015; pp. 450–454.

21. Marcelloni, F.; Vecchio, M. A Simple Algorithm for Data Compression in Wireless Sensor Networks. *IEEE Commun. Lett.* **2008**, *12*, 411–413. [CrossRef]

22. Marcelloni, F.; Vecchio, M. An Efficient Lossless Compression Algorithm for Tiny Nodes of Monitoring Wireless Sensor Networks. *Comput. J.* **2009**, *52*, 969–987. [CrossRef]

23. Szalapski, T.; Madria, S.; Linderman, M. TinyPack XML: Real time XML compression for wireless sensor networks. In Proceedings of the 2012 IEEE Wireless Communications and Networking Conference (WCNC), Shanghai, China, 1–4 April 2012; pp. 3165–3170.

24. Liang, Y.; Li, Y. An Efficient and Robust Data Compression Algorithm in Wireless Sensor Networks. *IEEE Commun. Lett.* **2014**, *18*, 439–442. [CrossRef]

25. Zou, Z.; Bao, Y.; Deng, F.; Li, H. An Approach of Reliable Data Transmission With Random Redundancy for Wireless Sensors in Structural Health Monitoring. *IEEE Sens. J.* **2015**, *15*, 809–818.

26. Hung, N.Q.V.; Jeung, H.; Aberer, K. An Evaluation of Model-Based Approaches to Sensor Data Compression. *IEEE Trans. Knowl. Data Eng.* **2013**, *25*, 2434–2447. [CrossRef]

27. Rubin, M.J.; Wakin, M.B.; Camp, T. Lossy Compression for Wireless Seismic Data Acquisition. *IEEE J. Sel. Top. Appl. Earth Obs. Remote Sens.* **2016**, *9*, 236–252. [CrossRef]

28. Long, S.; Xiang, P. Lossless Data Compression for Wireless Sensor Networks Based on Modified Bit-Level RLE. In Proceedings of the 2012 8th International Conference on Wireless Communications, Networking and Mobile Computing, Shanghai, China, 21–23 September 2012; pp. 1–4.

29. Koc, B.; Sarkar, D.; Kocak, H.; Arnavut, Z. A study of power consumption on MSP432 family of microcontrollers for lossless data compression. In Proceedings of the 2015 12th International Conference on High-capacity Optical Networks and Enabling/Emerging Technologies (HONET), Islamabad, Pakistan, 21–23 December 2015; pp. 1–5.

30. Chianphatthanakit, C.; Boonsongsrikul, A.; Suppharangsan, S. A Lossless Image Compression Algorithm using Differential Subtraction Chain. In Proceedings of the 2018 10th International Conference on Knowledge and Smart Technology (KST), Chiang Mai, Thailand, 31 January–3 February 2018; pp. 84–89.

31. Boonsongsrikul, A.; Lhee, K.S.; Hong, M. Securing data aggregation against false data injection in wireless sensor networks. In Proceedings of the 2010 The 12th International Conference on Advanced Communication Technology (ICACT), Phoenix Park, Korea, 7–10 February 2010; Volume 1, pp. 29–34.

32. Pop, S.; Ciascai, I.; Pitica, D. Statistical analysis of experimental data obtained from the optical pendulum. In Proceedings of the 2010 IEEE 16th International Symposium for Design and Technology in Electronic Packaging (SIITME), Pitesti, Romania, 23–26 September 2010; pp. 207–210.

33. Bhandari, S.; Bergmann, N.; Jurdak, R.; Kusy, B. Time Series Data Analysis of Wireless Sensor Network Measurements of Temperature. *Sensors* **2017**, *17*, 1221. [CrossRef] [PubMed]

34. CC1310 SimpleLink^{TM} Ultra-Low-Power Sub-1 GHz Wireless MCU. Available online: www.ti.com/product/CC1310 (accessed on 13 February 2019).

35. MSP430FR5969 LaunchPad Development Kit. Available online: http://www.ti.com/tool/MSP-EXP430FR5969 (accessed on 17 January 2019).

36. Code Composer Studio (CCS) Integrated Development Environment (IDE). Available online: http://www.ti.com/tool/CCSTUDIO (accessed on 30 February 2019).

Review

Mathematical Methods and Algorithms for Improving Near-Infrared Tunable Diode-Laser Absorption Spectroscopy

Tianyu Zhang, **Jiawen Kang**, **Dezhuang Meng**, **Hongwei Wang**, **Zhengming Mu**, **Meng Zhou**, **Xiaotong Zhang and Chen Chen ***

Key Laboratory of Geophysical Exploration Equipment, Ministry of Education, College of Instrumentation & Electrical Engineering, Jilin University, Changchun 130026, China; zty@jlu.edu.cn (T.Z.); kangjw6515@mails.jlu.edu.cn (J.K.); mengdz6515@mails.jlu.edu.cn (D.M.); wanghw6515@mails.jlu.edu.cn (H.W.); muzm17@mails.jlu.edu.cn (Z.M.); zhoumeng17@mails.jlu.edu.cn (M.Z.); zxt18@mails.jlu.edu.cn (X.Z.)
* Correspondence: cchen@jlu.edu.cn; Tel.: +86-137-5606-4009

Received: 22 October 2018; Accepted: 27 November 2018; Published: 6 December 2018

Abstract: Tunable diode laser absorption spectroscopy technology (TDLAS) has been widely applied in gaseous component analysis based on gas molecular absorption spectroscopy. When dealing with molecular absorption signals, the desired signal is usually interfered by various noises from electronic components and optical paths. This paper introduces TDLAS-specific signal processing issues and summarizes effective algorithms so solve these.

Keywords: TDLAS; signal processing; gas sensor; denoise; interference fringe; background correction

---

1. Introduction

Sensors based on Tunable Diode Laser Absorption Spectroscopy (TDLAS) have the advantages of high sensitivity, high stability, high selectivity and fast response, and have been widely applied in atmospheric environmental monitoring [1–3], medical health [4], industrial production [5,6], military surveying [7,8] and other fields. Given that the absorption spectrum is primarily determined by the atomic and molecular composition of the measured sample, it is a useful tool to determine the presence of a particular substance in a sample [9–12]. Nevertheless, the performance of TDLAS systems can be limited by many factors [13,14], especially, the measurement signal incorporates numerous contributions from optical components (interference fringe) and electronic components [15,16]. Background errors, which are caused by background extraction, also limit the detection precision of the system. Therefore, signal preprocessing is necessary to improve the accuracy of TDLAS-based analytical instruments.

In the development of TDLAS, many methods have been proposed to improve system accuracy and to measure resolution [17–22], which can be divided into two classes: software processing and hardware-based processing. Some hardware-based approaches such as multi-pass cells, differential schemes, and wavelength modulation techniques have been proved to ameliorate signal quality effectively [23–26]. Besides, with the continuous evolution of data processing technology, many data analysis algorithms have emerged in some fields [27,28]. Some of these methods have been introduced and applied in TDLAS systems to enhance the accuracy and resolution performance. Thereby the current review focuses more on software-based methods. Some algorithms have also been proposed to address pertinent issues in TDLAS signals, and the applicability of these algorithms have been experimentally demonstrated.

2. Factors Affecting the Accuracy of TDLAS

Many factors can affect the accuracy of a TDLAS system [29], such as white noise in electronic units, interference fringes in optical components [30,31], errors in background extraction [32,33], signal drift, and atmospheric pressure changes [34,35]. Some errors even arise from dust in light paths [36,37] and mechanical jamming when the device is used. Among these errors, the first three interferences are universal, and this article mainly introduces research works on these three issues. Examples of these polluted waveforms are shown in Figure 1.

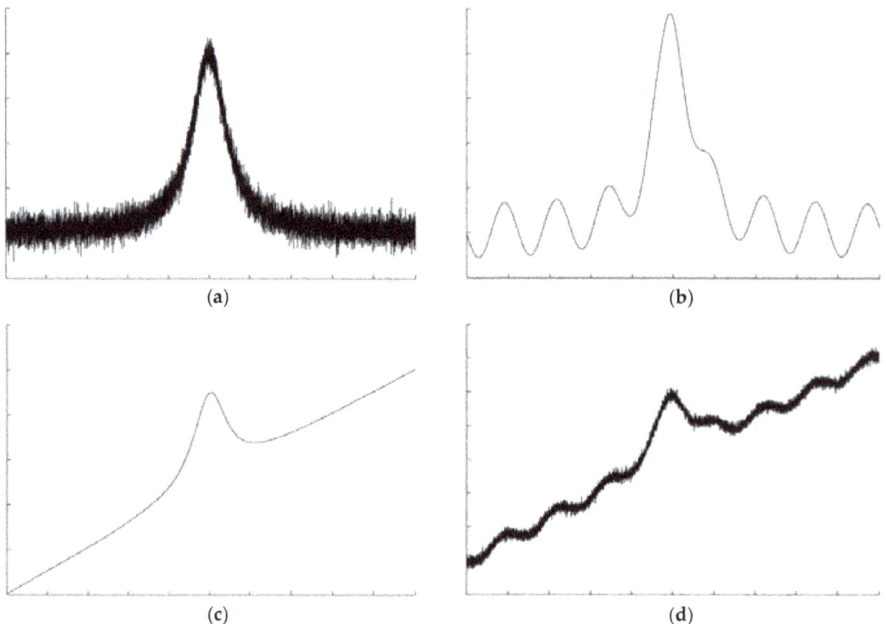

Figure 1. Signal simulation. (**a**) White noise; (**b**) interference fringe; (**c**) background; (**d**) multiple interference signal.

Among these problems, the most common issue is background error, which is easier to work out but should be given priority to deal with. Interference fringes are relatively difficult to remove, and are usually the main constraint on the accuracy of TDLAS. The noise, especially high-frequency white noise, should be handled after the first two problems are resolved.

2.1. Denoise

The signal obtained from the TDLAS system usually tends to exhibit a limited signal-to-noise ratio (SNR). The amplitude of absorption peak is smaller than that in other detection methods because the gas absorption is weaker in the near-infrared than in the mid-infrared region [38]. Especially in trace gas detection, the gas absorption signal is so weak that it can be easily submerged in various noise and spikes. These noises may originate from TEC or lock-in amplifier, the $1/f$ noise or power supply voltage jitter [39]. In this case, those algorithms that are most commonly used in general signal processing, like the least-squares method, cannot efficiently extract the precise desired signal. In recent years, some advanced algorithms have been proposed or introduced into TDLAS systems to process signals with low SNR, including wavelet transform (WT) [39,40], adaptive Savitzky–Golay algorithm [41], and empirical mode decomposition (EMD)-FCR algorithm [42]. These algorithms have been verified to be significant for practical issues.

2.2. Interference Fringe

The detection sensitivity of TDLAS technology is severely restricted by optical interference due to the strong coherence of the laser [43,44]. Especially at low concentration, this disturbance leads to baseline fluctuation and causes error in waveform extraction [45]. These optical interferences may arise from the multiple reflections on the reflecting or scattering surfaces in the light path [43,46], which then periodically fluctuates like a sine function in the measurement signal. Unlike the electrical noise, optical interference like the effect of interference fringes is in the low frequency segment and exhibits large amplitude. Distinguishing the location of signal peaks under some extreme cases may be difficult. Even if a comprehensive algorithm, such as the WT, can perform interference fringe removal, it cannot achieve high-precision extraction in many special situations. Some strategies for removing optical interference have been proposed [47,48], and these targeted algorithms give better results compared with common methods.

2.3. Background Correction

The carrier density inside the diode laser controls the light emission frequency and light emission intensity equally due to the nature of the semiconductor laser. Hence, when the modulation current drives the laser to sweep the absorption peak of the gas, the incident laser intensity will also be similarly modulated. As a result, the measuring signal peak will contain a ramp known as "background". Before extracting useful information, the raw spectral signal must be intensity normalized, that is, the raw spectrum is divided by the background line. For the information contained in the measurement signal, the error caused by the background removal also causes an error in the peak and outline area of the signal waveform, especially for the method of calculating the concentration information by using the signal area. The error is relatively large. Therefore, the background correction pretreatment is quite necessary. Like the removal of optical interference, background correction can be performed by some comprehensive algorithms, such as the EMD algorithm, which can obtain some background information by decomposing a final Intrinsic Mode Function (IMF) [49,50]. Some improvements that add an iteration process have been proposed to achieve an accurate background correction [51,52], and the ideas and strategies in these algorithms can be referenced to address the background problem before other signal processes.

3. Algorithms for TDLAS Signal Processing

Algorithms applied to denoising and signal processing have been further developed, and signal averaging is an early and simple method for signal processing [53]. Many kinds of algorithms have been developed to denoise from different perspectives, examples of which are the linear filters based on early filtering theory, such as the Wiener filter and Kalman filter [54,55]; the fitting algorithms based on nonlinear regression, such as least-squares method; and some other decomposition algorithms based on signal decomposition and reconstruction, such as empirical mode decomposition [49]. In recent years, some novel strategies have been developed to deal with the usual signal disturbances in TDLAS. For any denoising process, the ideal situation is to obtain a priori knowledge of a noise model before selecting an algorithm. Therefore, algorithm classification based on noise models has practical significance. This article will introduce some solutions addressing the three issues mentioned above. To compare the different methods, we present the experimental performance of each algorithm in each section.

3.1. Denoising

Noise reduction is required in various signal preprocessing techniques. In this section the particular strengths of the WT, adaptive Savitzky–Golay algorithm, and EMD-FCR algorithm will be introduced for their applications in TDLAS noise deduction. In addition to white noise removing, these methods can theoretically solve other interferences to some extent. In some studies, WT [56] is introduced into TDLAS sign analytical process, which has been widely used in other signal processing fields [16]. Wavelet analysis is a signal time-frequency analysis method for processing local or transient signals. It originates from Fourier algorithm transformation, combines the concepts of signal stretching and translation, and involves dual locality in the time and frequency domains. This variable resolution analysis method focuses on both the low-frequency trends and high-frequency details of signals.

3.1.1. Wavelet Transform (W-T)

The wavelet-based scenarios can be an effective approach to modeling the absorption and work out complicated signal situations because of the special characteristics of the time-frequency relationship. The amplitude of the absorption signal peak is small and contains both high-frequency white noise and low-frequency fluctuations which are usually caused by temperature drift and interference fringes. WT offers a window that varies with signal frequency band, allowing different scales of noise to be resolved into different sub-bands. As a result, time resolution improves at high frequency and frequency resolution improves at low frequency [57–59]. Thus, wavelet denoising is a powerful tool to extract desired signals from multiple noise pollution. However, one of the drawbacks of WT is it's the strong subjectivity of the choice of parameters. Human errors will greatly affect the decomposition performance of the algorithm. In this part, we mainly introduce the studies of WT to deal with the noise of low-SNR situation, in which the experimental process and data show significant reference for further research, and the background removal application will be introduced later in this article. In this paper, only the key concepts of WT are presented. Detailed mathematical treatment can be found in the cited references. Compared with e Fourier transform (FT), WT uses a finite-length, attenuating wavelet basis as a decomposition basis function, instead of an infinite-length trigonometric function. By selecting different scale functions and wavelet basis, it is possible to synthesize signals with time-domain scale discrepancy. Similarly, when decomposing a signal by wavelet basis and decomposition scale, its localization characteristics can be mapped into different frequencies. Thus, this method can be used to analyze non-stationary signals. Furthermore, by choosing an appropriate threshold to filter the decomposition result and reconstruct this signal, undesired noise can be removed, which is referred to as wavelet denoising (WD). The flow chart of the WT is shown in Figure 2.

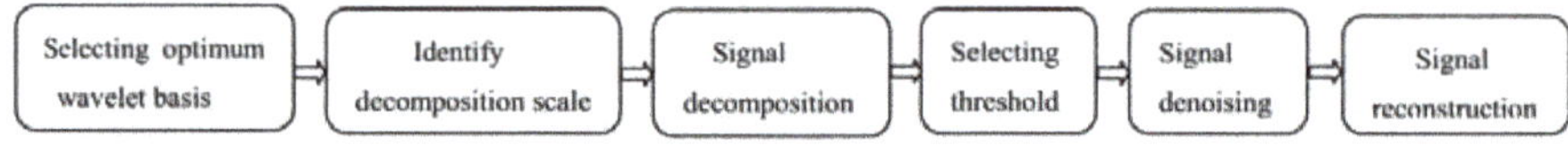

Figure 2. Flow chart of WT. This figure was obtained from reference [39].

Xia et al. [39] introduced WT to deal with low-SNR signal in TDLAS systems. In their experiments, wavelet basis symlet 6 (a kind of symmetric basis function) and decomposition scale six was used, and an approximation coefficient of a term of absorption signal was reserved and reconstructed only at a certain frequency. After signal reconstruction, the signal without and with WD was compared. Figure 3 shows that wavelet denoising can strikingly optimize the signal.

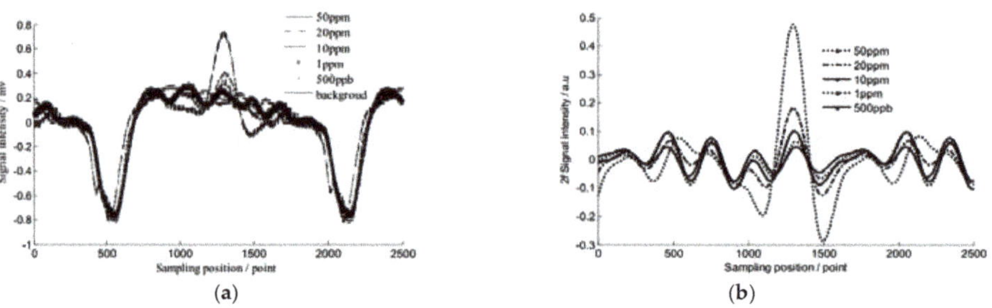

Figure 3. (**a**) signal without WD (**b**) signal with WD. This figure was obtained from reference [39].

Zheng et al. [40] also studied the application of wavelet-denoising-assisted wavelength modulation technique in a TDLAS-based near-infrared CH_4 detection device. Furthermore, detailed experimental data are provided to confirm the improvement of WD for polluted signals. A comparison between the sensing performances under the cases with and without WD use is shown in Table 1. Moreover, the sample gas was set up in two groups, a low-concentration group (scale of 0–1 kppm) and a high-concentration group (scale of 0–50 kppm). Experimental results demonstrated that the wavelet denoising method has great practical significance, and especially in low-concentration gas detection, the quality of the signal is enhanced significantly.

Table 1. Comparison between the sensing performances under the cases of with and without WD use. Accu: accuracy; MDL: minimum detection limit; RT: response time; AD: Allan deviation. This table was obtained from reference [40].

	MDL (ppm)	Non-Averaged Detection Range on 4 ppm Gas Sample (ppm)	Accu (%)	AD (ppm)
Using WD	1	3–5	<3.8% (C > 4 ppm)	0.08 (τ = 500 s) 0.35 (τ = 30 s)
Without using WD	4	2.6–5.5	<6.2% (C > 4 ppm)	0.13 (τ = 500 s) 0.46 (τ = 30 s)

3.1.2. Adaptive Savitzky–Golay (S–G) Algorithm

S–G filter is a classic smoothing denoising method [60,61] and is one of the most common pretreatment methods in spectrum analysis [62]. Li et al. proposed a simple but robust modified adaptive S–G algorithm for TDLAS signal processing [41], which shows unique superiority when temporal resolution and low system cost are priorities. This approach is developed from the S–G smoothing filter. The S–G filter using the least squares fitting coefficient as a filter response function is a smoothing filtering method for high frequency noises. As for the basic S–G filter, its effectiveness is strongly dependent on window size. As explained above, one of the difficulties in TDLAS signal processing is that the noise can originate from multi-frequency components. With a fixed window it is hard to match each of these signal segments. This new method presents a variable window and provides two additional criteria for TDLAS signal processing to determine the optimal window size. Compared with many preset parameters of WT, this adaptive algorithm reduces the subjective error.

The basic method of the S–G algorithm involves the following steps: (i) selecting window size (ii) selecting a polynomial function for the data point in window (iii) correcting the data point at the center of the selected interval by the polynomial coefficients as shown in Figure 4 and shifting the analysis window to the right by one data point.

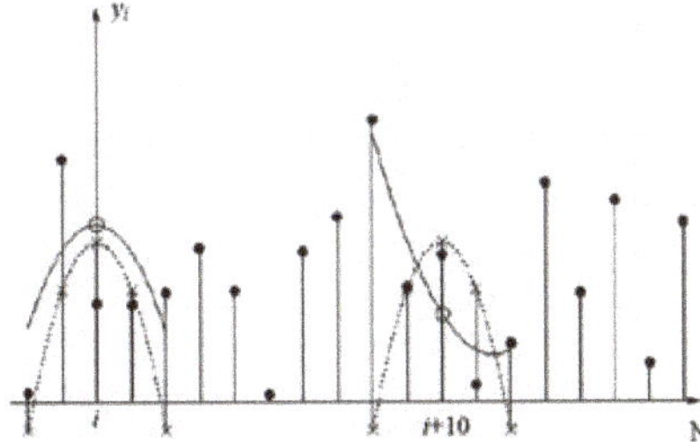

Figure 4. Illustration of least-squares smoothing by locally fitting a low-order polynomial (solid line) to five input samples: dot denotes the raw input samples, circle denotes the least-squares smoothed samples, and x denotes the effective impulse response samples. The dotted line denotes the polynomial approximation to centered unit impulse. This figure was taken from [41].

The above process is repeated. In this modified approach, two criteria are introduced to work out the optimal window size, namely, "PolyFit" and a threshold "Th". "PolyFit" is a signal segment in a polynomial function, which we regarded as noiseless. In a process of a segment of data, correlation coefficient R between the "PolyFit" and the same segment in the S–G-filter-smoothed data is utilized to assess the optimal filtering parameters instead of SNR. This condition is valid for noise reduction but is not credible for signal preservation. The threshold "Th", which is defined as the difference of peak heights between "PolyFit" and the S–G filtering smoothed data, is used to ensure filtering parameters without excessive signal distortion. Thus, each data interval can be modified under an optimal window size, and potential signal distortion can be alleviated in signal processing. The flow chart of the modified algorithm is shown in Figure 5.

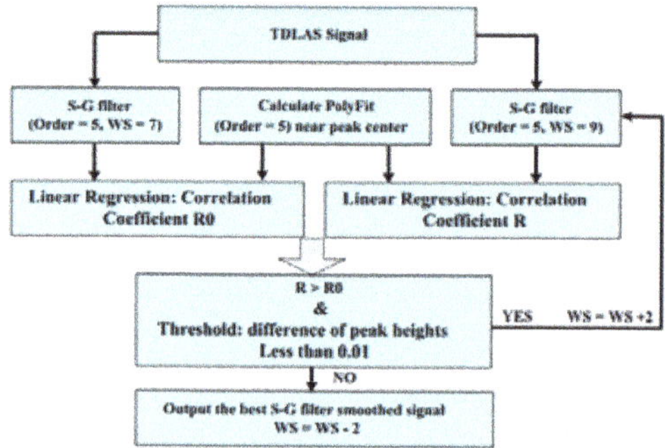

Figure 5. Flow chart of adaptive Savitzky–Golay algorithm. This figure was adopted from [41].

A series of experiments was performed to investigate the effectiveness of the algorithm and its applicability in various situations, for example, suitability evaluation for absorption spectra with different line shapes under the different pressures (between a few mbar and 1 bar). These experimental results indicated that the developed algorithm is reliable for practical application, and this method could also be used to construct an optimal calibration model for TDLAS spectra with different background structural characteristics (linear or nonlinear baseline drift). However, when applying the method to the simulated signals with different sampling points, one has to compromise between noise reduction and temporal resolution.

At a concentration of 1.5% of CO_2, the filter results of S–G algorithm compared with WT-based filter are shown in Figure 6a,b. The WT-based filter shows a strong noise reduction ability, the SNR enhancement factor is 5.5, and the S–G filter is 4.7. However, the WT-based filter requires more parameters and costs more time.

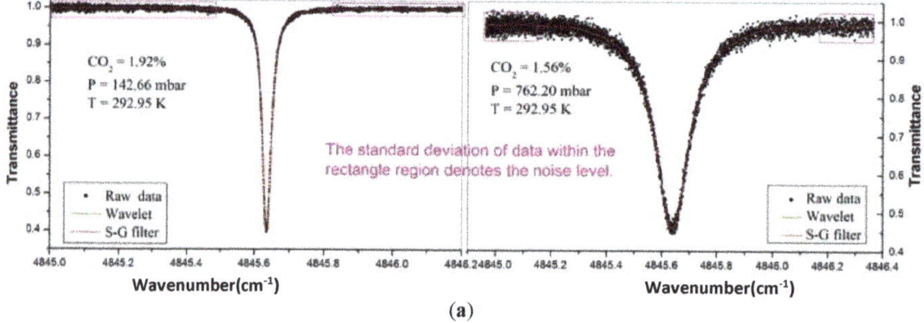

Figure 6. *Cont.*

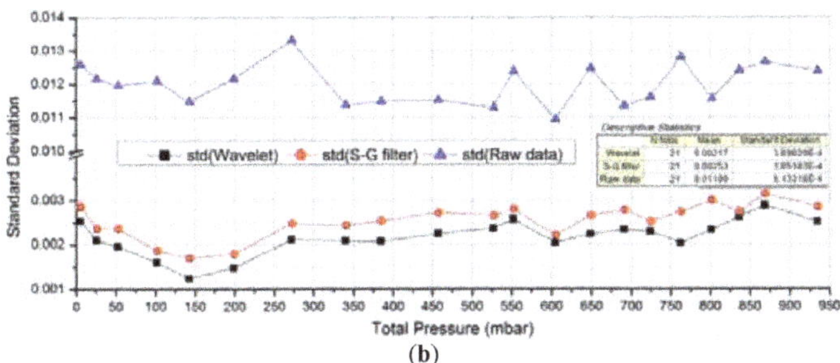

Figure 6. (**a**) Raw signal and processed signal by wavelet and S–G filter under different noise level. (**b**) Standard deviations of three sets of data under different pressure. This figure was obtained from [41].

3.1.3. EMD-FCR Algorithm

EMD algorithm is a time domain decomposition method based on the time scale features of the processed data [63]. EMD has been widely applied in many fields due to its excellent performance in processing non-stationary and non-linear signals [64–67]. In theory, the EMD algorithm can decompose any complicated signal into finite IMFs, and preset basis functions are not required. The signal decomposition depends only on the characteristics of signal itself, which is the essential difference from WT. Meng et al. [42] introduced the EMD algorithm into TDLAS signal processing, and proposed an improved algorithm that combines EMD, S–G filter, cross-correlation, and signal reconstruction (FCR), which is referred to as the EMD-FCR algorithm. This new method shows better applicability for second harmonic signal processing.

The essence of EMD is using the thought of stationary time series (STS) to decompose a frequency irregular wave into multiple regular waves and residual waves (original waveform = Σ IMFs + residual wave). Each IMF must meet two conditions: (1) in the whole data set, the number of extrema and the number of zero crossings must either equal or differ at most by one; (2) at any point, the mean value of the envelope defined by the local maxima and the envelope defined by the local minima are zero. In the EMD-FCR algorithm, each IMF requires being filtered by S–G filter (Figure 7b) and then cross-correlation calculations to obtain the cross-correlation coefficients between the original signal and each filtered IMFs.

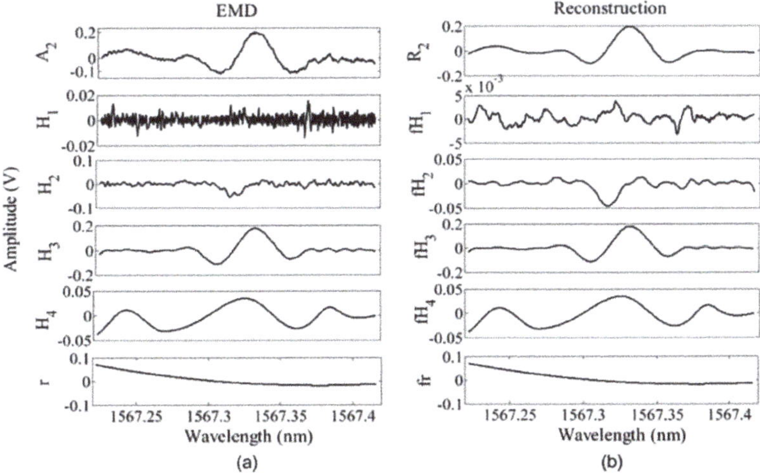

Figure 7. (**a**) Signal decomposition into IMF. (**b**) Reconstructed IMF after S–G filtering. This figure was obtained from reference [42].

Finally, each filtered IMF is weighed by its corresponding correlation coefficient and then added up to reconstruct a new signal. A portion of noise in original signal is removed by S–G filter. The remaining noise shows a low correlation with second harmonic signal so that it accounts for a small proportion of the reconstructed signal. Thus, the majority of the noise is suppressed.

The algorithm is assessed by simulation and experiment. In the two tests, EMD-FCR was compared with the Wiener filter, Kalman filter, and Wavelet filter. The results indicated that EMD-FCR performed best in both tests (Table 2 and Figure 8a). By this method, the SNR significantly improved from 7.32 dB to 14.31 dB, and the MDL decreased from 18 ppm to 2 ppm with SNR = 3 dB.

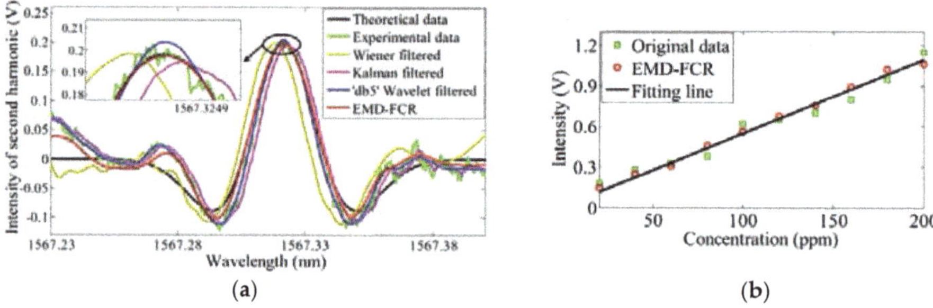

Figure 8. (**a**) Performance of different methods. (**b**) Relationship between the second harmonic intensity and gas concentration. This figure was obtained from [42].

Table 2. SNR and residual sum of squares (SSR) of different filters. This table was obtained from reference [42].

Filter	Wiener	Kalman	Wavelet	EMD-FCR
SNR_1 (dB)	14.17	11.21	13.26	14.82
$SSR_1 \left(V^2 \right)$	0.0287	0.3010	0.0027	0.0014
SNR_2 (dB)	14.02	11.81	13.96	14.31
$SSR_2 \left(V^2 \right)$	0.9859	0.7523	0.5943	0.2538

In further research, demodulation error experiments verified its reliability for extended (hour lomg) work. The errors of the second harmonic intensity after 50 min was only 2.113×10^{-5} V. Varying-concentration experiments indicated that the linear correlation coefficient of second harmonic intensity and gas concentration was improved from 0.93290 to 0.99297 by using the EMD-FCR algorithm.

3.1.4. Summary of Denoise Algorithm

For signal denoising, WT is a powerful tool, which can achieve a high SNR. However, the WT algorithm relies on many parameters, such as setting the wavelet base and degree of decomposition, which are prone to introducing subjective errors. Adaptive S–G filter is an improved version of the SG filter whose window size and polynomial order vary with the local features of signals with high precision. Given that most commercial software libraries include a function for the S–G filter, this algorithm is easy to implement. However, due to its nature as a smoothing filter and its specialization in Gaussian noise, some S–G filter experiments shows that the method may not work well if the signal contains large fluctuations. The EMD-FCR algorithm is an improved algorithm based on empirical mode decomposition. The filtering principle of EMD-FCR is signal decomposition and reconstruction, which can deal with non-stationary signals well. The algorithm is also self-adaptive because this decomposition depends on the characteristic of the signal itself. However, this algorithm is not ideal. When dealing with scale-mutative signal, the EMD algorithm may suffer mode-mixing problems [68–70].

3.2. Interference Fringe

In some practical case [71,72], white noise removal algorithms cannot effectively solve the interference fringe problem, especially when the signal is severely affected. Ensuring the accuracy of signal extraction is hard. Therefore, after the noise reduction, the following algorithms can be used to remove the interference fringes. The Levenberg-Marquardt (L-M) nonlinear fitting and the semi-parametric interference-immune algorithm can perform this task, respectively, from the perspective of the time and frequency domains.

3.2.1. L-M Nonlinear Fitting

L-M algorithm [73] is the most widely used nonlinear least-square algorithm, which uses gradient and iteration to find the largest or smallest value and then obtains the optimal solution of the requested parameters. Yan et al. [47] and Wagner et al. [74] used this algorithm for TDLAS curve-fitting. This algorithm converges fast and shows both the advantages of gradient method and Newton method [75–77]. However, one obvious drawback of L-M is that this iterative fitting requires a large amount of computation. Given the second-harmonic signal 2f, which is not a particularly complex function, has not too much parameter to be estimated, this method is usable for TDLAS signal processing. However, Yan et al. [47] mentioned that the L-M algorithm requires approximately 60,000 operations in a single iteration. Thus, adequate hardware support is indispensable to guarantee this algorithm will run well.

The specific mathematical principles of the L-M algorithm are not elaborated here because of its extensive use. However, we simply introduce the general iteration steps:

(a)   Select initial value x_0 and termination condition ε, and calculate $e_0 = ||Y - f(x_0)||$ and let step length $\lambda_0 = 10^{-3}$;

(b)   Compute the Jacobian matrix J^k_x, and construct the incremental equations;

(c)   Solve the incremental equation, and obtain Δ_{k+1};

(d)   If $||Y - f(x_0)||$ less than or equal to e_k, forward to e; else let $\lambda_{k-1} = 10\lambda_k$ and go back to a;

(e)   If $||\Delta_k||$ less than ε, stop iteration and output the result; else let $\lambda_{k-1} = 10\lambda_k$ and go back to b;

The baseline noise is measured first. As shown in Figure 9a, at a concentration of zero, the measured signal still shows pronounced fluctuation after smoothing filtration, which filtered out most of the high-frequency electronic noise.

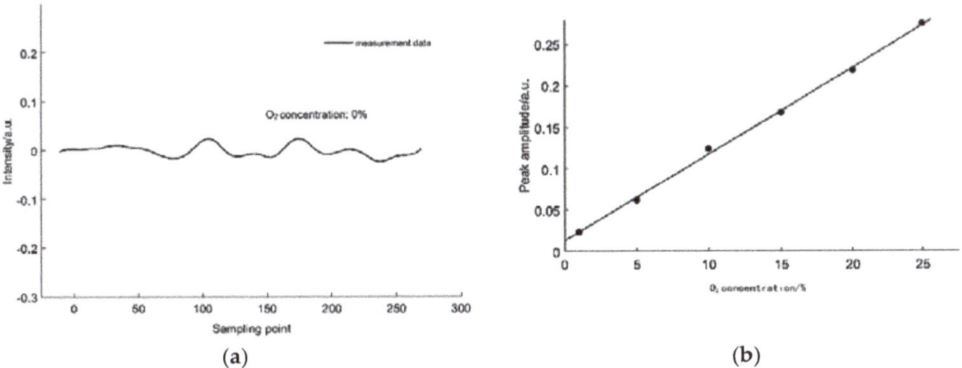

(a)   (b)

Figure 9. (a) Zero concentration fringe. (b) Relationship between the peak amplitude and gas concentration. This figure was adapted from [47].

In non-zero concentration experiments, the fitted and actual peak amplitudes show a 15% concentration error, which was the result of reducing the computational complexity of the L-M algorithm. After error correction, the measured nonlinearity between the gas concentration and the calculated concentration was 1.08%. This figure is 0.103% in the EMD algorithm experiment. However, because of the different measurement equipment and the different functional areas of the two algorithms, comparing these two values does not make much sense, that is, the two measurements may not use the same smoothing algorithm.

3.2.2. Semi-Parametric Interference-Immune Algorithm

Michelucci and Venturini proposed a novel semi-parametric algorithm to eliminate the signal distortion and background fluctuation caused by interference [36]. This algorithm shows a significant effect for dealing with strong interference signals. Compared with some of the above time-domain algorithms, this method starts from the frequency domain to solve the problem that the time domain algorithm is not good at. For severely polluted signals, even though the signal amplitude is ten times smaller than the fringes, the time-domain waveform has been severely disturbed, and the conventional time-domain fitting makes it difficult to restore the signal itself. However, these disturbances are easily distinguished in frequency, independently of the amplitude of the interferences. In general, the desired signal in TDLAS system, like absorption peak and second-harmonic waveform, can be modeled using known linear type. Calculating the DFT of model function is easy. Therefore, the DFT of the measurement signal can be fitted to obtain the line type parameters, and entering the parameter is not needed in this algorithm. However, for general spectrum fitting algorithms, when the signal interference is too weak, the contribution of the undesired factors on the frequency spectrum is not obvious and this method is hard to be applied.

This algorithm involves calculating the parameters by fitting the DFTs of model function and measured signal using the parameters to reconstruct the corresponding line shape. The steps of the algorithm are summarized in the figure. First, in order to improve the accuracy and reduce the window effect when measuring the signal DFT, the author chooses Tukey window (Figure 10a) is the compensation window [78,79] so that the signal decreases rapidly to zero on the sides. The next step is to determinate the optimal cut-off point i_0 to maximize the coefficient of determination R^2 obtained by fitting the DFT for $i > i_0$ to the functional form of the Fourier transform of the line shape. At every measurement, the i_0 is recalculated to guarantee that the algorithm will not be influenced by fringe changes in time, solving long-time stability problems arising from changes over time of the background, like thermal drift. Finally, DFT is fitted by using the DFT of model function to fit the DFT of the measured signal to obtain the parameters to determine the target signal. The algorithm flowchart is shown in Figure 10c.

In the simulation, the author simulates three background interferences to test the algorithm: periodic disturbance, weak disturbances of large FSR, and a complex disturbance with summing of 100 cosine functions. The result of simulation shows the discrepancy between the results obtained with the algorithm, and the expected values for the line parameters is less than 0.3%. As long as the background interference shows no fitting obstacle in the spectrum, the algorithm can perform signal extraction well, and this process is slightly affected by the interference amplitude.

Deliberately made interference fringes are utilized to test the practicality of the algorithm, and the measured signal is shown in Figure 10b. Two different windows interfere with two different intensity fringes. Despite the strong fluctuation, the extracted line shows a remarkable agreement with the expected curves from the HITRAN database [80], with deviation of the area of 0.1%. The experimental result is shown in Figure 10d,e.

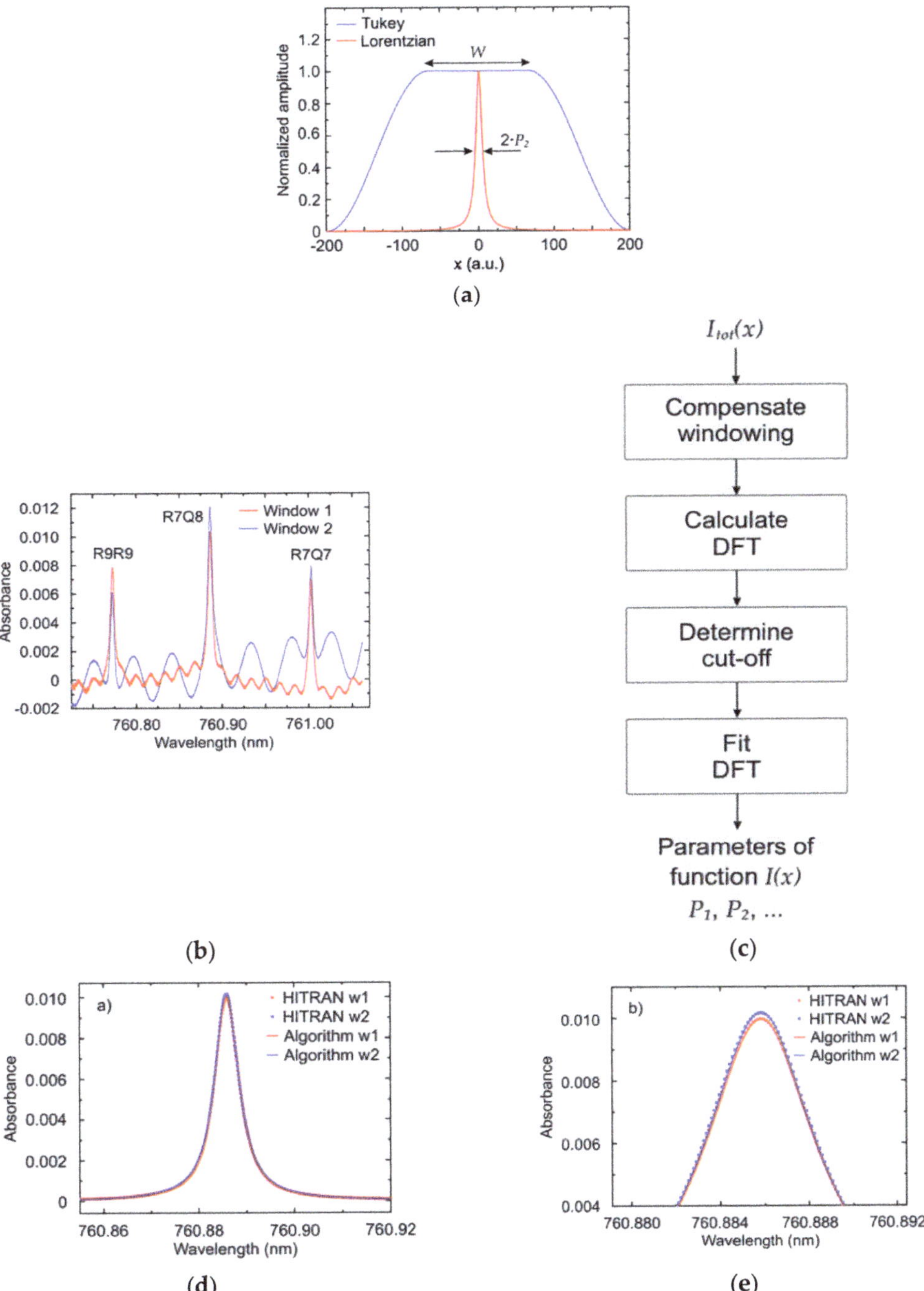

Figure 10. (**a**) Tukey window; (**b**) deliberately made interference fringe, (**c**) algorithm flowchart, (**d**) experimental result, (**e**) detail of the processed signal. This figure was adapted from [36].

These experiments show that the algorithm can effectively improve the system accuracy in the case of strong interference and solve the background fluctuation of the signal in a targeted manner. On the other hand, additional experiments remain to be performed to test the performance of this algorithm under the interference of other features.

3.2.3. Summary of Interference Fringe Processing Algorithm

For interference fringe problem, L-M nonlinear fitting and semi-parametric interference-immune algorithm are two solutions discussed in this paper. The L-M algorithm fits the signal in the time domain and is a widely utilized nonlinear least-squares method. This algorithm offers the advantages of both the Newton method and gradient method and fast convergence. Nevertheless, this iterative fitting requires the device to possess a high computational power. The semi-parametric interference-immune algorithm is a spectral fitting algorithm that can cope with the difficult situation of many time-domain analysis and presents strong immunity to strong optical interference signals. The signal extraction is independent from the amplitude of the interference. This method requires that the interference fringes of the measurement signal can be easily resolved in the spectrum, if the interference fringes are small, time-domain fitting can be performed directly and does not require spectrum analysis.

3.3. Baseline Drift

Background correction is required before signal fitting, otherwise the background will produce a large error for some fitting algorithms. In the following sections, two background correction strategies are introduced. These strategies adopt an iterative method to maximize the real baseline position. These strategies may be based on some algorithmic improvements with high reference value.

3.3.1. Advanced Integrative (AI) Algorithm

The AI algorithm proposed by Skrotzki et al. [51] is a modified fitting algorithm for the drawback of the integrative evaluation method, which calculates the molecular concentration by the integral area of the absorption line. Thus, the baseline error is made close to zero by fitting the no-absorption area and multiple iterations to improve the accuracy of the calculation results. An important feature is that the AI fitting algorithm is restricted to the evaluation of single absorption lines with precomputed line width. This feature suffers from limitations but exhibits a very fast reaction rate, and the fitting process does not dependent on appropriately chosen start values for the initialization, indicating its advantages in terms of robustness.

In particular, the authors compared it with the L-M algorithm and proved that the algorithm achieves similar accuracy as the L-M algorithm under proper application conditions, and the speed is 3–4 times faster than the L-M algorithm compared with the huge computational load of the latter. The AI method can be applied to embedded systems with limited computing power. In conclusion, this algorithm is an alternative for dealing with single absorption peak fitting in TDLAS systems.

Before introducing the principle, emphasizing the three assumptions and prerequisites for applying this algorithm is necessary: (1) the incident intensity I_0 (a parameter in Beer-Lambert law) is sufficiently known; (2) measurement signal is directly given in wavenumber domain; (3) only a single line absorption spectrum is considered.

Each iteration involves four steps (Figure 11). In Step 1, a polynomial fit is applied to the flanks of the absorption line signal to correct the background. In Step 2, the absorption line position m_0 is determined to retrieve the full absorption line profile. In Step 3, the line area obtained by integrate is corrected for the area within the flanks of the absorption line that are not covered by $[v_3]$ (shown in the figure). In Step 4, Voigt fit is used to obtain a good line shape that is approximate to the actual background.

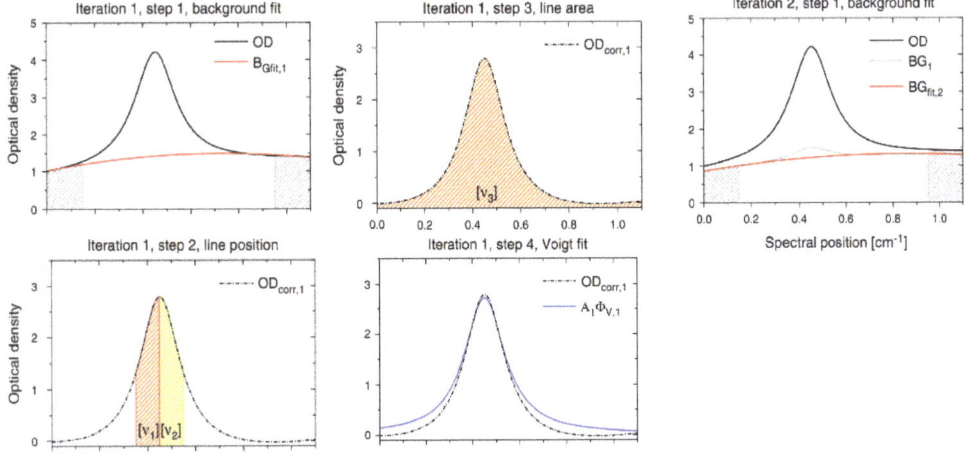

Figure 11. Steps of advanced integrative algorithm. This figure was obtained from [51].

In the second and subsequent iterations, Step 1 aims to fit the previous iteration by the same method, and the other steps are performed in the same manner discussed in the procedure described above. After multiple iterations, the precision of line area and line position continues to increase.

In this experiment, six iterations were necessary to fulfill the terminating conditions the author has chosen, yielding a fit precision of the line area A_6 of at least 10^{-3} and of the line position v_6 of at least 10^{-4}, respectively. Each iteration's results are shown in Table 3.

Table 3. Evolution of the relative deviation of absorption line area A_i and position mi from the prescribed 'true' values together with the signal-to-noise ratio $S/N_{3\sigma,i}$ for each iteration of the AI fit. This table is from [41].

Iteration i	Rel. Dev. Line Area	Rel. Dev. Line Position	$S/N_{3\sigma,i}$
1	0.118	4×10^{-4}	12
2	0.038	3×10^{-4}	38
3	0.012	3×10^{-4}	119
4	0.004	2×10^{-4}	339
5	0.001	2×10^{-4}	726
6	<0.001	1×10^{-4}	1060

The AI and L-M algorithms were used to compare the water vapor measurement experiments. The average relative deviation of the two algorithms was $0.1 \pm 0.2\%$, and the peak relative deviation was maintained within the range of $\pm 0.7\%$. Figure 12 shows the dynamic response of the relative deviation and SNR with the dynamic variation of H_2O concentration. Moreover, typical computational times obtained for the AI algorithm were 100–200 µs for the full evaluation of an absorption line profile. On the other hand, if the absorption line profile in the measurement signal was very unsatisfactory, it may not ideally converge. In the article, the authors summarize the characteristics of the two algorithms in terms of stability, speed, and flexibility, as shown in the figure, which can be a reference in practical applications.

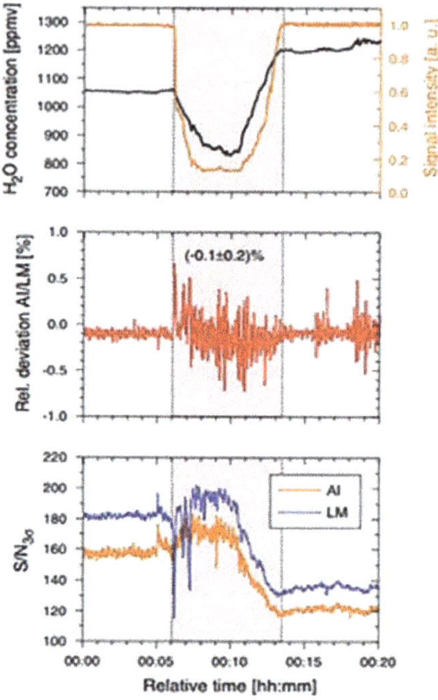

Figure 12. Comparison of AI fit and LM fit with relative time. This figure was obtained from [51].

3.3.2. Wavelet-Based Method for Baseline Drift

An AI algorithm calculates the background line by fitting the no absorption flanks. However, for poor-quality spectra, distinguishing the no absorption area using direct visual inspection (DVI) is hard. A new strategy was proposed by Li et al. [52], using wavelet decomposition and iteration to remove drift background. The application of WT for TDLAS signal denoising has been introduced, but the above studies tend to solve high-frequency noise, such as white Gaussian noise. This method uses WT based on the optimal wavelet pairs to find baseline and uses iteration to determine the precise location. In addition to the solution to the baseline drift, the strategy of separating the process of denoising and removing background is also meaningful. Unlike the block threshold strategy [81], this method uses different wavelet and decomposition levels to deal with noise and baseline. The characteristics of the two types of interference are considered and which of the two types shows a strong sense of reference is discussed.

In this method, denoising is separate from baseline removal regardless of their order. To remove the background, it is found that the wavelets bior2.2 or bior3.3 are good candidates for denoising TDLAS signals. A higher decomposition level than the optimal decomposition level for denoising was performed first. All detail coefficients were set to zero, and the approximation coefficients were used to reconstruct the signal. In this manner, a main background is obtained. The raw signal from the main background is subtracted, decomposed, and reconstructed. The iteration is repeated until the background reaches a precision calculated by root mean square error. Typically, this procedure is finished within 10 iterations. After removing the background, wavelet Daubechies7 is used for denoising. Conversely, an optimal decomposition level than the high decomposition level for denoising was first performed, and the best decomposition level between 5–7 was selected. The noise was then removed by decomposition and reconstruction. Figure 13 shows that the simulation results of the algorithm were very successful, the background was effectively corrected while preserving primary useful information, and the SNR was significantly improved.

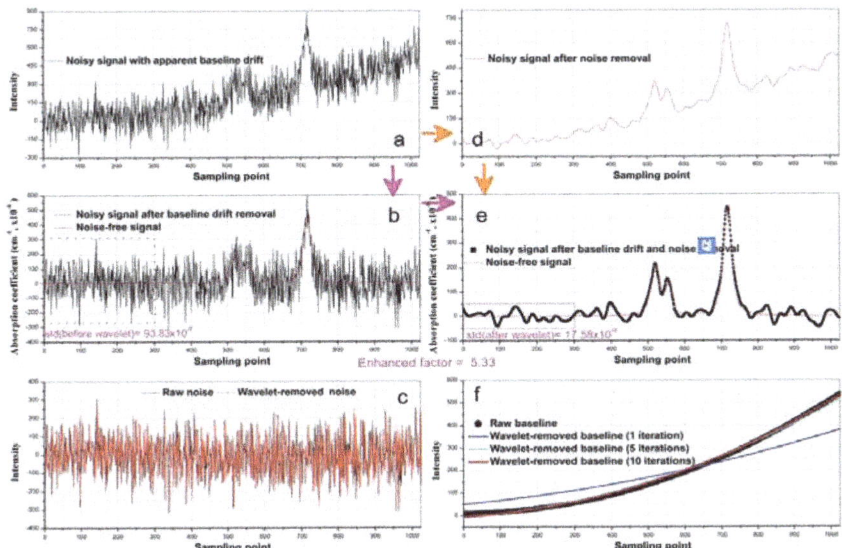

Figure 13. Nonlinear baseline correction and denoising using DWT. (**a**) Noisy signal with apparent baseline drift; (**b**) baseline drift removed signal from (**a**) and noise-free signal; (**c**) raw noise and wavelet-removed noise; (**d**) denoised signal from (**a**); (**e**) baseline drift removed signal from (**d**) and noise-free signal; (**f**) raw baseline and wavelet-removed baseline. This figure was obtained from [52].

The CO_2 absorption experiment, which is shown in Figure 14, demonstrates the effectiveness of this algorithm for solving baseline drift problems. The calculated SNR in DVI and DWT are 131.3 and 781.8, respectively.

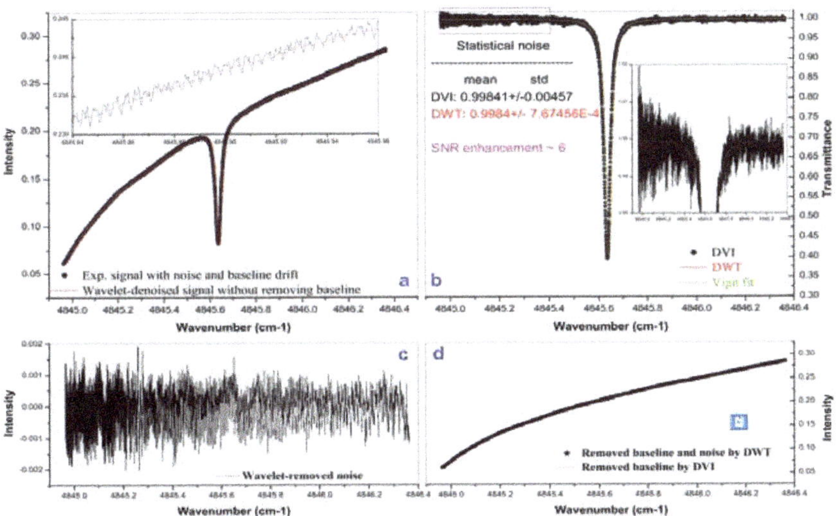

Figure 14. (**a**) Experimental spectrum of CO_2 (pressure = 140 mbar; path length = 4800 cm; mixing ratio = 1.0%; temperature = 292.95 K) and wavelet–denoised signal; (**b**) Baseline drift removed transmittance signals using DVI and DWT, as well as the Voigt fit; (**c**) DWT removed noise; (**d**) DWT removed baseline with noise and DVI removed baseline. Parts of the "baselines" are expanded for clarity and statistical noises are also provided in the inset. This figure was obtained from [52].

3.3.3. Summary of Background Removal Algorithms

In background correction, both the AI algorithm and the wavelet-based schemes improve the accuracy by introducing an iteration process. The AI algorithm corrects the background iteration through the absorption line area. The algorithm is very lightweight and suitable for solving the simple case of a single absorption peak, which poses the advantages of small calculation and fast speed. The latter uses decomposed wavelet background correction and iteration. In this strategy, the authors separate the background correction from denoising and use different wavelet and decomposition levels to process the background and the noise, respectively, which demonstrates flexible utilization of WT.

4. Conclusions

When dealing with TDLAS signals, optical factors, electronic factors, and the nature of the semiconductor lasers can cause disturbances. These common problems are summarized in three models: signal denoising, interference fringes, and background correction. In the above article, we have reviewed and compared some effective algorithms based on resent research works. Representative experiments were presented to evaluate the performance both qualitatively and quantitatively.

In essence, these signal problems are interferences superimposed on the original signal. These interferences are classified into three noise models for reduction due to the differences in spectral characteristic. Electrical white noise is a multiple frequency signal with a small amplitude, whereas interference fringes show concentrated frequency and large amplitude, and baseline drift is a ramp signal close to DC. Therefore, different strategies must be selected to deal with different signal models. For example, the adaptive S–G algorithm, which utilizes shift windows, can eliminate high-frequency/low-frequency well. By contrast, low-frequency and high-amplitude interference noise is difficult to remove with smoothing algorithms, but interference fringes are easy to process in the frequency domain by using a semi-parametric interference-immune algorithm. Nevertheless, the strategy involved in the algorithm must not be limited to the algorithm itself, like the schemes of correlation coefficient weight method, iteration, adaptive improvement, and problem decomposition, which can offer a foothold to solve any problem.

In comparison, signal decomposition and reconstruction-based algorithms, such as WT and EMD-FCR, can partly deal with all the three noise models because of their properties, such as multiscalability. When using WT, the selected parameters show remarkable effects. Therefore, when comparing algorithms, the details of experiments should be given particular importance. In future research, we expect that additional flexible strategies of signal decomposition and reconstruction algorithms will be developed to broaden their application for a wider variety of noise models.

Author Contributions: T.Z., J.K. and C.C. conceived and designed the research. T.Z., J.K., D.M., H.W., Z.M., M.Z. and X.Z. performed the research. T.Z., J.K., D.M., H.W. and C.C. wrote the paper.

Funding: This research was supported by Supported by the China Postdoctoral Science Foundation Funded Project (Grant Nos. 2013M530141, 2014T70290), the Project of Science and Technology Office of Jilin Province, China (No. 20150520096JH, 20170101200JC), the Project of Observation Instrument Development for Integrated Geophysical field of China Mainland (Grant No. Y201616, Y201706), the National Key R&D Program of China (No. 2016YFC0303902).

Conflicts of Interest: The authors declare no conflict of interest.

References

1. Cassidy, D.T.; Reid, J. Atmospheric pressure monitoring of trace gases using tunable diode lasers. *Appl. Opt.* **1982**, *21*, 1185–1190. [CrossRef] [PubMed]
2. Wang, W.; Lv, Y. The principal, preparation and application of quantum cascade laser. *Laser J.* **2018**, *39*, 7–11.
3. Nikodem, M.; Wysocki, G. Chirped laser dispersion spectroscopy for remote open-path trace-gas sensing. *Sensors* **2012**, *12*, 16466–16481. [CrossRef] [PubMed]

4.	Li, M.; Bai, F. Design of High Sensitivity Infrared Methane Detector Based on TDLAS-WMS. *Laser J.* **2018**, *39*, 75–79.

5.	Kluczynski, P.; Jahjah, M.; Nähle, L.; Axner, O.; Belahsene, S.; Fischer, M.; Koeth, J.; Rouillard, Y.; Westberg, J.; Vicet, A. Detection of acetylene impurities in ethylene and polyethylene manufacturing processes using tunable diode laser spectroscopy in the 3-μm range. *Appl. Phys. B* **2011**, *105*, 427. [CrossRef]

6.	Zhang, L.; Cui, X. The Design of Carbon Monoxide Detector Based on Tunable Diode Lasers Absorption Spectroscope. *Laser J.* **2014**, *35*, 54–56.

7.	Werle, P. A review of recent advances in semiconductor laser based gas monitors. *Spectrochim. Acta Part A Mol. Biomol. Spectrosc.* **1998**, *54*, 197–236. [CrossRef]

8.	Werle, P.; Slemr, F.; Maurer, K.; Kormann, R.; Mücke, R.; Jänker, B. Near-and mid-infrared laser-optical sensors for gas analysis. *Opt. Lasers Eng.* **2002**, *37*, 101–114. [CrossRef]

9.	Wang, F.; Cen, K.; Li, N.; Jeffries, J.B.; Huang, Q.; Yan, J.; Chi, Y. Two-dimensional tomography for gas concentration and temperature distributions based on tunable diode laser absorption spectroscopy. *Meas. Sci. Technol.* **2010**, *21*, 045301. [CrossRef]

10.	Kurtz, J.; Aizengendler, M.; Krishna, Y.; Walsh, P.; O'Byrne, S.B. Flight test of a rugged scramjet-inlet temperature and velocity sensor. In Proceedings of the 53rd AIAA Aerospace Sciences Meeting, Kissimmee, FL, USA, 5–9 January 2015; p. 0110.

11.	Wang, F.; Wu, Q.; Huang, Q.; Zhang, H.; Yan, J.; Cen, K. Simultaneous measurement of 2-dimensional H_2O concentration and temperature distribution in premixed methane/air flame using TDLAS-based tomography technology. *Opt. Commun.* **2015**, *346*, 53–63. [CrossRef]

12.	Buchholz, B.; Afchine, A.; Ebert, V. Rapid, optical measurement of the atmospheric pressure on a fast research aircraft using open-path TDLAS. *Atmos. Meas. Tech.* **2014**, *7*, 3653–3666. [CrossRef]

13.	Lins, B.; Zinn, P.; Engelbrecht, R.; Schmauss, B. Simulation-based comparison of noise effects in wavelength modulation spectroscopy and direct absorption TDLAS. *Appl. Phys. B* **2010**, *100*, 367–376. [CrossRef]

14.	Chen, K.; Mei, M. Detection of Gas Concentrations Based on Wireless Sensor and Laser Technology. *Laser J.* **2014**, *35*, 50–54.

15.	Mueller, H.G.; Weber, J.; Hornsby, B.W.Y. The effects of digital noise reduction on the acceptance of background noise. *Trends Amplif.* **2006**, *10*, 83–93. [CrossRef] [PubMed]

16.	Misiti, M.; Misiti, Y.; Oppenheim, G.; Poggi, J.-M. Wavelets and their applications. *Int. J. Imaging Syst. Technol.* **2010**, *7*, 151.

17.	Li, J.; Yu, B.; Zhao, W.; Chen, W. A review of signal enhancement and noise reduction techniques for tunable diode laser absorption spectroscopy. *Appl. Spectrosc. Rev.* **2014**, *49*, 666–691. [CrossRef]

18.	Zhang, K.; Zhang, L.; Zhao, Q.; Liu, S.; Chen, S.; Wu, Y.; Wang, K.; Yang, X. Application of digital quadrature lock-in amplifier in TDLAS humidity detection. *Opt. Spectrosc. Imaging Int. Soc. Opt. Photonics* **2017**, *10461*, 1046109.

19.	Mohammad, I.L.; Anderson, G.T.; Chen, Y. Noise estimation technique to reduce the effects of 1/f noise in Open Path Tunable Diode Laser Absorption Spectrometry. *Int. Soc. Opt. Photonics* **2014**, *9113*, 91130S.

20.	Chighine, A.; Fisher, E.; Wilson, D.; Lengden, M.; Johnstone, W.; McCann, H. An FPGA-based lock-in detection system to enable Chemical Species Tomography using TDLAS. In Proceedings of the 2015 IEEE International Conference on Imaging Systems and Techniques (IST), Macau, China, 16–18 September 2015; pp. 1–5.

21.	Tu, G.; Dong, F.; Wang, Y.; Culshaw, B.; Zhang, Z.; Pang, T.; Xia, H.; Wu, B. Analysis of random noise and long-term drift for tunable diode laser absorption spectroscopy system at atmospheric pressure. *IEEE Sens. J.* **2015**, *15*, 3535–3542. [CrossRef]

22.	He, Q.; Dang, P.; Liu, Z.; Zheng, C.; Wang, Y. TDLAS–WMS based near-infrared methane sensor system using hollow-core photonic crystal fiber as gas-chamber. *Opt. Quantum Electron.* **2017**, *49*, 115. [CrossRef]

23.	Frish, M.; Wainner, R.; Laderer, M.; Parameswaran, K.; Sonnenfroh, D.; Druy, M. Precision and accuracy of miniature tunable diode laser absorption spectrometers. *Proc. SPIE* **2011**, *8032*, 803209.

24.	Knight, J.; Birks, T.; Cregan, R.; Russell, P.S.J.; De Sandro, P. Large mode area photonic crystal fibre. *Electron. Lett.* **1998**, *34*, 1347–1348. [CrossRef]

25.	Dong, L.; Tittel, F.K.; Li, C.; Sanchez, N.P.; Wu, H.; Zheng, C.; Yu, Y.; Sampaolo, A.; Griffin, R.J. Compact TDLAS based sensor design using interband cascade lasers for mid-IR trace gas sensing. *Opt. Express* **2016**, *24*, A528–A535. [CrossRef] [PubMed]

26. Wang, F.; Chang, J.; Wang, Q.; Wei, W.; Qin, Z. TDLAS gas sensing system utilizing fiber reflector based round-trip structure: Double absorption path-length, residual amplitude modulation removal. *Sens. Actuators A Phys.* **2017**, *259*, 152–159. [CrossRef]

27. Shao, L.; Yan, R.; Li, X.; Liu, Y. From heuristic optimization to dictionary learning: A review and comprehensive comparison of image denoising algorithms. *IEEE Trans. Cybern.* **2014**, *44*, 1001–1013. [CrossRef] [PubMed]

28. Gupta, K.; Gupta, S. Image denoising techniques-a review paper. *IJITEE* **2013**, *2*, 6–9.

29. Werle, P.; Slemr, F. Signal-to-noise ratio analysis in laser absorption spectrometers using optical multipass cells. *Appl. Opt.* **1991**, *30*, 430–434. [CrossRef]

30. Masiyano, D.; Hodgkinson, J.; Tatam, R.P. Use of diffuse reflections in tunable diode laser absorption spectroscopy: Implications of laser speckle for gas absorption measurements. *Appl. Phys. B* **2008**, *90*, 279–288. [CrossRef]

31. Bomse, D.S.; Stanton, A.C.; Silver, J.A. Frequency modulation and wavelength modulation spectroscopies: Comparison of experimental methods using a lead-salt diode laser. *Appl. Opt.* **1992**, *31*, 718–731. [CrossRef]

32. Hennig, O.; Strzoda, R.; Mágori, E.; Chemisky, E.; Tump, C.; Fleischer, M.; Meixner, H.; Eisele, I. Hand-held unit for simultaneous detection of methane and ethane based on NIR-absorption spectroscopy. *Sens. Actuators B Chem.* **2003**, *95*, 151–156. [CrossRef]

33. Le Barbu, T.; Vinogradov, I.; Durry, G.; Korablev, O.; Chassefière, E.; Bertaux, J.-L. TDLAS a laser diode sensor for the in situ monitoring of H_2O, CO_2 and their isotopes in the Martian atmosphere. *Adv. Space Res.* **2006**, *38*, 718–725. [CrossRef]

34. Weibring, P.; Richter, D.; Fried, A.; Walega, J.; Dyroff, C. Ultra-high-precision mid-IR spectrometer II: System description and spectroscopic performance. *Appl. Phys. B* **2006**, *85*, 207–218. [CrossRef]

35. Buchholz, B.; Kühnreich, B.; Smit, H.; Ebert, V. Validation of an extractive, airborne, compact TDL spectrometer for atmospheric humidity sensing by blind intercomparison. *Appl. Phys. B* **2013**, *110*, 249–262. [CrossRef]

36. Michelucci, U.; Venturini, F. Novel semi-parametric algorithm for interference-immune tunable absorption spectroscopy gas Sensing. *Sensors* **2017**, *17*, 2281. [CrossRef] [PubMed]

37. Wang, J.; Yu, D.; Ye, H.; Yang, J.; Ke, L.; Han, S.; Gu, H.; Chen, Y. Applications of optical measurement technology in pollution gas monitoring at thermal power plants. *Proc. SPIE* **2011**, *8197*, 819702.

38. Reid, J.; Labrie, D. Second-harmonic detection with tunable diode lasers—Comparison of experiment and theory. *Appl. Phys. B* **1981**, *26*, 203–210. [CrossRef]

39. Xia, H.; Dong, F.-Z.; Zhang, Z.-R.; Tu, G.-J.; Pang, T.; Wu, B.; Wang, Y. Signal analytical processing based on wavelet transform for tunable diode laser absorption spectroscopy. *Proc. SPIE* **2010**, *7853*, 785311.

40. Zheng, C.-T.; Ye, W.-L.; Huang, J.-Q.; Cao, T.-S.; Lv, M.; Dang, J.-M.; Wang, Y.-D. Performance improvement of a near-infrared CH_4 detection device using wavelet-denoising-assisted wavelength modulation technique. *Sens. Actuators B Chem.* **2014**, *190*, 249–258. [CrossRef]

41. Li, J.; Deng, H.; Li, P.; Yu, B. Real-time infrared gas detection based on an adaptive Savitzky–Golay algorithm. *Appl. Phys. B* **2015**, *120*, 207–216. [CrossRef]

42. Meng, Y.; Liu, T.; Liu, K.; Jiang, J.; Wang, R.; Wang, T.; Hu, H. A modified empirical mode decomposition algorithm in TDLAS for gas detection. *IEEE Photonics J.* **2014**, *6*, 1–7. [CrossRef]

43. Hodgkinson, J.; Tatam, R.P. Optical gas sensing: A review. *Meas. Sci. Technol.* **2012**, *24*, 012004. [CrossRef]

44. Hartmann, A.; Strzoda, R.; Schrobenhauser, R.; Weigel, R. Ultra-compact TDLAS humidity measurement cell with advanced signal processing. *Appl. Phys. B* **2014**, *115*, 263–268. [CrossRef]

45. Cao, J.N.; Wang, Z.; Zhang, K.-K.; Yang, R.; Wang, Y. Etalon effects analysis in tunable diode laser absorption spectroscopy gas concentration detection system based on wavelength modulation spectroscopy. In Proceedings of the Symposium on Photonics and Optoelectronics, Chengdu, China, 19–21 June 2010; pp. 1–5.

46. Masiyano, D.; Hodgkinson, J.; Schilt, S.; Tatam, R.P. Self-mixing interference effects in tunable diode laser absorption spectroscopy. *Appl. Phys. B* **2009**, *96*, 863–874. [CrossRef]

47. Yan, J.; Zhai, C.; Wang, X.; Huang, W. The research of oxygen measurement by TDLAS based on Levenberg-Marquardt nonlinear fitting. *Spectrosc. Spectr. Anal.* **2015**, *35*, 1497–1500.

48. Webster, C.R. Brewster-plate spoiler: A novel method for reducing the amplitude of interference fringes that limit tunable-laser absorption sensitivities. *JOSA B* **1985**, *2*, 1464–1470. [CrossRef]

49. Huang, N.E.; Shen, Z.; Long, S.R.; Wu, M.C.; Shih, H.H.; Zheng, Q.; Yen, N.-C.; Tung, C.C.; Liu, H.H. The empirical mode decomposition and the Hilbert spectrum for nonlinear and non-stationary time series analysis. *Proc. Math. Phys. Eng. Sci.* **1998**, *454*, 903–995. [CrossRef]
50. Xie, Q.; Li, J.; Gao, X.; Jia, J. Real time infrared gas detection based on a modified EMD algorithm. *Sens. Actuators B Chem.* **2009**, *136*, 303–309. [CrossRef]
51. Skrotzki, J.; Habig, J.C.; Ebert, V. Integrative fitting of absorption line profiles with high accuracy, robustness, and speed. *Appl. Phys. B* **2014**, *116*, 393–406. [CrossRef]
52. Li, J.; Yu, B.; Fischer, H. Wavelet transform based on the optimal wavelet pairs for tunable diode laser absorption spectroscopy signal processing. *Appl. Spectrosc.* **2015**, *69*, 496–506. [CrossRef]
53. Werle, P.; Mücke, R.; Slemr, F. The limits of signal averaging in atmospheric trace-gas monitoring by tunable diode-laser absorption spectroscopy (TDLAS). *Appl. Phys. B* **1993**, *57*, 131–139. [CrossRef]
54. Werle, P.W.; Scheumann, B.; Schandl, J. Real-time signal-processing concepts for trace-gas analysis by diode-laser spectroscopy. *Opt. Eng.* **1994**, *33*, 3093–3106.
55. Leleux, D.; Claps, R.; Chen, W.; Tittel, F.; Harman, T. Applications of Kalman filtering to real-time trace gas concentration measurements. *Appl. Phys. B* **2002**, *74*, 85–93. [CrossRef] [PubMed]
56. Coifman, R.R.; Meyer, Y.; Wickerhauser, V. Wavelet analysis and signal processing. In *Wavelets and Their Applications*; Citeseer: State College, PA, USA, 1992.
57. Tsatsanis, M.K.; Giannakis, G.B. Multirate filter banks for code-division multiple access systems. In Proceedings of the International Conference on Acoustics, Speech, and Signal Processing, ICASSP-95, Detroit, MI, USA, 9–12 May 1995; pp. 1484–1487.
58. Li, J.; Parchatka, U.; Fischer, H. Applications of wavelet transform to quantum cascade laser spectrometer for atmospheric trace gas measurements. *Appl. Phys. B* **2012**, *108*, 951–963. [CrossRef]
59. Duan, H.; Gautam, A.; Shaw, B.D.; Cheng, H.H. Harmonic wavelet analysis of modulated tunable diode laser absorption spectroscopy signals. *Appl. Opt.* **2009**, *48*, 401–407. [CrossRef] [PubMed]
60. Savitzky, A.; Golay, M.J. Smoothing and differentiation of data by simplified least squares procedures. *Anal. Chem.* **1964**, *36*, 1627–1639. [CrossRef]
61. Madden, H.H. Comments on the Savitzky-Golay convolution method for least-squares-fit smoothing and differentiation of digital data. *Anal. Chem.* **1978**, *50*, 1383–1386. [CrossRef]
62. Czarnecki, M.A. Resolution enhancement in second-derivative spectra. *Appl. Spectrosc.* **2015**, *69*, 67–74. [CrossRef]
63. Boudraa, A.-O.; Cexus, J.-C. EMD-based signal filtering. *IEEE Trans. Instrum. Meas.* **2007**, *56*, 2196–2202. [CrossRef]
64. Pines, D.; Salvino, L. Structural health monitoring using empirical mode decomposition and the Hilbert phase. *J. Sound Vib.* **2006**, *294*, 97–124. [CrossRef]
65. Poungponsri, S.; Yu, X.-H. An adaptive filtering approach for electrocardiogram (ECG) signal noise reduction using neural networks. *Neurocomputing* **2013**, *117*, 206–213. [CrossRef]
66. Lingfang, S.; Yechi, W. Soft-sensing of oxygen content of flue gas based on mixed model. *Energy Procedia* **2012**, *17*, 221–226. [CrossRef]
67. Kopsinis, Y.; McLaughlin, S. Development of EMD-based denoising methods inspired by wavelet thresholding. *IEEE Trans. Signal Process.* **2009**, *57*, 1351–1362. [CrossRef]
68. Tang, B.; Dong, S.; Song, T. Method for eliminating mode mixing of empirical mode decomposition based on the revised blind source separation. *Signal Process.* **2012**, *92*, 248–258. [CrossRef]
69. Hu, X.; Peng, S.; Hwang, W.-L. EMD revisited: A new understanding of the envelope and resolving the mode-mixing problem in AM-FM signals. *IEEE Trans. Signal Process.* **2012**, *60*, 1075–1086.
70. Torres, M.E.; Colominas, M.A.; Schlotthauer, G.; Flandrin, P. A complete ensemble empirical mode decomposition with adaptive noise. In Proceedings of the IEEE International Conference on Acoustics, Speech and Signal Processing (ICASSP), Prague, Czech, 22–27 May 2011; pp. 4144–4147.
71. Werle, P. Laser excess noise and interferometric effects in frequency-modulated diode-laser spectrometers. *Appl. Phys. B* **1995**, *60*, 499–506. [CrossRef]
72. Hansen, P.; Pereyra, V.; Scherer, G. Nonlinear Least Squares Problems. 2004. Available online: http://www.imm.dtu.dk/pcha/LSDF/NonlinDataFit.pdf (accessed on 29 November 2018).
73. Ranganathan, A. The levenberg-marquardt algorithm. *Tutoral LM Algorithm* **2004**, *11*, 101–110.

74. Wagner, S.; Klein, M.; Kathrotia, T.; Riedel, U.; Kissel, T.; Dreizler, A.; Ebert, V. Absolute, spatially resolved, in situ CO profiles in atmospheric laminar counter-flow diffusion flames using 2.3 μm TDLAS. *Appl. Phys. B* **2012**, *109*, 533–540. [CrossRef]

75. Bertsekas, D.P. *Nonlinear Programming*; Athena Scientific Belmont: Belmont, MA, USA, 1999.

76. Wedderburn, R.W. Quasi-likelihood functions, generalized linear models, and the Gauss—Newton method. *Biometrika* **1974**, *61*, 439–447.

77. Hartley, H.O. The modified Gauss-Newton method for the fitting of non-linear regression functions by least squares. *Technometrics* **1961**, *3*, 269–280. [CrossRef]

78. Tukey, J.W. An introduction to the calculations of numerical spectrum analysis. *Spectra Anal. Time* **1967**, 25–46.

79. Harris, F.J. On the use of windows for harmonic analysis with the discrete Fourier transform. *Proc. IEEE* **1978**, *66*, 51–83. [CrossRef]

80. Rothman, L.S.; Gordon, I.E.; Babikov, Y.; Barbe, A.; Benner, D.C.; Bernath, P.F.; Birk, M.; Bizzocchi, L.; Boudon, V.; Brown, L.R. The HITRAN2012 molecular spectroscopic database. *J. Quant. Spectrosc. Radiat. Transf.* **2013**, *130*, 4–50. [CrossRef]

81. Ramos, P.M.; Ruisánchez, I. Noise and background removal in Raman spectra of ancient pigments using wavelet transform. *J. Raman Spectrosc.* **2005**, *36*, 848–856. [CrossRef]

MDPI
St. Alban-Anlage 66
4052 Basel
Switzerland
Tel. +41 61 683 77 34
Fax +41 61 302 89 18
www.mdpi.com

Sensors Editorial Office
E-mail: sensors@mdpi.com
www.mdpi.com/journal/sensors